OPTICAL COMMUNICATION SYSTEMS

John Gowar

Lecturer in Electronic Engineering, University of Bristol

Prentice/Hall  **International**

Englewood Cliffs, NJ London New Delhi Rio de Janeiro
Singapore Sydney Tokyo Toronto Wellington

Library of Congress Cataloguing in Publication Data

Gowar, John, 1938–
Optical communication systems.

(Prentice-Hall international series in optoelectronics)
Includes bibliographies and index.
1. Optical communications. I. Title. II. Series.
TK5103.59.G68 1983 621.38'0414 83-8694
ISBN 0-13-638056-5
ISBN 0-13-638156-1 (pbk.)

British Library Cataloguing in Publication Data

Gowar, John
Optical communication systems.
1. Optical communications
I. Title
621.38'0414 TK5103.59

ISBN 0-13-638056-5
ISBN 0-13-638156-1 (pbk.)

ISBN 0-13-638056 5

ISBN 0-13-638156 1 {PBK}

Prentice-Hall International, Inc., *London*
Prentice-Hall of Australia Pty, Ltd., *Sydney*
Prentice-Hall Canada, Inc., *Toronto*
Prentice-Hall of India Private Ltd., *New Delhi*
Prentice-Hall of Japan, Inc., *Tokyo*
Prentice-Hall of Southeast Asia Pte., Ltd., *Singapore*
Prentice-Hall Inc., *Englewood Cliffs, New Jersey*
Prentice-Hall do Brasil Ltda., *Rio de Janeiro*
Whitehall Books Ltd., *Wellington, New Zealand*

10 9 8 7 6 5 4 3 2

Typeset by Pintail Studios Ltd., Ringwood, Hants., UK.
Printed in the United States of America

Contents

Foreword

With the appearance of this second volume in Prentice-Hall's new series on optoelectronics, a few words of editorial comment seem appropriate. The accent of the series will be on texts written with an eye to teaching rather than as monographs intended primarily for research workers already well versed in the general principles of their subject. A series on optoelectronics seems especially timely in view of the rapid expansion of interest in this subject. This expansion is being driven primarily by the burgeoning developments in a variety of applications fields, especially in electronic communications. In the past few years it has become widely recognized that the rapidly growing need for high speed, high information capacity communications systems will strain the capacity of traditional techniques involving all-electrical networks. These needs, driven particularly by the widespread introduction of links between computers and also by the desire to distribute video information over complex networks, have validated the requirement for optical communication techniques. Conveniently, such optical communication systems are just emerging from the status of laboratory prototypes mainly developed in the 1970s. A variety of Government-supported schemes have recently been launched in the UK to assist in the introduction of these new methods in the communications industry. Foremost are the JOERS and the Alvey schemes, both of which include the promotion of optoelectronic research and development work between industry and government, university or technical college R&D groups. These initiatives will appreciably increase the already strong demand for personnel trained in optoelectronics. Development of a more general awareness of the subject amongst technically trained people, and even for the general public, will also become increasingly appropriate.

The first book in the Prentice-Hall Optoelectronics series, by John Wilson and John Hawkes, was published in Spring 1983. It is a general text, set at a level appropriate for a final year undergraduate course on the basic principles of the three areas into which the subject can conveniently be divided. These three elements are optical (often called lightwave) communications, optical signal processing, and displays. This second text in the series, by John Gowar, treats the first of these elements, the core topic for the majority of those interested in the subject, in significantly greater depth. Starting with basic concepts, the reader is taken to a stage appropriate for an MSc degree examination course. Both texts emphasize their basically tutorial approach by the inclusion of many useful examples in every chapter. Gowar also ends each chapter with a concise summary of the basic principles which have been illustrated. I believe that these texts, taken in combination, will enable a student in science and engineering to build up a fully comprehensive appreciation of optical communications. Both texts provide ample references to original research publications for those requiring detail on specific aspects.

Further volumes in the series are in preparation or are being planned to treat other aspects of optoelectronics in comparable detail. A number of texts will also cover specific areas of specialization in rather greater depth.

P. J. Dean
Series Editor
October 1983

Preface

It has come to be generally accepted that during the last two decades of the 20th century optical fiber communication techniques will become firmly established in most of the developed countries of the world. It follows that during this period the emphasis in the teaching of the principles of telecommunications and data transmission will move from being concerned entirely with electrical and radio systems to dealing with optical, electrical and radio techniques on more or less equal terms.

Until recently the subjects associated with optical communications have been discussed mainly in review articles and in advanced, usually multi-author, specialist texts. One or two books set at a much simpler technical level and aimed at the nonspecialist user of optical fiber technology have also appeared. Many of these are excellent, however, their main purpose has been to review the subject rather than to teach it. The present book is intended to be a general tutorial introduction set at a level suitable for final-year undergraduate students of Electronic Engineering, Physics or other related subjects. It is hoped that it may equally meet the needs of graduate engineers and scientists who are contemplating research or development work in this field or who have a general interest and seek a comprehensive introduction. To a large extent the book has been developed from a third-year undergraduate course given for several years by the author to Electrical and Electronic Engineering and Physics students in the University of Bristol. He is grateful for the interest and enthusiasm shown by these students.

One of the nice features of a new technology is the way it brings together in common interest people with widely varying academic backgrounds, and cuts across existing, artificial subject boundaries. However, this also gives rise to one of the biggest problems in a book of this kind, and that is the level of prior knowledge to assume in each of the traditional subject areas covered. Here, a basic knowledge of electromagnetic theory and of solid-state physics is assumed, as is some familiarity with the elementary theory of analogue and digital communication systems and electronic circuits. Naturally, the level of approach will reflect the inclination and interest of the author. What has to be avoided, though, is making the treatment of subjects with which the reader is familiar appear trivial (or worse, misguided!) without rendering them unintelligible for the reader with only a limited background. Time will tell to what extent these traps have been avoided.

It may be noticed that, unlike most books concerned in some way with electromagnetic waves, no attempt has been made here to derive, justify or explain Maxwell's equations. There seemed to be nothing that could usefully be added to the enormous literature which already exists on this subject. On the other hand, consideration of semiconductor properties has been developed here from a fairly basic level in the hope that some readers may find helpful the largely diagrammatic approach based on electron energy levels.

The end-of-chapter references have deliberately been kept small in number and are intended to serve two purposes. Some are sources of material, knowledge of which is assumed in the text. Others extend the subject beyond the level of the treatment given here. It must be emphasized that this book is not intended to be a review of current research activity and no attempt has been made to reference the enormous literature. Several other texts provide a ready access to this, and in particular the 1980 review paper of Botez and Herskowitz* contains no fewer than 640 references. In particular, *IEEE Transactions on Communications*, **COM–26** (7), (July 1978); *Proceedings of the IEEE*, **68** (10), (October 1980); *IEEE Journal on Selected Areas in Communications*, **SAC–1** (3), (April 1983), give good reviews of the progress and activities in the subject at each of their respective dates.

The present text has benefited greatly from the comments and suggestions made by the reviewers, and the author has been grateful for them. The many remaining deficiencies are, of course, his own responsibility. Thanks must go to many past and present colleagues whose influences have helped in various direct and indirect ways. Special thanks are due to Geoff Reed and Ted Thomas for giving up their time to read and criticize drafts of the manuscript, to Ken Sander and others for help at various times with mathematical proofs, and to Carrie Pharoah and Tina Watson for their valuable assistance in the preparation of the figures. The author owes particular debts of gratitude to Angela Tyler for typing the manuscript (some parts many times!) and to his wife, Ann, for her help in checking the completed text and for her support throughout its preparation.

J. Gowar
University of Bristol
October 1983

* D. Botez and G. J. Herskowitz, Components for Optical Communications Systems: A Review, *Proc. I.E.E.E.* **68** (6) 689–731 (1980).

1

A General Introductory Discussion

1.1 HISTORICAL PERSPECTIVE

There is, of course, nothing new in the use of optical frequencies for the transmission of information. Visual methods of communicating are widely used in the animal kingdom, obvious examples being courtship displays and warnings of aggressive intent, and from the time of the most primitive human civilizations man has employed artificial optical techniques to send information over long distances. He has, for example, used smoke signals or reflected sunlight during the day and fire beacons at night. Perhaps we may indulge briefly in two references from classical antiquity:

> In Aeschylus's play *Agamemnon*, written in the 5th century B.C. and concerned with events from mythical times nearly a millennium earlier, Clytemnestra explains how she had learned of the fall of Troy the previous night:
>
> > 'Hephaistos the fire-god, beaconing the news,
> > Flashed a courier-chain of fires from peak-to-peak.'†
>
> She then gives a graphic description of the eight stages used to relay the news by beacon from Asia Minor to Argos.
>
> More prosaically, but possibly more reliably, Herodotus writing a little later tells how in 480 B.C. the Persian leader Mardonius dreamed of sending a similar message by similar means in the other direction across the Aegean, to inform his emperor (Xerxes) of the fall of Athens. His dream was unfulfilled.

Each of these examples is concerned with the transmission of a single item of information by means of a prearranged signal. Although more sophisticated methods of signalling were discussed in classical times, only the use of beacons

† Line 281 of the Oresteia. In the translation by Frederick Raphael and Kenneth McLeish: *The Serpent Son*, C.U.P. (1979).

seems to have continued over the intervening centuries until the development of the semaphore towards the end of the 18th century. In due time this was overtaken by the electric telegraph on land and had to compete with signalling flags and shuttered signalling lamps at sea. Such techniques were in turn largely replaced by the telephone and by wireless telegraphy. At this point a major change took place in the form in which information was transmitted. The early systems were all what we would now call *digital* systems, but the telephone and 'radio' have allowed the transmission of *analogue* information in analogue form, that is as a continuous time-varying electrical waveform.

It is clear that as we enter the 21st century, the era of 'information technology', telecommunication will again revert to an optical frequency carrier and will be predominantly digital in form. It is true that some specialist users may wish to preserve analogue transmission and some to use unguided waves in the atmosphere or in space, but it may confidently be predicted that most information will be transmitted as a sequential stream of light pulses guided along silica optical fibers, the pulses representing binary digital data. It is with the techniques that are being used in the first generation of such systems, and the limitations that are imposed on them, that this book is concerned. The technology and techniques to be described are all well established; nothing new needs to be invented for the wholesale introduction of silica fibers (rather than copper cable) as the principal medium of transmission. Legitimate delays in making this change arise from the need to establish the reliability and life of components and to define suitable standards, but the country which fails to invest in an adequate telecommunication system will be left behind in an age in which the electronic storage, processing and transmission of information will be one of the main activities of business life.

The novelty of modern optical communication systems is that the optical signal is usually guided and the signalling rates are very high. But it needs to be stressed that they are, as yet, no higher than those used in modern, electronic, digital, trunk-telephone equipment. The main incentive for changing from coaxial cable and microwave radio to optical fiber is that of minimizing the overall system cost. It is, therefore, essential that the information capacity of fibers is used to the maximum and that the signalling rates employed should be able to match those of alternative electrical and radio channels. Thus most interest concerns optical pulse repetition frequencies in the range 1 Mb/s to 1000 Mb/s. It has been established, as we shall see later, that the spacing between the repeaters of an optical system can be made sufficiently large for most to be sited in existing telephone exchange buildings. The need for remote, underground repeaters with their associated problems of power supply and maintenance is then largely eliminated. Further advantage derives from the small size, the overall compatibility, and the relative simplicity of the fiber system. The main bulk material, silica, is more abundant and should be cheaper than the copper needed for guided electrical systems.

The modern era of optical communication may be said to have originated with the invention of the laser in 1958, and the first developments soon followed the realization of the first lasers in 1961. Compared to conventional sources of optical frequency, laser radiation is highly monochromatic and coherent, and very intense. In fact it is much more like the radiation generated by a normal microwave transmitter. It is thus a natural step to think of its potential as a carrier for telecommunications. Initially the principal motivation was the enormous bandwidth that would be available, if it were possible to modulate the laser light at only a few percent of its fundamental frequency. The importance of this will be familiar to most readers and is further discussed in Section 1.2. We find that in a system using a helium–neon laser (free space wavelength 0.63 μm, frequency 4.7×10^{14} Hz) a channel bandwidth of 1% of the laser frequency would be 4700 GHz—enough to carry nearly a million simultaneous television channels! The frequencies and wavelengths corresponding to the radio, microwave and optical regions of the electromagnetic spectrum are illustrated in Fig. 1.1. It should be noted that a logarithmic scale has been used. This means that each decade of frequency represents a range, or bandwidth, nine times greater than the whole range of frequencies below it.

The 1960s saw many ideas for laser transmitters using different modulation techniques (frequency modulation, phase modulation, intensity modulation, amplitude modulation, polarization modulation, pulse frequency modulation) and a number of systems using unguided transmission in free space were developed. Some of these will be discussed further in Chapter 16.

During this period a number of people experimented with guided systems in which the laser beam was confined to a transmission channel by lenses spaced at 10 m or 100 m intervals. It is to the credit of K. C. Kao and his fellow workers at the Standard Telecommunications Laboratories in Harlow, England, who were active in this work, that they came to realize that a much simpler guidance system, using a continuous glass fiber of a kind that had already been employed in short lengths for endoscopes and other applications, might be used for telecommunication. The 1966 paper by Kao and Hockham

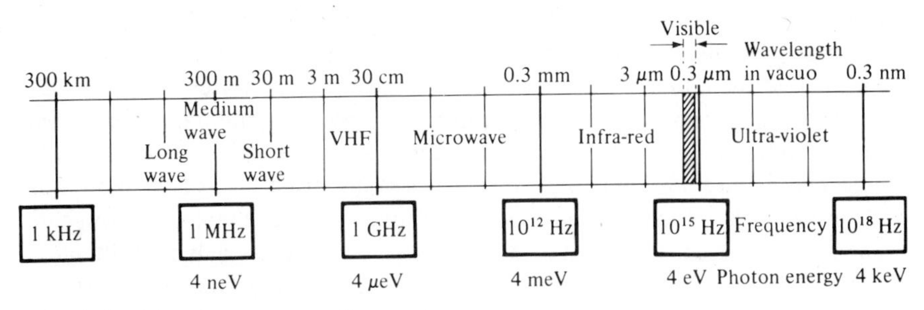

Fig. 1.1 The electromagnetic spectrum.

[1.1] may be said to have laid the foundation for the subject of fiber-optic communications.

The immediate problem was that of optical attenuation in the fiber. Whereas, on a clear day, atmospheric attenuation might be a few dB/km, the best glass then available gave rise to a minimum attenuation of about 1000 dB/km at visible wavelengths. Kao and Hockham's main thesis was that if this could be reduced to 20 dB/km at a convenient wavelength in the visible or near infrared, then practical fiber-optic communication systems could be developed. At such a level of attenuation, the transmitted power would be reduced by a factor of 10^6 after 3 km. By seeking ways to eliminate the absorbing impurities from the fiber, glass manufacturers led by Corning in the USA succeeded in reducing fiber attenuation to this level by 1970 and to 2 dB/km by 1975. More recently Japanese workers have held the published 'records' for minimum attenuation with 0.5 dB/km in 1976 and 0.2 dB/km in 1979. If this last figure could be maintained over long lengths of fiber, the transmitted power would be reduced by a factor of only two after travelling a distance of 15 km. However, it should be emphasized that these later results have involved longer wavelengths (1.55 μm) and laboratory conditions. They are achieved mainly by the elimination of hydroxyl ions (OH^-) from the fiber. Fibers have been shown to be robust and can be formed into cable by conventional manufacturing processes which we shall discuss in Chapter 4.

By 1980 many manufacturers in a number of countries were producing low-loss fiber with attenuation less than 10 dB/km, reliable semiconductor optical sources (GaAs) and detectors (Si) had been developed, and full-scale trials of fiber-optic links integrated into the conventional telephone network were in progress in all countries having a developed communications industry. Some of these trials will be described in Chapter 17. They have, without exception, been entirely successful. So much so that the telephone authorities plan that as the majority of high-capacity trunk telephone links become digital during the 1980s, they will at the same time become optical. There is every indication that optical fibers will be used in the transatlantic submarine cable (TAT 8) scheduled to be laid in 1988.

As we shall find in later chapters, the semiconductor sources that are used are invariably broad-band and cover a range of wavelengths of about 30 nm in the case of light-emitting diodes and about 3 nm in the case of lasers. This means that these first-generation optical communication systems are relatively crude in comparison with a sophisticated modern radio communication system and consist of little more than a broad-band noise source that is simply switched on and off. Some of the earliest radio telegraphy was achieved on the same principle, before the introduction of tuned circuits enabled narrow-band carriers to be used. The result is that the enormous theoretical bandwidth of the optical system is lost, but a simple and cheap system has been realized instead.

Other than in Chapter 16, which deals with *unguided* optical transmission, and the remainder of this chapter, which addresses more general telecommunication problems, our principal purpose will be to make a detailed examination of the various components of optical fiber communication systems and in particular to assess the limitation that each part may place on the bandwidths and the transmission distances that can be obtained.

From what has been said so far it might be thought that it is only communicating over long distances that is of interest to anyone. But, of course, a great deal of data has to be transmitted over relatively short lengths of the order of meters or tens of meters, for example, in telemetry, or within a digital computer or control system, or between computers or other systems. The basic theory which we shall be discussing holds just as well for such links, and optical fiber systems have a number of additional merits which will encourage their use.

We shall conclude this introductory section without further detailed comment by referring to Table 1.1, which lists some of the advantages and disadvantages of optical communication systems in comparison with their electrical counterparts. Where more detailed discussion is to be found in future chapters, forward references are given.

1.2 THE MEASUREMENT OF INFORMATION AND THE CAPACITY OF A TELECOMMUNICATION CHANNEL

The block diagram of a generalized optical communication channel given in Fig. 1.2 shows it to be identical in form with other types of communication system. The difference is that the carrier frequency used is several orders of magnitude higher than those used in radio and microwave systems. The purpose of any communication channel is to send information over a required distance, and we may assess its performance in terms of the amount of information it is able to carry and the distance this information can be sent without intermediate repeaters. In order to do this we need to consider the nature of the information and the way in which it may be quantified. This will then give us a measure of the information-carrying capacity of any proposed communication channel and we shall be able to identify features of the channel that may limit its capacity. For these purposes, some basic ideas from communications theory are introduced in this section, and in Section 1.4 they will be applied in a general way to optical communication channels. This will enable us to identify the likely merits of the optical system and to seek out those components within it that will limit its performance. We shall be using a number of standard results and readers to whom these are unfamiliar are referred to general telecommunications texts such as [1.2]–[1.4] for detailed derivations.

Table 1.1 Advantages and disadvantages of optical communication links in comparison with radio and microwave systems, and with guided electrical systems

Unguided systems (see Chapter 16)	
Advantages	*Disadvantages*
1. Higher ratio of received power to transmitted power, with smaller transmitter and receiver apertures.	1. Less suited to broadcasting because of narrow beam.
2. Better directional resolution with smaller transmitter and receiver apertures.	2. Requires accurate pointing.
3. Very compact transmitting and receiving modules possible for short range applications (~ 1 km).	3. Generation efficiency of optical carrier is low.
4. Good security.	4. There is a significantly higher level of noise in the receiver, due in part to the quantum limited nature of the detection process at optical frequencies (see Chapter 14).
5. Exploits unused part of electromagnetic spectrum.	5. Atmospheric effects.
6. No communications licence required!	6. Possible hazard.

continued

Guided systems

Advantages

1. Low attenuation and low dispersion possible, allowing repeater spacing to be high (10–50 km) (see Chapter 3).
2. Small diameter of individual fiber channel.
3. Tight bending radii possible (see Chapter 3).
4. Low weight of high capacity channel.
5. Bulk material costs low.
6. Cables may be nonconducting and noninductive. This means that the transmission is not subject to electromagnetic interference, that the terminal equipments may be electrically isolated from one another, and that in the event of damage to the cable the hazard of unsecured live conductors is avoided. These advantages are particularly valuable for transmission between high voltage equipment and in environments suffering from excessive electromagnetic noise or from potentially explosive atmospheric conditions.
7. Negligible crosstalk.
8. High security: tapping possible only by tampering with individual fiber and then likely to be detectable.
9. Flexibility of bandwidth: fiber types give possibility of replacing all levels of the digital hierarchies as discussed in Section 1.3.
10. System flexibility gives possibility of 'graceful growth' as improved sources, fibers and detectors become available or as new requirements are demanded, whilst the overall system compatibility is retained. Thus the system may at different times be upgraded in a number of different ways.

Disadvantages

1. Difficulty of splicing fibers (see Chapter 4).
2. Need for additional conducting members in cable when electrical supplies are required for remote terminals.
3. Susceptibility of fiber to ingress of water into cable.
4. Susceptibility to ionizing radiation (see Chapter 3).
5. Poor source efficiency and limited power (see Chapter 9).
6. Laser and LED nonlinearities limit analogue use.
7. Not suited to ternary signalling (compare the use of positive as well as negative levels on digital telephone lines).
8. Fibers not directly suited to multiple-access use, therefore difficult to implement multi-access data bus (see Chapter 17).
9. Higher receiver noise (see Chapter 14).

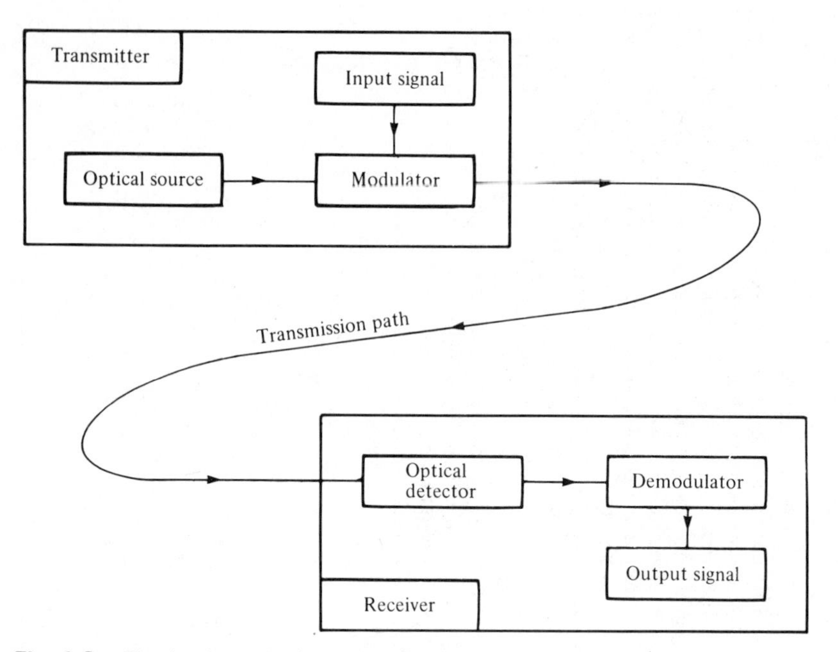

Fig. 1.2 The basic optical communication system.

We note first of all that most information in its original form involves some physical parameter that varies continuously with time and which may take on any of a continuous range of possible values. Familiar examples are the sound pressure waves characterizing speech, and the continuously varying, two-dimensional pattern of light intensity representing an optical image, perhaps one which we may wish to transmit by television. The transducers that convert this information into an electrical signal (the microphone and the camera tube, respectively, in our two examples) leave it as a continuous and continuously time-varying waveform. It remains an *analogue* signal.

By contrast, some information is *discrete*. For example, letters on a page are each individual entities and each has its own meaning. By means of suitable encoding the letters may be transmitted as discrete entities at discrete times. Figure 1.3 shows a familiar optical communication system that does just that!

A third possibility is that the information takes on a discrete form, but one that varies continuously with time. An example of this is the signal derived from a facsimile machine being used to scan a printed page. The output is a continuous function of time but at any instant it is at one of two levels corresponding to black or white.

The fourth and final combination is that of a sampled waveform in which the information may assume any of a continuous range of values, but only at discrete times.

Fig. 1.3 A familiar digital optical communication technique.

However, information in any of these four forms may be quantified by considering what we should have to do in order to convert it to a sequence of discrete binary digits (bits) that would represent it fully. The number of bits that we should require for it to be possible in principle to recover the information in its original form is then a measure of the amount of information we are handling. It is important to be clear that we are not, here, making any judgement about the value of the information—it may be gibberish! We are only assessing the effort that will be required to store or transmit it. We shall first go through this procedure for an arbitrary analogue waveform, and then apply the result to compare the information contained in various familiar forms.

Figure 1.4(a) represents part of an analogue waveform. The first step in converting this to a sequence of binary levels is to sample the waveform at regular discrete intervals of time as shown in Fig. 1.4(b). According to the *sampling theorem* it is necessary that the sampling frequency ($f_s = 1/T$) should be more than twice the highest frequency contained in the original waveform, f_m, if the

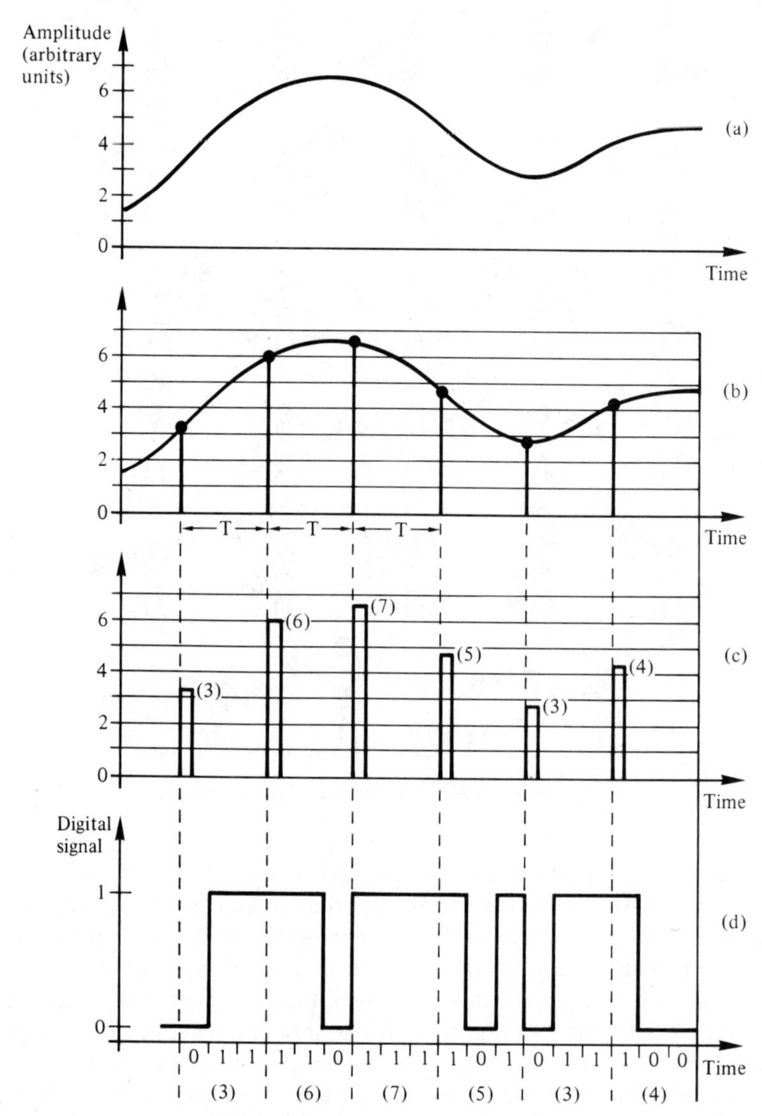

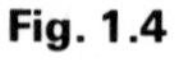
Fig. 1.4 Steps in the conversion of an analogue waveform into a binary digital signal: (a) part of an analogue waveform; (b) analogue waveform sampled at time intervals of T; (c) the sampled waveform showing allocation of amplitude to one of $2^3 = 8$ levels; (d) sampled information converted to binary digital form with 3 bits/sample.

sampled waveform of Fig. 1.4(c) is to represent it fully. Provided this condition, first derived by Nyquist, is met, the original waveform may be recovered simply by putting the sampled waveform through a low-pass filter which lets through frequencies below f_m. The frequency range 0 to f_m is the *bandwidth* of the signal, which in general we shall write as Δf. In this case $\Delta f = f_m$. Thus $f_s > 2\Delta f$.

The next step is to allocate each sampled level to one of a finite number of amplitude levels. In principle the sampled amplitudes may take on any of a continuous range of values. In practice we must remember that there is always a random fluctuation present, superimposed on the intended signal waveform. This is the system *noise* and it makes it unprofitable to attempt to distinguish between levels whose amplitudes are much closer together than the r.m.s. value of the fluctuations. Indeed it is the ratio of the maximum signal amplitude (A_S) to the r.m.s. noise amplitude (A_N) that determines the number of levels needed to give a sufficiently precise representation of the original waveform. Let the chosen number of levels be m. Then each sampled value requires

$$N = \log_2 m$$

binary digits to encode it. When this binary digital signal is decoded back to reproduce the original waveform, the errors produced by the *quantization* of the sampled amplitudes leads to additional noise, known as *quantization noise.* It can be shown that the quantization noise is comparable to or less than the original noise on the waveform provided that m is greater than $[1 + (A_S/A_N)^2]^{1/2}$. So, in order to represent our original analogue waveform, which covered a bandwidth, Δf[Hz], and had a *dynamic range*, A_S/A_N, we shall require a minimum of B binary digits per second [b/s], where

$$B = 2\,\Delta f \log_2 [1 + (A_S/A_N)^2]^{1/2} = \Delta f \log_2 [1 + (A_S/A_N)^2] \quad (1.2.1)$$

We may look at this result in another way. A communication channel which is able to carry analogue signals covering a bandwidth, Δf, and in which it is possible to maintain a ratio of peak signal to r.m.s. noise at the receiver (where this ratio is usually least) of A_S/A_N, is said to have a channel capacity, B[b/s], which is also given by eq. (1.2.1) This equation is known as Shannon's formula and its correct interpretation has been the subject of a great deal of discussion. Here we are seeking no more than to use it as a basis for the comparison of the information content of different types of signal waveform, and for the comparison of the information capacity of different types of communication channel. In each case the information rate is directly proportional to the range of frequencies being sent (Δf) and depends logarithmically upon the minimum (receiver) signal-to-noise ratio. When it comes to practical channels, as we shall see in the examples which follow, the digital encoding of an analogue waveform normally requires a substantially greater bit rate than eqn. (1.2.1) indicates.

Similarly, the amount of information that it is possible to transmit along a communication channel is normally significantly less than predicted by eqn. (1.2.1).

In practical systems A_S/A_N is much greater than unity and is usually quoted in dB. As a result eqn. (1.2.1) can be simplified as follows. Let us say that the value of A_S/A_N is X dB, that is,

$$\frac{X}{[\text{dB}]} = 20 \log_{10}(A_S/A_N)$$

Then,

$$\frac{B}{[\text{b/s}]} = 0.332 \frac{X}{[\text{dB}]} \frac{\Delta f}{[\text{Hz}]} \tag{1.2.2}$$

since $\log_2 10 = 3.32$ and we may neglect 1 in comparison with $(A_S/A_N)^2$.

By way of example, let us now consider the amount of information stored in familiar forms such as a book, a gramophone record, and a videotape of a TV film. We will also examine at the same time the capacity of communication channels needed for the live transmission of audio and video information.

1. *A book.* A printed book of 100,000 words, say 250 pages, may contain an average of 5 letters per word. According to the code used most widely for converting text into digital form, ASCII, each letter is represented by seven binary digits. With $2^7 = 128$, there is ample allowance for upper- and lower-case letters, numerals, spaces and punctuation. The total information content on this basis is therefore 3.5 Mbits.
2. *A speech channel.* The bandwidth required for an intelligible speech channel is about 3 kHz: 300 Hz to 3.4 kHz is the normal specification for a telephone circuit. The overall signal-to-noise ratio should exceed about 30 dB ($A_S/A_N = 31.6$) to be acceptable. Direct application of the Nyquist and Shannon criteria indicates an information rate of 30 kb/s with a minimum sampling frequency of about 6 kHz and at least 5 bits per sample.

   In practice a single, digital, pulse-code-modulated (PCM) telephone voice channel operates at 64 kb/s. The analogue waveform is sampled at 125 μs intervals ($f_s = 8$ kHz) and each sample is encoded into an 8-bit word.
3. *A gramophone record.* Assume that a double-sided, long playing, gramophone record plays for 50 min and produces a signal covering a bandwidth of 20 kHz with a dynamic range of 80 dB. In theory this could be converted to PCM form using 14 bits per sample and a 40 kHz sampling frequency. The information rate of such a signal would be 560 kb/s and the total information content 1.68 Gbits.

Laser-read audio digital disks (Compact Discs) do employ 14 bits per sample, but use a sampling frequency of 44.33 kHz and thus require a channel capable of carrying 620 kb/s.

4. *A film.* A 625-line, PAL, color television, video signal occupies a frequency range of 5.5 MHz. For good quality the ratio of peak signal to r.m.s. noise should be at least 50 dB. Using eqn. (1.2.2) the information rate is then $0.332 \times 50 \times 5.5 \times 10^6 = 91$ Mb/s, and the information transmitted during a 100-minute film would comprise $91 \times 10^6 \times 6000 = 550$ Gbits.

   In practice it is possible to encode the video waveform satisfactorily using 8 bits per sample, but the sampling frequency is governed by the waveform characteristics. It is desirable that f_s should be a multiple of the color subcarrier frequency, $f_{sc} = 4.43$ MHz. Thus $f_s = 13.3$ MHz ($3f_{sc}$) has been used quite widely, but $f_s = 17.7$ MHz ($4f_{sc}$) is regarded as preferable. In the latter case the bit rate required for a digital video channel is 142 Mb/s. Much effort has been expended to try to obtain satisfactory digital transmission more economically. More sophisticated encoding techniques can reduce the required number of bits per sample to 5, and experiments have shown that a sampling frequency of 8.86 MHz, which is twice the color subcarrier frequency but below the Nyquist limit, can give adequate results, provided that certain, special signal processing techniques are used. By such techniques the bit rate required is reduced to 44.3 Mb/s. It must be emphasized that these techniques depend entirely upon the particular characteristics of the PAL video waveform and could not be applied to just any signal of similar overall specification. However, there is considerable redundancy in most live visual images (only a small fraction of the picture changes with each frame) and taking advantage of this in the encoding process can also reduce the bit-rate requirement very considerably.

We need to make one final point and that concerns *multiplexing*, that is the simultaneous transmission of a number of independent signals along one communication channel of high capacity. There are two principal ways in which this may be carried out. The first way, which is more appropriate for analogue signals, is to multiplex in the frequency domain. Each signal will modulate its own carrier, and the carrier frequencies are spaced out within the bandwidth of the high-capacity channel. For example, in a radio transmission system the television signals discussed just now in Example 4 would normally be allocated a channel bandwidth of 8 MHz. If several are to be transmitted simultaneously, each will be allocated a carrier and the carrier frequencies will be spaced at least 8 MHz apart. The second way, which is only suitable for digital signals, is

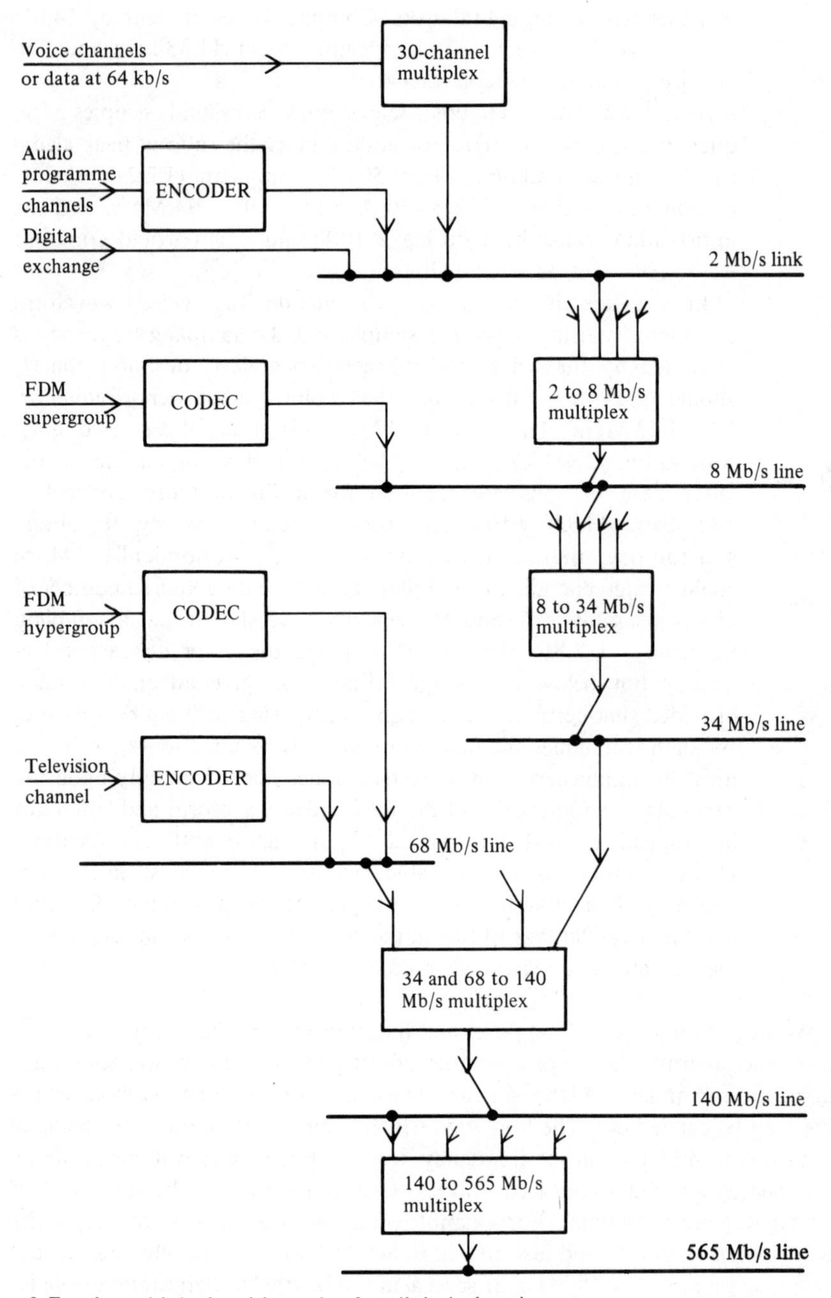

Fig. 1.5 A multiplexing hierarchy for digital signals.

Table 1.2 International systems of time-division-multiplexed digital hierarchies

		Bit rates in Mb/s			Number of multiplexed voice channels	Typical electrical system		Likely optical system	
		USA	Japan	Europe		Usual guided channel type	Typical repeater spacing	Fiber type	Estimated repeater spacing
Level 1	(T1)	1.544	1.544		24	0.63 mm		Step	
						Twisted pair	1.8 km		
				2.048	30	Voice frequency line			10–50 km
								or	
Level 2	(T2)	6.312	6.312		96	1.27 mm			
							3.0 km		
				8.448	120	Screened pair		Graded-index	
Level 3			32.064		480	2.9 mm			
							2.0 km	Graded-index	10–20 km
				34.368	480	Coaxial line			
	(T3)	44.736			672				
Level 4			97.728		1440	1.2/4.4 mm		Graded-index	
							2 km	or	5–50 km
				139.264	1920 (2 TV channels)	Coaxial line		Single-mode	
	(T4)	274.176			4032	2.6/9.5 mm			
							1 km	Single-mode	
Level 5			397.20			Coaxial line			10–40 km
		565		564.992	7680 (8 TV channels)				

to multiplex in the time domain. Thus, pulse code modulated signals from several sources may be interleaved with one another and sent along a single, high bit-rate channel. Of course, means must be found to identify and separate them at their destination and additional bits are required for this.

Figure 1.5 gives an indication of the way in which signals from various sources may be converted to digital form and transmitted at appropriate bit rates in a telecommunication system. It also shows how the levels in the system may be multiplexed together to form a coherent hierarchy. At the first level on the European standard, thirty 64 kb/s voice channels are *time-division-multiplexed* onto a single channel carrying 2.048 Mb/s. Four of these channels are multiplexed to form the second level, and so on. In the system shown, TV signals are digitized at 68 Mb/s and two TV channels may be multiplexed onto the fourth level of the system.

Some of the essential characteristics of three of the digital hierarchies in use in trunk telecommunication systems are set out in Table 1.2. An indication is given of the type of guided electrical and optical channels that would be suitable for the different levels in the hierarchies and the repeater spacings that can be achieved with each. The types of fiber referred to are described in detail in Chapter 2. It should be noted that the T3 level is not in use at the present time in the US except for some experimental optical fiber links. The 'Level 5' bit rates are at the stage of being proposals rather than fully adopted and implemented standards.

1.3 COMMUNICATION SYSTEM ARCHITECTURE

Although we shall chiefly be concerned with the physical nature of optical communication systems and their components, it is as well to be aware of the various structural forms the systems may take, as dictated by different requirements. For example, the following four systems may be identified clearly.

1. Single, independent, point-to-point channels, e.g. telemetric links, field telephones.
2. Broadcast systems, in which information from one or more sources is simultaneously available at many receivers.
3. Point-to-point links with limited switching requirements, either with many receivers connected to relatively few sources, e.g. subscriber access to data banks or TV film libraries; or with many sources connected to a few receivers, e.g. telemetric control systems in which the controllers may wish to select and limit the data they receive.
4. Fully switched point-to-point links in which every terminal can be

connected at choice to every other terminal for one-way or two-way communication, e.g. telephone systems using an exchange or a hierarchy of exchanges, data networks using a multi-access data bus.

In each case transmission may be by means of guided or unguided channels.

Now, it is important to recognize that the electromagnetic spectrum represents a limited natural resource. As such, it needs to be conserved and it should be fully exploited. The use for communications of unguided electromagnetic waves of visible or infra-red frequencies will indeed exploit a hitherto unused part of the spectrum. There is a further advantage in that the lateral spread of an optical channel can be more easily confined to a small space than its radio or microwave counterpart. Thus physically adjacent channels may use the same spectral frequencies with less probability of crosstalk and interference. A similar space advantage obtains with guided optical waves because fibers can be made very much smaller than equivalent electrical transmission links. This is illustrated by Fig. 1.6 and has led some writers to talk of *space-division multiplexing*!

This benefit causes, as a direct consequence, two serious problems for an optical link using unguided transmission. They are the need for accurate tracking and the need for a method of spreading the optical beam when a signal is to be broadcast over a wide area.

In regions where the electromagnetic spectrum is congested it is reasonable to try to restrict the use of unguided transmission to those applications for which it is essential. The only situation which meets this condition is independent of the four requirements we have identified; it is when one or more of the terminals in the system has to be *mobile*. However, the type of system for which unguided radio transmission has shown the greatest economic advantages over guided systems is in broadcasting. The limitations this imposes should not be overlooked. In the UK there is provision for four independent television channels. In parts of the USA it is possible to receive up to eight or ten separate stations. It is hard to see how either of these numbers could be increased significantly using unguided broadcast transmission alone. The hidden costs should not be overlooked either. The broadcasting frequencies occupy the vast majority of the readily usable radio spectrum up to 1 GHz. The amount left for other, especially mobile, users such as police, taxis, citizen-band, air traffic control, radio-telephone, is minimal and quite inadequate.

It would be easy to digress further into these important political matters, but what we need to do here is to relate this discussion to the possible future use of optical communication channels. It is important to recognize, and it may already be clear, that optical links, whether guided or unguided, are most directly suited to independent, point-to-point channels, the first of our four

Fig. 1.6 Space division multiplexing. [© Standard Telecommunication Laboratories. Reprinted by permission of STL Ltd.]

The photograph shows standard $3\frac{1}{2}''$ dia. ducts carrying, respectively, 320 twisted-wire pairs, 20 coaxial cables, and 5 optical fibers. Each has approximately the same information carrying capacity.

system types. In so far as routing and switching may be required between multiple users in a more complex system, this has to be provided by conventional electrical or electronic exchange techniques, for which the signals are converted back from optical to electrical form. When optical channels are introduced into the typical telephone system, the fourth of the system types listed earlier, then they are best regarded as point-to-point links between exchanges or between exchange and subscriber.

The extent to which optical fibers will penetrate the various levels of telephone networks remains a matter of debate. It is clear that their greatest economic advantage lies as high-capacity channels linking the upper levels of the exchange hierarchy. But it seems likely that as optical technology becomes more commonplace and cheaper, their use will be extended down to the lowest levels—the local loop lines between exchanges and subscribers. The fact that the exchange hierarchy. But it seems likely that as optical technology becomes more commonplace and cheaper, their use will be extended down to the lowest levels—the local loop lines between exchanges and subscribers. The fact that these optical fiber lines will, as we shall see, have a much greater information-carrying capacity than their electrical counterpart, perhaps by a factor of a thousand, will leave the telephone authorities in something of a dilemma. With the widespread use of optical fibers the information capacity of transmission lines would by present standards become unlimited. However, the provision of adequate exchange capacity would be much more difficult and very expensive. This, together with the high historic capital investment in existing electrical local loop lines, represent two serious obstacles to the widespread provision of local, high-bandwidth services integrated into the existing telephone system. It may well be that some organizations such as the railroad or the electricity supply authorities, which may have a particular need for such services, will set up their own independent optical networks.

1.4 THE BASIC OPTICAL COMMUNICATION SYSTEM

Let us return to the generalized block diagram of an optical communication system shown in Fig. 1.2. The main components are:

1. The optical source.
2. A means of modulating the optical output from the source with the signal to be transmitted.
3. The transmission medium.
4. The photodetector which converts the received optical power back into an electrical waveform.
5. Electronic amplification and signal processing required to recover the signal and present it in a form suitable for use.

The block diagram may be used equally to represent analogue or digital, guided or unguided systems.

When optical fibers are used as the transmission medium, very few combinations of possible types of source and detector are sensibly compatible. A semiconductor source is called for, and for the receiver a semiconductor junction photodiode detector is required. A great merit of semiconductor light-emitting diode (LED) and laser sources is that the optical power produced can be modulated directly and very easily by controlling the electric current flowing.

Many more combinations of source, detector and modulation method are possible when unguided transmission is used, and a brief discussion of some possible systems is given in Chapter 16. The first requirement is for an optical source of high intensity, and this usually means that a laser is called for. Then, except in the case of semiconductor lasers, an external means of modulating the light is usually needed. Attenuation along the transmission path is often variable, leading to a fluctuating level of received power. In analogue systems this makes it impractical to modulate the optical power directly and a sub-carrier normally has to be used. At the receiver, either semiconductor photodetectors or photomultipliers may be used, but the choice of detector depends upon the wavelength to be used and upon the physical size of the system.

The two most important technical parameters of a communication channel are its information-carrying capacity and the maximum unrepeatered distance the signal can be sent. As we shall see, these two parameters are often closely interrelated. We showed in Section 1.2 that the channel capacity is determined by the signal bandwidth that can be accommodated and by the signal-to-noise ratio at the receiver. We shall examine each of these factors in turn.

Signal bandwidth may be limited at almost any point in the system:

1. By the rate at which the source can be modulated.
2. By the modulator itself.
3. By the transmission medium: if the medium is dispersive it will distort the signal waveform during propagation.
4. By the detector.
5. In the receiver electronics.

In practice LED sources can be modulated at frequencies up to about 100 MHz and laser sources up to about 1 GHz without undue difficulty. Semiconductor p-i-n and avalanche photodiodes are available that are able to respond to optical power modulation frequencies above 1 GHz. However, the use of these highest frequencies does require quite advanced and sophisticated receiver amplifier design.

Most important is the fact that in general an optical fiber acts as a dispersive transmission medium. Optical pulses spread out as they propagate, and phase distortion of analogue signals occurs. In effect the fiber acts as a low-pass filter in which the upper cutoff frequency is inversely proportional to the propagation distance. A given fiber type can be characterized by a *bandwidth–distance product*. This may be lower than 10 MHz.km or higher than 10 GHz.km depending upon the type of fiber and the characteristics of the source used. We discuss fiber dispersion in detail in Chapters 2, 5 and 6.

In unguided optical systems the transmission medium, air or space, is not significantly dispersive and does not, therefore, limit the system bandwidth. But with so many system combinations possible, general comments on the rest of the system are impractical at this stage.

The other important parameter, the signal-to-noise ratio, is determined by the effective noise level at the input to the receiver amplifier, and by the useful optical power reaching the detector. One feature that distinguishes optical communication systems from conventional ones is that the receiver noise contains a component proportional to the received optical power. This is the *shot noise* that is characteristic of the quantum-limited detection process. Thus, in the most usual type of system, where it is the optical power that is modulated, the noise level is signal dependent. It is essential that receiver noise is minimized, but in general it would be expected to increase in proportion to the signal bandwidth.

The received optical power depends upon the power transmitted and the attenuation in the channel. We have already made the point that the level of attenuation that could be attained was one of the key parameters that determined the feasibility of optical fiber systems. It is desirable that the efficiencies of the energy conversion processes at the transmitter (electrical to optical) and the detector (optical to electrical) should be as high as possible. Source efficiencies are particularly poor.

In analogue systems the signal-to-noise ratio directly determines the quality of the channel; in digital systems it determines the probability of error in deciding whether or not a pulse has been transmitted. These are matters which are discussed in detail in Chapter 15, but the following figures may help to give a guide to the sort of performance which might be expected from a digital optical fiber link. It is convenient here to express the various optical power levels in dBm, that is, as a ratio of 1 mW. This notation is customary in communications engineering.

A typical optical power level that can be launched onto a telecommunications grade fiber using an LED source is about 50 μW (–13 dBm). With a laser source it would be in the region of 1 mW (0 dBm). The minimum received optical power needed to ensure a satisfactorily low error rate is typically in the order of 0.1 nW/(Mb/s). Consider a system operating at 10 Mb/s as an example. Then the power level required at the receiver is in the order of 1 nW (–60 dBm). We have to make an allowance for fiber attenuation and leave a system power margin. A generous figure for this might be 10 dB, that is, a factor of ten. The power balance may then be set out as in Table 1.3.

With a fiber attenuation coefficient of 5 dB/km the LED would permit a repeater spacing of 7.4 km and the laser 10 km. Of course, the dispersive properties of the fiber must also be adequate to carry the data at 10 Mb/s over these distances, so its dispersion must be low enough to support bit rate × distance products of 74 (Mb/s) km and 100 (Mb/s) km in the two cases. In

Table 1.3

Source		LED	Laser
Transmitter power		−13 dBm	0 dBm
Minimum required power at receiver	−60 dBm		
Power margin	10 dB		
		−50 dB	−50 dB
Allowed fiber attenuation		37 dB	50 dB

comparison with the standards set by some of the more advanced types of optical fiber, an attenuation of 5 dB/km and a dispersion limit of 70–100 (Mb/s) km represent very modest demands, as we shall see. Much longer repeater spacings at much higher bit rates have been obtained in many of the fully engineered systems described in Chapter 17.

PROBLEMS

1.1 Calculate the transmission frequency and the photon energy of each of the following optical sources:
(a) He–Ne laser, $\lambda = 0.6328\ \mu m$
(b) Nd^{3+} laser, $\lambda = 1.059\ \mu m$
(c) CO_2 laser, $\lambda = 10.6\ \mu m$.

1.2 Calculate the spread of frequencies between the halfpower points of the following optical sources:
(a) A particular GaAlAs laser with a spectral halfwidth of 3 nm about a wavelength of 0.82 μm.
(b) A particular InGaAsP/InP light-emitting diode source having a spectral halfwidth of 110 nm about a peak power wavelength of 1.55 μm.

1.3 Calculate the distance over which the optical power in a beam will be attenuated by a factor of ten when propagating in fibers having the following attenuation coefficients:
(a) 2000 dB/km
(b) 20 dB/km
(c) 0.2 dB/km.

1.4 The attenuation of optical power $P(x)$ with propagation distance x may be expressed in terms of an attenuation coefficient α defined from $P(x) = P(0) \cdot \exp(-\alpha x)$. Obtain the relationship between α expressed in units of $[m^{-1}]$ and the attenuation coefficient expressed in units of [dB/km]. Hence calculate the values of α for the three types of glass of Problem 1.3.

1.5 The information content of a 40-minute, 10,000 word lecture may be communicated: (a) by telegraphing the words, (b) by telephoning the speech waveform, (c) by transmitting a televized videogram. Using the conversion standards discussed on pp. 12 and 13 compare the channel capacity required to carry the lecture by each of these three methods, assuming an average of five letters per word.

REFERENCES

1.1 K. C. Kao and G. A. Hockham, Dielectric-fibre surface waveguides for optical frequencies, *Proc. I.E.E.* **113**, 1151–8 (1966).

1.2 J. Brown and E. V. D. Glazier, *Telecommunications*, 2nd edn, Chapman & Hall (1974).

1.3 M. Schwartz, *Information Transmission, Modulation and Noise*, 3rd edn., McGraw-Hill (1980).

1.4 H. Stark and B. Tuteur, *Modern Electrical Communications Theory and Systems*, Prentice-Hall (1979).

SUMMARY

Table 1.1 lists advantages and disadvantages of optical frequency communication systems.

Although the hypothetical signal bandwidth associated with optical frequencies cannot yet be exploited, and unguided propagation has very limited application, optical fibers offer telecommunications a new medium, particularly suited to the needs of long-distance, high-capacity, point-to-point digital links.

Table 1.2 shows the way in which fibers may be deployed at various levels in the digital communication hierarchies.

Development of low-loss fibers, attenuation <5 dB/km, and easily modulated, high-radiance, semiconductor light sources have made optical fiber links superior in performance to those using electrical transmission lines.

2

Elementary Discussion of Propagation in Optical Fibers

2.1 INTRODUCTION TO FIBER PROPAGATION USING A RAY MODEL

2.1.1 Introduction

In Chapters 2 to 6 we shall consider all aspects of the optical fiber as a transmission medium for optical signals, being particularly concerned to identify those fiber properties that may limit the overall information capacity of an optical fiber communication system. The present chapter will treat the propagation of light rays according to the laws of geometrical optics. The effect of the fiber material on the propagation will be summarized in the material refractive index, n, and initially we shall assume that n is not a function of the optical wavelength. Of course, light is an electromagnetic wave phenomenon and in Section 2.2, a brief summary is given of the theory of the propagation of electromagnetic waves in bulk dielectric media. This provides a basis for understanding both the observed manner in which the refractive index does vary with wavelength in fiber materials and one of the fundamental causes of optical attenuation. We return to consider attenuation fully in Chapter 3, having first given an elementary treatment of total fiber dispersion in Section 2.3 and introduced the idea of the root-mean-square pulse width in Section 2.4. Chapter 4 is concerned with some of the techniques used in the manufacture of fibers and fiber cables.

The ray approximation represents the limit in which the wavelength of the light, λ, tends to zero in comparison with the dimensions of the medium. It supposes that locally the electromagnetic field is that of a plane wave, and the ray trajectory is then normal to the planes of constant phase of the wave. As we shall see, optical fibers may have core diameters as large as 1 mm or as small as a few μm. In some of the most common types of fiber the diameter is in the region of 50 μm. With such fibers the ray approximation might be thought to be reaching the limit of its applicability.

In Chapters 5 and 6 we shall examine the behavior of fibers as dielectric guides for electromagnetic waves of optical frequency, and we shall find that for most types of fiber the assumptions made in the waveguide approach are equivalent to those of the ray approximation. Only when the fiber core diameter becomes very small, as in the type of fiber known as single-mode or monomode fiber, is there a significant difference. The level of the theoretical treatment given in Chapters 5 and 6 is necessarily appreciably above that of the rest of the book. It may well be considered to be beyond the scope of a purely undergraduate text and many readers may wish to skip over the detailed theory. Other readers may find the necessarily brief treatment of the basics of electromagnetic wave propagation and the theory of dielectrics given in the present chapter to be inadequate in its detail or rigor. Such readers are respectfully referred to those many excellent texts which deal more specifically with such matters.

2.1.2 Step-index Fibers: Numerical Aperture and Multipath Dispersion

The fact that a transparent, dielectric medium having a refractive index higher than its surroundings would act as a *lightguide* was demonstrated by Tyndall at a Royal Institution Lecture in 1870 using a water jet.

Refraction at an interface between uniform media is governed by Snell's law, formulated in 1621, and is illustrated in Fig. 2.1(a). There a ray of light is shown, passing from a medium of higher refractive index, n_1, into a medium of

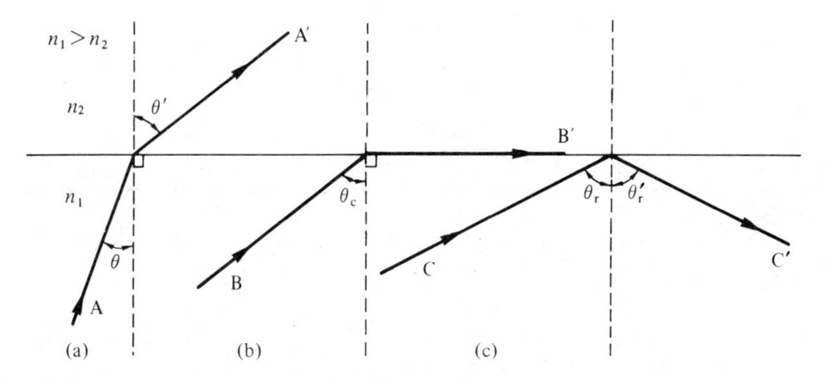

Fig. 2.1 Refraction and total internal reflection at a dielectric interface: (a) ray AA′ is refracted according to Snell's law: $n_1 \sin\theta = n_2 \sin\theta'$; (b) ray BB′ is the critical ray: $n_1 \sin\theta_c = n_2$; (c) ray CC′ is totally reflected at the interface $\theta_r = \theta_r'$.

lower refractive index, n_2. For $0 < \theta < \theta_c$ and $0 < \theta' < \pi/2$,

$$n_1 \sin \theta = n_2 \sin \theta' \tag{2.1.1}$$

where θ and θ' are the angles of incidence and refraction as defined on the diagram, and $\theta = \theta_c$ is the *critical angle*, at which $\theta' = \pi/2$, as shown in Fig. 2.1(b). Thus,

$$n_1 \sin \theta_c = n_2 \tag{2.1.2}$$

For $\theta > \theta_c$ *total internal reflection* occurs with no losses at the boundary, as in Fig. 2.1(c).

Now consider a cylindrical glass fiber consisting of an inner *core* of refractive index n_1, and an outer *cladding* of refractive index n_2, where again $n_1 > n_2$. The end face of the fiber is cut at right angles to the fiber axis and Fig. 2.2 shows a ray entering the end face from the air outside (refractive index n_a). The ray will propagate unattenuated along the fiber by means of multiple internal reflections provided that the angle of incidence onto the core-cladding interface, θ, is greater than the critical angle, θ_c. This requires that the angle of obliqueness to the fiber axis, $\phi = \pi/2 - \theta$, be less than $\phi_m = \pi/2 - \theta_c$, and that the angle of incidence, α, of the incoming ray onto the end face of the fiber be less than a certain value, α_m.

In order to calculate α_m and ϕ_m we assume $n_a = 1$ and apply Snell's law:

$$\sin \alpha = n_1 \sin \phi = n_1 \cos \theta \tag{2.1.3}$$

For the critical rays,

$$\sin \alpha_m = n_1 \sin \phi_m = n_1 \cos \theta_c \tag{2.1.4}$$

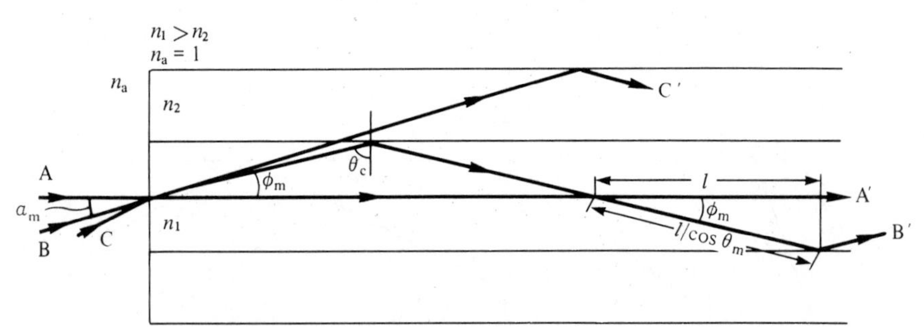

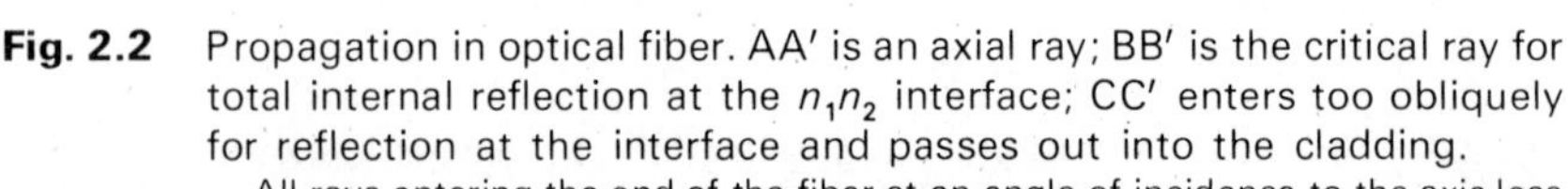

Fig. 2.2 Propagation in optical fiber. AA′ is an axial ray; BB′ is the critical ray for total internal reflection at the $n_1 n_2$ interface; CC′ enters too obliquely for reflection at the interface and passes out into the cladding.

All rays entering the end of the fiber at an angle of incidence to the axis less than α_m will propagate within the core of the fiber.

Clearly the rays propagating within the core travel different path lengths depending on their obliqueness. Over an axial distance l, these will range from l for the axial ray to $l/\cos \phi_m$ for the most oblique ray (critical ray BB′).

But from eqn. (2.1.2)

$$n_1 \sin \theta_c = n_2 \quad \therefore \quad \cos \theta_c = \frac{(n_1^2 - n_2^2)^{1/2}}{n_1} \tag{2.1.5}$$

$$\therefore \quad \sin \alpha_m = (n_1^2 - n_2^2)^{1/2} \tag{2.1.6}$$

If we write

$$\Delta n = n_1 - n_2 \tag{2.1.7}$$

and

$$n = \frac{n_1 + n_2}{2} \tag{2.1.8}$$

then

$$\sin \alpha_m = (2n \, \Delta n)^{1/2} \tag{2.1.9}$$

The greater the value of α_m, the greater is the proportion of the light incident onto the end face that can be collected by the fiber and be propagated by total internal reflection. By analogy with the term used to define the light-gathering power of microscope objectives, $n_a \sin \alpha_m$ is known as the *numerical aperture* (NA) of the fiber. Thus, putting $n_a = 1$,

$$(\mathrm{NA}) = \sin \alpha_m = (2n \, \Delta n)^{1/2} \tag{2.1.10}$$

We shall show first that of the light emitted by a small diffuse source situated on the fiber axis near to the end face, only a fraction $(\mathrm{NA})^2$ can be collected by and propagated along the fiber.

Consider a small diffuse light source such as the isotropic (Lambertian) radiator shown in Fig. 2.3, in which the power radiated per unit solid angle in a

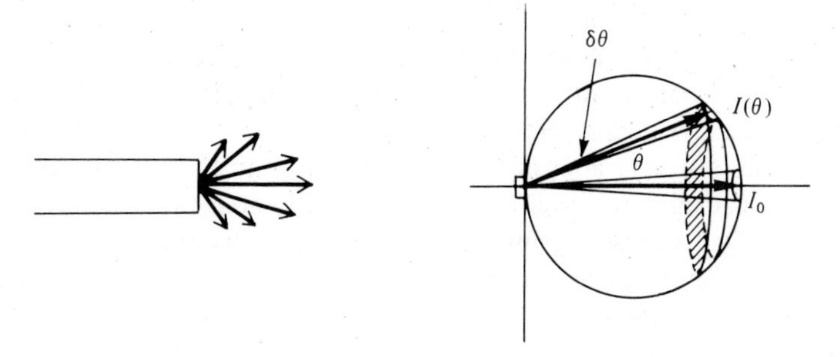

Fig. 2.3 Diffuse light source.

The power radiated into a small solid angle $\delta\Omega$ in a direction θ to the perpendicular is $I(\theta)\,\delta\Omega = I_0 \cos \theta \, \delta\Omega$. The elementary annular ring, whose radius subtends an angle θ and whose width subtends an angle $\delta\theta$ at the source, itself subtends a solid angle $\delta\Omega = 2\pi \sin \theta \, \delta\theta$.

direction at an angle θ to the normal to the surface is given by

$$I(\theta) = I_0 \cos\theta \tag{2.1.11}$$

The total power Φ_0 emitted by such a source is obtained by integrating $I(\theta)$ over all forward directions.

$$\Phi_0 = \int_0^{\pi/2} (I_0 \cos\theta)(2\pi)(\sin\theta)\, d\theta$$
$$= -2\pi I_0 [\tfrac{1}{2} \cos^2\theta]_{\theta=0}^{\pi/2} = \pi I_0 \tag{2.1.12}$$

But the power from such a source that can be collected by an adjacent fiber whose core diameter is greater than the diameter of the source is given by Φ, where

$$\Phi = \int_0^{\alpha_m} (I_0 \cos\theta)(2\pi)(\sin\theta)\, d\theta$$
$$= -\pi I_0 [\cos^2\theta]_{\theta=0}^{\alpha_m}$$
$$= \pi I_0 \sin^2\alpha_m = \Phi_0 (\mathrm{NA})^2$$
$$\therefore \quad \frac{\Phi}{\Phi_0} = (\mathrm{NA})^2 = 2n\,\Delta n \tag{2.1.13}$$

Clearly, in order to collect as much light as possible it is necessary to make n and Δn large, and the best we can do is to use a glass with high refractive index without any cladding. Then the total internal reflection will occur at the glass–air surface just as with Tyndall's water jet it occurred at the water–air surface. Some short-range optical data transmission systems do use bundles of unclad glass fibers in which this condition is satisfied. There are two problems with this arrangement. First, when light suffers total internal reflection an electromagnetic disturbance, known as an *evanescent wave*, does penetrate the reflecting interface. The amplitude of the disturbance decays exponentially with distance away from the surface and cannot normally propagate in the medium of lower refractive index. However, any variation or non-uniformity in the region of the reflecting interface may cause the conversion of the evanescent wave into a propagating wave. In an unclad fiber bundle, the surface conditions are necessarily variable and uncontrolled as the individual fibers come into contact with each other and with their surroundings. As a result a significant fraction of the propagating power is coupled out of the fibers, and high attenuation results.

The second problem is that any short pulse of light launched onto the fiber consists of some rays which propagate along the fiber axis and some which have the trajectories of the most oblique rays. These two extremes are illustrated in Fig. 2.2. Now the refractive index of a medium is simply a measure of the velocity of propagation, v, of the light in the medium,

$$v = c/n \tag{2.1.14}$$

An axial ray, then, will travel an axial distance, l, along the fiber in a time $n_1 l/c$ while the most oblique ray that can be propagated will cover the same axial distance in a time given by

$$\frac{n_1 l}{c \cos \phi_m} = \frac{n_1 l}{c \sin \theta_c} = \frac{n_1^2 l}{n_2 c} \tag{2.1.15}$$

If the two rays are launched together they will be separated on arrival by ΔT where

$$\Delta T = \frac{n_1}{n_2} \frac{l}{c} \Delta n \tag{2.1.16}$$

and a pulse containing rays at all possible angles will spread out during propagation by an amount given by

$$\frac{\Delta T}{l} = \frac{n_1}{n_2} \frac{\Delta n}{c} \tag{2.1.17}$$

This is known as the *multipath time dispersion* of the fiber. For an unclad glass fiber with, say, $n_1 = 1.5$, $n_2 = 1$, and taking $c = 3 \times 10^8$ m/s, eqn. (2.1.17) gives

$$\frac{\Delta T}{l} = 2.5 \times 10^{-9} = 2.5 \text{ ns/m} = 2.5 \text{ } \mu\text{s/km}$$

In this case light incident on the end face of the fiber at all angles would be expected to propagate.

Cladding the core with glass having a slightly lower refractive index gives rise to three effects:

(i) if the cladding is of high quality and of sufficient thickness to contain the evanescent wave it will greatly reduce attenuation;
(ii) time dispersion will be reduced;
(iii) the light gathering power of the fiber will be reduced.

When $\Delta n \ll n$, eqn. (2.1.17) for the time dispersion of the fiber may be approximated by

$$\Delta T/l \simeq \Delta n/c \tag{2.1.18}$$

A *step-index* fiber of this type is shown in Fig. 2.4. Cables based on such fibers are widely available. If we take as typical values, $n = 1.5$ and $\Delta n = 0.01$, then on the basis of the results derived so far, the numerical aperture is $(\mathrm{NA}) = 0.173$, the acceptance angle is $\alpha_{\mathrm{m}} = 10°$ and the fraction of the optical power from a diffuse source that the fiber will propagate is $(\mathrm{NA})^2 = 0.03 = 3\%$. Finally the time dispersion of the fiber is

$$\frac{\Delta T}{l} = 3.4 \times 10^{-11} = 34 \text{ ns/km}$$

The relation between the time dispersion of a length of fiber, ΔT, and its signal bandwidth, Δf, and the maximum bit rate, B, that it will support will be discussed more rigorously in Section 2.4 and Chapter 15. To an approximation that is sufficient for most purposes, however, we may take

$$B \simeq 2\Delta f \simeq \frac{1}{\Delta T} \tag{2.1.19}$$

Then

$$(\Delta f)l \simeq \frac{1}{2}\frac{c}{\Delta n} \tag{2.1.20}$$

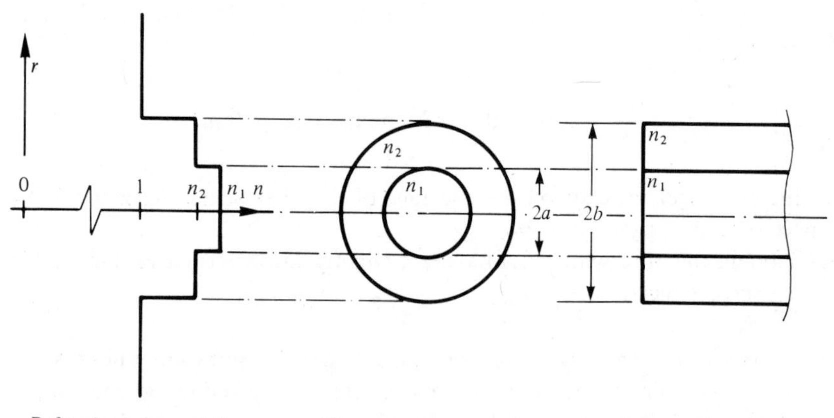

Fig. 2.4 Step-index fiber.
Core and cladding diameters, $2a$ and $2b$, are tending to standardize on 50 μm and 125 μm respectively, but many other sizes are made. In some applications larger diameters are required, for example core diameters of 100 μm to 300 μm and cladding diameters in the range 200 μm to 500 μm. Such fibers tend to be rather stiff.

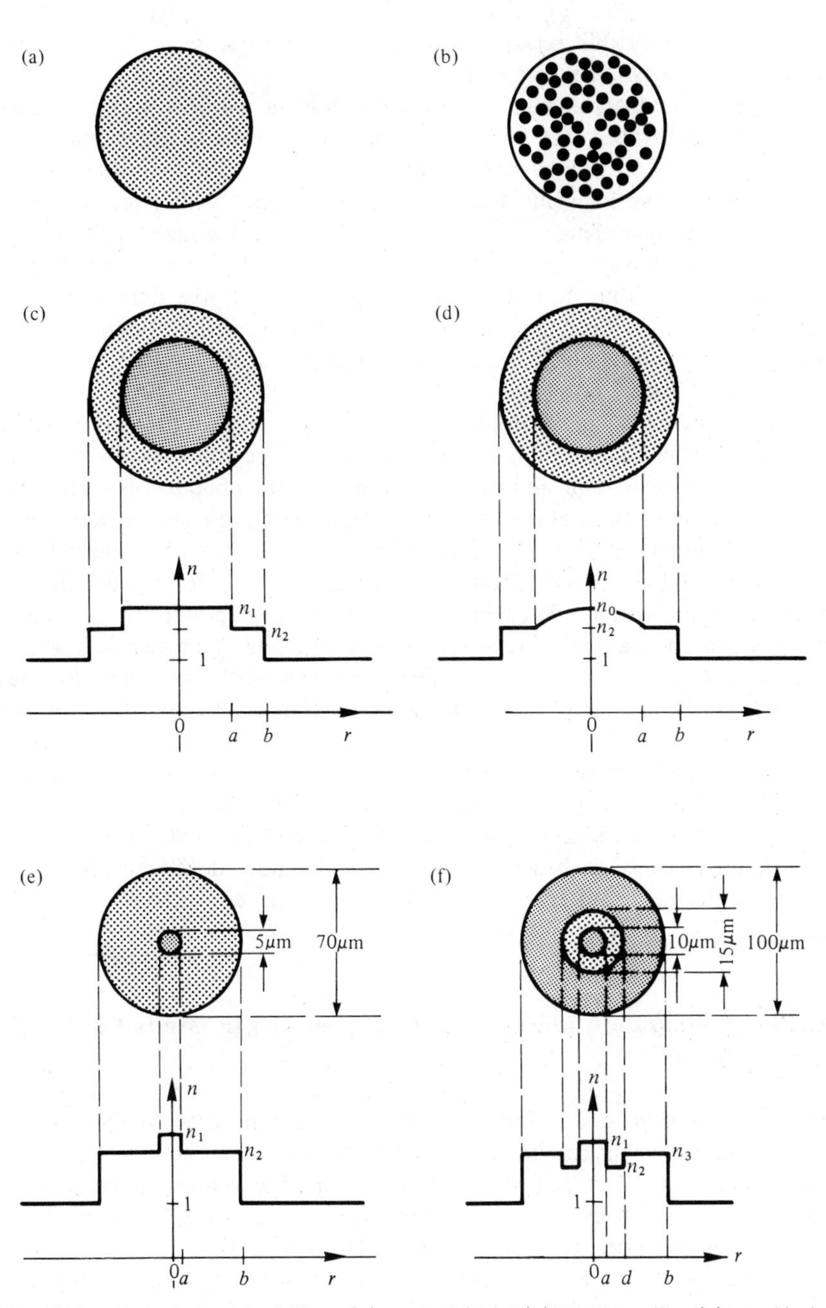

Fig. 2.5 Types of optical fiber: (a) unclad fiber; (b) fiber bundle; (c) step-index fiber; (d) graded-index fiber; (e) single-mode fiber; (f) W-profile fiber.

Hence the fiber of this example may be said to have a *bandwidth–distance product* of approximately 15 MHz.km.

So far we have considered only rays which enter the fiber such that they pass through the fiber axis, the so-called *meridional rays*. In general there are also rays which propagate without satisfying this condition: these are *skew rays*. It is possible for some skew rays to be confined within the core even though they are travelling at very oblique angles to the fiber axis. It seems that in practice such rays are soon scattered out of the core at bends and irregularities, and thus do not contribute significantly to the time dispersion, but a proper analysis is complex. The question of the amount of optical power that can usefully be coupled into a fiber from an extended source is a matter we return to in Chapter 8.

The bandwidth–distance product indicated by eqn. (2.1.20) is in practice over-pessimistic. As a result of scattering within the fiber, the most oblique rays suffer higher attenuation and over long distances the obliqueness of the trajectories of the more axial rays tends to average out. These are matters which we shall return to in Section 6.6, but as a result of such effects dispersion is reduced and in long lengths of fiber may increase as $(\text{distance})^{1/2}$. Nevertheless, dispersion represents a severe constraint on the use of step-index fibers, confining them to applications of relatively short range and of modest bandwidth. The example given at the end of Chapter 1 illustrates this. There are two types of fiber which seek to overcome this limitation, and these, together with the step-index fiber and the fiber-bundle, are illustrated by means of section diagrams and refractive index profiles in Fig. 2.5. One of these, the *graded-index* fiber shown in Fig. 2.5(d) has come to dominate the early phase of fiber development and we shall deal with this next. The *single-mode* fiber shown in Fig. 2.5(e) is likely to become important with future developments and is discussed further in Section 2.3 and in Chapter 5 where the possible advantages of the W-profile shown in Fig. 2.5(f) are also mentioned.

2.1.3 Propagation and Multipath Dispersion in Graded-index Fibers

The principle of the graded-index fiber is readily described qualitatively, but a proper theory soon results in considerable analytical complexity. With reference to Fig. 2.6, it can be seen that the axial ray follows the most direct path through the fiber but travels through the most refractive part of the medium and therefore travels most slowly. The oblique rays necessarily travel further but much of their path lies in a medium of lower refractive index, in which they travel faster. With the correct refractive index profile it might be

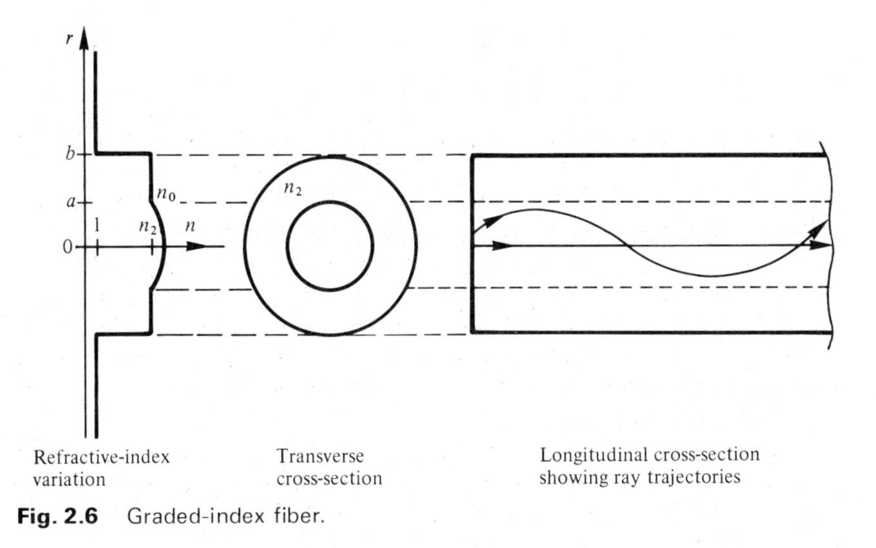

Fig. 2.6 Graded-index fiber.

thought that all possible rays entering at one point could be brought to a focus at a succession of periodically spaced points along the fiber. By Fermat's principle this would imply that their axial velocities were all the same, and that time dispersion was zero.

It can be shown (see, e.g., Section 3.2.1 of Ref. [2.1]) that the trajectory of a ray passing through a medium of nonuniform refractive index is given by

$$\frac{\mathrm{d}}{\mathrm{d}s}\left(n\frac{\mathrm{d}\mathbf{r}}{\mathrm{d}s}\right) = \nabla n \tag{2.1.21}$$

where $\mathbf{r}$ is the general position vector of a point on the ray path, and $\mathrm{d}s$ is an elementary distance measured along the trajectory.

We shall apply eqn. (2.1.21) to the case of a cylindrical fiber in which the refractive index has radial symmetry. For the moment we shall restrict the discussion to meridional rays, and furthermore to those meridional rays that remain always nearly parallel to the fiber axis. This is the paraxial ray approximation, and it allows us to approximate $\mathrm{d}s$ to the axial distance $\mathrm{d}z$. Then eqn. (2.1.21) reduces to

$$\frac{\mathrm{d}^2 r}{\mathrm{d}z^2} = \frac{1}{n}\frac{\mathrm{d}n}{\mathrm{d}r} \tag{2.1.22}$$

where r is now the radial distance of the ray from the axis and z is a distance measured along the axis. It is easy to show that a parabolic profile will give rise to a sinusoidal variation of r with z. Let

$$n(r) = \begin{cases} n_0 \left[1 - \Delta' \left(\dfrac{r}{a} \right)^2 \right] & \text{for } r < a \\ n_0(1 - \Delta') = n(a) & \text{for } r \geqslant a \end{cases} \tag{2.1.23}$$

where n_0 is the refractive index on the axis, a is the core radius, and

$$\Delta' = \frac{n_0 - n(a)}{n_0} \tag{2.1.24}$$

is the fractional total change of the core refractive index.

Differentiating eqn. (2.1.23) gives

$$\frac{\mathrm{d}n}{\mathrm{d}r} = -\frac{2n_0 r}{a^2} \cdot \Delta' \tag{2.1.25}$$

By further restricting the discussion to rays that remain close to the axis we may assume that $n_0/n \simeq 1$, so that eqn. (2.1.22) becomes

$$\frac{\mathrm{d}^2 r}{\mathrm{d}z^2} \simeq -\frac{2r}{a^2} \cdot \Delta' \tag{2.1.26}$$

If we consider the ray which enters the fiber such that $r = r_0$ and $\mathrm{d}r/\mathrm{d}z = r_0'$ at $z = 0$, integration of eqn. (2.1.26) shows that its trajectory is

$$r = r_0 \cos (2\Delta')^{1/2} \frac{z}{a} + r_0' \frac{a}{(2\Delta')^{1/2}} \sin (2\Delta')^{1/2} \frac{z}{a} \tag{2.1.27}$$

The trajectories of two such groups of rays are illustrated in Fig. 2.7; those for which $r_0 = 0$ and those for which $r_0' = 0$. All are non-dispersive.

If we attempt to relax the paraxial ray approximations, the equations become considerably more complicated. However, it may be shown (see Ref. [2.2]) that *all* meridional rays are nondispersive when the refractive index profile takes the form

$$n(r) = n_0 \operatorname{sech} \alpha r \simeq n_0 \left[1 - \tfrac{1}{2} \alpha^2 r^2 + \tfrac{5}{24} \alpha^4 r^4 + \cdots\right] \tag{2.1.28}$$

The polynomial expansion shows that the parabolic index is a first approximation to this, with $\Delta' = (\alpha a)^2/2$. Skew rays are another matter and there is no theoretical profile that will make them nondispersive, independent of initial position and angle, and nondispersive with respect to meridional rays.

The practical problems involved in producing graded-index fibers will be examined in Chapter 4 and we shall return again to the theory of wave and ray

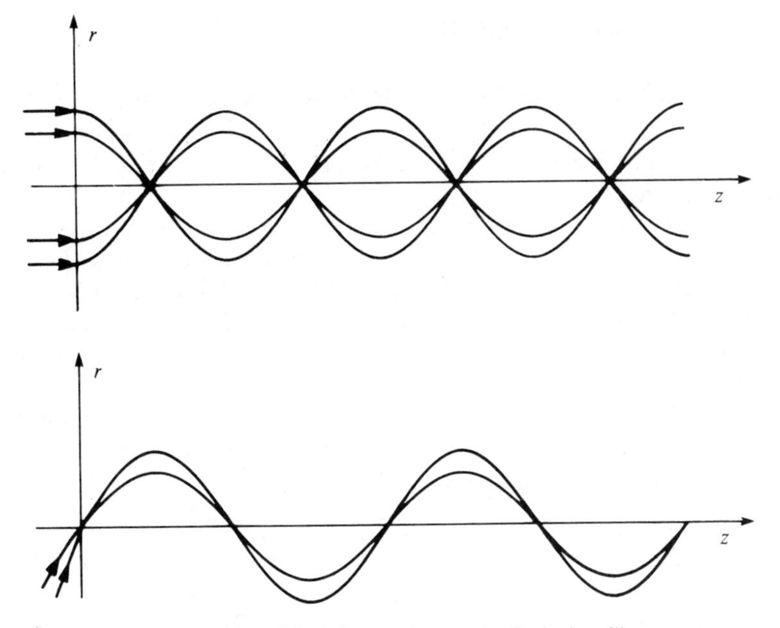

Fig. 2.7 Trajectories of meridional rays in parabolic index fiber. The refractive index profile

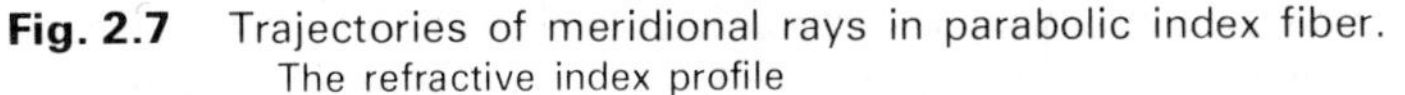

$$n(r) = n_0 \left[1 - \Delta' \left(\frac{r}{a} \right)^2 \right]$$

is assumed to encompass all the ray trajectories shown. The diagrams are intended to imply circular symmetry about the horizontal axes.

propagation in them in Chapter 6 and Appendix 3, respectively. There we shall find that with an ideal profile multipath dispersion would be reduced below 0.1 ns/km. In practice, well graded fibers can be readily manufactured so that their multipath dispersion is less than 1 ns/km. However, even a crude grading of the index profile can be beneficial. The time dispersion of the step-index fiber used as an example in Section 2.1.2 might be reduced from 34 ns/km to something less than 10 ns/km simply by smoothing out the refractive index change between core and cladding.

Before returning to these matters, it is necessary to consider a further, and quite unrelated, source of time dispersion in optical fibers: that arising from the fact that the refractive index is indeed a function of wavelength. This might well be called chromatic dispersion but it is most commonly referred to as *material dispersion.*

2.2 MATERIAL DISPERSION

2.2.1 The Refractive Index of Bulk Media: Theory

The speed of transmission of electromagnetic waves through transparent materials is influenced by the interaction of the waves with the molecules of the material. Because the interaction is a function of the wave frequency, so too is the velocity of propagation of the waves: the material is said to be *dispersive*. One consequence of this is that short pulses of light tend to spread out as they propagate. The spread is proportional to the range of optical frequencies in the pulse and constitutes another important factor limiting the bandwidth of optical fibers.

In optics it is normal to talk, as we have so far, about the refractive index, n, of the medium. This tells us the factor by which the phase velocity, v_p, of the waves is reduced below their velocity in vacuum, c:

$$v_p = \frac{c}{n} \tag{2.2.1}$$

Another customary usage which derives from the 17th century scientific origins of optics is to talk not of the frequency, f, of optical sources but of their wavelength, λ. This refers to the free space wavelength, and $\lambda = c/f$. Within a refractive medium the wavelength is reduced locally to λ_m, such that,

$$\lambda_m = \lambda/n \tag{2.2.2}$$

and

$$v_p = \lambda_m f \tag{2.2.3}$$

We will represent an electromagnetic wave of frequency, f, propagating in the z-direction through a refractive medium by writing the amplitude of its electric field component in the x-direction as the real part of E_x, where

$$E_x(z, t) = E_0 \exp\{-j(\omega t - \beta z)\} \tag{2.2.4}$$

and where E_0 is a field constant; $\beta = 2\pi/\lambda_m$ is the propagation constant in the medium; $\omega = 2\pi f$ is the angular frequency of the wave; and $j^2 = -1$.

Equation (2.2.4) represents a plane wave travelling through bulk material and we may imagine it to be plane-polarized so that the electric field vector is confined to the x–z plane. The phase velocity of this wave is $v_p = \omega/\beta$, thus

$$v_p = \frac{\omega}{\beta} = \lambda_m f = \frac{c}{n} \tag{2.2.5}$$

and

$$n = \frac{c}{\omega}\beta \tag{2.2.6}$$

If the wave is attenuated as it passes through the medium, then we may introduce an attenuation coefficient, a, so that:

$$\begin{aligned} E_x(z, t) &= E_0 \exp(-az) \exp\{-j(\omega t - \beta z)\} \\ &= E_0 \exp\{-j[\omega t - (\beta + ja)z]\} \end{aligned} \tag{2.2.7}$$

This behavior may equally well be represented by defining a complex refractive index, n^*, such that

$$n^* = n + jn' = \frac{c}{\omega}(\beta + ja) \tag{2.2.8}$$

Thus the real part of the refractive index is still given by eqn. (2.2.6) while the imaginary part becomes

$$n' = \frac{c}{\omega} a \tag{2.2.9}$$

We shall show that the same physical processes that give rise to a variation of refractive index with frequency also give rise to attenuation of electromagnetic waves. Thus the refractive index of a dispersive medium is both a function of frequency and complex. These physical processes can easily be described qualitatively for a typical dielectric medium, but quantitative theoretical treatment for any but the simplest materials is impossibly complicated.

The electric field component of the propagating optical electromagnetic wave produces polarization in the molecules of the material such that they or their electronic structure oscillate at the wave frequency. The effect of the oscillating charge is to radiate new waves at the same frequency and these interfere with the original wave in such a way that the resultant wave has a net phase shift with respect to the original. Because this is happening continuously, the total phase shift is proportional to the propagation distance and the resultant wave appears to travel with a lower phase velocity.

The interaction between the wave and the molecules takes the form of a series of damped harmonic resonances. Above each resonant frequency the particular atomic or electronic charge motion that gives rise to the interaction is unable any longer to follow the electric field oscillations. The material no longer polarizes in that particular way, with the result that the refractive index is reduced below its value at frequencies below the resonance.

The effect of an electrostatic field in polarizing a dielectric material is normally expressed in terms of the relative dielectric constant or permittivity, ε_r, of the material. The refractive index results from the polarizing action in a high-frequency field and can easily be related to the high-frequency permittivity of the material. It is known from electromagnetic theory that the phase velocity of electromagnetic waves in a medium having a relative permeability, μ_r, and a relative permittivity, ε_r, is

$$v_p = \frac{1}{\sqrt{(\mu_0 \mu_r \varepsilon_0 \varepsilon_r)}} = \frac{c}{\sqrt{(\mu_r \varepsilon_r)}} \tag{2.2.10}$$

where μ_0 and ε_0 are respectively the permeability and permittivity of free space. Thus $n = \sqrt{(\mu_r \varepsilon_r)}$, and since magnetic effects are normally negligible in nonmagnetic materials, we may put $\mu_r = 1$, so that in practice,

$$n = \sqrt{\varepsilon_r} \tag{2.2.11}$$

See Appendix 1.

What we need then is a theory that will give ε_r and hence n as a function of frequency at optical frequencies. Many physics texts which deal with electromagnetism, optics and the solid state give elementary treatments, and readers are referred to them for greater detail.

The usual starting point is to take the polarizability of an individual molecule in the material as being α. This means that the electric dipole moment, p_x generated in a local electric field in the x-direction, E_x, is

$$p_x = \alpha E_x \tag{2.2.12}$$

In a dilute (i.e. gaseous) material containing N molecules per unit volume, the volume polarization, P_x, is

$$P_x = N p_x = N \alpha E_x \tag{2.2.13}$$

Now, the relative permittivity, ε_r, is defined as

$$\varepsilon_r = \frac{\varepsilon_0 E_x + P_x}{\varepsilon_0 E_x} = 1 + \frac{P_x}{\varepsilon_0 E_x} \tag{2.2.14}$$

Thus

$$\varepsilon_r = 1 + \frac{N\alpha}{\varepsilon_0} \tag{2.2.15}$$

In a solid dielectric an allowance has to be made for the polarizing effect on any one molecule of the polarization of all the surrounding molecules. In the simplest approximation, which is exact for an ideal cubic lattice, each polarizable molecule is assumed to be in a spherical cavity in otherwise uniform dielectric. The local field is thereby increased by a factor $(1 + P_x/3\varepsilon_0 E_x)$ over the average field, E_x. The polarization thus becomes

$$P_x = N\alpha E_x(1 + P_x/3\varepsilon_0 E_x) = N\alpha E_x + N\alpha P_x/3\varepsilon_0 = \frac{N\alpha E_x}{1 - (N\alpha/3\varepsilon_0)} \tag{2.2.16}$$

The resulting relative permittivity is obtained by substituting eqn. (2.2.16) into eqn. (2.2.14):

$$\varepsilon_r - 1 = \frac{N/\varepsilon_0}{(1/\alpha - N/3\varepsilon_0)} \tag{2.2.17}$$

This result is sometimes expressed in the alternative form derived by Clausius and Mossotti:

$$\frac{\varepsilon_r - 1}{\varepsilon_r + 2} = \frac{N\alpha}{3\varepsilon_0} \tag{2.2.18}$$

Now for a typical dielectric the mean molecular polarizability, α (and hence the average polarization per unit volume, P_x) varies with the frequency of the exciting electric field in the manner illustrated in Fig. 2.8. The 'radio frequency' transition derives from heavily damped, orientational effects and plays no part at frequencies of interest to us. The other transitions are the result of the resonances described earlier. The higher-frequency effect derives from the response of the electronic structure of the molecules to the optical field. In practice there are a number of such resonances in the ultra-violet part of the spectrum. The lower frequency transition is just one shown out of several that

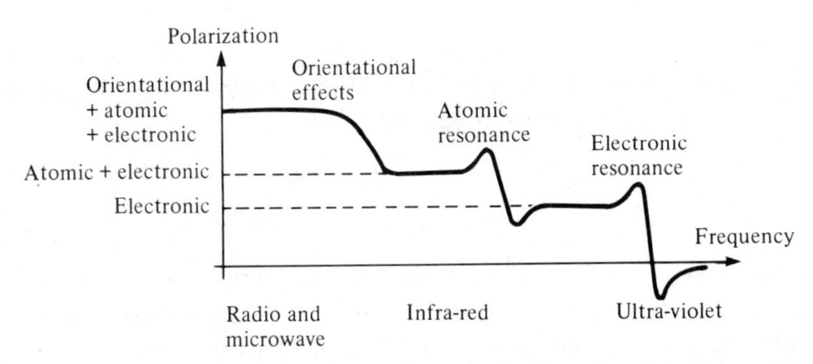

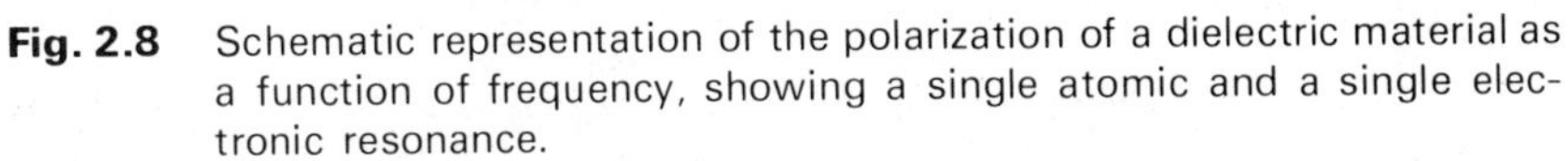

Fig. 2.8 Schematic representation of the polarization of a dielectric material as a function of frequency, showing a single atomic and a single electronic resonance.

result from the motion of the molecules as a whole in response to the optical field. These are lattice vibrations and are excited by the electric field at frequencies in the infra-red region of the spectrum. In both sets of resonances, the charge which is displaced in the interaction with the electric field and which thus gives rise to polarization is subject to a restoring force which is proportional to the displacement. The charge motion, then, is that of a harmonic oscillator. The electric field in the x-direction resulting from the electromagnetic wave at a given point in the material is obtained by setting z = constant in eqn. (2.2.7) and may be expressed as the real part of $E_x = E_1 \exp(-j\omega t)$ where E_1 is a constant field.

Then the differential equation relating the displacement, x, of an electronic charge, e, of mass, m, subject to this electric field is

$$\ddot{x} + \gamma_k \dot{x} + \omega_{0k}^2 x = \frac{e}{m} E_1 \exp(-j\omega t) \tag{2.2.19}$$

where $\omega_{0k}/(2\pi)$ is the resonant frequency of the particular interaction represented by the subscript k, and γ_k is a damping factor representing the dissipative effects associated with this interaction and resulting from radiation losses and collisions.

The solution for a forced, damped oscillation of this type gives

$$x = \frac{(eE_1/m)\exp(-j\omega t)}{\omega_{0k}^2 - \omega^2 - j\gamma_k \omega} \tag{2.2.20}$$

The molecular polarizability is now seen to become a complex function of frequency. We shall represent it as α^*, where

$$\alpha^* = \frac{p_x}{E_x} = \frac{xe}{E_x} = \frac{(e^2/m)}{\omega_{0k}^2 - \omega^2 - j\gamma_k \omega} \tag{2.2.21}$$

The relative permittivity is similarly a complex function of frequency, $\varepsilon_r^*(\omega)$, which may be obtained by substituting the complex form for the atomic polarizability, eqn. (2.2.21), into eqn. (2.2.17):

$$\varepsilon_r^*(\omega) = 1 + \frac{Ne^2/m\varepsilon_0}{\omega_{0k}^2 - \omega^2 - j\gamma_k \omega - Ne^2/3m\varepsilon_0} \tag{2.2.22}$$

If we take account of all possible resonances and represent the strength of each by the factor g_k (a term which arises in a quantum-mechanical formulation of the problem), then ε_r^* becomes, as a function of frequency,

$$\varepsilon_r^*(f) = 1 + K \sum_k \frac{g_k}{f_{1k}^2 - f^2 - j\gamma_k f/(2\pi)} \tag{2.2.23}$$

where

$$f_{1k}^2 = \frac{1}{4\pi^2}\left(\omega_{0k}^2 - \frac{Ne^2}{3m\varepsilon_0}\right) \quad \text{and} \quad K = \frac{Ne^2}{4\pi^2 m\varepsilon_0}$$

Clearly the refractive index, too, is now complex, and with

$$n^* = n + jn' \tag{2.2.8}$$

we have

$$(n^*)^2 = [n^2 - (n')^2] - 2jnn' = \varepsilon_r^* \tag{2.2.24}$$

In materials of interest to us attenuation must be very small so we shall be concerned only with frequencies well away from the resonant frequencies, where we can assume that $n' \ll n$. Then

$$n^2 \simeq \mathrm{Re}\,(\varepsilon_r^*) \tag{2.2.25}$$

and

$$2nn' = \mathrm{Im}\,(\varepsilon_r^*) \tag{2.2.26}$$

The real and imaginary parts of n^* for an idealized dielectric are shown in Fig. 2.9. Away from the resonances the effect of the imaginary part of ε_r^* may be neglected and the refractive index may then be expressed as

$$n^2 - 1 = \sum_k \frac{Kg_k}{f_{1k}^2 - f^2} = \sum_k \frac{G_k \lambda^2}{\lambda^2 - \lambda_{1k}^2} \tag{2.2.27}$$

where

$$\lambda_{1k} = c/f_{1k} \quad \text{and} \quad G_k = \frac{Kg_k \lambda_{1k}^2}{c^2}$$

The variation of the refractive index of optical materials is usually expressed in the form of eqn. (2.2.27), which is known as the *Sellmeier dispersion formula*. It is one of the humbling facts of science history that the essential features of this analysis were first developed by Maxwell in an *examination question* he set for the Cambridge Mathematical Tripos in 1869. W. Sellmeier independently arrived at eqn. (2.2.27) in a series of papers published in *Annalen der Physik und Chemie* in 1872.† It has subsequently been derived on a number of occasions using apparently more sophisticated models for dielectric materials, but the essential mechanism of dispersion theory has

† W. Sellmeier, Concerning the forced vibrations of particles of matter excited by vibrations of the aether [i.e. *electromagnetic waves*] and their reaction on the latter, particularly for the purpose of explaining dispersion and its anomalies. *Annalen der Physik und Chemie* (ed. J. C. Poggendorff), (5th Series) **145**, 399–421 & 520–49 (1872); **147**, 386–403 & 525–54 (1872).

remained unchanged in these models, namely the response of elastically bound charge in the high-frequency electric field.

An excellent fit to experimental data can usually be obtained by taking three terms in the dispersion formula, two corresponding to electronic resonances in the u.v. and one deriving from an atomic resonance in the i.r.

The form in which eqn. (2.2.27) is presented is rather cumbersome for purposes of analysis, and because we are concerned with values of λ that are far removed from any of the λ_{1k} values, we can approximate the Sellmeier equation by a polynomial series in λ^2. We have

$$\frac{G_k \lambda^2}{\lambda^2 - \lambda_{1k}^2} = \frac{G_k}{1 - \lambda_{1k}^2/\lambda^2} = -\frac{G_k \lambda^2/\lambda_{1k}^2}{1 - \lambda^2/\lambda_{1k}^2} \tag{2.2.28}$$

For those terms where $\lambda > \lambda_{1k}$, put $k = k'$ and assume $\lambda \gg \lambda_{1k'}$. Then, using a binomial expansion,

$$\frac{G_{k'} \lambda^2}{\lambda^2 - \lambda_{1k'}^2} \simeq G_{k'}(1 + \lambda_{1k'}^2/\lambda^2 + \lambda_{1k'}^4/\lambda^4 + \cdots) \tag{2.2.29}$$

For those terms where $\lambda < \lambda_{1k}$, put $k = k''$ and assume $\lambda \ll \lambda_{1k''}$. Then, again using a binomial expansion, we may put

$$\frac{G_{k''} \lambda^2}{\lambda^2 - \lambda_{1k''}^2} \simeq -G_{k''}(\lambda^2/\lambda_{1k''}^2)(1 + \lambda^2/\lambda_{1k''}^2 + \cdots) \tag{2.2.30}$$

Thus eqn. (2.2.27) becomes

$$n^2 - 1 = \sum_k \frac{G_k \lambda^2}{\lambda^2 - \lambda_{1k}^2} \simeq \cdots + \frac{A}{\lambda^4} + \frac{B}{\lambda^2} + C + D\lambda^2 + E\lambda^4 + \cdots \tag{2.2.31}$$

where

$$A = \sum_{k'} G_{k'} \lambda_{1k'}^4$$

$$B = \sum_{k'} G_{k'} \lambda_{1k'}^2$$

$$C = \sum_{k'} G_{k'}$$

$$D = \sum_{k''} - G_{k''}/\lambda_{1k''}^2$$

$$E = \sum_{k''} - G_{k''}/\lambda_{1k''}^4$$

It can be seen from Fig. 2.9 that in the optical regions away from resonances we should expect n to increase slowly with frequency and thus to decrease slowly with increasing wavelength. Thus $dn/d\lambda$ is small and negative in those regions of interest to us. We can also see from Fig. 2.9 that there is a close association between dispersion (regions where n varies with frequency) and attenuation (regions where n' is significant). This association is quite fundamental in origin. In any invariant, causal, linear system in which a finite excitation produces a finite response, the imaginary part of the response function can always be uniquely determined from a knowledge of the real response function and vice versa. In physics the equations expressing this are known as the Kramers–Kronig relations while they are known to electrical engineers from the work of Bode.

This section has examined the theoretical basis of dispersion in dielectric media. In the next section we shall present empirical results for materials used

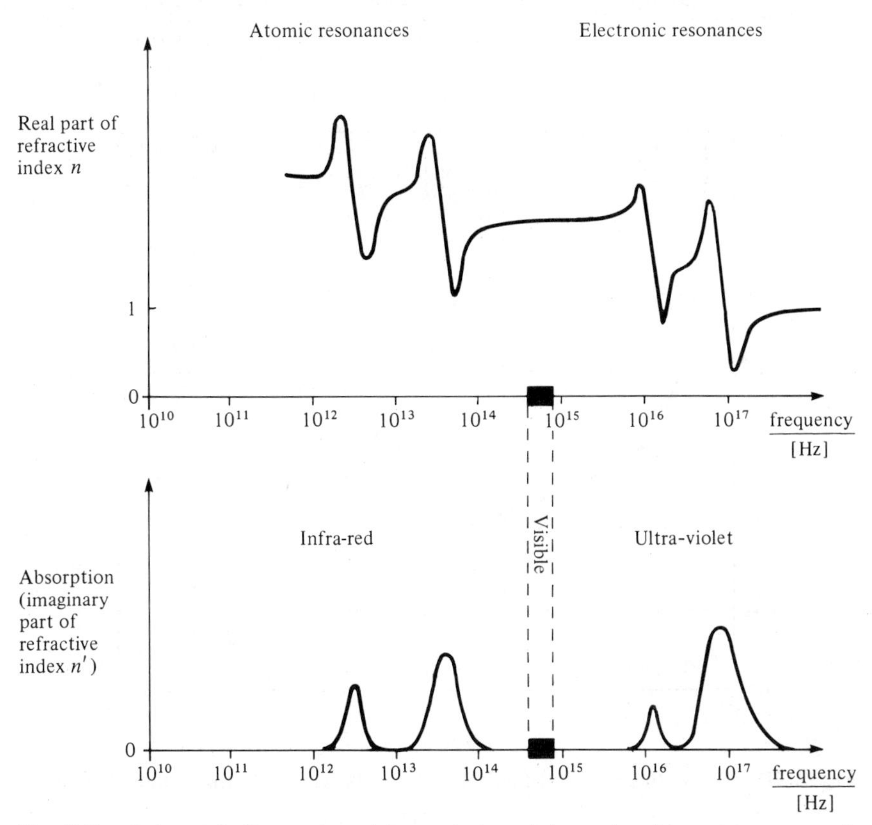

Fig. 2.9 Schematic illustration of the variation of the real and imaginary parts of the refractive index of a dielectric material showing atomic and electronic resonances.

in the manufacture of optical fibers and we shall find them to be in broad agreement with this theory. In later sections we shall explore the consequences of this variation of refractive index with frequency as it affects the signalling rates that can be achieved in optical fibers using different optical frequencies.

2.2.2 The Refractive Index of Bulk Media: Experimental Values

The starting material for most high-quality optical fibers is pure silica (SiO_2) in the form of fused quartz. The refractive index of bulk specimens of pure silica in the wavelength range 0.2 μm to 4.0 μm was determined with high precision by Malitson of the National Bureau of Standards in 1965.† He fitted his results to a Sellmeier dispersion equation having three terms, two in the u.v. and one in the i.r. and, with λ in μm, his results took the form:

$$n^2 - 1 = \frac{0.6961663\lambda^2}{\lambda^2 - (0.0684043)^2} + \frac{0.4079426\lambda^2}{\lambda^2 - (0.1162414)^2} + \frac{0.8974794\lambda^2}{\lambda^2 - (9.896161)^2} \quad (2.2.32)$$

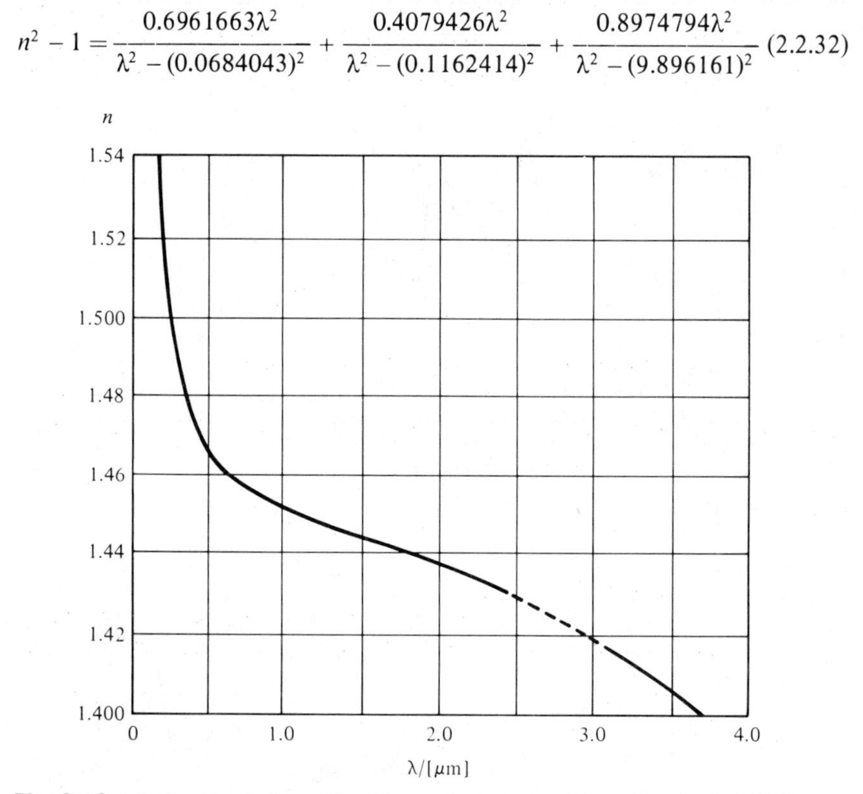

Fig. 2.10 Refractive index of fused quartz as derived from the data of Malitson.

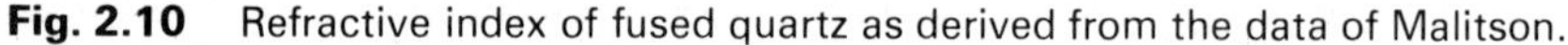

† I. H. Malitson, Inter-specimen comparison of the refractive index of fused silica, *Jnl. Optical Soc. America*, **55**, 1205–1209 (1965).

This equation for $n(\lambda)$ is plotted in Fig. 2.10.

In order to make clad or graded-index fibers means have to be found to vary the refractive index. This is normally achieved by adding substantial concentrations of oxide impurities to the silica. Clearly these will introduce further resonances in which either the oscillator strengths or the resonant frequencies or both will be changed. Thus while n may be varied as required, care has to be taken that this does not introduce extra dispersion and at the same time increase attenuation by introducing a resonance closer to the working wavelength.

At wavelengths of interest the refractive index of pure silica is in the region of 1.45. This can be lowered by the addition of impurities such as boria (B_2O_3) and fluorine (F), and can be raised by doping with oxides such as titania (TiO_2), caesia (Cs_2O), alumina (Al_2O_3), zirconia (ZrO_2), germania (GeO_2) and phosphorus pentoxide (P_2O_5), as shown in Fig. 2.11.

At first sight it is attractive to think of using undoped silica for the core material on the grounds that this material is immediately and readily available

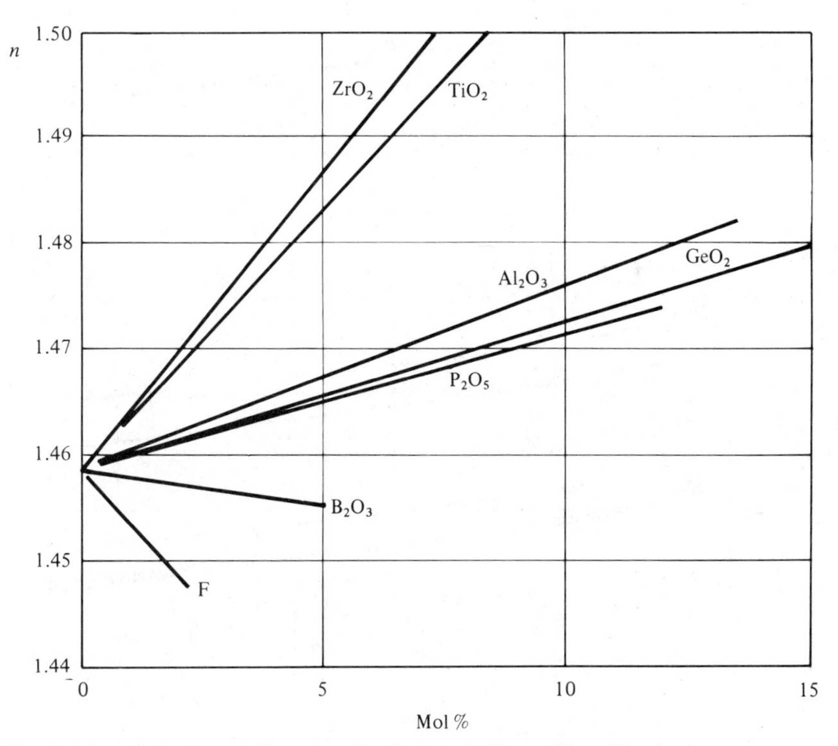

Fig. 2.11 Variation of the refractive index of silica with oxide doping concentration (the results are based on measurements made at wavelengths in the region of 0.6 μm).

in a very pure form. Silica doped with B_2O_3 or F is then required for the cladding, unless some quite different material such as a low refractive index polymer can be used. In practice the development of the vapor deposition techniques described in Chapter 4, as well as making index grading possible, has also enabled the formation of doped silica layers with an even lower level of unwanted contamination than is normally obtained with the best available synthetic silica. Thus index-raising dopants can be used for the fiber core as well as, or instead of index-lowering dopants for the cladding.

It is found that the best results can be obtained when the fiber core is made from silica doped with both GeO_2 and P_2O_5 and a cladding of pure silica or silica doped with B_2O_3 or F is used. The reasons are complex, and each of the other suggested dopants has its disadvantages. For example, TiO_2 is found to suffer chemical reduction during the drawing process leaving Ti^{3+} ions in the silica matrix. These cause optical absorption. The large, high-temperature mobility of P_2O_5 in silica makes difficult the control of the index profile of

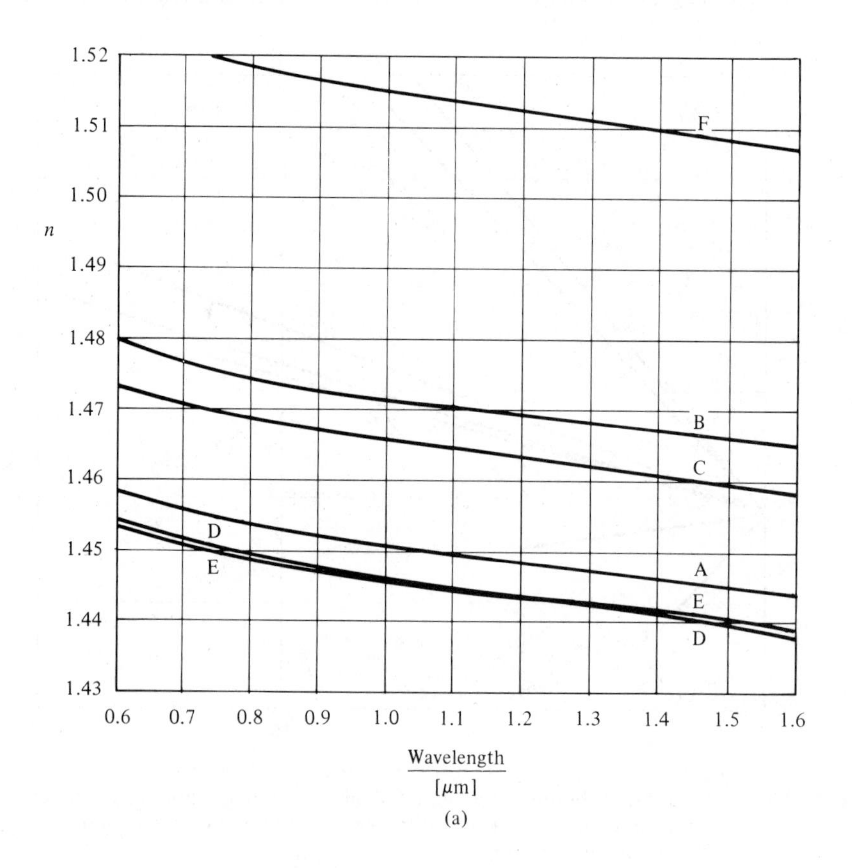

(a)

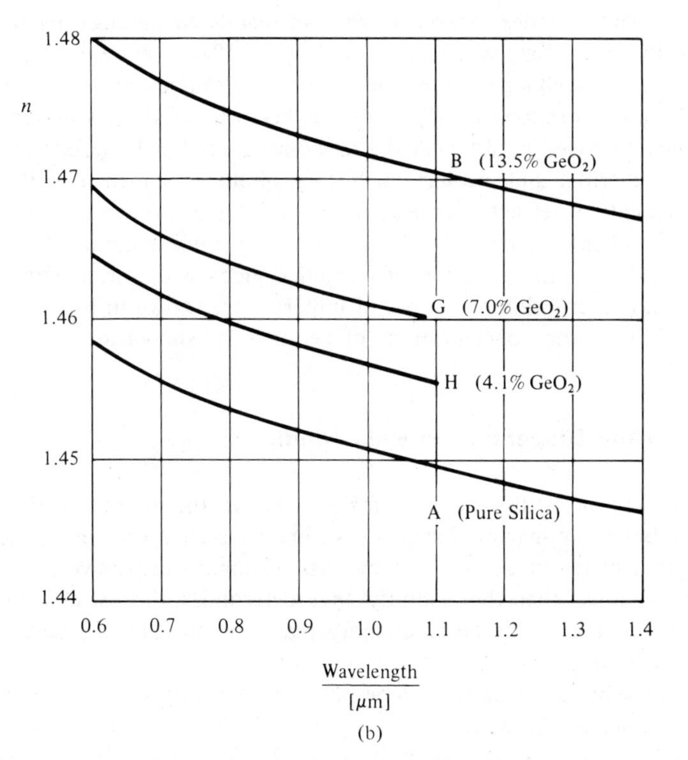

(b)

Fig. 2.12 Variation of the refractive indices of silica glasses with wavelength:

(a) Glass compositions (mol %)

A	pure silica
B	13.5% GeO_2; 86.5% SiO_2
C	9.1% P_2O_5; 90.9% SiO_2
D	13.3% B_2O_3; 86.7% SiO_2
E	1.0% F; 99.0% SiO_2
F	16.9% Na_2O; 32.5% B_2O_3; 50.6% SiO_2

[Data derived from J. W. Fleming, *Ets. Lett.* **14**, 326–8 (25 May 1978).]

(b) Germania–silica glasses (mol %)

A	Pure silica
B	13.5% GeO_2; 86.5% SiO_2
G	7.0% GeO_2; 93.0% SiO_2
H	4.1% GeO_2; 95.9% SiO_2

[Data derived from J. W. Fleming, *Jnl. Am. Ceramic Soc.* **59**, 503–7 (1976).]

graded-index fibers. But when P_2O_5 is used together with GeO_2 it has the effect of lowering the deposition temperature and a better profile is obtained.

The maximum index difference that can readily be obtained using the germania–phosphosilicate system is about 4%, but 1% is a more typical figure. At these levels the changes in n and $\mathrm{d}n/\mathrm{d}\lambda$ are roughly proportional to the impurity concentration and Fig. 2.11 has been drawn on this assumption. In borosilicate glasses in particular the refractive index has been found to depend upon the thermal and mechanical history of the specimen. For this reason measurements of refractive index and dispersion made on sections taken from the drawn fiber may differ from those made on bulk material. Bulk material measurements of the variation of refractive index with wavelength for silica doped with several different types of impurity are shown in Fig. 2.12(a). The effect of increasing concentrations of germania is shown in Fig. 2.12(b).

2.2.3 Time Dispersion in Bulk Media

In optics the word dispersion normally refers to the quantity $\mathrm{d}n/\mathrm{d}\lambda$, but in optical fiber communication systems we are concerned with the tendency for pulses of light to spread out as they pass through a dispersive medium. We shall now show that the quantity that determines this is not $\mathrm{d}n/\mathrm{d}\lambda$ but $\lambda \mathrm{d}^2 n/\mathrm{d}\lambda^2$, and it is this latter quantity that is implied by the term material dispersion throughout the rest of this book.

Any disturbance or signal superimposed onto a wave propagates not at the phase velocity of the wave,

$$v_p = \frac{\omega}{\beta} \tag{2.2.5}$$

but at the *group velocity*, v_g, which is given by

$$v_g = \frac{\mathrm{d}\omega}{\mathrm{d}\beta} = \frac{1}{\mathrm{d}\beta/\mathrm{d}\omega} \tag{2.2.33}$$

Justification for this assertion can be found in most texts on wave propagation, in particular, in Refs. [1.2] and [2.1]. Now in a nondispersive medium the phase velocity is independent of the wave frequency, and the group velocity and the phase velocity are the same:

$$\beta = \frac{\omega}{v_p} \qquad \text{and} \qquad v_g = \frac{1}{\mathrm{d}\beta/\mathrm{d}\omega} = v_p$$

But in a dispersive medium, where by definition the phase velocity is a function of frequency, v_g and v_p differ:

$$v_g = \frac{1}{d\beta/d\omega} = \frac{v_p}{1 - (\omega/v_p)(dv_p/d\omega)} \tag{2.2.34}$$

This is important since it is the signal velocity that we are invariably concerned with in communications. A light pulse, for example, travels through a dispersive medium with a speed v_g. What complicates the issue is that as a result of the dispersion the pulse is necessarily attenuated and to some extent distorted during its travel. We may, nevertheless, define a *group index*, N, such that

$$N = \frac{c}{v_g} \tag{2.2.35}$$

and in a dispersive medium N will be different from the ordinary refractive index, or *phase index*, n.

It will be apparent already that we have taken an over-simplistic view in Section 2.1.2 in expressing the multipath time dispersion by means of eqn. (2.1.17). The velocity of propagation of pulses of light in the ray model of Fig. 2.2 is c/N_1, so that the transit time difference between the axial pulses and those travelling by the most oblique route should be given by

$$\frac{\Delta T}{l} = \frac{N_1}{n_2}\frac{\Delta n}{c} \tag{2.2.36}$$

where N_1 is the group index of the core. However, eqn. (2.1.17) remains an adequate approximation for normal step-index fibers. The effect of material dispersion on propagation in graded-index fibers requires more careful treatment and is discussed further in Chapter 6.

As we have seen, the dispersive properties of optical materials are traditionally expressed in terms of the variation of the refractive index with free-space wavelength: $n(\lambda)$. We therefore need to be able to express v_g and N in terms of n and λ. Note first that

$$N = \frac{c}{v_g} = c\frac{d\beta}{d\omega} = c\frac{d}{d\omega}\left(\frac{\omega n}{c}\right) = \frac{d(\omega n)}{d\omega} = n + \frac{\omega\, dn}{d\omega} \tag{2.2.37}$$

Now

$$\frac{dn}{d\omega} = \frac{dn}{d\lambda}\frac{d\lambda}{d\omega}$$

and with $\omega = 2\pi c/\lambda$,

$$\frac{d\omega}{d\lambda} = -\frac{2\pi c}{\lambda^2}$$

Substituting these expressions into eqn. (2.2.37) gives

$$N = n + \frac{2\pi c}{\lambda}\frac{dn}{d\lambda}\left(\frac{-\lambda^2}{2\pi c}\right) = n - \lambda\frac{dn}{d\lambda} \tag{2.2.38}$$

Thus

$$v_g = \frac{c}{N} = \frac{c}{[n - \lambda\, dn/d\lambda]} \tag{2.2.39}$$

An impulse of light, then, travels a distance l through a medium in a time, t, given by

$$t = \frac{l}{v_g} = \frac{Nl}{c} = \left[n - \lambda\frac{dn}{d\lambda}\right]\frac{l}{c} \tag{2.2.40}$$

If the light contains a range of free-space wavelengths $\Delta\lambda$ about λ and if the medium is dispersive, the impulse spreads out as it propagates and arrives over a range of times, Δt, given by

$$\Delta t = \frac{dt}{d\lambda}\Delta\lambda = \frac{l}{c}\frac{dN}{d\lambda}\Delta\lambda = \frac{l}{c}\left[\frac{dn}{d\lambda} - \frac{dn}{d\lambda} - \lambda\frac{d^2n}{d\lambda^2}\right]\Delta\lambda = -\frac{l}{c}\lambda\frac{d^2n}{d\lambda^2}\Delta\lambda \tag{2.2.41}$$

It is customary to define the spectral width, $\Delta\lambda$, of an optical source as the range of wavelengths over which the spectral power exceeds 50% of the peak spectral power. It is also convenient to define a relative spectral width, γ, given by

$$\gamma = \left|\frac{\Delta\lambda}{\lambda}\right| = \left|\frac{\Delta\omega}{\omega}\right| \tag{2.2.42}$$

Thus an impulse, after propagating for a distance, l, through a dispersive medium, will have broadened into a pulse of half-power width, τ, where

$$\tau = \frac{l}{c}\gamma\left|\lambda^2\frac{d^2n}{d\lambda^2}\right| \tag{2.2.43}$$

That is,

$$\frac{\tau}{l} = \frac{\gamma}{c} | Y_{\mathrm{m}} | \tag{2.2.44}$$

where

$$Y_{\mathrm{m}} = \lambda^2 \frac{\mathrm{d}^2 n}{\mathrm{d}\lambda^2} \tag{2.2.45}$$

represents a material dispersion coefficient. If we write $\Delta f \simeq 1/4\tau$ as the approximate signal bandwidth of the fiber (see Section 2.4), this gives

$$(\Delta f) l \simeq \frac{c}{4\gamma | Y_{\mathrm{m}} |} \tag{2.2.46}$$

We have introduced modulus signs into the definitions of γ and τ because in general what is of interest is the absolute value of the wavelength spread $\Delta\lambda$ or the pulse width τ. We are not normally concerned whether it is the longer or the shorter wavelengths that arrive first.

We emphasize again that τ is the half-power pulse width, and that putting $\Delta f = 1/4\tau$ is an approximation, just as was the relationship suggested in Section 2.1.2 between bandwidth and the total multipath dispersion: $\Delta f = 1/2\Delta T$. These matters will be the subject of further discussion and analysis, particularly in Section 2.4.

Graphs showing the variation with wavelength of Y_{m} and $(\lambda/c)(\mathrm{d}^2 n/\mathrm{d}\lambda^2)$ for bulk samples of pure and doped silica are presented in Fig. 2.13. These data have been derived from the results shown in Fig. 2.12 and it needs to be stressed that they may not apply exactly to material of similar composition that has been drawn into fiber. From these curves we may deduce that at 0.85 μm (a typical wavelength for gallium arsenide sources) doping silica with germania has the effect of increasing both refractive index and dispersion; doping with boria will decrease both; doping with phosphorus pentoxide increases n but has little effect on the dispersion.

For pure silica at 0.85 μm, $Y_{\mathrm{m}} = 0.021$. Thus $\tau/l = 7.2 \times 10^{-11}\gamma$ [s/m] and $(\Delta f)l = (3.5 \times 10^9)/\gamma$ [m/s]. We shall see in Chapter 8 that a typical range of wavelengths produced by GaAs light-emitting diode sources is some 30 nm about 850 nm, so that $\gamma = 0.035$, and in bulk silica the rate at which pulses spread is then

$$\frac{\tau}{l} = \frac{\gamma}{c} | Y_{\mathrm{m}} | = \frac{0.035 \times 0.021}{3 \times 10^8} = 2.5 \times 10^{-12} = 2.5 \text{ ns/km},$$

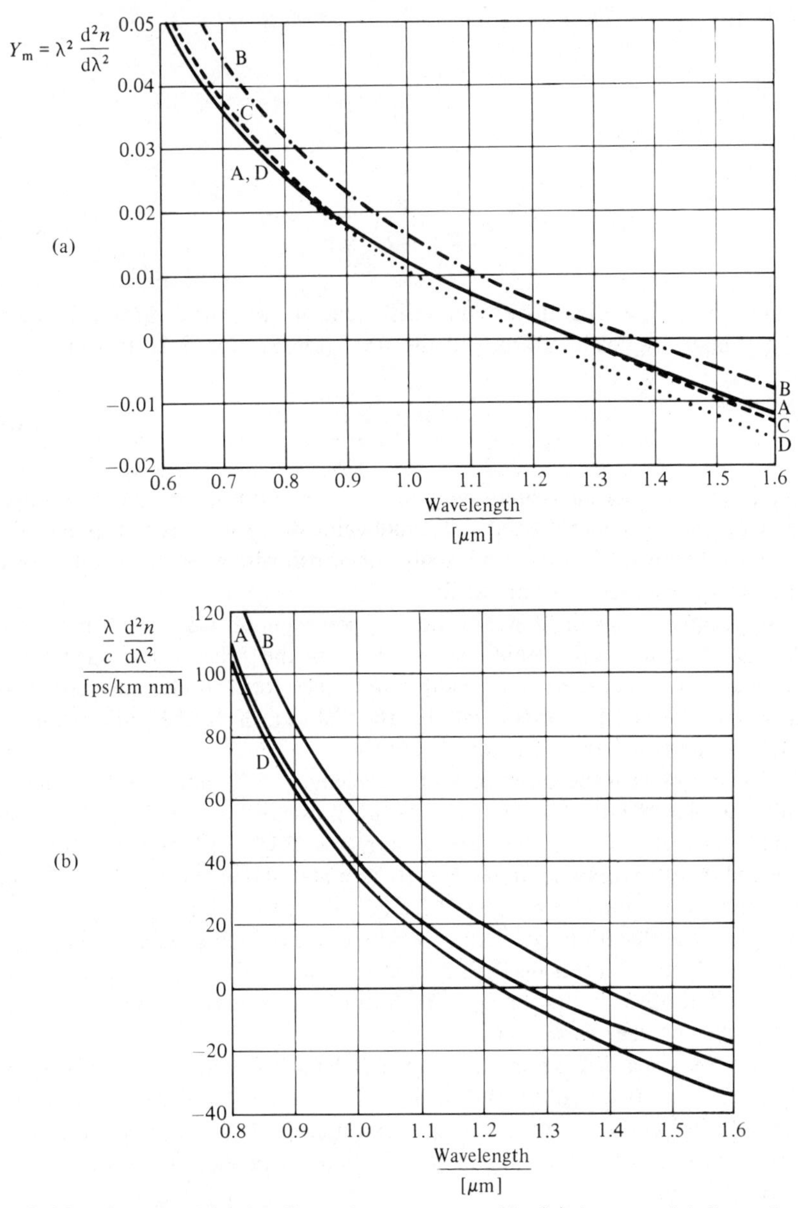

Fig. 2.13 Graphs of (a) the dispersion parameter Y_m; (b) the material dispersion, $(\lambda/c)\, d^2n/d\lambda^2$, in units of ps/(km.nm).

Both are for pure and doped silica, based on the data for bulk samples shown in Fig 2.12. The letters A–D refer to the compositions given in Fig. 2.12.

and the bandwidth–distance product is $(\Delta f)l = 100$ MHz.km. Laser sources tend to emit over a narrower spectral range, something like 3 nm, so that for them $\gamma = 0.0035$, and in silica $\tau/l = 0.25$ ns/km and $(\Delta f)l = 1$ GHz.km. These values should be compared with the (ΔT) values quoted earlier for multipath dispersion, namely $\Delta T/l \simeq 2\tau/l \simeq 34$ ns/km for a typical step-index fiber and 2500 ns/km for an unclad fiber, which in terms of bandwidth become approx. 15 MHz.km and 0.2 MHz.km respectively.

It will be seen in Fig. 2.13 that in the case of pure silica $d^2n/d\lambda^2$ changes sign at a wavelength $\lambda = \lambda_0 = 1.276$ μm. This is the point of inflection in the $n(\lambda)$ vs. λ curve and it has frequently been referred to in the literature as the 'wavelength of zero material dispersion'. Such an expression is misleading from a practical point of view because a real pulse contains a spread of wavelengths and these propagate with a range of group velocities even though the longest and shortest wavelengths go at the same speed. This situation is illustrated in Fig. 2.14(a). Whether $d^2n/d\lambda^2$ is greater or less than zero is of no consequence to the question of pulse spreading.

Material dispersion is minimal for sources which emit at wavelengths near to λ_0 and such sources would permit the maximum information capacity of the fiber to be exploited. It is clearly important to know what this limit is and for this reason we shall derive the correct expressions for time dispersion in the region of the dispersion minimum.

For a source which spreads over a range of wavelengths $\Delta\lambda$ about a center wavelength, λ_m such that it straddles λ_0, that is:

$$\left(\lambda_0 - \frac{\Delta\lambda}{2}\right) < \lambda_m < \left(\lambda_0 + \frac{\Delta\lambda}{2}\right), \tag{2.2.47}$$

the pulse spreading can be obtained by means of a Taylor expansion of the pulse propagation time, t, about λ_0. We have defined

$$N = \frac{ct}{l} = n - \lambda\frac{dn}{d\lambda} \tag{2.2.38}$$

Let

$$N = N_0 \text{ at } \lambda = \lambda_0 \text{ and}$$

$$\lambda - \lambda_0 = x \tag{2.2.48}$$

Then

$$N(\lambda) = N_0 + x\left(\frac{dN}{d\lambda}\right)_{\lambda_0} + \frac{x^2}{2!}\left(\frac{d^2N}{d\lambda^2}\right)_{\lambda_0} + \cdots \tag{2.2.49}$$

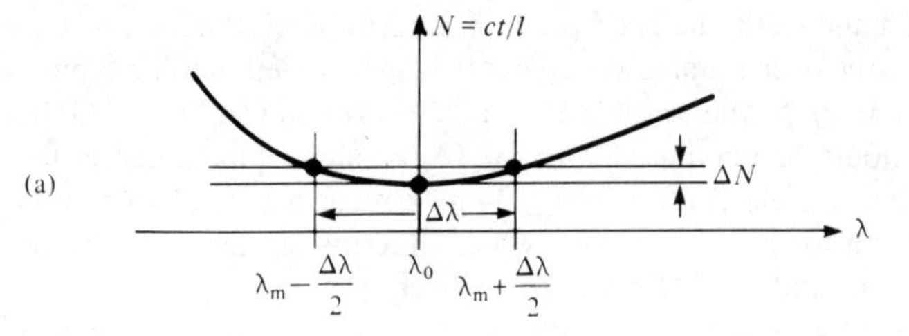

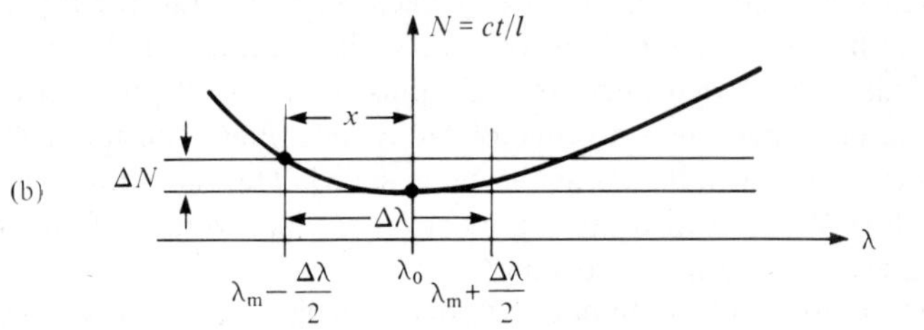

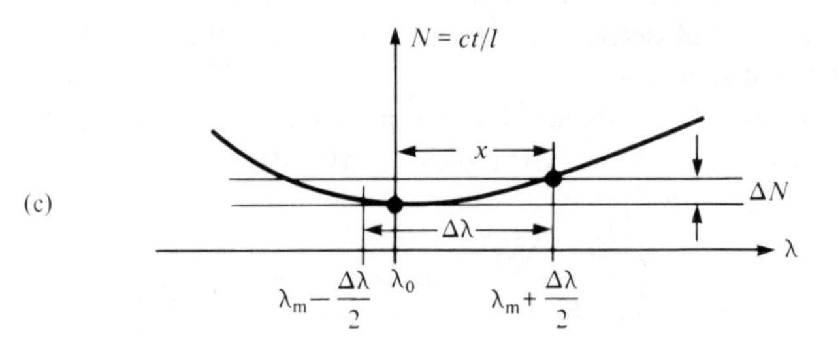

Fig. 2.14 Region of minimum spread of propagation times.

(a) Minimum spread of propagation times, $\lambda_m = \lambda_0$;

(b) $(\lambda_0 - \frac{1}{2}\Delta\lambda) < \lambda_m < \lambda_0$;

(c) $\lambda_0 < \lambda_m < (\lambda_0 + \frac{1}{2}\Delta\lambda)$.

$N = ct/l$ is the normalized propagation time.

$N = c/v_g = n - \lambda\, dn/d\lambda$ and has a minimum value of N_0 at $\lambda = \lambda_0$.

A Taylor expansion about λ_0, writing $x = \lambda - \lambda_0$, gives

$$\Delta = N - N_0 = x\left(\frac{dN}{d\lambda}\right)_{\lambda_0} + \frac{x^2}{2!}\left(\frac{d^2N}{d\lambda^2}\right)_{\lambda_0} + \frac{x^3}{3!}\left(\frac{d^3N}{d\lambda^3}\right)_{\lambda_0} + \cdots$$

in which the first term is clearly zero. In the derivation given in the text, the third and higher terms have been neglected.

Now

$$\left(\frac{\mathrm{d}N}{\mathrm{d}\lambda}\right)_{\lambda_0} = -\lambda_0\left(\frac{\mathrm{d}^2 n}{\mathrm{d}\lambda^2}\right)_{\lambda_0} = 0$$

and

$$\left(\frac{\mathrm{d}^2 N}{\mathrm{d}\lambda^2}\right)_{\lambda_0} = -\left(\frac{\mathrm{d}^2 n}{\mathrm{d}\lambda^2}\right)_{\lambda_0} - \lambda_0\left(\frac{\mathrm{d}^3 n}{\mathrm{d}\lambda^3}\right)_{\lambda_0} = -\lambda_0\left(\frac{\mathrm{d}^3 n}{\mathrm{d}\lambda^3}\right)_{\lambda_0} \tag{2.2.50}$$

Therefore,

$$[N(\lambda) - N_0] = -\lambda_0\left(\frac{\mathrm{d}^3 n}{\mathrm{d}\lambda^3}\right)_{\lambda_0}\frac{x^2}{2!} + \cdots \tag{2.2.51}$$

N_0 represents the minimum transit time, and $[N(\lambda) - N_0]$ will represent the transit time spread $(c\tau/l)$ if we evaluate it at the value of λ which corresponds to the greater of $[\lambda_0 - (\lambda_m - \frac{1}{2}\Delta\lambda)]$ and $[(\lambda_m + \frac{1}{2}\Delta\lambda) - \lambda_0]$. The two cases are:

(i) $(\lambda_0 - \frac{1}{2}\Delta\lambda) < \lambda_m < \lambda_0$, as shown in Fig. 2.14(b), when

$$\frac{c\tau}{l} = \frac{\lambda_0}{2}\left(\frac{\mathrm{d}^3 n}{\mathrm{d}\lambda^3}\right)_{\lambda_0}\left[\lambda_0 - \left(\lambda_m - \frac{\Delta\lambda}{2}\right)\right]^2 \tag{2.2.52}$$

(ii) $\lambda_0 < \lambda_m < (\lambda_0 + \frac{1}{2}\Delta\lambda)$, as shown in Fig. 2.14(c), when

$$\frac{c\tau}{l} = \frac{\lambda_0}{2}\left(\frac{\mathrm{d}^3 n}{\mathrm{d}\lambda^3}\right)_{\lambda_0}\left[\left(\lambda_m + \frac{\Delta\lambda}{2}\right) - \lambda_0\right]^2 \tag{2.2.53}$$

The neglected higher-order terms would become significant if the spectral width of the source, $\Delta\lambda$, approached 100 nm.

At $\lambda_m = \lambda_0$ the minimum value of the bulk material dispersion is

$$\frac{\tau}{l} = -\lambda_0\left(\frac{\mathrm{d}^3 n}{\mathrm{d}\lambda^3}\right)_{\lambda_0}\frac{(\Delta\lambda)^2}{8c} = (-)\frac{\gamma^2\lambda_0^3}{8c}\left(\frac{\mathrm{d}^3 n}{\mathrm{d}\lambda^3}\right)_{\lambda_0} \tag{2.2.54}$$

For pure silica at $\lambda = \lambda_0 = 1.276\ \mu\mathrm{m}$,

$$\lambda_0^3\left(\frac{\mathrm{d}^3 n}{\mathrm{d}\lambda^3}\right)_{\lambda_0} = -0.048$$

so that $\tau/l = 2 \times 10^{-11}\gamma^2$ s/m.

Consider a possible LED source which is centered on λ_0 and has a value of γ of 0.04. This implies a wavelength spread of 51 nm about 1.276 μm and gives

$\tau/l = 3.2 \times 10^{-14} = 32$ ps/km, and $(\Delta f)l \simeq 8$ GHz.km. The values with a typical laser source would be some two orders of magnitude better than this. In either case material dispersion becomes very small, so there has been considerable pressure for sources and detectors to be developed to exploit this region. Two further points may also be noted from Fig. 2.13. First, Y_m remains quite small at wavelengths beyond λ_0. For example, at $\lambda = 1.55$ μm, in pure silica, which as we shall see in Section 2.5 is a region of minimum attenuation,

$$Y_m = -0.01, \qquad \text{giving} \qquad \tau/l = 3.4 \times 10^{-11}\gamma$$

Then, with a source for which $\gamma = 0.04$, $\tau/l = 1.3$ ns/km and $(\Delta f)l =$ 200 MHz.km, while with $\gamma = 0.004$, $\tau/l = 0.13$ ns/km and $(\Delta f)l =$ 2 GHz.km. Secondly, the value of λ_0 may be varied by the use of dopants. Boria will enable it to be decreased below 1.22 μm and germania can cause it to be raised above 1.37 μm as Fig. 2.13 shows.

2.3 THE COMBINED EFFECT OF MATERIAL DISPERSION AND MULTIPATH DISPERSION

Up to now we have examined two independent effects which give rise to time dispersion in optical fibers: multipath dispersion and material dispersion. Under normal circumstances we should expect them to be present together, and the question arises as to how we should combine them in order to determine the total fiber dispersion.

If we were concerned only with the time difference on arrival between the fastest and the slowest waves, it would be perfectly proper simply to add the two effects. The first light to arrive would be that having the wavelength with the fastest group velocity and it would have travelled the shortest optical distance. The last light to arrive would have the wavelength with the slowest group velocity and would have travelled the longest optical distance. For practical purposes, however, this is an unsophisticated and over-pessimistic view.

In estimating the bandwidth or maximum bit rate of an optical communication channel we need to consider the shape of the received pulses. The shape on arrival of an impulse that has been broadened by material dispersion will reflect the power distribution among the wavelengths making up the pulse. We shall find in general that most optical sources have a distribution of power with wavelength that is roughly Gaussian in form. We should then expect the received pulse shape to take on a similar form about the mean arrival time, t_0, as shown in Fig. 2.15(a). Although we have as yet no theoretical basis for predicting the distribution of optical power among the various ray trajectories that

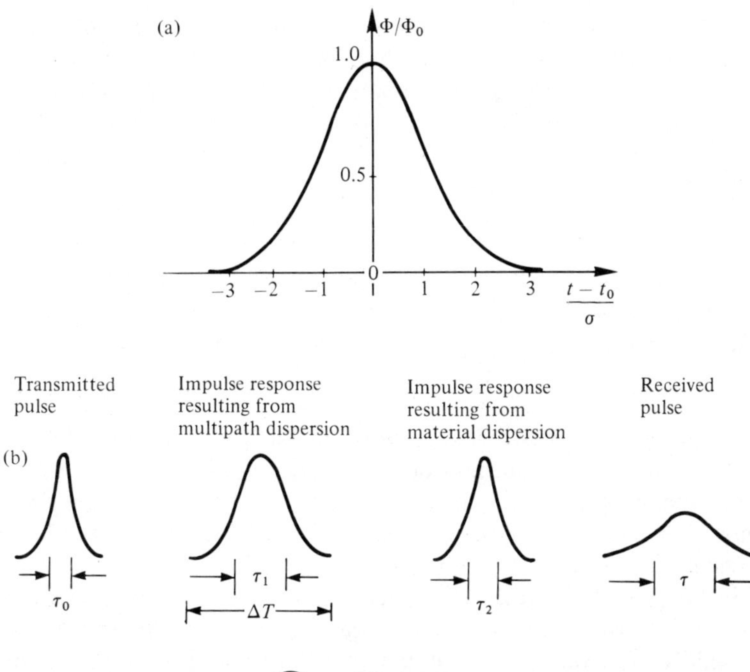

Fig. 2.15 Dispersive effects in combination:
(a) Theoretical pulse shape

$$\text{Gaussian, } \Phi(t) = \Phi_0 \exp\left[-\frac{(t-t_0)^2}{2\sigma^2}\right]$$

halfwidth, $\tau = 2.355\sigma$.

(b) Real pulses will of course fall to zero in a finite time. Nevertheless, the mechanisms causing broadening can in practice be treated as though they were giving rise to independent Gaussian pulses so that

$$\tau \simeq (\tau_0^2 + \tau_1^2 + \tau_2^2)^{1/2}.$$

propagate, it is intuitively reasonable to suppose that most power will propagate in those rays which travel an average optical path length rather than in those which traverse either the longest or the shortest optical distance. And if this is so, multipath dispersion, also, will cause an impulse to broaden into a pulse of approximately Gaussian shape.

Now let us assume that an impulse is broadened by both multipath and material dispersion, that the two mechanisms are uncorrelated and that they independently lead to approximately Gaussian pulses having half-height pulse widths τ_1 and τ_2 respectively. The two mechanisms will then combine to produce a pulse which remains roughly Gaussian in shape and which will have a width, τ, given by:

$$\tau = (\tau_1^2 + \tau_2^2)^{1/2} \tag{2.3.1}$$

When the transmitted pulse is not an impulse to start with but is also roughly Gaussian with a full-width at half-height value of τ_0, then we may extend this argument as illustrated in Fig. 2.15(b) to suggest that τ becomes:

$$\tau = (\tau_0^2 + \tau_1^2 + \tau_2^2)^{1/2} \tag{2.3.2}$$

where τ_0 is the initial pulse width; τ_1 is the half-height pulse-width that would result from multipath broadening alone (for any given fiber its value would be rather less than the values for the total pulse width, ΔT, quoted in Section 2.1, perhaps by a factor of two); and τ_2 represents the broadening due to material dispersion alone as given by eqns (2.2.44) or (2.2.54).

In Section 2.1 we showed that a typical step-index fiber would give rise to an overall pulse width of $\Delta T/l = 34$ ns/km as a result of multipath time dispersion. This might be equivalent to a half-power width somewhere in the region of 15 ns/km. With graded-index fibers this figure can be reduced to something in the region of 0.5 ns/km. In Section 2.2.3 we have indicated the level of material dispersion that might be expected with light-emitting diode and laser sources operating at different wavelengths with silica fibers. In Table 2.1 we show how these results can be combined. By a simple extension of eqn. (2.3.2) we may write

$$\tau = \left[\frac{\tau_0^2}{l^2} + \left(\frac{\tau_1}{l}\right)^2 + \left(\frac{\tau_2}{l}\right)^2\right]^{1/2} l \tag{2.3.3}$$

As before τ_0 represents the half-power width of the transmitted pulse, (τ_1/l) the effect of multipath dispersion and (τ_2/l) the effect of material dispersion.

In Table 2.1 we have taken $\tau_0 = 0$ and $\Delta\lambda = 30$ nm and 3 nm for the 0.9 μm LED and laser respectively. At the longer wavelengths we have used $\gamma = 0.04$ and 0.004. It is clear that in the step-index fiber, multipath dispersion totally dominates in all cases. With graded-index fiber we have assumed 0.5 ns/km to be a typical value for multipath dispersion. This will always dominate material dispersion when laser sources are used. With light-emitting diodes material dispersion is dominant except for wavelengths around 1.3 μm.

It is clear that in order to reap the full benefit of the lower material dispersion that can be obtained at wavelengths around 1.3 μm, it will be necessary to

Table 2.1 The combined effects of multipath and material dispersion in step-index (SI) and graded-index (GI) silica fibers at different wavelengths

Wavelength (μm)	Source	Material dispersion (τ_2/l) [ns/km]	Total dispersion for SI fiber (τ/l) [ns/km] [multipath dispersion $\tau_1/l = 15$ ns/km]	Total dispersion for GI fiber (τ/l) [ns/km] [multipath dispersion $\tau_1/l = 0.5$ ns/km]
0.9	LED	2.1	15	2.2
	Laser	0.2	15	0.5
1.3	LED	0.1	15	0.5
	Laser	0.01	15	0.5
1.55	LED	1.2	15	1.3
	Laser	0.1	15	0.5

reduce multipath dispersion below the quoted value of 0.5 ns/km. There are two ways in which this may be achieved. In the first the core diameter is reduced until 'single-mode' operation is achieved. A laser source is then necessary but multipath dispersion is totally eliminated: it may be thought that only one ray trajectory is possible—however, the ray model is quite inappropriate under these conditions. Total dispersion can then be very small indeed, perhaps as low as 10 ps/km. Single-mode fibers are discussed further in Chapter 5. The second method involves a very careful profiling of the refractive index of a graded-index fiber, and the levels of total dispersion that could in principle be obtained are discussed in Chapter 6. In practice the best graded-index fibers show a multipath dispersion of 0.2–0.3 ns/km, giving bandwidths in the region of 1 GHz.km.

2.4 ROOT-MEAN-SQUARE PULSE WIDTHS AND FREQUENCY RESPONSE

2.4.1 RMS Pulse Widths

Another measure of pulse width, which we shall use in later chapters and which is particularly valuable when the precise shape of the pulse is unknown, is the root-mean-square (r.m.s.) pulse width, σ. This is defined as follows: the received power distribution as a function of time is $\Phi(t)$ and the total energy in the pulse is

$$\mathcal{E} = \int_{-\infty}^{\infty} \Phi(t)\, \mathrm{d}t \tag{2.4.1}$$

The mean pulse arrival time is

$$t_0 = \frac{1}{\mathcal{E}} \int_{-\infty}^{\infty} t\Phi(t)\, \mathrm{d}t \tag{2.4.2}$$

and the r.m.s. pulse width, σ, is given by

$$\sigma^2 = \frac{1}{\mathcal{E}} \int_{-\infty}^{\infty} t^2 \Phi(t)\, \mathrm{d}t - t_0^2 \tag{2.4.3}$$

We may also characterize the spectral distribution of the light making up the pulse by means of an r.m.s. spectral width defined by direct analogy with eqns (2.4.1–2.4.3). Further, we may define the spectral width either in terms of the spread of wavelengths contained in the light or in terms of the spread of angular frequencies. Let the total power radiated by a given source at some instant be Φ_0 where

$$\Phi_0 = \int_0^{\infty} \Phi_\omega(\omega)\, \mathrm{d}\omega \tag{2.4.4}$$

$\Phi_\omega(\omega)$ is then the power spectral density expressed as a function of the optical angular frequency. Then the mean optical angular frequency of the light is given by $\bar{\omega}$ where

$$\bar{\omega} = \frac{1}{\Phi_0} \int_0^{\infty} \omega\Phi_\omega(\omega)\, \mathrm{d}\omega \tag{2.4.5}$$

And finally the r.m.s. line width, σ_ω, is given by:

$$\sigma_\omega^2 = \frac{1}{\Phi_0} \int_0^{\infty} \omega^2 \Phi_\omega(\omega)\, \mathrm{d}\omega - \bar{\omega}^2 \tag{2.4.6}$$

It is equally possible to define an average source wavelength, $\bar{\lambda}$, and to express the r.m.s. line width in terms of wavelength, σ_λ. These alternative definitions then become:

$$\Phi_0 = \int_0^{\infty} \Phi_\lambda(\lambda)\, \mathrm{d}\lambda \tag{2.4.7}$$

where $\Phi_\lambda(\lambda)$ is the power spectral density expressed as a function of wavelength. (The units of $\Phi_\lambda(\lambda)$ are [W/m] whereas the units of $\Phi_\omega(\omega)$ are

[W s].) Then,

$$\bar{\lambda} = \frac{1}{\Phi_0}\int_0^\infty \lambda \Phi_\lambda(\lambda)\, d\lambda \tag{2.4.8}$$

and

$$\sigma_\lambda^2 = \frac{1}{\Phi_0}\int_0^\infty \lambda^2 \Phi_\lambda(\lambda)\, d\lambda - \bar{\lambda}^2 \tag{2.4.9}$$

It is important to appreciate that σ_λ is a wavelength spread and σ_ω an angular frequency spread.

Now let us again assume that an impulse is broadened by both multipath and material dispersion, that the two mechanisms are uncorrelated and that they independently lead to approximately Gaussian pulses having r.m.s. pulse widths of σ_1 and σ_2 respectively. The two mechanisms will again combine to produce a pulse which remains approximately Gaussian in shape and will have an r.m.s. width, σ, given by

$$\sigma = (\sigma_1^2 + \sigma_2^2)^{1/2} \tag{2.4.10}$$

When the transmitted pulse is not an impulse to start with but is roughly Gaussian in shape and of r.m.s. width σ_0, then again,

$$\sigma = (\sigma_0^2 + \sigma_1^2 + \sigma_2^2)^{1/2} \tag{2.4.11}$$

The value of σ_1 in comparison with the total multipath pulse width discussed in Section 2.1 depends upon the way in which the optical power is distributed among the ray trajectories. The value of σ_2 may be obtained by reference to eqn. (2.2.41), which becomes:

$$\sigma_2 = -\frac{l}{c}\frac{dN}{d\lambda}\sigma_\lambda = \frac{l}{c}\left|\lambda^2 \frac{d^2 n}{d\lambda^2}\right|\frac{\sigma_\lambda}{\lambda}$$

$$= \frac{l}{c}\left|\omega \frac{dN}{d\omega}\right|\sigma_\omega/\omega = \frac{l}{c}|Y_m|\gamma_\sigma \tag{2.4.12}$$

where

$$\gamma_\sigma = \frac{\sigma_\omega}{\omega} = \frac{\sigma_\lambda}{\lambda}$$

We will next relate the root-mean-square pulse width, σ, to the full width at halfheight, τ, and other characteristic parameters of some hypothetical pulses having the different analytical shapes illustrated in Fig. 2.16. These

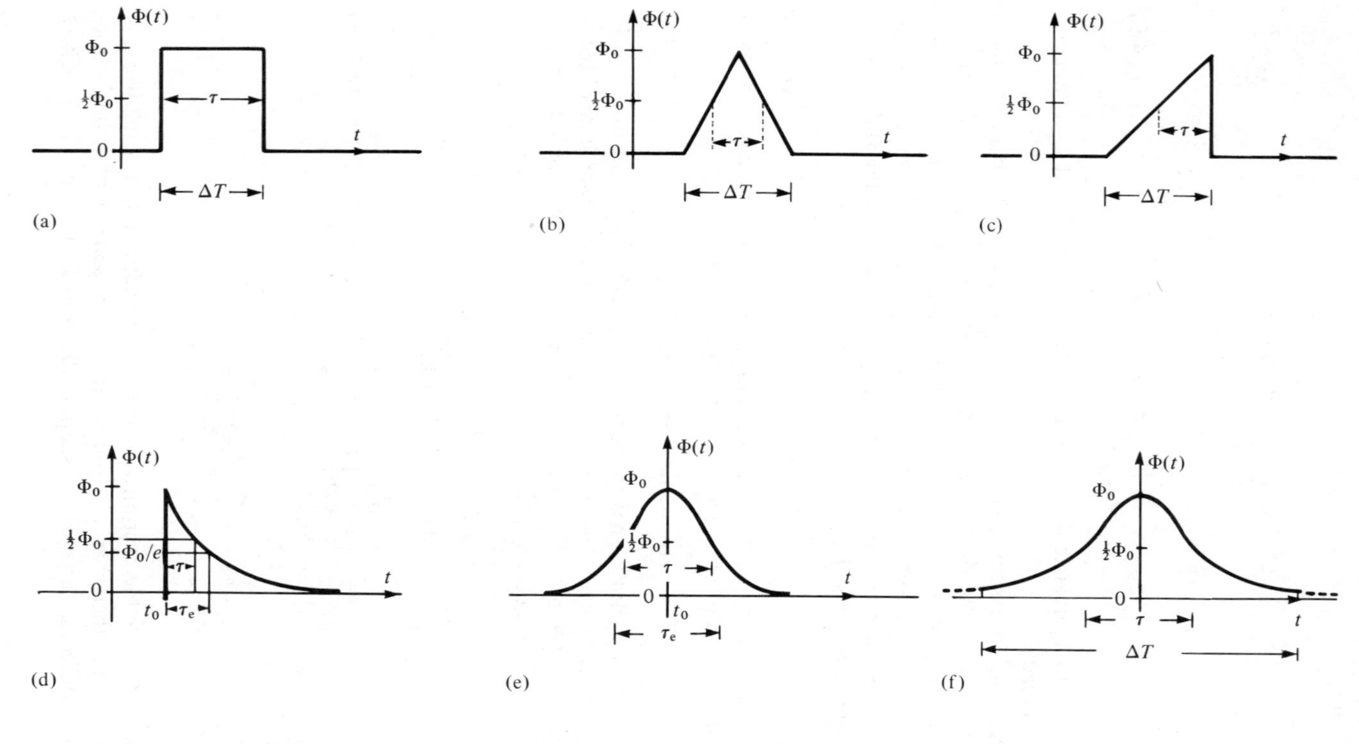

Fig. 2.16 R.M.S. pulse widths. (a) square pulse—$\sigma = \tau/\sqrt{12}$; (b) triangular pulse—$\sigma = \tau/\sqrt{6}$; (c) triangular pulse—$\sigma = \sqrt{2}\tau/3$; (d) exponential pulse—$\Phi(t) = \Phi_0 \exp[-(t - t_0)/\tau_e]$, $\sigma = 1.44\tau$; (e) Gaussian pulse—$\Phi(t) = \Phi_0 \exp[-(t - t_0)^2/2\sigma^2]$, $\sigma = 0.425\tau$; (f) Lorentzian pulse—$\Phi(t) = \Phi_0/[1 + 4(t - t_0)^2/\tau^2]$.

relationships are obtained straightforwardly by an evaluation of eqn. (2.4.3) in each case.

(a) *Rectangular*

$$\tau = \Delta T$$

$$\sigma = \frac{\Delta T}{2\sqrt{3}} = 0.289\Delta T = 0.289\tau \tag{2.4.13}$$

(b) *Triangular, 1*

$$\tau = 0.5\Delta T$$

$$\sigma = \frac{\Delta T}{2\sqrt{6}} = 0.204\Delta T = 0.408\tau \tag{2.4.14}$$

(c) *Triangular, 2*

$$\tau = 0.5\Delta T$$

$$\sigma = \frac{\Delta T}{3\sqrt{2}} = 0.236\Delta T = 0.471\tau \tag{2.4.15}$$

(d) *Exponential*

$$\tau = 0.693\tau_e$$

$$\sigma = \tau_e = 1.44\tau \tag{2.4.16}$$

(e) *Gaussian*

$$\tau = (\log 2)^{1/2}\tau_e = 0.833\tau_e$$

$$\sigma = \frac{\tau_e}{2\sqrt{2}} = 0.354\tau_e = 0.425\tau \tag{2.4.17}$$

(f) *Truncated Lorentzian*

$$\sigma = \frac{\tau}{2}\left|\frac{X}{\tan^{-1}X}\right|^{1/2} \tag{2.4.18}$$

where $X = \Delta T/\tau$. Note that $\sigma \to \infty$ as $X \to \infty$.

Some of these relationships will be used in later analysis. Their verification is left as an exercise for the reader.

2.4.2 Frequency Response

As we shall find in later chapters, an optical fiber communication system may be treated as a band-limited linear system. This arises because the signal is represented in the receiver by the current generated there. This is proportional to the received optical power, and in turn to the optical power transmitted. Subject to source linearity this is proportional to the signal current in the transmitter, giving overall linearity. We have seen that material and multipath dispersion will cause an optical impulse launched onto the fiber to spread out as it propagates. The received pulse then represents the impulse response of the fiber. Fourier transformation may be used to convert this impulse response into a corresponding, equivalent frequency response. However, because during transmission the signal amplitude is represented by the optical power, an ambiguity arises in the definition of the fiber bandwidth.

A normalized fiber impulse response $h(t)$ may be defined as follows. If an optical impulse of energy $\mathcal{E}_{\mathrm{T}}$ is launched onto the fiber and it gives rise to a received pulse of optical power $\Phi_{\mathrm{R}}(t)$, then

$$h(t) = \Phi_{\mathrm{R}}(t)/\mathcal{E}_{\mathrm{R}} \tag{2.4.19}$$

where

$$\mathcal{E}_{\mathrm{R}} = \int_{-\infty}^{\infty} \Phi_{\mathrm{R}}(t)\, \mathrm{d}t \tag{2.4.20}$$

is the received energy. Thus $\mathcal{E}_{\mathrm{R}}/\mathcal{E}_{\mathrm{T}}$ represents the fiber attenuation. The corresponding frequency response, $H(f)$, of the fiber may be obtained in the usual way by taking the Fourier transform of $h(t)$:

$$H(f) = \int_{-\infty}^{\infty} h(t) \exp(-\mathrm{j}2\pi f t)\, \mathrm{d}t \tag{2.4.21}$$

In general $H(f)$ is a complex function representing the variation of both amplitude and phase with frequency.

It is customary to define the bandwidth of a linear system as the range of frequencies over which $|H(f)|$ exceeds $1/\sqrt{2}$ of its maximum value. That is, the bandwidth represents the range of frequencies between the '–3 dB' or 'half-power' points on the frequency characteristic. In optical fiber systems this has become known as the *electrical bandwidth* of the system, and we shall designate it as $(\Delta f)_{\mathrm{el}}$. The terms '–3 dB' and 'half-power' then refer to the amplitude of the electrical signal generated in the receiver. We are laboring this elementary point because it is also customary to measure optical power levels in dBm and ratios of optical power in dB, and the two sets of units differ by a factor of two. Thus, if at some frequency the received modulated optical power were reduced to half its maximum value, $|H(f)| = \frac{1}{2}$, we would say that the optical power was 3 dB down. But after conversion back to an electric current the electrical signal would be 6 dB below its maximum value. It is therefore also possible to define an *optical bandwidth* for a fiber, $(\Delta f)_{opt}$, based on the frequency range over which $|H(f)|$ exceeds $\frac{1}{2}$.

A frequency parameter that is probably of more interest than either $(\Delta f)_{el}$ or $(\Delta f)_{opt}$ is the maximum bit rate, B, that can be usefully transmitted through the fiber. It will be shown in Chapter 15 that for a wide range of pulse shapes B should not exceed $1/4\sigma$. If it does, then the received optical power level required to ensure a certain minimum error rate in the regeneration process is found to increase rapidly. Using this relationship between the bit-rate capacity of the fiber and the r.m.s. pulse width gives us a means of linking both these parameters to the bandwidth as represented by $(\Delta f)_{el}$ or $(\Delta f)_{opt}$ and to the time dispersion as represented by the full width at half-height pulse length, τ, or the overall pulse length, ΔT. To demonstrate this we will evaluate the frequency responses corresponding to four of the hypothetical pulse shapes discussed in Section 2.4.1, on the unlikely assumption that they represent the impulse responses of particular fibers. The results justify, with some reservations, the use of the relationships

$$(\Delta f)_{opt} \simeq 2(\Delta f)_{el} \simeq B = \frac{1}{4\sigma} \simeq \frac{1}{2\tau} \simeq \frac{1}{\Delta T} \tag{2.4.22}$$

as a working guide. It will be noted that the fiber bandwidth values quoted earlier in this chapter refer to the electrical rather than the optical bandwidth. Once again it will be left to the reader to verify these results in detail.

Case (a) Rectangular Pulse

$$\begin{aligned} h(t) &= 1/\tau, \quad |t| < \tau/2 \qquad H(f) = \frac{\sin \pi f \tau}{\pi f \tau} \\ &= 0, \quad |t| > \tau/2 \end{aligned} \tag{2.4.23}$$

$$1.44(\Delta f)_{opt} = 1.96(\Delta f)_{el} = B = \frac{1}{4\sigma} = \frac{1}{1.15\tau} = \frac{1}{1.15\Delta T}$$

Case (b) Triangular Pulse 1

$$h(t) \begin{cases} = \dfrac{1}{\tau}\left(1 - \dfrac{|t|}{\tau}\right), & |t| < \tau \\ = 0, & |t| > \tau \end{cases} \qquad H(f) = \left(\frac{\sin \pi f \tau}{\pi f \tau}\right)^2 \tag{2.4.24}$$

$$1.01(\Delta f)_{opt} = 1.92(\Delta f)_{el} = B = \frac{1}{4\sigma} = \frac{1}{1.64\tau} = \frac{1}{0.82\Delta T}$$

Case (d) Exponential Pulse

$$h(t) = (1/\tau_e)\exp(-t/\tau_e), \quad t > 0 \qquad H(f) = (1 + j2\pi f\tau_e)^{-1}$$
$$= 0, \quad t < 0 \tag{2.4.25}$$

with $\tau = 0.693\tau_e$,

$$|H(f)| = (1 + 4\pi^2 f^2 \tau_e^2)^{-1/2}$$

$$0.91(\Delta f)_{opt} = 1.57(\Delta f)_{el} = B = \frac{1}{4\sigma} = \frac{1}{5.8\tau}$$

Case (e) Gaussian Pulse

$$h(t) = \frac{1}{\sqrt{(2\pi)}\sigma}\exp(-t^2/2\sigma^2) \qquad H(f) = \exp(-2\pi^2 f^2 \sigma^2) \tag{2.4.26}$$

$$1.34(\Delta f)_{opt} = 1.89(\Delta f)_{el} = B = \frac{1}{4\sigma} = \frac{1}{1.69\tau}$$

2.4.3 Total R.M.S. Pulse Width

When an optical fiber can be treated as a linear system in the way we have just discussed, then equation (2.4.11) is true irrespective of the pulse shape. This may be shown as follows.† With reference to Fig. 2.17, we will show that any pulse $h_0(t)$ having an r.m.s. pulse width σ_0 which passes through a linear system whose impulse response $h_1(t)$ has an r.m.s. pulse width σ_1 will emerge as a pulse of r.m.s. width σ_2, where

$$\sigma_2^2 = \sigma_0^2 + \sigma_1^2 \tag{2.4.27}$$

For convenience we shall assume $h_0(t)$ and $h_1(t)$ to be normalized, that is

$$\int_{-\infty}^{\infty} h_0(t)\,dt = 1 \tag{2.4.28}$$

$$\int_{-\infty}^{\infty} h_1(t)\,dt = 1 \tag{2.4.29}$$

† The author is grateful to Dr S. D. Personick for demonstrating this proof.

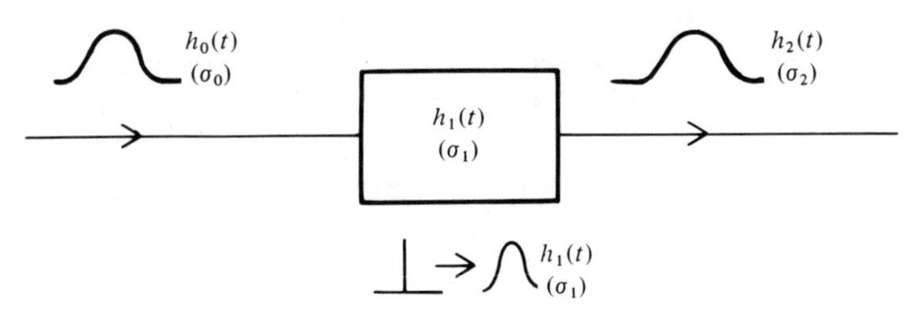

Fig. 2.17 R.M.S. pulse width.

and take the time zero to be centered on the mean time of $h_0(t)$, that is

$$\int_{-\infty}^{\infty} t h_0(t)\,\mathrm{d}t = 0 \tag{2.4.30}$$

From the convolution integral for a linear network,

$$h_2(t) = \int_{-\infty}^{\infty} h_0(t-\tau)h_1(\tau)\,\mathrm{d}\tau \tag{2.4.31}$$

Using the definition of r.m.s. pulse length,

$$\sigma_2^2 = \int_{t=-\infty}^{\infty} t^2 h_2(t)\,\mathrm{d}t - \left[\int_{t=-\infty}^{\infty} t h_2(t)\,\mathrm{d}t\right]^2 \tag{2.4.32}$$

We will evaluate the second term in equation (2.4.32) first by writing $x = t - \tau$, $\mathrm{d}x = \mathrm{d}t$, and using equation (2.4.31)

$$\int_{t=-\infty}^{\infty} t h_2(t)\,\mathrm{d}t = \int_{x=-\infty}^{\infty} (x+\tau)\left[\int_{\tau=-\infty}^{\infty} h_0(x)h_1(\tau)\,\mathrm{d}\tau\right]\mathrm{d}x$$

$$= \int_{\tau=-\infty}^{\infty} h_1(\tau)\left[\int_{x=-\infty}^{\infty} (x+\tau)h_0(x)\,\mathrm{d}x\right]\mathrm{d}\tau$$

$$= \int_{\tau=-\infty}^{\infty} h_1(\tau)[0+\tau]\,\mathrm{d}\tau = \int_{t=-\infty}^{\infty} t h_1(t)\,\mathrm{d}t \tag{2.4.33}$$

The first term in equation (2.4.32) may be evaluated using a similar substitution, namely $x = t - \tau$ so that $t^2 = x^2 + 2x\tau + \tau^2$, then

$$\int_{t=-\infty}^{\infty} t^2 h_2(t)\,\mathrm{d}t = \int_{t=-\infty}^{\infty} (x^2 + 2x\tau + \tau^2)\left[\int_{\tau=-\infty}^{\infty} h_0(x)h_1(\tau)\,\mathrm{d}\tau\right]\mathrm{d}x$$

$$=\int_{\tau=-\infty}^{\infty} h_1(\tau)\left[\int_{x=-\infty}^{\infty} (x^2 + 2x\tau + \tau^2)h_0(x)\,\mathrm{d}x\right]\mathrm{d}\tau$$

$$=\int_{\tau=-\infty}^{\infty} h_1(\tau)[\sigma_0^2 + 0 + \tau^2]\,\mathrm{d}\tau$$

$$=\sigma_0^2 + \int_{t=-\infty}^{\infty} t^2 h_1(t)\,\mathrm{d}t \qquad (2.4.34)$$

By analogy with equation (2.4.32) we may write

$$\sigma_1^2 = \int_{t=-\infty}^{\infty} t^2 h_1(t)\,\mathrm{d}t - \left[\int_{t=-\infty}^{\infty} t h_1(t)\,\mathrm{d}t\right]^2 \qquad (2.4.35)$$

Thus, putting together equations (2.4.32–35) we obtain

$$\sigma_2^2 = \sigma_0^2 + \sigma_1^2 \qquad (2.4.27)$$

This result is not dependent on the assumption of equation (2.4.30), which merely serves to simplify the algebra, and clearly it may be extended indefinitely if the output pulse is passed into further band-limiting systems. Thus, provided that material and multipath dispersion may be taken to be independent linear processes, it follows that equation (2.4.11) will apply for pulses of any physically realizable shape.

PROBLEMS

2.1 With reference to Fig. 2.2, calculate the values of $(\mathrm{NA})^2$, the angles α_m and ϕ_m, and the dispersion parameters $(\Delta T/l)$ and $(B\,.\,l)$ for the following step-index fibers:

(a) $n_1 = 1.470$, $n_2 = 1.455$, $n_a = 1$.
(b) $n_1 = 1.46$, $n_2 = 1.40$, $n_a = 1$.
(c) $n_1 = 1.46$, $n_2 = n_a = 1$ (unclad fiber).

2.2 Show that eqn. (2.1.21) may be approximated by eqn. (2.1.22) for paraxial rays.

2.3 A fiber has a core refractive index that varies with radius according to eqn. (2.1.23) with $a = 30$ μm and $\Delta' = 0.01$. Show that paraxial rays from an axial, point source will be brought to a focus at periodic axial distances of 0.67 mm.

2.4 Using the Sellmeier coefficients for pure silica given in eqn. (2.2.32), derive the coefficients A, B, C, D, E required in the polynomial expansion for the refractive index, eqn. (2.2.31). Hence obtain values for the refractive index of pure silica at 0.82, 1.27 and 1.55 μm.

(Note that it is a straightforward matter to write a short computer program which will give values for n and its derivatives, such as $Y_m = \lambda^2 d^2n/d\lambda^2$, using either the Sellmeier or the polynomial coefficients. A listing of measured Sellmeier coefficients may be found in Ref. [5.3], Table 7.3.)

2.5 The amplitude of a carrier wave of angular frequency ω_c is modulated sinusoidally at an angular frequency ω_m. Show that the propagation velocity of the *modulation* is given by $(d\beta/d\omega)^{-1}_{\omega c}$ where $\beta = 2\pi/\lambda$ is the propagation constant of the carrier wave.

2.6 Show that for a source centered on the wavelength of minimum dispersion, λ_0, the next most significant term in eqn. (2.2.54) is

$$\pm [2\lambda_0^3(d^3n/d\lambda^3)_{\lambda_0} + \lambda_0^4(d^4n/d\lambda^4)_{\lambda_0}](\gamma^3/48)$$

2.7 For pure silica $\lambda_0 = 1.28$ μm, $\lambda_0^3(d^3n/d\lambda^3)_{\lambda_0} = -0.048$ and $\lambda_0^4(d^4n/d\lambda^4)_{\lambda_0} = 0.2274$. Calculate the dispersion caused by sources having spectral halfwidths of 50, 100 and 150 nm about λ_0, noting the significance of the term in γ^3, and noting also whether the longer or the shorter wavelengths give rise to the greater dispersion. Hence explain whether the true minimum dispersion would occur with sources of slightly shorter or slightly longer wavelengths.

2.8 Verify the relationships between the r.m.s. and halfheight widths of the pulses shown in Fig. 2.16.

2.9 Verify the frequency responses given by eqns (2.4.23)–(2.4.26).

2.10 Verify the relationship between:
(a) $(\Delta f)_{opt}$, $(\Delta f)_{el}$ and B
(b) σ, τ and ΔT,
for the four pulse shapes quoted on pp. 65 and 66.

REFERENCES

2.1 M. Born and E. Wolf, *Principles of Optics*, 5th Ed., Pergamon (1975).
2.2 D. B. Keck, 'Optical fibre waveguides', in M. K. Barnoski (Ed.) *Fundamentals of Optical Fiber Communications*, Academic Press (1976).

SUMMARY

Using the concept of rays (valid in the limit $\lambda \to 0$) light in step-index (SI) fibers is guided by total internal reflection at the core–cladding interface. The fiber captures and propagates a fraction $\Phi/\Phi_0 = (NA)^2 = 2n\,\Delta n$ of the light emitted by a diffuse source.

Multipath dispersion, $\Delta T/l = \Delta n/c$, limits the bandwidth and bit-rate capacity to $2(\Delta f)_{el} \simeq B \simeq 1/\Delta T \simeq 1/2\tau_1 \simeq c/l\,\Delta n$, in theory. Capacity is higher in practice.

In graded-index (GI) fibers multipath dispersion can be reduced below $\tau_1/l = 1$ ns/km. It is totally absent from single-mode fibers.

Material dispersion (τ_2/l) is least at frequencies where fundamental attenuation is least, and may be modified by the dopants used to vary the refractive index. $(\tau_2/l) = (\gamma/c)|Y_m|$, where $\gamma = |\Delta\lambda/\lambda|$ and $Y_m = \lambda^2 d^2 n/d\lambda^2$.

In silica $Y_m = 0$ at $\lambda = 1.276$ μm. Then $(\tau_2/l) = -(\gamma^2/8c)\lambda^3 d^3 n/d\lambda^3$.

Total dispersion, $\tau = (\tau_1^2 + \tau_2^2)^{1/2}$ is dominated by multipath dispersion in SI fibers and in laser-driven GI fibers, but by material dispersion in LED-driven GI fibers (except near the dispersion minimum). See Table 2.1.

The impulse response of a fiber as measured by its r.m.s. pulse width, σ, may be related to the bit-rate capacity: $B = 1/4\sigma$. Frequency response can be derived by Fourier transformation (and vice versa). Both electrical and optical bandwidths, $(\Delta f)_{el}$ and $(\Delta f)_{opt}$, may be defined using the -3 dB and -6 dB points on the electrical response curve, respectively.

3

Attenuation in Optical Fibers

3.1 ATTENUATION MECHANISMS

3.1.1 Introduction

For a material to be considered as a possible optical fiber material it has to be highly transparent to electromagnetic radiation of wavelengths in the region of 1 μm. We are thus concerned here with any mechanism that may give rise to attenuation of light over the wavelength range, say, of 0.5–2.0 μm.

We recall that in their original proposal in 1966 Kao and Hockham suggested that an overall attenuation figure of 20 dB/km would be needed before optical fibers could be of use in trunk telecommunications, and that by 1980 the lowest attenuation recorded under laboratory conditions in a single, unspliced length of fiber was as little as 0.2 dB/km. This attainment was due entirely to the understanding that had been gained of the causes of attenuation and to the high quality control over the materials used which enabled the main causes to be eliminated.

The mechanisms of attenuation may be classified broadly into two principal groups:

> *Absorption* is basically a material property and at the wavelengths we are concerned with it takes place when electronic transitions, the resonances discussed in Section 2.2.1, are excited within the material and are followed by nonradiative relaxation processes. As a result there is an increase in thermal energy in the material.
>
> *Scattering* may be partly a material property but may also be caused

by imperfections in the fiber geometry. It occurs when the mode of propagation of the light is changed such that some of the optical energy leaves the fiber. There is no conversion of the radiant energy into other forms.

At the present time nearly all optical fibers are made from high silica content glasses doped with oxides such as boria, titania, germania, or phosphorus pentoxide. As in the rest of the book, these are the materials we shall concentrate on, discussing in turn the principal causes of absorption and scattering. However, it should be said that a great many materials have been suggested for use as optical fibers and quite a number have been tried experimentally. For example, in the early stages of development, before it was established that fibers could be drawn from multicomponent glasses, fibers having a liquid core enclosed in a glass cladding were made successfully (tetrachlorethylene was one liquid used—bubble-free, of course!). A number of workers have experimented with sodium silicate and calcium silicate glasses which have quite low melting points, in the region of 1100°C, and are quite easy to handle. Others have used lead silicate glasses which offer the possibility of large index differences. Glasses can be formed from sulphides and selenides as well as from oxides and the possible use of these and even of single-crystal materials for transmission at longer wavelengths is a matter that has caused some speculation. However, single crystals are unlikely ever to have the mechanical properties required in practical fibers and these other glassy materials are so far from being realizable as useful light guides that we will not consider them further. A group of transparent materials that we will consider further are polymers. They are discussed separately in Section 3.4.

Attenuation does not depend only upon the quality of the core material. The cladding material may also be significant. In the process of total internal reflection the electromagnetic fields penetrate the core-cladding interface and extend into the cladding. Thus a small fraction of the total optical power propagates in the cladding. If this is of poor quality or is highly absorbing it will contribute to the overall fiber attenuation. Fibers required to have minimal attenuation are made with the cladding material as high in quality and as carefully controlled as the core. It is then necessary to ensure that any light scattered into the cladding does not propagate and enter the detector because this is likely to increase the range of propagation velocities and make the fiber dispersion worse. Two steps can be taken to avoid this: the outer layers of the cladding can be made absorbing so that the scattered rays are attenuated there but the propagating light is unaffected; the cladding itself can be surrounded by a protecting polymer layer of higher refractive index into which the scattered rays pass and are absorbed.

3.1.2 Absorption

We have already seen in Section 2.2.1 that the electronic and atomic resonances which are responsible for the dispersive properties of a dielectric material also give rise to absorption in the vicinity of the resonant frequencies. For materials of interest to us these are u.v. resonances associated with the electronic structures of the crystal atoms, and i.r. resonances associated with lattice vibrations of the atoms themselves. Although these resonant frequencies are well away from the optical frequencies we wish to use, the absorption they produce is very strong and the tails of their absorption bands do extend into this region at the very low levels of attenuation we are concerned with. An estimate of the attenuation produced by the absorption band edges in a germania-doped silica fiber is shown in Fig. 3.1.

The window between the two edges should be greatest at about 1.5 μm, but it is considerably reduced because, above about 0.3 μm, another fundamental loss mechanism dominates the u.v. absorption band edge. This is Rayleigh scattering, which we shall consider in the next section.

The i.r. band edge becomes significant at wavelengths longer than 1.5 μm. This edge is associated with the characteristic stretching vibrations of the oxide

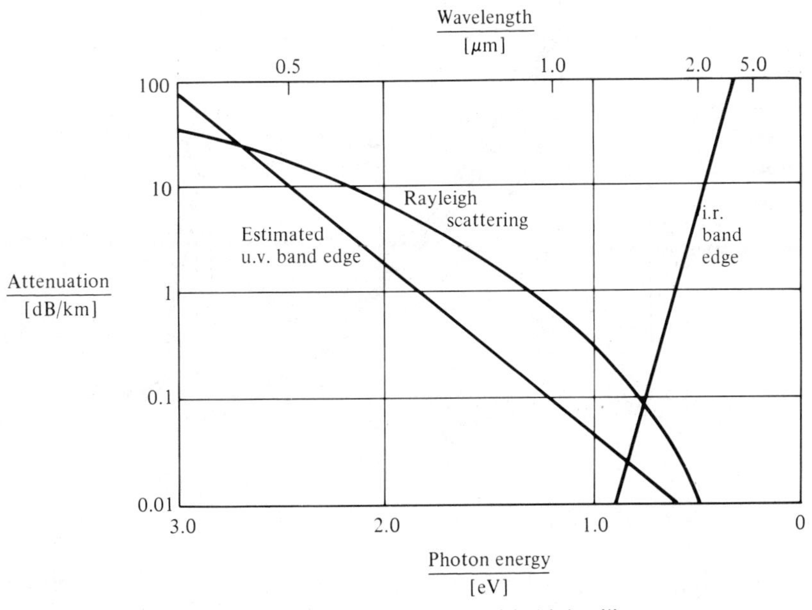

Fig. 3.1 Fundamental loss limits in glasses with high silica content.

bonds which have the following fundamental frequencies:

Si–O	9.0 μm
Ge–O	11.0 μm
P–O	8.0 μm
B–O	7.3 μm

From this point of view germania should be the most favorable impurity because of the longer wavelength associated with the Ge–O stretching vibration. This is borne out by the measurements plotted in Fig. 3.2. These show that the i.r. band edge does indeed move to shorter wavelengths when P_2O_5 and B_2O_3 are used as dopants. While this has little effect on the attenuation level at 0.85 μm, it rules out the use of these materials as principal dopants in fibers designed to be used at longer wavelengths.

It may be noted that the figure of 9.0 μm for the Si–O resonance wavelength does not agree with the value of 9.9 μm used in eqn. (2.2.32) in the Sellmeier equation for the dispersion of silica. There are two reasons for this: the first is that eqn. (2.2.32) is an empirical relation giving a best fit to data from a

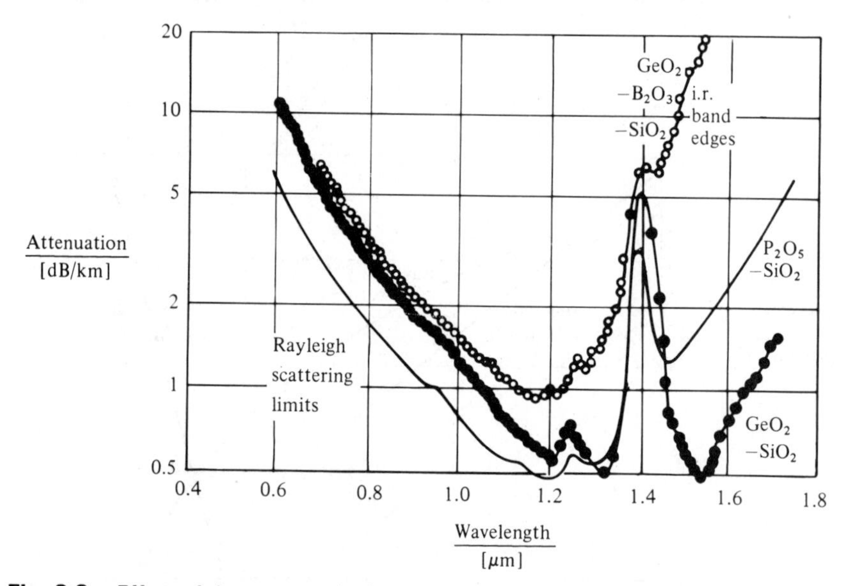

Fig. 3.2 Effect of dopants on the i.r. band edge and the Rayleigh scattering loss. [Data taken from H. Osanai *et al.*, Effect of dopants on transmission loss of low-OH-content optical fibers, *Ets. Lett.* **12**, 549–50 (14 Oct. 1976).]

The absorption peaks near 1.4 μm and 1.25 μm are caused by residual water vapor.

restricted region well away from the supposed resonances. The quality of the fit to the data is not very sensitive to the precise values of λ_k used, and eqn. (2.2.32) is, therefore, not a good measure of the resonant frequencies. The second reason is that the data have been approximated by a Sellmeier formula having three terms when there are in fact many resonances whose effect on dispersion is being summarized in this way. The λ_k values, then, represent some crude average of the wavelengths of these resonances. It may also be noted in passing, just to put things into perspective, that the absorption coefficient at the Si–O resonance wavelength would be about 10 dB/μm!

The band edge absorption is inherent in the material used to make the fiber. However, this material may also contain impurity atoms and molecules which can cause absorption at wavelengths of interest. In practice it is found that water vapor and the first row of transition metals (vanadium, chromium, manganese, iron, cobalt and nickel) are the most pernicious of these. The metals occur in the glass as ions, which by virtue of their electronic structure give rise to broad-band absorption at wavelengths which may depend upon the state of oxidation of the ion. As a very approximate indication, the concentrations of these transition-metal ionic impurities have to be kept below about 1 part in 10^9 for their contribution to absorption not to exceed 1 dB/km at wavelengths in the region of 1 μm.

The absorption caused by water vapor arises from the basic stretching vibration of the O–H bond. Its fundamental frequency (f_0) is centered at 2.73 μm but it gives rise to harmonics and to combination frequencies with the Si–O bending resonance at 12.5 μm (frequency f_s). Table 3.1 lists some of these absorption bands together with their assignment and their absorption strength. Most may be observed on the attenuation curves of Figs 3.2, 3.3 and 3.5. There is considerable unharmonicity, which means that the harmonics are not exact multiples of the fundamental frequency. The absorption peaks tend to be quite broad and slightly asymmetric with respect to wavelength, giving relatively greater amounts of absorption at shorter wavelengths. There is also

Table 3.1 OH^- absorption bands

Approximate resonance wavelength [μm]	Origin	Approximate absorption caused by 1 p.p.m. of OH^- [dB/km]
1.39	$2f_0$	65
1.24	$2f_0 + f_s$	2.3
1.13	$2f_0 + 2f_s$	0.1
0.95	$3f_0$	1
0.88	$3f_0 + f_s$	0.1
0.72	$4f_0$	0.05

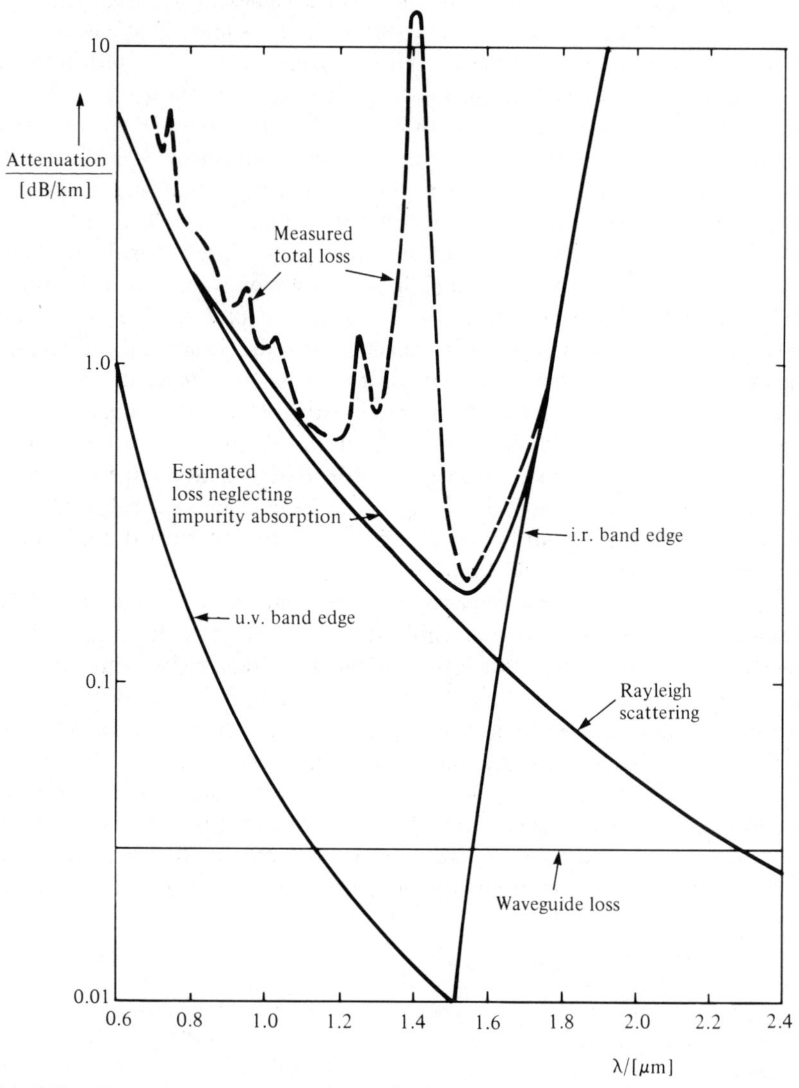

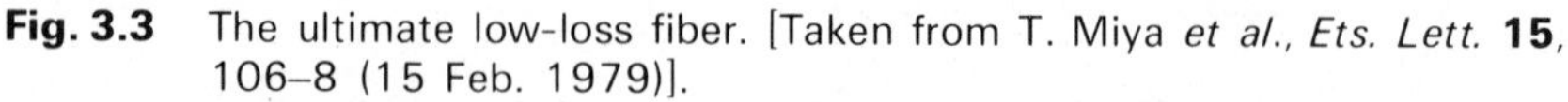

Fig. 3.3 The ultimate low-loss fiber. [Taken from T. Miya *et al.*, *Ets. Lett.* **15**, 106–8 (15 Feb. 1979)].

These data show the experimentally determined attenuation of a single, 2.2 km length of single-mode, germania-doped fiber having $\Delta = 0.0019$. Estimates of the contributing loss factors are shown.

evidence that they are broader in borosilicate glasses than in others. When P_2O_5 is present the absorption bands are complicated by a P–OH resonance at 3.05 μm which has a first harmonic between 1.5 μm and 1.6 μm.

For fibers designed to be used in the wavelength range 0.8 μm to 0.9 μm, a water vapor content between 1 part in 10^6 and 1 part in 10^7 is small enough to have negligible effect, but fibers designed to use the windows around 1.2, 1.3 or 1.6 μm require a concentration of less than 1 part in 10^8. This is exceedingly difficult to achieve. In Fig. 3.3 we have reproduced the measured total attenuation curve for one of the lowest loss fibers produced up to 1980. The estimated contributions from various absorption and scattering processes are shown. It is only when impurity absorption is reduced to the levels obtained here that these other contributions can be identified with any confidence.

3.1.3 Scattering

By its nature glass is a disordered structure in which there are microscopic variations around the average material density, and local microscopic variations in composition. Each of these gives rise to fluctuations of refractive index on a scale that is small compared to optical wavelengths, that is, on a submicron scale. This is fundamental to any glassy material, however carefully made, and it causes the light to be scattered in the manner known as Rayleigh scattering (see, e.g., Ref. [2.1], Section 2.13.5). The light is then lost from the fiber.† Indeed, if visible light from a laser is coupled into a long, unsheathed fiber coil and the fiber is viewed from the side in a darkened room, the scattered light is clearly seen, diminishing in intensity along the length of the fiber.

The loss caused by this mechanism can be minimized by cooling the melt from which the fiber is drawn in as carefully controlled a manner as possible. The loss is likely to be higher in multicomponent glasses because of compositional variations. Its essential characteristic is that the scattered power and hence the attenuation is inversely proportional to the fourth power of the wavelength. Figure 3.1 shows that it is Rayleigh scattering rather than the u.v. absorption band edge that is the principal cause of loss in silica fibers at wavelengths less than 1.5 μm. For glasses with high silica content, a typical value for this attenuation mechanism is 1 dB/km at 1 μm, with germania and boria doping giving a slight increase and phosphorus pentoxide a slight decrease. This effect can be seen in Fig. 3.2. For sodium borosilicate glasses a typical value at 1 μm is in the region of 2 dB/km.

† Some light may be back-scattered so that it propagates in the reverse direction. Observation of this effect is a useful way of assessing fiber attenuation. However, it seriously limits bidirectional transmission, and even double back-scattering, in which some light ends up delayed but travelling in the original direction, can cause problems.

Up to the present we have tacitly assumed that the geometry of any fiber is perfect from end to end and that the fiber itself is straight. In practice, of course, this is not the case and the bends and imperfections that do occur lead to the guided rays being scattered into rays which are transmitted away from the core-cladding interface. Major disturbances to the geometry of this interface (steps, offsets, inclusions) and large imperfections within the fiber core (bubbles, impurities) each give rise to a large local scattering loss. Such imperfections show up quite clearly as locally bright regions in the experimental arrangement which demonstrates Rayleigh scattering. This enables unsatisfactory lengths of fiber to be identified easily and discarded.

Tight bends similarly cause some of the light not to be internally reflected but to propagate into the cladding and be lost. In theory the optical power scattered out of a fiber at a major bend depends exponentially on the bend radius, R. The *bending* loss is then proportional to $\exp(-R/R_C)$ where the critical radius $R_C \simeq a/(\mathrm{NA})^2 = a/2n\,\Delta n$, a being the radius of the fiber core. At a bend of radius R_C the loss would be very considerable, but because of the form of the exponential function the losses decrease rapidly for less tight bends.

In practice, however, it is the mechanical properties of the fiber rather than bending losses that determine the minimum allowable bending radius. If the fiber is bent so tightly that the surface strain exceeds about 0.2%, then significant stress cracking is likely to develop in the fiber during life. This must be prevented by enclosing the fiber in a suitably rigid cable. Consider a fiber of core radius $a = 30\ \mu\mathrm{m}$, cladding diameter $2b = 125\ \mu\mathrm{m}$ and having $n = 1.5$, $\Delta n = 0.01$, so that $\mathrm{NA} = 0.17$. Let the fiber be wound on a drum of radius $(R - b)$ so that the neutral axis of the fiber is bent into a curve of radius R as shown in Fig. 3.4. The compression strain at the inner fiber surface and the tension strain at the outer surface are both given by b/R. For this strain to be less than 0.2%, R must be greater than $b/0.002 = 500b$. In the example given this requires $R > 31$ mm. On the other hand $R_C = a/(2n\,\Delta n) = a/0.03 = 2$ mm. Clearly a mechanically acceptable bending radius gives rise to negligible bending losses.

Although the losses from gross bends may be unimportant, a continuous succession of very small bends may cause a very significant rise in fiber attenuation. The effect is known as *microbending* loss, and is particularly noticeable when an unsheathed fiber is wound in tension onto a drum. The microbends arise as the fiber is distorted by imperfections on the drum surface. A similar effect can easily be caused by the pressures exerted on a fiber by adjacent members inside a cable. Continuous, small variations in core diameter, which can easily arise during manufacture, may give rise to a similar scattering mechanism and cause what are known as *waveguide* losses.

Bending, microbending and waveguide losses have been the subject of considerable theoretical analysis. This is too complex and too detailed for us to be able to pursue here but it is considered further in Chapter 5. Suffice it to say

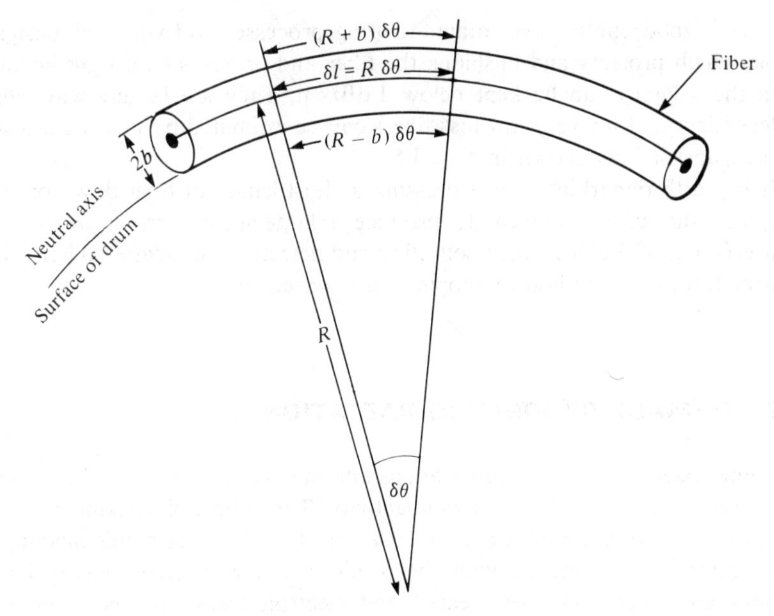

Fig. 3.4 Surface strain produced by bending.

$$\text{Strain at outer surface (tension)} = \frac{(R+b)\delta\theta - R\delta\theta}{R\delta\theta}$$

$$= \frac{b}{R} = \text{Strain at inner surface (compression)}$$

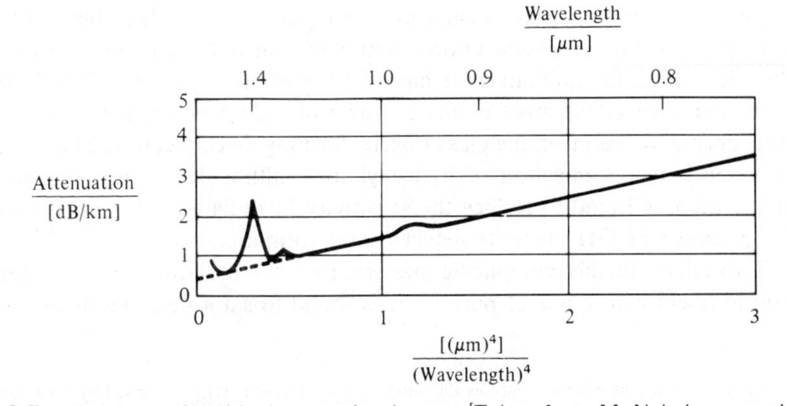

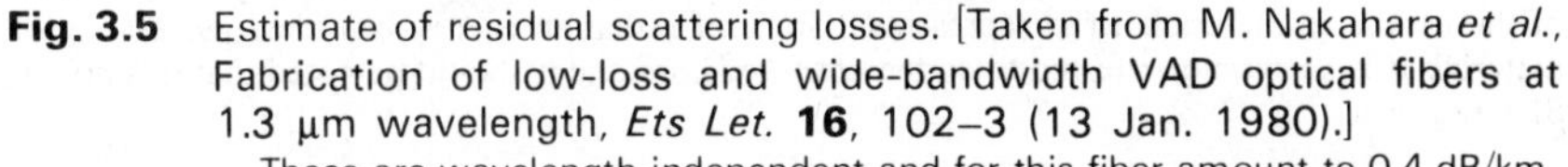
Fig. 3.5 Estimate of residual scattering losses. [Taken from M. Nakahara *et al.*, Fabrication of low-loss and wide-bandwidth VAD optical fibers at 1.3 μm wavelength, *Ets Let.* **16**, 102–3 (13 Jan. 1980).] These are wavelength-independent and for this fiber amount to 0.4 dB/km.

that with good control over manufacturing processes and with well designed cable which protects and cushions the fiber and prevents over-tight bending, then these losses can be kept below 1 dB/km. They are largely wavelength independent and for very low loss fibers can be estimated by plotting attenuation against λ^{-4} as shown in Fig. 3.5.

It is worth remarking that if pressing a short length of fiber down onto an irregular surface is sufficient to produce a large local increase in the light scattered out of the fiber, then collecting and detecting this scattered light gives a straightforward method of tapping an exposed fiber.

3.2 DAMAGE BY IONIZING RADIATION

In some space and military applications communication links may be required to be resistant to high levels of radioactivity. The effects of ionizing radiation on glasses are very complex and quite severe. The chemical bonds binding the glass matrix are disrupted with the result that new electron energy levels, donors and acceptors, are created and electron transitions between them become possible. Many of these transitions give rise to absorption in the visible and near infra-red.

At low levels of irradiation the additional radiation-induced attenuation rises in direct proportion to the irradiation. But the sensitivity to radiation varies for different types of fiber from 0.1 to 10(dB/km)/rad.† These figures are based on attenuation measurements made at an optical wavelength of 0.82 μm. There is some evidence that the attenuation increase is less for light of longer wavelengths.

The bonds in multicomponent glasses are particularly vulnerable and fibers having cores made from silica doped with GeO_2 or B_2O_3 are more sensitive to the effects of radiation than those having cores of pure synthetic silica. In these latter the chief effect involves the removal of individual oxygen atoms from their normal positions in the glass matrix, forming defect centers. The presence of a modest concentration of hydroxyl ions, although it raises the normal attenuation, is found to reduce the sensitivity to radiation. It is assumed that the presence of OH^- inhibits defect center formation.

Two effects further complicate matters. First the radiation-induced attenuation in fibers with cores of pure silica is found to saturate at levels of several

† The *rad* is a unit which quantifies the amount of radiation energy absorbed by a given mass of the irradiated material. 1 rad represents 1 joule of radiation energy absorbed per 100 kg of material. It is sometimes referred to as the *absorbed dose* and it should be noted that it is not an SI unit.

hundred to several thousand dB/km. Secondly, in all types of fiber the defects produced anneal out with time. This process may be accelerated by heat treatment and by flooding the material with white light. Recovery is more complete in pure silica. Germania-doped fibers in particular exhibit an extremely high transient attenuation which, for a few seconds following a radiation flash, may disable a system designed to withstand the long-term effects.

Most fibers required to be radiation hard are used in relatively short links so that a fairly high level of normal attenuation is acceptable, and additional radiation-induced attenuation is less critical. A numerical example will illustrate this. Consider two systems in each of which the ratio between the minimum required receiver power and the power generated by the transmitter is 50 dB. Say that of this 30 dB is allowed for normal fiber attenuation and 20 dB is a generous, unallocated safety margin. Let the first system use low-loss fiber of normal attenuation 5 dB/km in order to obtain a repeater spacing of 6 km, whereas the second only requires a repeater spacing of 150 m and thus uses fiber with a normal attenuation of 200 dB/km. The effect of 1000 rad irradiation on the first system would be disastrous. If it increased attenuation by 100 dB/km, then even a repeater spacing reduced from 6 to 0.5 km would still need the full 50 dB of available loss. The second system, however, would survive. Fiber attenuation would increase from 200 to 300 dB/km and give a loss of 45 dB over 150 m. This remains within the safety margin.

A fiber which may be of particular use in radiation-resistant systems is one that has a pure synthetic silica core with a small remaining water content, and a polymer cladding. Polymer and polymer-clad-silica (PCS) fibers are discussed further in Section 3.4.

It is worth remarking that a fiber designed to withstand a 1000 rad radiation dose would not be part of a manned system. A dose of 500 rads is sufficient to cause approximately 50% human mortality within 30 days of exposure.

3.3 THE OPTIMUM WAVELENGTH FOR SILICA FIBERS

Before passing on to the question of fiber manufacture, it is worth pausing for a moment to review, from the point of view of fiber properties, the relative merits of different source wavelengths in an optical communication system. We do this on the assumption that suitable semiconductor sources and detectors could be produced if required.

The key system parameters are bandwidth and repeater spacing, and the key fiber properties which determine them are dispersion and attenuation. Figure 3.3 shows that in a low-loss fiber, attenuation minima occur at 0.9, 1.0, 1.2, 1.3 and 1.55 μm. If water vapor can be eliminated and germania-doped silica is used, attenuation at 1.3 and 1.55 μm can be significantly lower than it is around 0.85 μm, the wavelength of the most common, gallium arsenide, sources. It may be lowest of all at 1.55 μm. Clearly there is a strong incentive for the use of longer wavelengths. This is reinforced by the variation of material dispersion with wavelength. We recall from Section 2.2.3 that a minimum occurs in the region of 1.3 μm. The question is, then, which is the better wavelength, 1.3 or 1.55 μm.

In order to illustrate this discussion we have calculated, using some simplifying assumptions, the repeater spacing that might be obtained at different bit rates with a number of hypothetical systems. The results are plotted in Figs. 3.6 and 3.7. They include step-index (SI), graded-index (GI) and single-mode (SM) fibers, and three source wavelengths, 0.9, 1.3 and 1.55 μm. Both laser and LED sources are considered, and for completeness results are also given for the polymer-clad-silica (PCS) fibers discussed in Section 3.4. These last are based on an attenuation figure of 20 dB/km and a dispersion of 100 ns/km. For the other fibers, multipath dispersion has been taken as 10 ns/km for SI fiber, 0.5 ns/km for GI fiber and zero for single-mode fiber. Representative figures for material dispersion and attenuation are based on the values for high-quality germania-doped silica shown in Figs. 2.13(b), 3.2 and 3.3. With a small

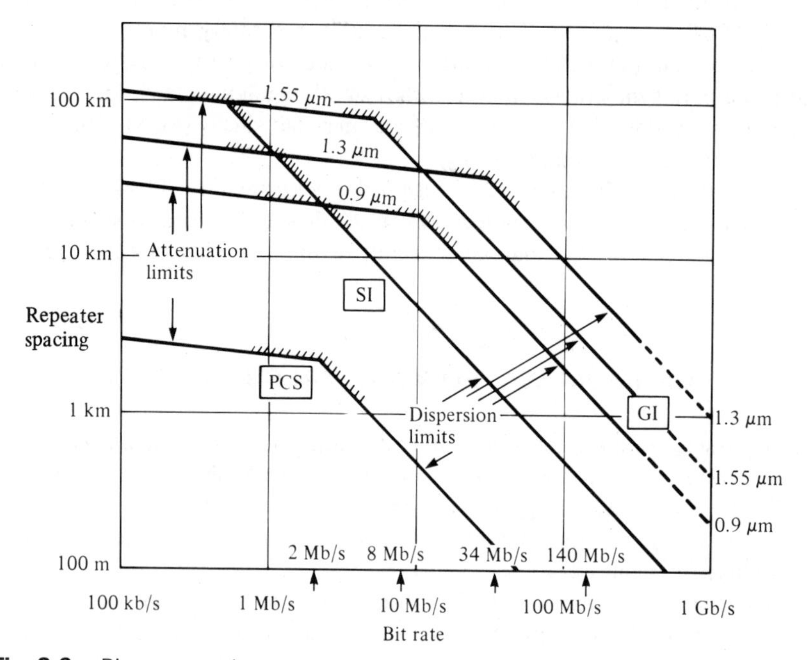

Fig. 3.6 Bit rates and repeater spacings obtainable in optical communication systems using LEDs.

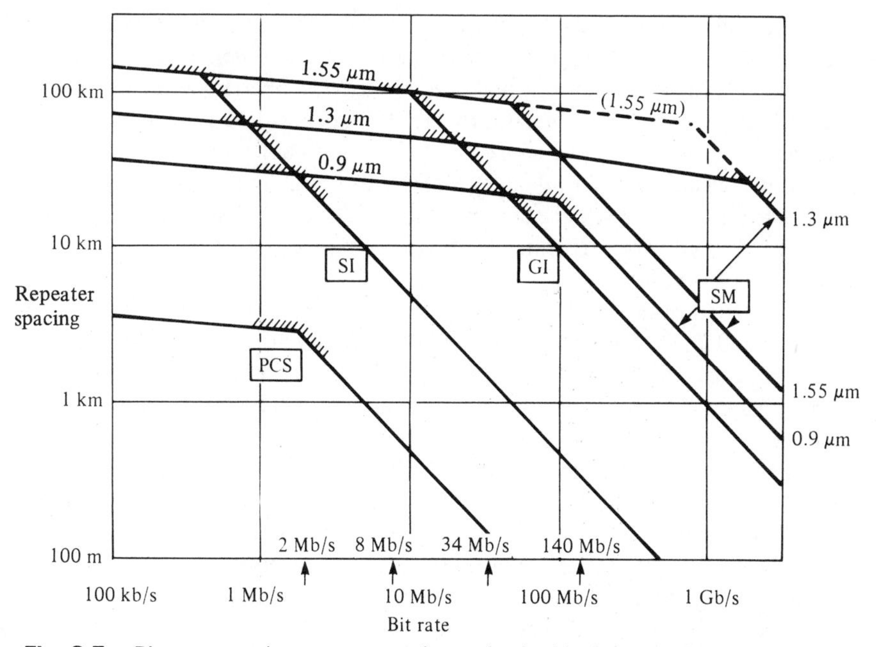

Fig. 3.7 Bit rates and repeater spacings obtainable in optical communication systems using lasers.

allowance for cabling losses and splices we have taken attenuation at 0.9 μm to be 2.0 dB/km and material dispersion to be 70 ps/(km.nm). At 1.3 μm the chosen figures are 1.0 dB/km and 2 ps/(km.nm), and at 1.55 μm they are 0.5 dB/km and 20 ps/(km.nm). Total dispersion is obtained by adding together the mean square values of the multipath and the material dispersion as in Table 2.1. The dispersion figures are based on full-width-at-half-height values, τ, which are assumed to be approximately twice the r.m.s. pulse widths, σ. The dispersion-limited bit rate is taken as $B = 1/4\sigma = 1/2\tau$. No account has been taken of the more complex interactions between the various causes of dispersion which are discussed in Chapter 6 or of the possible trade-off between dispersion and signal power which is discussed in Chapter 15. A 10 dB safety factor has been allowed for connectors and component degradation. In all cases we have assumed that the minimum power required at the detector is 0.1 nW/(Mb/s). This is likely to be optimistic at higher bit rates and for p–i–n photodiodes, but pessimistic at lower bit rates and for avalanche photodiodes. It is assumed in each case that LED sources launch 50 μW (–13 dBm) onto the fibers with a fractional spectral spread of $\gamma = 0.04$, and that laser sources launch 1 mW (0 dBm) with $\gamma = 0.004$.

Under the imposed conditions Fig. 3.6 indicates that with LED sources and step-index fiber, longer wavelengths are only advantageous at bit rates lower than about 2 Mb/s. At higher bit rates performance is limited by multipath dispersion. With LED sources and well-graded GI fiber material dispersion reduces the performance at high bit rates at 0.9 μm and 1.55 μm, but is not significant at 1.3 μm. If graded-index fiber with still better index-grading can be produced consistently, the advantage of 1.3 μm is increased further. Then a particularly cheap and reliable system using a 1.3 μm LED source and a p–i–n photodiode could permit repeater spacings of up to 20 km at bit rates possibly as high as 300 Mb/s.

With laser sources 1.55 μm is again advantageous at low bit rates, as can be seen in Fig. 3.7. With both SI and GI fibers multipath dispersion dominates at higher bit rates, independent of wavelength. The situation is more complicated when single-mode fibers are used, for reasons that will be discussed further in Chapter 5. At 0.9 μm and 1.55 μm with normal laser sources ($\gamma = 0.004$) the system will be dispersion limited when the bit rate exceeds 50–100 Mb/s. This gives the advantage above 100 Mb/s to 1.3 μm, which is always attenuation limited. However, lasers can be made to operate in a single longitudinal mode, in which case γ may be less than 0.0001. Then, material dispersion is small even at 1.55 μm, and full advantage can be taken of the minimal attenuation at this wavelength (dashed curves on Fig. 3.7).

Not surprisingly there is intense world-wide activity aimed at developing longer wavelength sources that will match the power and reliability of gallium arsenide sources and long wavelength detectors that will match the efficiency and noise performance of silicon diodes. Some of this work is discussed in later chapters. There is every indication that a high proportion of future optical communication systems will use light having free-space wavelengths of 1.55 and 1.3 μm.

3.4 ALL-PLASTIC AND POLYMER-CLAD-SILICA (PCS) FIBERS

For communicating at low transmission rates over short distances a very cheap kind of optical fiber made entirely of plastic may be worth considering. These are likely to be limited to data rates of a few Mb/s and to distances of a few hundred meters. The two transparent plastic materials most commonly used for the core of the fiber are polymethylmethacrylate (PMMA, better known as Perspex) and polystyrene. Their refractive indices are 1.49 and 1.59 respectively. PMMA is suitable as a cladding material for polystyrene-cored fibers; otherwise a fluorocarbon polymer or a silicone resin may be used.

The great advantage of plastic fibers lies in their cheapness and in the ease with which they can be manipulated and connected. They can be made to have a large numerical aperture and a range of convenient diameters, say 0.5–1.0 mm. The fact that plastic fibers are soft and nonbrittle means that even these relatively large-diameter fibers can be bent into tight corners. They can often be satisfactorily cut to length with a razor blade and precision couplers are no longer essential.

Disadvantages arise out of the high attenuation and dispersion inherent in polymer materials and from the rapid variation of material properties with temperature. There is an upper working limit of 80–100°C. In some cases the temperature coefficients of the core and cladding refractive indices are quite different, with the result that the numerical aperture varies with temperature. In some systems as the temperature is reduced it is possible for the cladding refractive index to become greater than that of the core, so that the fiber ceases to act as a light guide at all below some limiting temperature.

The high attenuation is caused by absorption and scattering. There are strong absorption bands associated with modes of stretching vibration of the various types of C–H bond in the material. Important harmonic resonances are centered around 1.09, 1.02, 0.91 and 0.74 μm, but there are many other combination tones and harmonics at shorter wavelengths, as can be seen in Fig. 3.8. The large-chain molecules that make up the polymer together with the effect of dust-particle inclusions are responsible for the high Rayleigh scattering losses. The wavelength of minimum attenuation is usually between 0.5 and 0.7 μm and minimum values between 200 and 2000 dB/km are typical. It has been suggested that the substitution of deuterium for hydrogen in the polymer

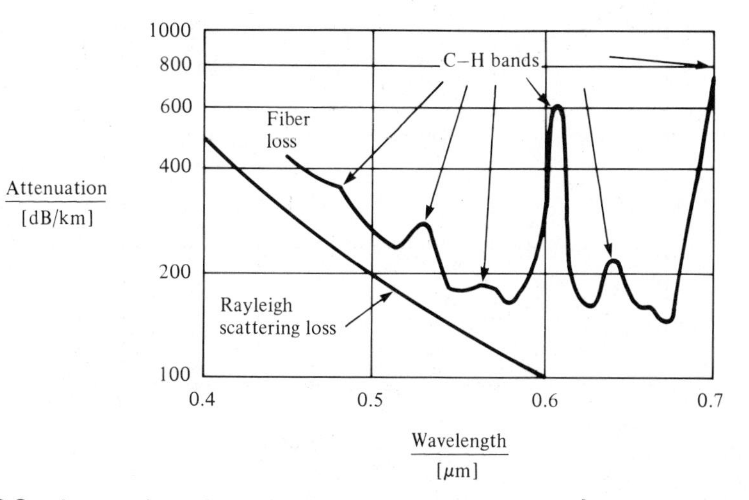

Fig. 3.8 Attenuation of a polystyrene-cored plastic fiber. [Data derived from S. Oikawa *et al.*, *Ets Lett.* **15**, 829–30 (6 Dec. 1979).]

would move the absorption bands to longer wavelengths and further reduce the minimum attenuation levels obtainable.

The proximity of the C–H absorption bands gives rise to high values of material dispersion. However, plastic fibers are step-index fibers and one of their merits is that they can be made with a large numerical aperture. Thus, in practice, total dispersion is dominated by the multipath dispersion.

A type of fiber whose properties lie between those of the high quality, doped silica fibers and the all-plastic fiber is one made with a core of pure silica and a cladding of transparent polymer. This is the PCS, polymer-clad-silica fiber, whose typical operating properties were included in Figs. 3.6 and 3.7. Several successful versions of this type of fiber have been reported. The problem is to find a cladding polymer with a sufficiently low refractive index. Fluorocarbon polymers and some proprietary silicone resins are two types that have been used. The variation of refractive index with wavelength for a particular silicone resin is shown in Fig. 3.9. In Fig. 3.10 we show how the optical attenuation of this resin varies with wavelength, and how this affects the total loss of a PCS fiber clad with the resin.

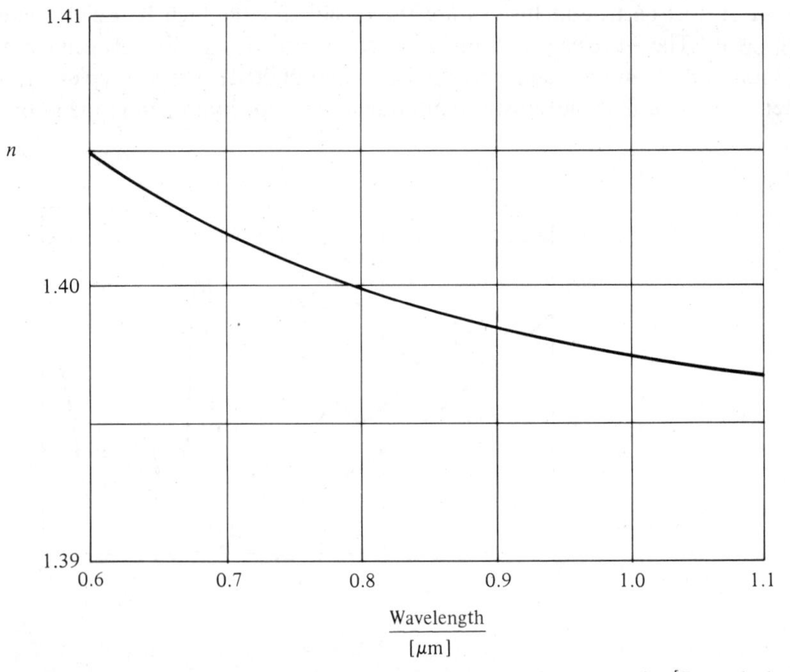

Fig. 3.9 Bulk refractive index of a proprietary, cured silicone resin. [Data derived from J. W. Fleming, *Applied Optics* **18**, 4000–2 (1979).]

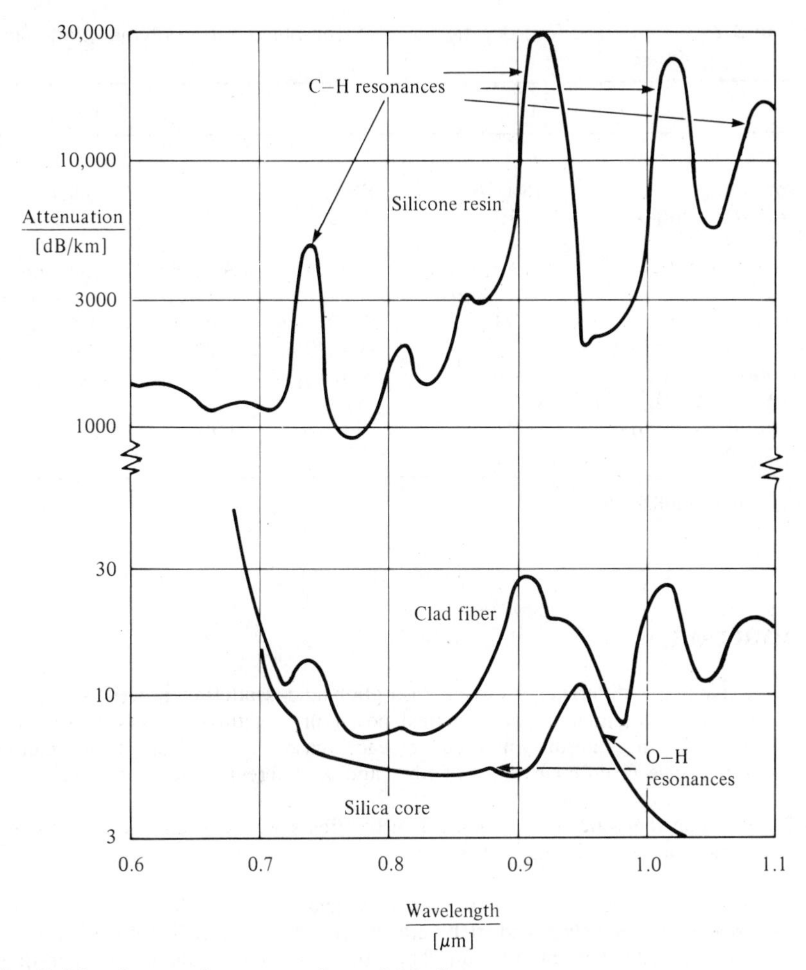

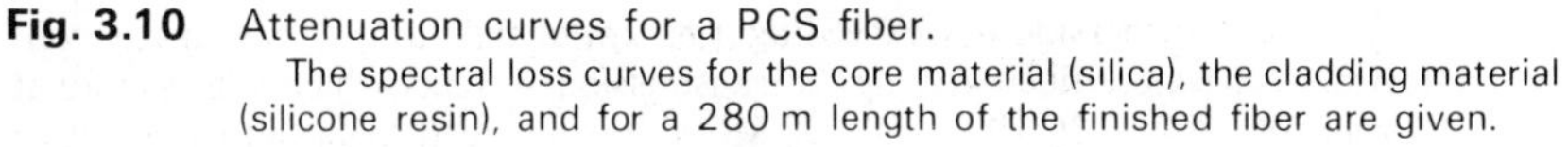
Fig. 3.10 Attenuation curves for a PCS fiber.
The spectral loss curves for the core material (silica), the cladding material (silicone resin), and for a 280 m length of the finished fiber are given.

The properties of some typical examples of all-plastic and PCS fiber are listed in Table 3.2.

It is found in practice that the half-height pulse widths, τ, are often 5–10 times smaller than the theoretical values given for ΔT in the table. This is because preferential attenuation of the more oblique rays may reduce the effective numerical aperture of the fiber and thus the multipath dispersion as well.

Table 3.2 Characteristic properties of some all-plastic and polymer-clad-silica fibers

	1	2	3	4
Core:				
material	PMMA†	PS‡	PS‡	Silica
refractive index	1.49	1.59	1.59	1.46
Cladding:				
material	Fluorocarbon polymer	PMMA†	Silicone resin	Silicone resin
refractive index	1.39	1.49	1.40	1.40
Numerical aperature	0.54	0.55	0.75	0.41
Theoretical dispersion ($\Delta T/l = \Delta n/c$) [ns/km]	340	340	640	200
Minimum attenuation [dB/km]	~100	—	150	8

† Polymethylmethacrylate.
‡ Polystyrene.

PROBLEMS

3.1 Explain the difference between absorption and scattering processes as they determine the attenuation of optical power propagating in fibers. Outline the principal mechanisms involved in each case and indicate their relative importance in different types of fiber and at different optical wavelengths.

3.2 Estimate the water (OH^-) content in the fibers whose attenuation curves are plotted in Figs 3.2, 3.3 and 3.5.

3.3 A manufacturer offers two grades of optical fiber. One has attenuation not greater than 8 dB/km and multipath dispersion not exceeding 10 ns/km (full-width-at-halfheight) at 0.85 μm. The other has attenuation not greater than 4 dB/km and dispersion not exceeding 1 ns/km at 0.85 μm. A designer is considering using these fibers for digital transmission systems which will operate at bit-rates of 2 Mb/s, 20 Mb/s and 100 Mb/s. It is intended to use an LED source which will launch 150 μW onto the fibers and which has a spectral halfwidth of 35 nm about 850 nm. For satisfactory reception the received power should exceed 1 nW/(Mb/s). Assuming a value of 0.025 for the material dispersion parameter Y_m, and allowing a loss of 1 dB/km for splices, calculate in each case the maximum repeaterless transmission distance that could be obtained. Indicate in each case whether transmission would be limited by attenuation or by dispersion.

3.4 Discuss the advantages and disadvantages of the use of different optical frequencies (over the range 0.5–2.0 μm) for transmission through silica-based fibers.

3.5 With reference to the analysis of Section 2.2.1 explain why materials of greater atomic weight (such as selenides rather than oxides) might be thought to offer advantages as possible optical fiber materials.

SUMMARY

Attenuation in optical fibers is caused by absorption and scattering. After elimination of impurities, especially transition metals and water (OH^-), the fundamental absorption limit is set by the ultra-violet and infra-red band-edges of the intrinsic material.

Gross imperfections, bends, microbends and waveguide scattering may all give rise to scattering losses. If these are minimized, there remains a fundamental scattering loss which results from the disordered nature of the glassy material. This is Rayleigh scattering. It varies as λ^{-4} and causes a loss of about 1 dB/km at 1 μm in the best fibers.

Lowest reported loss figures are 0.2 dB/km at 1.55 μm, 0.5 dB/km at 1.3 μm and 2.0 dB/km at 0.85 μm.

Attenuation is increased when the fiber is exposed to ionizing radiation.

Considering attenuation and dispersion properties together, there are two preferred wavelengths: 1.55 μm, the attenuation minimum, and 1.3 μm, the dispersion minimum. The former is used with single-mode fiber and frequency-stabilized laser sources to give systems with the highest performance. The latter is used with LED sources and well-graded multimode fiber.

All-plastic and polymer-clad-silica (PCS) fibers may have application for low capacity, short range data transmission.

4

The Manufacture and Assessment of Silica Fibers and Cables

4.1 FIBER PRODUCTION METHODS

The methods used to prepare highly transparent materials for optical fibers may be divided into two broad groups: crucible or liquid phase methods, and vapor deposition methods. In both cases the achievement of low attenuation requires the most stringent control over the purity of the starting materials and in the avoidance of contamination during the whole manufacturing process. Only an outline will be given here, but more details will be found in Refs. [4.1]–[4.4]. As with many products the general principles are easy to describe, but success then depends on a great deal of detailed technique which is not amenable to theoretical analysis and which is normally acquired by experience and kept confidential to the manufacturers. Thus within each of these two groups of production methods there are many variants. Cutting across them is the important question of whether the fiber is produced directly in a continuous process, or a preform is first made and the fiber drawn from it in a subsequent process. Of the vapor deposition processes only one (vapor axial deposition) can readily be adapted for the continuous production of fiber and this is not normally done. Of the crucible methods the double crucible arrangement shown schematically in Fig. 4.1 has become important in the production of inexpensive fiber by a continuous process.

In general crucible processes may be used to produce fibers from glasses with low melting points. The highly purified, powdered components are heated together in a platinum or silica crucible. An electric furnace may be used in which case the components are heated by radiation from the furnace walls, which themselves have to be maintained free from contaminants. Alternatively radio-frequency induction heating may be used. With a metal crucible the r.f. coupling may be made to the crucible and the heat transferred by conduction. But in order to use a silica crucible, direct coupling to the melt has to be obtained, so the powdered glass components have to be pre-heated and a higher radio frequency used. The melt is then at a higher temperature than the

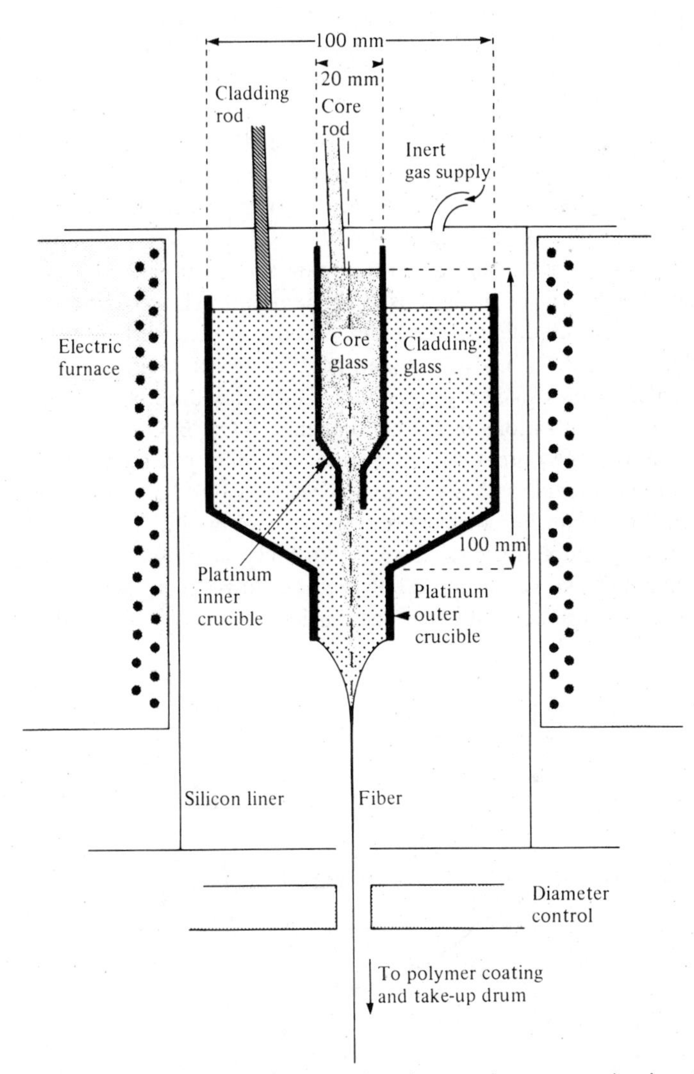

Fig. 4.1 Double crucible arrangement for the continuous production of clad fibers.

The crucible and furnace are shown in section with typical dimensions. Control of the core and cladding diameters is maintained by the fiber pulling speed and the head of molten glass in each crucible. With a suitable choice of materials and control of the melt temperature at the nozzle, index grading can be obtained by ionic diffusion.

crucible and so is less susceptible to contamination from it. Silica crucibles cannot normally be used more than once, since they do not withstand thermal cycling. Traditionally a rod of core glass and a tube of cladding glass are produced, the rod is assembled inside the tube and the two are drawn down together to produce clad fiber.

With the double crucible arrangement of Fig. 4.1, the starting materials may be fed in either in powder form or by means of high purity, preformed rods. The double crucible is mounted inside a vertical, silica-lined muffle furnace capable of raising the melt to 1000–1200°C. An inert gas atmosphere is maintained inside the furnace. When a dopant such as thallium with a relatively high rate of diffusion in silica is used to create the refractive index difference between core and cladding, some index grading at the interface occurs during the pulling process. It has been found possible to produce the same effect in sodium/calcium borosilicate glass fibers. The index difference is obtained by varying the concentrations of the components (SiO_2, B_2O_3, Na_2O, CaO) and diffusion across the core–cladding boundary within the melt gives sufficient grading to reduce multipath dispersion to between 1 and 5 ns/km. The levels of attenuation that can be achieved are shown in Fig. 4.2.

The melting temperature of glasses with high silica content is too high for crucible methods, so vapor deposition methods have to be used. Some of these use flame hydrolysis to synthesize fine particles of glass from halide vapors of

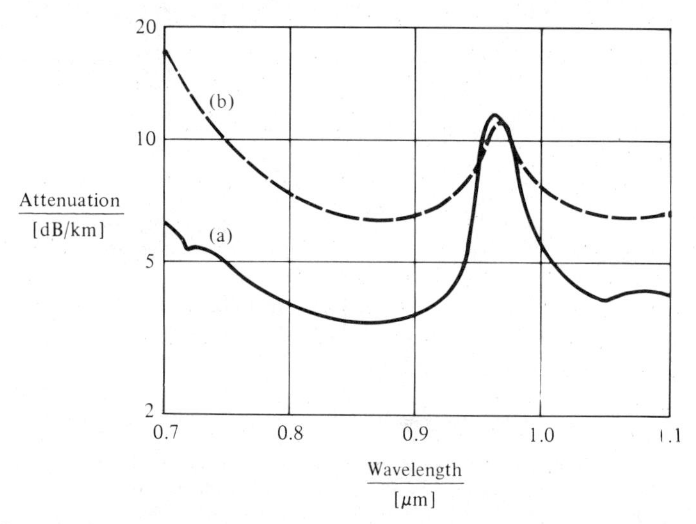

Fig. 4.2 Attenuation level attained with fibers drawn using the double crucible process: (a) sodium borosilicate glass fiber; (b) sodium calcium borosilicate glass fiber. [Data derived from K. J. Beales *et al.*, *Ets. Lett.* **13**, 755–6 (24 Nov. 1977).]

the constituents, in reactions such as:

$$SiCl_4\uparrow + 2H_2O\uparrow = SiO_2 + 2H_2\uparrow + 2Cl_2\uparrow$$
$$GeCl_4\uparrow + 2H_2O\uparrow = GeO_2 + 2H_2\uparrow + 2Cl_2\uparrow$$
$$2POCl_3\uparrow + 3H_2O\uparrow = P_2O_5 + 3H_2\uparrow + 3Cl_2\uparrow$$
$$2BBr_3\uparrow + 3H_2O\uparrow = B_2O_3 + 3H_2\uparrow + 3Br_2\uparrow$$

In the technique that first yielded fibers with lower than 20 dB/km loss a glass 'soot' was deposited in this way onto an alumina rod. The halide vapors were introduced into a methane–oxygen flame directed onto the rod. Many layers were built up, their composition being varied so as to produce either a step or a graded variation of refractive index. The rod was then removed, leaving a porous glass preform. This was sintered to form a clear glass and this in turn was drawn into fiber. Preforms large enough to be drawn into 40–50 km of fiber can be made in this way. The main difficulty is the removal of the excess water vapor, left after the hydrolysis.

Another hydrolysis method that also enables large preforms to be made is vapor axial deposition (VAD) illustrated in Fig. 4.3. Core and cladding glasses are deposited simultaneously onto the end of a seed rod which is rotated to ensure azimuthal homogeneity and is drawn up into an electric furnace at about 2.5 mm per minute. There it is heated to about 1500°C in an atmosphere of oxygen and thionyl chloride vapor. This fulfills two functions:

(a) Any water is removed by chemical reaction:

$$SOCl_2\uparrow + H_2O \rightarrow SO_2\uparrow + Cl_2\uparrow + H_2\uparrow$$
$$2SOCl_2\uparrow + 2OH^- \rightarrow 2SO_2\uparrow + 2Cl_2\uparrow + H_2\uparrow.$$

(b) At the same time the rod, which is porous and initially may be about 60 mm in diameter and 200 mm long, is consolidated into a transparent, glassy preform of some 20 mm diameter.

Reasonably well controlled index grading can be obtained with germania-doped cores by carefully maintaining a temperature profile across the porous preform during deposition. The concentration of GeO_2 in the synthesized glass particles is found to increase steadily with deposition temperature over the range 300–800°C. Long graded-index fibers having an attenuation less than 0.5 dB/km at 1.55 μm and a dispersion less than 1 ns/km have been produced by this process. The attenuation results shown in Fig. 3.5 were obtained with a fiber made by this method. A failure to maintain adequate control over the core diameter has given rise to a residual waveguide loss of 0.4 dB/km in this particular fiber.

More recently improved drying methods have enabled fibers with a water

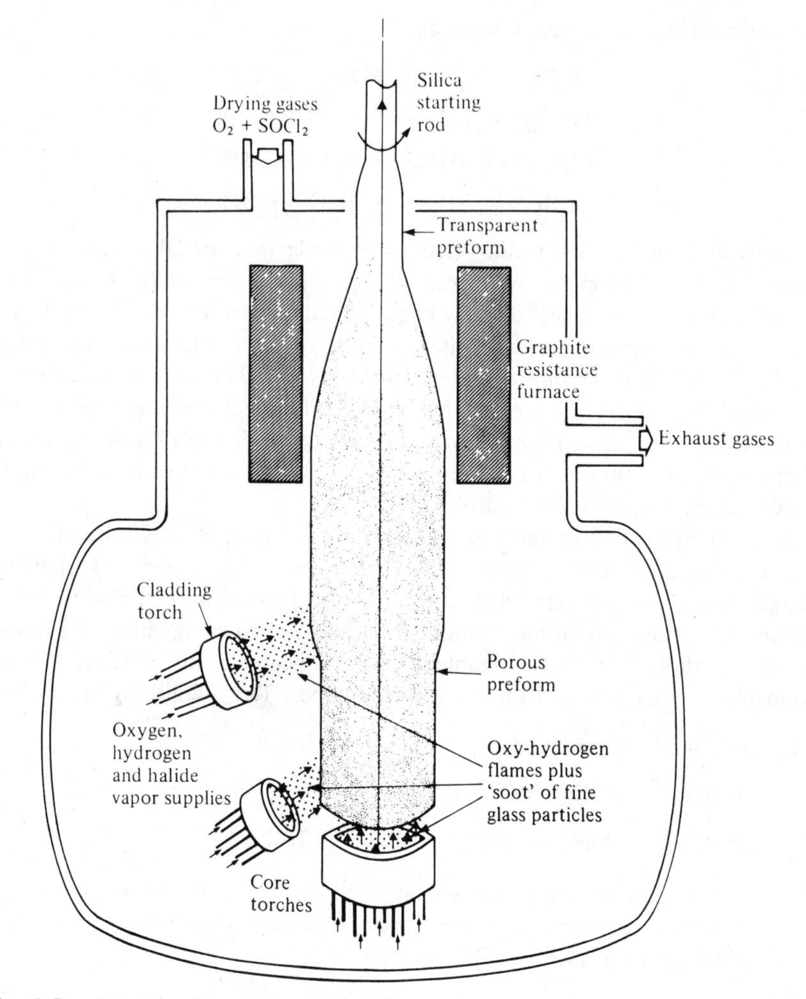

Fig. 4.3 A schematic diagram of preform manufacture using the vapor-phase axial deposition (VAD) process.

vapor content lower than one part in 10^9 to be produced by this method. A typical attenuation curve for such a fiber is shown in Fig. 4.4. The OH^- absorption peak at 1.39 μm is the only one that is detectable, and the range of wavelengths over which the attenuation remains below 1 dB/km extends from 1.1 μm to 1.7 μm. It has been found possible to introduce a wider range of doping materials into the flame in the form of a mist which is atomized from aqueous solution. The process is then no longer limited to those materials whose halides or hydrides have suitable vapor pressures.

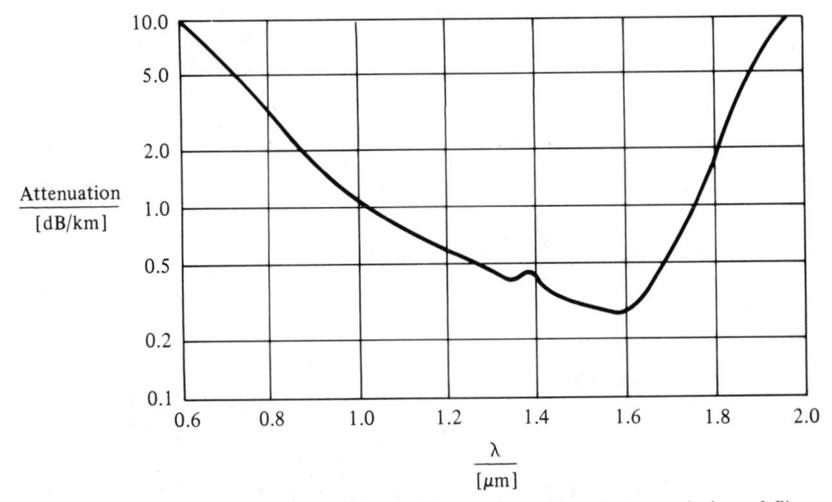

Fig. 4.4 Experimental determination of the attenuation characteristics of fibers made by the VAD process with improved drying.

In the technique that first yielded the highest quality fibers, those giving the results shown in Figs. 3.2 and 3.3, the vapor deposition takes place by thermal oxidation on the inside of a hollow tube of pure synthetic silica. Typically the tube is about 1 m long, some 15 mm in diameter and has a wall thickness of about 1 mm. It is first thoroughly cleaned and the inside tube surface etched and washed. It is then mounted horizontally and rotated on a glass-working lathe equipped with an oxy-hydrogen torch. This heats the complete outside circumference of a short length of the tube and can be translated at a controlled rate from one end of the tube to the other. Provision is made for passing vapors of silicon tetrachloride ($SiCl_4$) and of the chlorides or bromides of any required dopants such as $GeCl_4$, $POCl_3$ and BBr_3 along the inside of the tube together with oxygen. The flow rates of all of these are carefully controlled, and the highest purity is ensured by distillation of the source materials. A schematic diagram of the arrangement is shown in Fig. 4.5.

The normal preform production schedule starts with several passes of the flame with only oxygen flowing through. This heats the tube to about 1500°C and effectively polishes the inside surface. Then a number of passes are made with oxygen and the vapor of the cladding dopant, perhaps $SiCl_4$ and BBr_3, flowing. These are followed by more passes of the flame with the vapor of the core dopant, perhaps $SiCl_4$ and $GeCl_4$, passing through. When preforms for graded-index fibers are being made, some 50–100 such passes may be made in forming the core, with the concentration of the core dopant being gradually increased from pass to pass. Each of these may occupy 4 or 5 minutes.

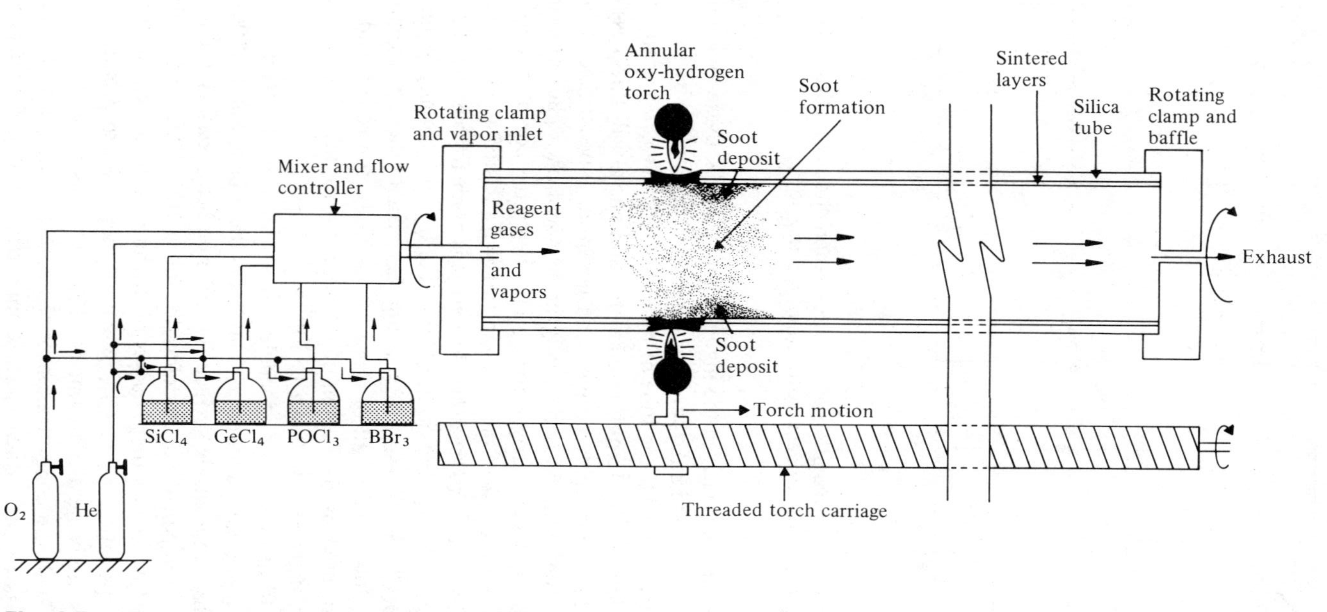

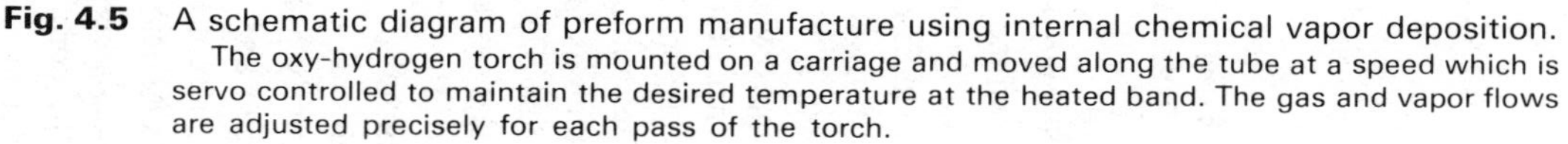

Fig. 4.5 A schematic diagram of preform manufacture using internal chemical vapor deposition. The oxy-hydrogen torch is mounted on a carriage and moved along the tube at a speed which is servo controlled to maintain the desired temperature at the heated band. The gas and vapor flows are adjusted precisely for each pass of the torch.

An interaction between the vapors and the oxygen takes place on the inside silica surface in the heated band with the result that a layer of SiO_2 plus dopants (GeO_2, B_2O_3) is deposited and the halogen gas that it liberates is carried off in the vapor stream. The form of the deposit may be a 'soot' or a more or less glassy layer, depending on the wall temperature. If 'sooty' layers are formed, subsequent heat treatments are needed to fuse them into glassy layers of varying composition to a total thickness of some 200–300 μm.

Typical reactions are:

$$\left.\begin{array}{l} SiCl_4\uparrow + O_2\uparrow \rightarrow SiO_2 + 2Cl_2\uparrow \\ GeCl_4\uparrow + O_2\uparrow \rightarrow GeO_2 + 2Cl_2\uparrow \\ 4POCl_3\uparrow + 3O_2\uparrow \rightarrow 2P_2O_5 + 6Cl_2\uparrow \end{array}\right\} \text{Core}$$

$$\left.\begin{array}{l} SiCl_4\uparrow + O_2\uparrow \rightarrow SiO_2 + 2Cl_2\uparrow \\ 4BBr_3\uparrow + 3O_2\uparrow \rightarrow 2B_2O_3 + 6Br_2\uparrow \end{array}\right\} \text{Cladding}$$

The temperature and vapor pressures are normally chosen so that these reactions take place to some extent in the vapor phase in the volume of the tube, a technique that is sometimes referred to as 'modified chemical vapor deposition'—MCVD. In earlier CVD processes lower temperatures and pressures were used and the reactions took place only at the surface. MCVD permits much faster deposition rates and these can be increased still further if a microwave frequency plasma is created in the reaction zone. The use of halide vapors rather than hydrides and the elimination of water vapor from the reaction zone enables a very low hydroxyl ion concentration to be achieved in the deposited layers.

At the end of these forming processes the tube is collapsed into the rod preform of 4–5 mm diameter by making further translations of the flame at a still slower rate so that the tube temperature is raised to around 1770°C. The torch speed can be controlled by means of a servo which monitors the tube temperature pyrometrically. The final pass during which the rod is formed may occupy as much as thirty minutes. Great care is required since it is this step that determines the circularity of the preform and the concentricity of core and cladding.

The fiber itself is drawn in a pulling machine like the one shown schematically in Fig. 4.6. With continuous production methods this is of course integrated with the rest of the process, but when preforms are made it may be separated, as shown. The hot zone has to be maintained at 1900–2000°C. These temperatures are normally obtained with a small, zirconia-lined, electric furnace, but other systems, including one which uses carbon dioxide laser radiation to heat the rod, have been shown to be successful.

Maintenance of the outside diameter of the fiber to sub-micron tolerance is

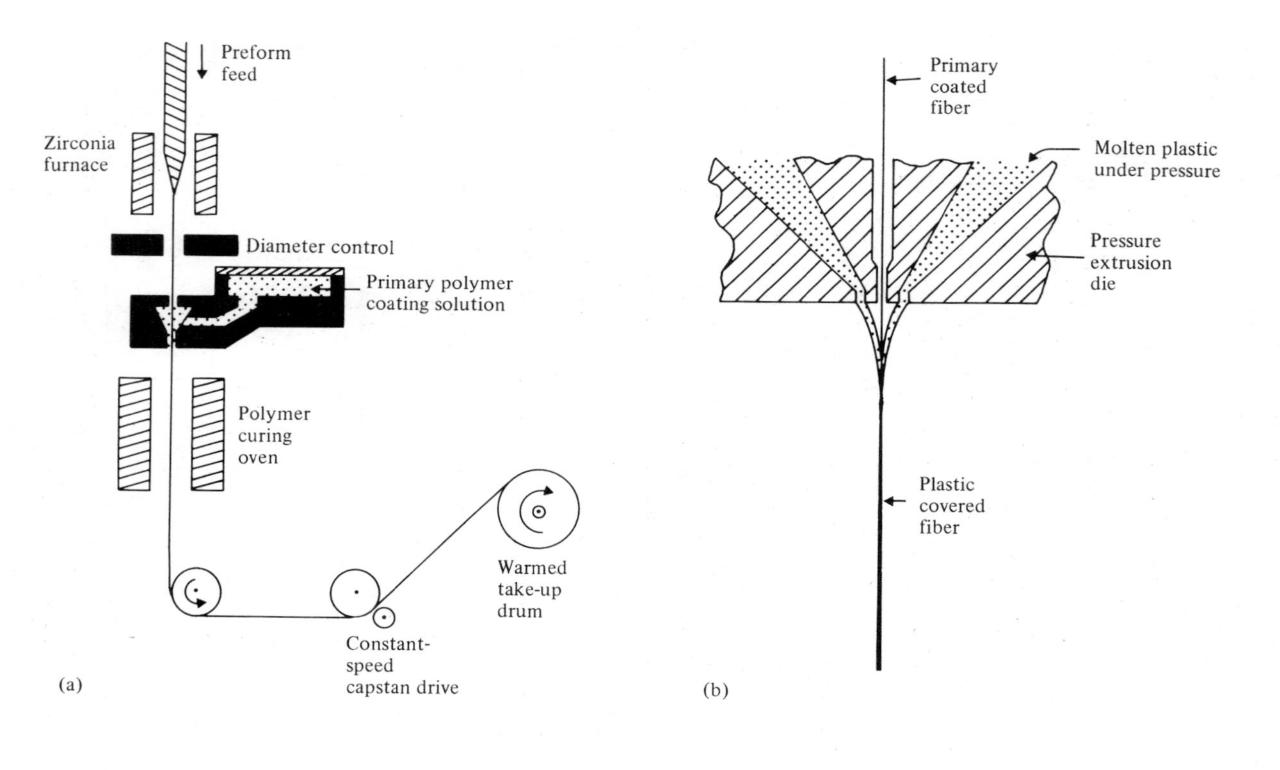

Fig. 4.6 Fiber drawing and coating: (a) schematic diagram of apparatus for drawing fiber from a rod preform and applying a primary polymer coating; (b) section of extrusion die for putting a plastic sheath onto the primary-coated fiber.

essential because when fibers are to be joined together it is the outside surfaces that are aligned. Any variation in the diameter gives rise to an offset of the fiber cores. The most common system used to monitor the diameter of the drawn fiber consists of a laser which illuminates the fiber and a photodetector placed within the diffraction pattern produced. This changes with any variation of fiber diameter and causes the photodiode current to change. This in turn acts as a signal which controls servomechanisms determining the winding rate of the fiber and the rate at which the preform is fed into the puller. A stability better than 0.1% can be achieved. The winding rate is best controlled by a precision capstan drive. Then, lengths of fiber can be wound onto reels as required without the necessity of stopping and restarting the drawing process at each changeover.

It is important to keep in mind the sizes and quantities of fiber that are involved. A 1 km length of fiber wound onto a drum of 30 cm diameter with a pitch of 5 fibers to the millimeter would occupy a length of 20 cm. At 100 μm diameter its weight would be less than 1 oz (about 20 g).

Winding rates may vary from about 0.2 m/s to about 5 m/s. Thus a preform for a high quality, graded-index fiber which may take 10 hours to make takes a further 5–10 hours to draw down to some 5–10 km of fiber. Even with the high degree of automation that is necessary, fiber produced in this way is not cheap, perhaps remaining over $100 per km. In comparison a fiber made in a continuous process and drawn at the faster rates indicated could become considerably cheaper, perhaps less than $10 per km.

The biggest obstacle to higher production rates is not the pulling process itself but the subsequent polymer coating operation. It has been found essential to give the drawn fiber a primary coating of polymer immediately it leaves the diameter monitoring cell. This is in order to protect the silica surface. Otherwise, under normal atmospheric conditions surface microcracks develop and cause a dramatic loss of fiber breaking strength. A layer of about 40 μm thickness of a polymer such as Sylgard® which adheres well to the fiber surface is found to inhibit the development of microcracks. Sometimes a thin (~10 μm) primary layer of high refractive index and then a thicker (~100 μm) buffer layer of lower refractive index are used. The primary polymer coating is usually deposited out of solution as shown in Fig. 4.6(a).

A further hazard that fibers may face during life is the ingress of water, which also weakens the fiber surface. To maintain fiber strength during service it is necessary to prevent the fiber from coming into contact with water which would diffuse through the polymer and into the silica. Some groups have experimented with ceramic coatings put onto the fiber, and others with metallic coatings put onto the polymer layer in order to make a water-tight seal. Otherwise reliance has to be placed on the outer cable produced for the fiber.

The final step in the pulling processes is to form a high-strength plastic extrusion over the fiber. This provides some protection against longitudinal and lateral mechanical forces, whether the fiber is to be used as a single fiber or is to be made up into a multifiber cable. The pulled fiber may be taken straight into a plastic extrusion die like the one shown in Fig. 4.6(b) or else this process may form part of a subsequent cabling operation. The finished fiber is normally wound onto a warmed aluminum drum. On cooling to room temperature this shrinks and releases any tension in the fiber, thereby minimizing microbending effects.

In describing the MCVD and VAD processes we have emphasized the production of graded-index fibers. However, both of these processes may easily be adapted to the production of single mode fibers having small core diameter simply by adjusting the quantities of core and cladding material in the preform.

We conclude this section by making a comparison between the MCVD and VAD processes. The need to maintain control over the deposition processes originally limited the size of an MCVD preform to one that would normally pull into 3–5 km of graded-index fiber. With higher vapor pressures and careful control over the deposition conditions larger preforms can be made. Furthermore, after the preform has been collapsed into a rod, it may be sleeved with a pure silica tube, which will form the outer part of the cladding, prior to being drawn into fiber. This is particularly appropriate for the production of monomode fibers. By means of such techniques MCVD preforms can be made big enough to pull into more than 10 km of graded-index fiber and more than 30 km of monomode fiber, whilst still retaining very high quality. VAD preforms are normally larger than MCVD preforms. Making use of subsequent sleeving, it has been shown that a single preform can yield more than 30 km of multimode fiber and up to 100 km of monomode fiber.

The refractive-index profiles of VAD fibers tend to be smoother than those of MCVD fibers, but it is more difficult to obtain a precise profile shape and a sharp cut-off at the core–cladding boundary. Figure 4.7(a) shows the way in which a VAD fiber profile may be varied by controlling the flame temperature during deposition by means of the oxygen–hydrogen ratio. The refractive-index profiles of MCVD preforms show two defects which tend to carry through into the drawn fiber and adversely affect its dispersion properties. These are a periodic variation of the index, layer by layer, and a dip on the axis to a value that is almost that of undoped silica. Both result from the diffusion and evaporation of dopant during the heat cycles that follow the initial deposition of the layers. The addition of a small concentration of phosphorus oxychloride to the germanium chloride vapor lowers the deposition temperature and reduces the periodic ripple. The central dip results from evaporation of dopant during the collapse stage. Maintaining an enhanced pressure of O_2, Cl_2 and dopant vapor during collapse mitigates this. The profile of a typical germania-doped, MCVD fiber is shown in Fig. 4.7(b).

In the MCVD process the reagents must be of the highest purity and should

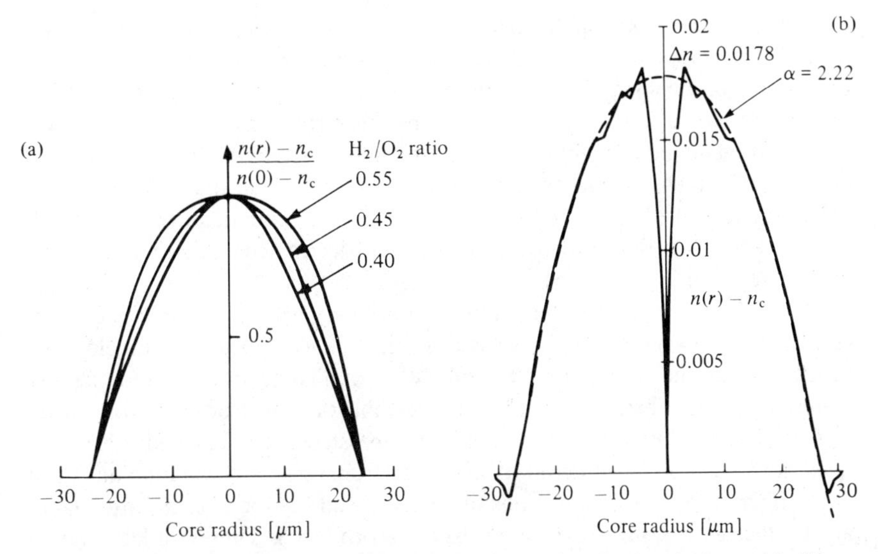

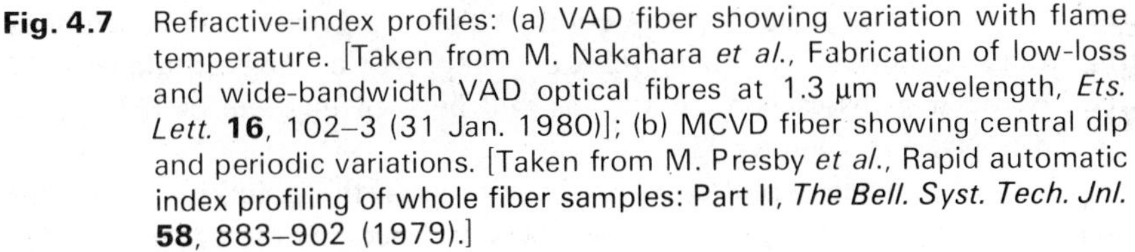

Fig. 4.7 Refractive-index profiles: (a) VAD fiber showing variation with flame temperature. [Taken from M. Nakahara *et al.*, Fabrication of low-loss and wide-bandwidth VAD optical fibres at 1.3 μm wavelength, *Ets. Lett.* **16**, 102–3 (31 Jan. 1980)]; (b) MCVD fiber showing central dip and periodic variations. [Taken from M. Presby *et al.*, Rapid automatic index profiling of whole fiber samples: Part II, *The Bell. Syst. Tech. Jnl.* **58**, 883–902 (1979).]

be specially dried. With VAD the drying of the porous preform is an essential part of the process. In both cases water vapor may diffuse out of the silica jacket during collapse. In order to prevent this adversely affecting attenuation, it is desirable to have several microns of deposited silica doped with P_2O_5 and B_2O_3 or F to form the core–cladding interface and the inner-cladding layers. These will then act as a buffer against the excess water vapor in the jacketing silica tube.

4.2 CABLES

For normal handling and use optical fibers will require additional protective layers after they emerge from the puller. Depending upon their intended use they may be made up into multifiber cable or be packaged individually. In the latter case it may be sufficient simply to run the polymer-coated fiber from the puller straight into the plastic extrusion plant to produce the final product: a coated and sheathed fiber of perhaps 0.5–1 mm total diameter.

Such fiber remains liable to excessively tight bending, which will cause microcracks to develop, and to external pressures and distortions, which will produce microbending losses. The purpose of more elaborate cabling is to minimize these effects and to protect the fiber from chemical and physical deterioration in hostile environments. It enables strength members of steel, polymer (Kevlar®) or carbon fiber to be incorporated, so that long lengths of cable can be hauled through ducts without the fiber being damaged by tension, and copper wires may also be included in order to feed electrical power to remote repeaters.

Naturally the designs of optical fiber cables vary widely and only one or two representative examples are illustrated in Fig. 4.8. The fiber may run along the neutral axis of the tube or on a helical path around a central strength member. It may lie loosely inside its duct within the cable or it may be held more rigidly in place. In the latter case the cable must be strong enough and adequately filled with cushioning material to minimize transverse and longitudinal strain at the fiber. The act of cabling has often been found to increase the attenuation of the fiber as a result of the microbending produced. The extra loss may be 0.5–2 dB/km initially, but there is evidence that it decreases subsequently as the strains introduced by cabling relax out. Improved cabling techniques have reduced cabling losses considerably as will be seen in some of the system examples given in Chapter 17.

A further comment about cost may be in order at this point. Except for the simplest arrangements such as single fibers and perhaps the cable shown in Fig. 4.8(a), cabling costs totally dominate the overall cost of cabled fiber. And the cost of cabling is much the same as the cost of making electrical cables of comparable complexity, say something in the order of $1–10/m. The cost advantage of optical fibers lies, then, in the increased capacity that can be obtained from a given cable and from the reduced cost of repeaters. To realize the benefit this extra capacity has to be usable without incurring additional complexity and cost at the terminals, for example in providing multiplexing and demultiplexing equipment. Consider a common requirement. A room housing some twenty or thirty computer graphics terminals has to be connected to a central computer situated several hundred meters away. The maximum information transmission rate to and from each terminal is 9.6 kb/s. The choice lies between laying a conventional cable containing a twisted-wire pair for each terminal and a two-fiber optical cable with multiplexing equipment at each end. At the present time, total cost would strongly favor the conventional solution, and only external constraints such as immunity from electromagnetic interference could lead to a demand for the use of the optical fiber. If the data transmission rates or the transmission distance increase, the balance of advantage will swing towards the optical fiber.

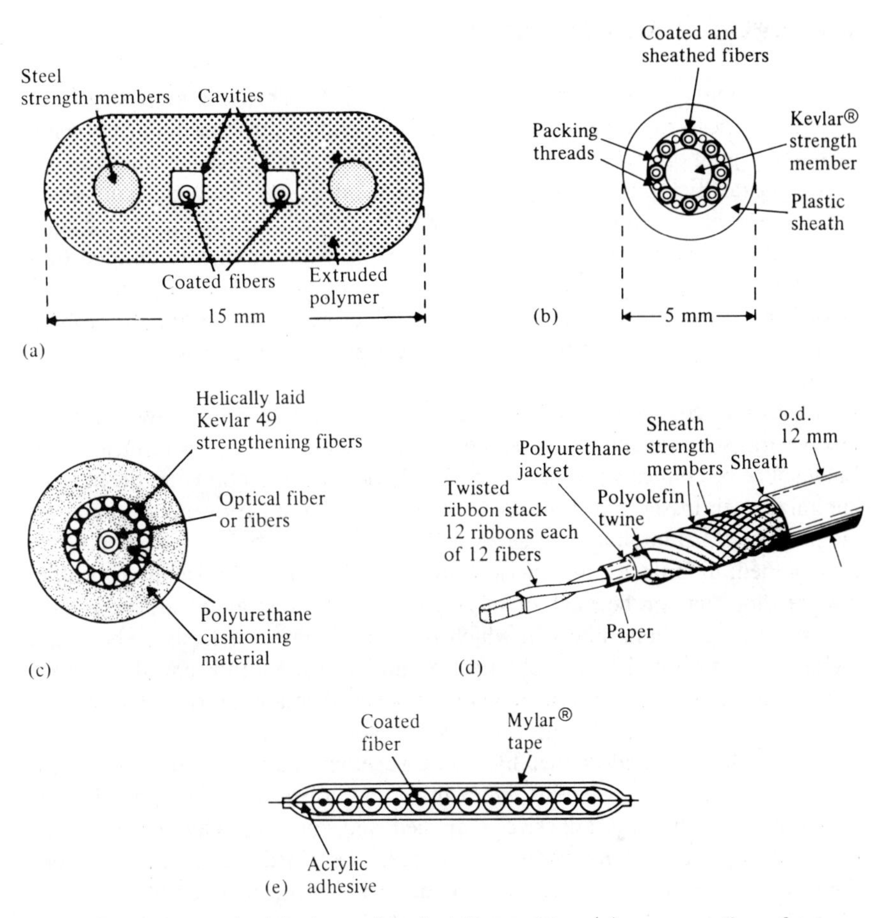

Fig. 4.8 Some typical designs of optical fiber cables: (a) cross-section of a two-fiber cable* made by BICC plc using Corning fibers [reproduced by permission of BICC plc]; (b) cross-section of an eight-fiber cable having no metallic components; (c) cross-section of a possible design of cable with the optical fiber on the cable axis and an outer strength member; (d) illustration of the 144-fiber cable designed and made by Bell Labs. The fibers are arranged in 12 ribbons each containing 12 fibers, and splicing techniques and connectors for the cable have been designed [from M. J. Buckler and C. M. Miller, Optical crosstalk evaluation for two end-to-end lightguide system installations, *The Bell Syst. Tech. Jnl.* **57** (1978)]; (e) cross-section of one of the 12-fiber ribbons used in the Bell Labs cable [from M. J. Buckler *et al.*, Lightguide cable manufacture and performance. *The Bell Syst. Tech. Jnl.* **57** (1978)].

* This optical fiber cable is the subject of British and foreign patents and patent applications owned by BICC plc.

4.3 SPLICES AND CONNECTORS

However long individual lengths of fiber may become, no system will be without the need for splices and connectors. We distinguish at once between a permanent joint or *splice* and a demountable *connector*. Splices will be required during installation and during service when cables are damaged and fibers become broken. Demountable connectors will normally be used in the terminal equipment. Sources and detectors will probably be permanently bonded to a short length of fiber, a 'pig-tail', so that they can be patched onto the main fiber link of the system using a standard fiber-to-fiber connector. Such an arrangement permits individual components to be tested separately and replaced if necessary.

Splices and connectors may be required between individual single fibers or involve the simultaneous multiple coupling of a number of fibers in a cable. Each splice or connector will give rise to additional attenuation and the need to minimize such losses sets a maximum tolerance on the permitted mismatch at the junction. Mismatch may arise because of differences between the two fibers to be joined: differences in numerical aperture (Δn), in refractive index profile, in core diameter, or because of misalignment. The tolerances are very tight indeed for single mode fibers in which the core diameter is likely to be in the region of 5–10 μm. In general offset is much more important than either angular misalignment or (in the case of connectors) end separation. This can be seen in Fig. 4.9, which is based on measurements made using GI fibers.

Splices may be made either by fusing together the two fiber ends, or by gluing them with a transparent, index-matching adhesive. A jig is required to hold the ends while the adhesive is applied and cured or while the ends are melted and joined. Many designs have been described. These can be more elaborate when the splices are to be made in the controlled conditions of the laboratory or factory, but must be safe, reliable and easy to use when splices have to be made in the field. In one of the more successful fusing methods an electric discharge is struck across the two fiber ends which melt in the heat generated. When they are brought into contact, surface tension forces tend to self-align the two fibers. Equipment has been developed in a number of places that allows splices to be made by nonspecialist staff in operational environments. The resulting losses may be of the order of 0.2 dB/splice for multimode fibers and no more than 0.5 dB/splice for monomode fibers.

Glued splices and demountable connectors give low loss only if the fiber end faces are clean, smooth and perpendicular to the fiber axis. Fibers can be cleaved reliably in a controlled manner by bending them in tension over a curved mandrel and scoring the outer surface. The fiber then breaks in the required manner as a result of the propagation of a stress crack. This technique will be familiar to anyone who has worked with glass.

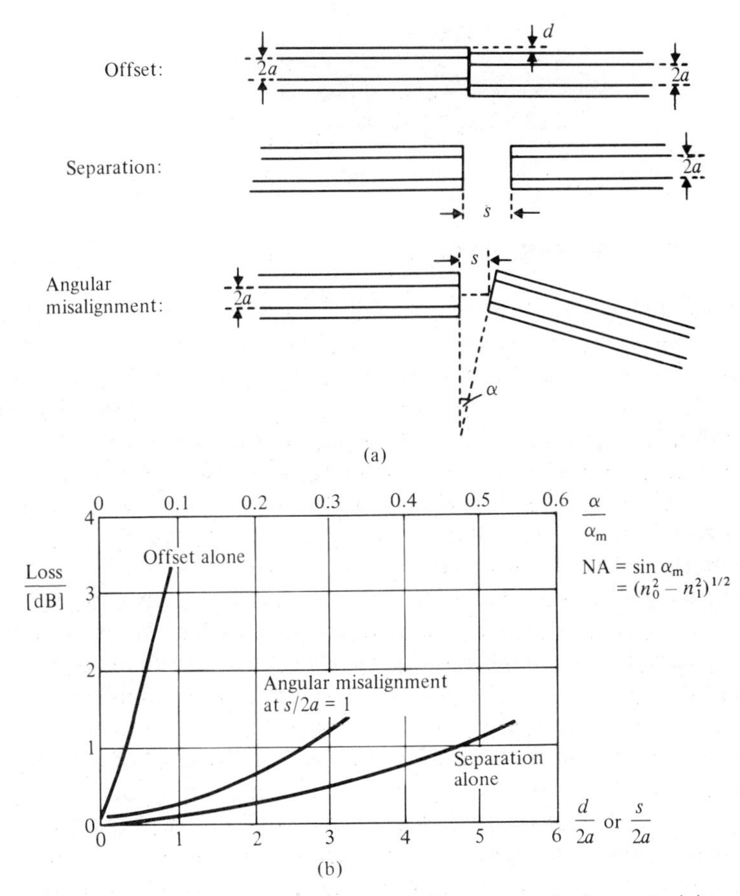

Fig. 4.9 Connector losses resulting from various types of mismatch: (a) types of mismatch; (b) measured connector losses. [Data derived from T. C. Chu and A. R. McCormick, Measurements of loss due to offset, end-separation, and angular misalignment in graded-index fibers excited by an incoherent source. *Bell Syst. Tech. Jnl.* **57**, 595–602 (1978).]

Connectors such as the one shown in Fig. 4.10 depend upon the precision of the mechanical location to preserve the required alignment. Initial insertion losses less than 0.5 dB/connector can be obtained. But mechanical wear and the likelihood of the fiber end surfaces becoming contaminated and abraded in working environments during a life which may involve hundreds or thousands of disconnections and reconnections will cause this to deteriorate. System designers differ on the margins that should be allowed, but the more conservative suggest that –3 dB/connector may be needed under more adverse conditions.

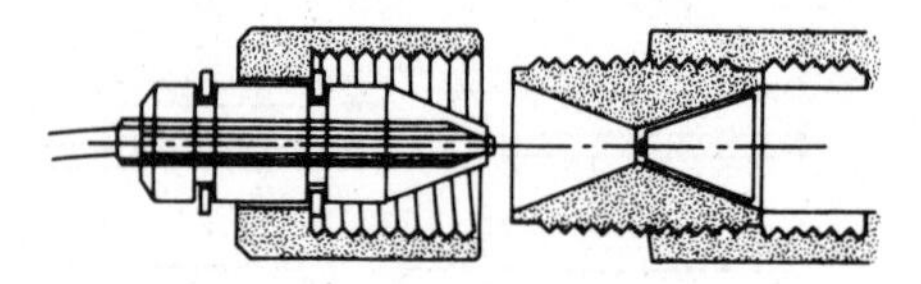

Fig. 4.10 A cone-centered connector for a single-fiber cable.

4.4 FIBER ASSESSMENT

The ability to measure accurately such fiber characteristics as the cladding and core diameters, the numerical aperture and the index profile, the attenuation and the pulse dispersion is essential both for the fiber manufacturer, who will wish to exercise some control over them, and for the optical system designer, who will wish to choose the fiber most suited to the requirement. Many techniques have been developed and much sophisticated equipment has been produced in order to facilitate these measurements. Some are designed to be used on-line during manufacture, some for use in the field after installation, while others may only be appropriate in the laboratory. Quite subtle techniques have been devised for the determination of the refractive index profiles of fibers and for the measurement of numerical aperture as a function of wavelength. Good descriptions of many of these techniques may be found in the more general reviews, such as Refs. [4.1]–[4.3], while more detailed and specialist treatments are given in Refs. [4.5] and [4.6]. Here we shall not attempt an exhaustive or detailed discussion, but simply describe some of the variations that are possible on a straightforward laboratory arrangement for the measurement of attenuation and dispersion, and comment on some of the problems of interpretation that may arise.

Once the fiber is made its important parameters are its attenuation and dispersion, and knowledge of these may be required over a range of optical wavelengths. The difficulty that arises is that neither may remain uniform along the length of the fiber, and both may depend upon the type of optical source that is used. The reason is that it is the more oblique rays rather than the axial rays that are more likely to be scattered out of the fiber by microbending and waveguide scattering. This effect is known as *mode stripping* and gives rise to differential attenuation among the various propagating rays (modes) and to a reduction in time dispersion as the more oblique rays are lost. However, microbends are also likely to cause scattering amongst the guided rays themselves and thus to convert axial into oblique rays and vice versa. After a certain distance an equilibrium distribution is set up among the propagating rays. This reduces multipath time dispersion and pulse spreading becomes proportional to the square root of distance. At the same time, because of the higher attenuation suffered by rays scattered into more oblique trajectories, fiber attenuation as a whole increases. As the quality of fibers and cables has improved, this effect, which is known as *mode conversion* and is discussed further in Chapter 5, has become less significant.

Clearly, when making measurements of attenuation or dispersion on a fiber it is desirable to replicate as closely as possible the type of optical source and the mechanical conditions that will be used in practice. The use of a very diffuse rather than a collimated source leads to additional attenuation and dispersion, certainly over the first section of the fiber.

The basic laboratory test equipment shown schematically in Fig. 4.11 enables attenuation, material dispersion and multipath dispersion to be measured over a wide range of wavelengths. The various possible forms taken by the individual elements in the system—the optical source and detectors, and the four optional components shown dotted in the diagram—will be discussed in the context of the various measurements.

Attenuation is best measured by mechanically chopping a continuous source. The reference detector ensures that the input optical power remains constant. A destructive measurement of attenuation as a function of fiber length is obtained simply by starting with the full length and cutting back progressively. A less destructive estimate of the attenuation can be made by measuring the power (Φ_1), transmitted by the full length fiber, (l_1) and then cutting back to about 5 m (l_2) for comparison (Φ_2). This removes some of the effects of the coupling and launching conditions. Then, with l_1 and l_2 measured

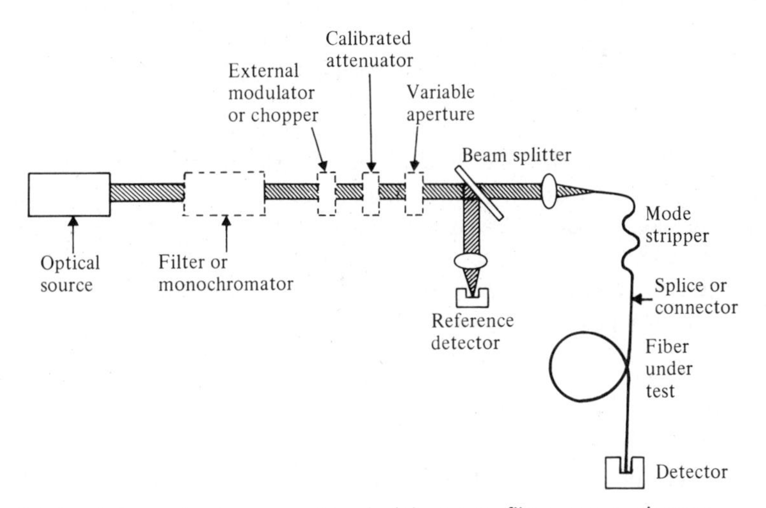

Fig. 4.11 Schematic diagram of basic laboratory fiber test equipment.

in km,

$$\frac{\text{Attenuation}}{[\text{dB/km}]} = \frac{10}{(l_1 - l_2)} \log_{10}(\Phi_1/\Phi_2) = \frac{10 \log_{10}(V_1/V_2)}{(l_1 - l_2)} \quad (4.4.1)$$

where V_1 and V_2 are the output voltages from the detector amplifier. This assumes linearity in the detector and to ensure this over a wide dynamic range it is better to make use of the calibrated attenuator for the second measurement. An LED or laser source at the optical frequency of interest may be used. Alternatively a wide-band source such as a tungsten halogen lamp or a xenon arc lamp may be used in conjunction with a monochrometer or a set of interference filters to cover a range of wavelengths. For wavelengths shorter than about 1 μm a silicon photodiode or even a photomultiplier tube may be used as a detector. For longer wavelengths germanium, cooled indium antimonide or some of the newer semiconductor detectors discussed in Chapter 12 have to be used.

In order to determine fiber dispersion some means has to be found to modulate the output power from the source. For measurements at the optical wavelength of principal interest, direct modulation of a semiconductor light-emitting diode or laser source may be used. This may take the form of pulse modulation, in which case the total dispersion may be estimated by observing the broadening of the optical pulses as a function of fiber length. This is a 'time domain' measurement. Alternatively a swept frequency sinusoidal modulator may be used to give a direct measurement of the fiber bandwidth.

The 'shuttle pulse' technique is a modification of the pulse method that has been used to study propagation in fairly short (<1 km) lengths of fiber. Partially silvered mirrors are placed at each end of the fiber and pulses may be observed, at either end, after multiple reflections.

The 'pulse delay' technique enables the effect of material dispersion to be identified separately. Here the variation with wavelength of the total delay time, t, over a given length of fiber is measured. Since the delay resulting from multi-path dispersion is independent of wavelength (something that may be checked by observation of the pulse shape which should not change from one wavelength to another), the variation of t is a direct consequence of the variation of the group index, N, with wavelength:

$$t(\lambda) = \frac{N(\lambda) l}{c} \quad (2.2.40)$$

And the slope of the graph of t vs. λ is a direct measure of material dispersion:

$$\frac{dt}{d\lambda} = \frac{l}{c}\frac{dN}{d\lambda} = -\frac{l}{c}\lambda\frac{d^2 n}{d\lambda^2} \quad (2.2.41)$$

The type of optical source most commonly used for these measurements is based on a pulsed laser that is sufficiently powerful to generate nonlinear optical effects in a suitable medium. In this way short pulses covering a range of wavelengths may be obtained. A narrow band of these wavelengths may be selected by means of a monochromator and, if necessary, further modulation may be imposed using an external modulator. Dye lasers and neodymium lasers have been used. One possible nonlinear medium is none other than a single-mode fiber. Illuminated with about 1 kW pulses of optical power at 1.06 μm from a Q-switched and mode-locked Nd–YAG laser, this acts as a wide-band source, emitting light from 1.1 to 1.6 μm. Another wide-band source makes use of a lithium niobate optical parametric amplifier. Both dye and neodymium lasers have been used as excitors, and output wavelengths covering the range 0.56–3.5 μm have been generated.

It needs to be stressed that these are advanced and developing techniques and that there remain some unresolved discrepancies among different measurements made directly on fibers, and between these and measurements made on bulk material.

4.5 A COMPARISON BETWEEN OPTICAL FIBERS AND CONVENTIONAL ELECTRICAL TRANSMISSION LINES

This final section of Chapter 4 attempts to sum up the practical consequences of the theory set out in the previous parts, and to assess the effectiveness of the various possible types of optical fiber link in comparison with more conventional communication channels. As much as possible of the relevant information has been gathered into Fig. 4.12, which plots attenuation against frequency for some representative transmission media.

From the point of view of the user an important difference between optical fibers and more conventional communication channels, such as twisted-wire or open-wire pairs, coaxial cable and waveguide, is that the fiber attenuation is independent of the signal bandwidth. This is partly because all optical transmitters are little more than broad-band noise sources, which we have already likened to the spark-gap radio transmitters that were in use for radio communication at the turn of the century. In digital transmission they are just being crudely switched on and off! But the important point here is that the signal bandwidth of an optical transmitter is small in comparison with the frequency spread of the source.

For open-wire (1) and twisted-pair (2) transmission lines attenuation is proportional to the square root of frequency. (The parenthesized numbers refer to the curves in Fig. 4.12.) At high frequencies this result is a consequence of the skin effect.

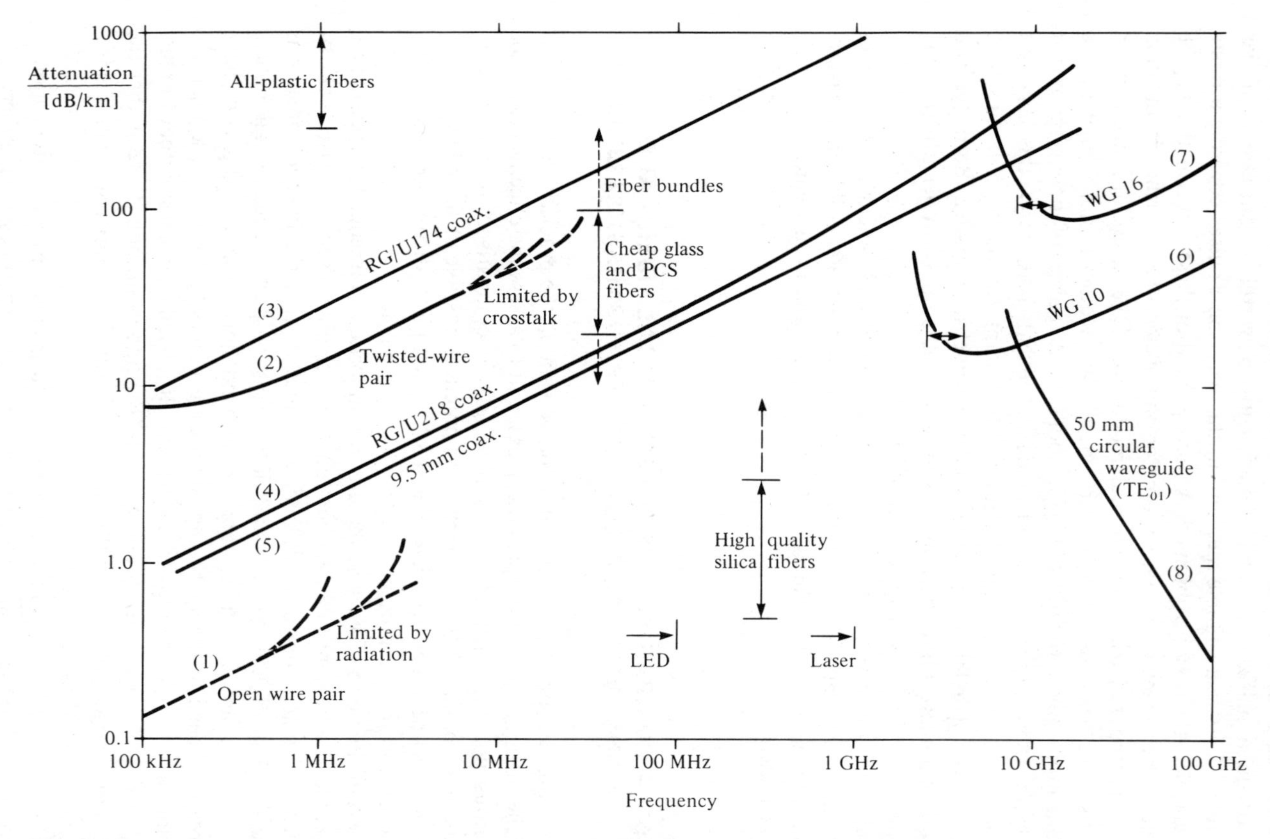

Fig. 4.12 Attenuation vs. frequency for some representative transmission media.

In coaxial cables two effects contribute to attenuation: losses in the dielectric are proportional to frequency but are not usually significant except at the highest frequencies; ohmic losses in the conductors are proportional to the square root of the frequency and are usually dominant. The full expression for attenuation is:

$$\frac{A}{[\mathrm{dB/m}]} = 20 \log_{10} \left[\frac{\pi f}{c} (\mu_r \varepsilon_r)^{1/2} \tan \delta + \tfrac{1}{2}(\pi f \varepsilon_r \varepsilon_0 / \sigma)^{1/2} \left(1 + \frac{b}{a}\right) \Big/ b \ln\left(\frac{b}{a}\right) \right] \tag{4.5.1}$$

where f, μ_r, ε_r, ε_0 and c have their usual meanings; $\tan \delta$ is the dielectric loss term; σ is the electrical conductivity of the conductors; a is the o.d. of the inner conductor; b is the i.d. of the outer conductor. SI units are assumed throughout.

With the second term dominating, attenuation is inversely proportional to the diameter of the cable. There is an indication of this in Fig. 4.12 where the specifications for two 50 Ω polyethylene cables are given, (3) and (4), together with a curve for a 75 Ω cable widely used for trunk telecommunications, (5). The dimensions of the cables are given in Table 4.1. The telecommunications cable achieves its relatively low attenuation by tight specification and the use of high quality materials.

An indication of the attenuation to be expected in two sizes of rectangular copper waveguide, WG10 and WG16, is given by curves (6) and (7). WG10 has internal dimensions of 72.14 by 34.04 mm and is normally operated in the frequency range 2.6–4 GHz. Within this range overmoding and its associated dispersion are avoided. Notice that this gives a maximum overall *signal* bandwidth of 1.4 GHz. WG16 has internal dimensions of 22.86 by 10.16 mm. Its normal operating frequency range is 8.2–12.4 GHz giving an overall *signal* bandwidth of 4.2 GHz.

Finally the remarkable properties of the TE_{01} mode in circular, 50 mm dia., copper waveguide are shown by curve (8).

For comparison the various types of optical fiber have attenuation levels which are independent of signal frequency. Depending upon the kind of fiber,

Table 4.1

Cable type	Curve	a (mm)	b (mm)	Cable diameter (mm)
RG/U174	(3)	0.48	1.5	2.54
RG/U218	(4)	4.9	17.3	22
Telecommunications cable	(5)	2.64	9.5	12

they may be less than 1 dB/km or more than 1000 dB/km. High quality, water-free, silica fibers may be used in the wavelength range 1.5–1.6 μm with overall attenuation levels less than 1 dB/km. At 0.85 μm, practical attenuation levels for long runs are more likely to lie in the range 2–5 dB/km. Cheaper fibers, made from sodium borosilicate or lead glasses or with silica core and polymer cladding, may achieve attenuation levels in the range 20–200 dB/km. For short-range links these have the benefit of being easier to handle and of having higher numerical aperture. The extra dispersion that this brings is unimportant over short distances. For very short links and reasonable environmental temperatures, all-plastic fibers offer a convenient and very cheap alternative.

The small size of an optical fiber in comparison with coaxial cable and waveguide requires emphasis. A single fiber cable need be no more than 1 mm in diameter, and a 12.5 mm dia. cable may, as we have seen, contain over 100 fibers. So when space is at a premium, far more information can be conducted through a given region using optical fibers than by any other means. However, the small size brings problems in handling and terminating the fibers. In applications where handling convenience is more important than range or bandwidth polymer fibers and glass fiber bundles of 0.5–1 mm dia. may be considered. Single core fibers of this size are also available but tend to become inconveniently stiff and susceptible to bends.

In Figure 4.12 an attempt has been made to indicate the likely upper operating signal frequencies or bit rates that can be achieved by the various types of fiber: polymer fibers up to a few megahertz; cheap glass fibers and fiber bundles up to tens of megahertz; multimode silica fibers using LEDs up to hundreds of megahertz and graded-index or monomode fibers using lasers reaching a gigahertz or so.

It becomes clear that for low bandwidth and short-range applications, where data are carried within an equipment or between units in a system, the merits of optical fibers lie in such matters as cost, weight, size, electrical isolation and electromagnetic compatibility. Attenuation and dispersion are relatively unimportant. At medium and high bandwidths and in long-range applications the low attenuation of high-quality silica fibers gives them a further strong advantage over all competitors, except for the long-haul circular waveguide using the TE_{01} mode of propagation. This has been the subject of extensive trials, and has an interesting case history. Although it requires an elaborate design and great care and precision in construction and installation in order to prevent the transmitted radiation being coupled out of the TE_{01} mode into others which are highly attenuated, this was successfully achieved in the trials. It was expensive, but with the enormous capacity fully utilized the cost per telephone circuit compared very favorably with other techniques. The problem was utilization: the change in the available capacity of a single link was too great for the system to accommodate. In consequence the trials were abandoned in favor of a more gradual introduction of optical fiber links.

PROBLEMS

4.1 Describe the various vapor deposition techniques that are used to fabricate high quality silica fibers. Estimate the length of a 125 μm diameter fiber that might be produced from a preform tube 1 m long, 25 mm o.d., 3 mm wall thickness.

4.2 Assume that the optical power density propagating along a fiber is uniformly distributed over the cross-section of the core, so that at a connector the power coupled into the next section depends only on the overlapping core area of the two fibers.

In a given connector between two fibers of core diameter d the lateral offset is x.

Show that the fraction of the propagating power transmitted through the connector is given by

$$(2/\pi)\{\cos^{-1}(x/d) - (x/d)(1 - x^2/d^2)^{\frac{1}{2}}\}$$

Express this as a loss in dB and compare it to the loss measured experimentally for graded-index fibers which is shown in Fig. 4.9.

Explain why the calculated loss may be higher than the loss observed in practice.

REFERENCES

4.1 J. E. Midwinter, *Optical Fibers for Transmission*, John Wiley (1979).
4.2 S. E. Miller and A. G. Chynoweth (Eds.), *Optical Fibre Telecommunications*, Academic Press (1979).
4.3 C. P. Sandbank (Ed.), *Optical Fiber Communication Systems*, John Wiley (1980).
4.4 K. J. Beales and C. R. Day, A review of glass fibres for optical communication, *Phys. and Chem. of Glasses*, **21** (1), 5–21 (1980).
4.5 D. Marcuse and H. M. Presby, Index profile measurements of fibers and their evaluation, *Proc. I.E.E.E.*, **68**, 666–88 (1980).
4.6 D. Marcuse, *Principles of Optical Fiber Measurements*, Academic Press (1981).

SUMMARY

Glass for fibers may be prepared by liquid phase (crucible) or vapor deposition techniques.

The double crucible method allows a continuous production of step- or graded-index fibers with attenuation less than 5 dB/km at 0.85 μm and graded-index bandwidths higher than 300 MHz.km.

With other liquid phase and most vapor deposition methods, preforms are first produced and lengths of fiber drawn from them.

The outside vapor deposition (OVD) and the vapor axial deposition (VAD) processes permit large preforms to be made quite rapidly. In principle the VAD method could be adapted to continuous production. Attenuation of 1 dB/km or less can be obtained from 1–1.7 μm with a minimum of 0.3 dB/km at 1.6 μm. Graded-index fibers with bandwidths exceeding 500 MHz.km can be made.

The modified chemical vapor deposition (MCVD) method has produced fibers with the lowest loss and the most carefully controlled index-grading. These include graded-index fibers with an attenuation minimum of 0.34 dB/km at 1.55 μm and bandwidths in excess of 1 GHz.km, and single-mode fibers with an attenuation minimum of 0.2 dB/km at 1.55 μm.

Drawn fiber must immediately be coated with polymer and then sheathed with extruded plastic.

Fibers must be protected from water, should not be subjected to excessive bending or put under tension. These considerations dominate cable design.

Average losses from fusion splices, which may be made in the field, are in the range 0.1–0.5 dB. Demountable connectors are needed at terminal equipment and may introduce a loss of 0.5–3.0 dB.

Many techniques for measuring important fiber parameters such as index profile, attenuation, and bandwidth have been developed.

Figure 4.12 compares the characteristics of different types of optical fiber transmission line with those of different types of electrical transmission line.

5

Electromagnetic Wave Propagation in Step-index Fibers

5.1 MODES AND RAYS

The picture of light propagation in optical fibers that we have used so far, that is of rays of light reflected back and forth from the core-cladding boundary and thereby travelling zig-zag fashion along the core of the fiber, is adequate for most purposes in optical communication systems that use multimode fibers. But it has to be recognized that light is an electromagnetic wave phenomenon and that we really ought to be thinking in terms of *modes of guided wave propagation*, rather than rays. The physical concept of a ray is a representation of a pencil of plane electromagnetic waves in the limit $\lambda \to 0$, something that can often be realized quite effectively in practice by using a narrow laser beam. In optical fibers it is not in general true that the optical wavelength is small in comparison with the fiber radius, so the ray model needs to be used with caution and the fiber itself should really be considered to be a dielectric waveguide.

In treating guided electromagnetic wave propagation in dielectric media in this chapter and Chapter 6 we shall, as ever, be attempting to tread a middle path. On the one hand, we shall assume some familiarity with basic electromagnetic theory. On the other, a detailed and rigorous discussion is beyond the scope of this book and for this the interested reader is referred to more advanced texts, e.g. Refs. [5.1]–[5.3]. The solution of Maxwell's equations even for the simplest case, which is that of a cylindrical, step-index fiber with an infinitely thick cladding, is difficult, and it is interesting to note that the further assumptions and simplifications that are usually made in order to treat the more complex types of fiber turn out in any case to be formally equivalent to those made in the ray model. We shall deal with step-index fiber first, and then with a class of graded-index fibers in Chapter 6. Because many readers may be familiar with the theory of guided waves in metal waveguides we have started out by putting the wave solutions in a form that is analogous to the one normally used in that theory. An approximation can be made which enables a simpler notation to be used for fiber guides, and this we will derive. Some readers may be familiar with the WKB (Wentzal, Kramers, Brillouin) approximation which can be used in the solution of wave equations. It will be found that the graded-index fiber provides a particularly nice example of its application, and this will be developed in Chapter 6.

We shall be seeking solutions to Maxwell's electromagnetic field equations for a nonconducting and charge-free dielectric. These are

$$\begin{aligned} \mathbf{curl\ E} &= -\frac{\partial \mathbf{B}}{\partial t} \qquad & \mathbf{curl\ H} &= \frac{\partial \mathbf{D}}{\partial t} \\ \text{div } \mathbf{D} &= 0 \qquad & \text{div } \mathbf{B} &= 0 \end{aligned} \tag{5.1.1}$$

where $\mathbf{D} = \varepsilon_r \varepsilon_0 \mathbf{E}$ and $\mathbf{H} = \mathbf{B}/\mu_r \mu_0$

- $\mathbf{E}$ is the electric field vector,
- $\mathbf{D}$ the electric displacement vector,
- $\mathbf{B}$ is the magnetic induction vector,
- $\mathbf{H}$ the magnetic field vector,
- μ_0 is the permeability of free space,
- ε_0 is the permittivity of free space,
- μ_r is the relative permeability of the material and is assumed always to be 1.
- ε_r is the relative permittivity of the material.

We have already discussed the plane wave solutions of these equations which apply in regions unconstrained by boundaries. Such waves are *transverse electromagnetic* (TEM) waves in which the electric and magnetic field vectors are mutually perpendicular, and both are perpendicular to the direction of propagation. We shall now be seeking guided wave solutions that are subject to the radially symmetric variation of $\varepsilon_r(r)$ in the fiber. It is found that such waves cannot be pure TEM waves but always have an axial component of **E** or **H**. Because of the radial symmetry, the solutions are conveniently expressed in cylindrical polar coordinates and have the general form

$$\begin{aligned} \mathbf{E} &= \mathbf{E}(r, \phi) \exp \{-j(\omega t - \beta z)\} \\ \mathbf{H} &= \mathbf{H}(r, \phi) \exp \{-j(\omega t - \beta z)\} \end{aligned} \tag{5.1.2}$$

where, as before $\omega = 2\pi f$ is the angular frequency and β is a propagation constant. Since both electric and magnetic field vectors exhibit the same functional relationship, we shall use the following notation:

$$\psi = \psi(r, \phi) \exp \{-j(\omega t - \beta z)\} \tag{5.1.3}$$

where ψ stands for the local value of either the electric field **E** or the magnetic field **H**. The boundary condition imposed by the fiber is the prescribed variation of permittivity, $\varepsilon_r(r)$, or refractive index, $n(r) = \sqrt{\varepsilon_r(r)}$. The solutions are also subject to the requirements that the fields be finite on the axis and zero at infinity. That is: $\psi(0, \phi) \neq \infty$ and $\psi(\infty, \phi) = 0$. These conditions give rise to eigenvalue solutions for $\psi(r, \phi)$, each of which has a particular β value. It is these discrete, propagating, electromagnetic field patterns that we have referred to as *modes*.

Both of the types of fiber that we have discussed so far—the step-index and the graded-index fibers—are able to support many such modes; in fact, many hundred in a fiber of typical dimensions. They are examples of *multimode* fibers. To some extent we may associate the different modes with the different ray trajectories we have referred to previously. Because the propagation constant, β, varies from mode to mode, each propagates with its own characteristic phase and group velocities. So the property we have hitherto referred to as multipath dispersion is perhaps better called *intermode* dispersion. In the literature the term is usually abbreviated just to *mode* dispersion.

As well as providing clearer physical insights into the optical propagation characteristics of fibers, the mode theory enables fibers with very small core diameter to be treated, it enables the power distribution within the fiber to be calculated and it demonstrates a further source of time dispersion. This arises because the velocity of propagation of any particular mode varies with frequency, independent of any material dispersion that may be present. This effect is known as *intra-mode* dispersion, or more commonly as *waveguide* dispersion, and whilst it is normally small compared to material dispersion, it can have the effect of shifting the wavelength of minimum dispersion to longer wavelengths, as we shall show in Section 5.5.

5.2 WAVE PROPAGATION MODES IN AN IDEAL STEP-INDEX FIBER

Here we shall consider a fiber which consists of a uniform core of radius, a, and refractive index, n_1, and an infinitely thick cladding of refractive index, n_2.

Solutions to eqns. (5.1.1), subject only to the discontinuity in ε_r at $r = a$, take the form of well known functions. These are Bessel functions, $J(x)$, within the core, and modified Hankel functions, $K(x)$, in the cladding.† The axial field

† The solution of Maxwell's wave equation in cylindrical coordinates is discussed in Appendix 1, Section A1.2. The origin and properties of the Bessel and Hankel functions will be familiar to many readers. They are discussed and tabulated in (for example) Chapter 9 of M. Abramowitz and I. A. Stegun, *Handbook of Mathematical Functions* (Dover, 1965).

components in the core ($r < a$) are given by

$$\psi_z = \psi_1 J_k(ur) \cos k\phi \tag{5.2.1}$$

where ψ_1 is a constant electric or magnetic field; k is an integer; and u is a parameter that is related to the as yet undetermined propagation constant of the guided wave, β, by

$$u^2 = \left(\frac{2\pi n_1}{\lambda}\right)^2 - \beta^2 = \beta_1^2 - \beta^2 \tag{5.2.2}$$

Note that

$$\beta_1 = \frac{2\pi n_1}{\lambda} = \frac{n_1 \omega}{c}$$

would be the propagation constant for plane transverse electromagnetic (TEM) waves in the material of the fiber core.

In the cladding ($r > a$) the axial field components are given by

$$\psi_z = \psi_2 K_k(wr) \cos k\phi \tag{5.2.3}$$

where ψ_2 is a constant electric or magnetic field; k is again an integer; and w is another parameter related to β:

$$w^2 = \beta^2 - \left(\frac{2\pi n_2}{\lambda}\right)^2 = \beta^2 - \beta_2^2 \tag{5.2.4}$$

Now

$$\beta_2 = \frac{2\pi n_2}{\lambda} = \frac{n_2 \omega}{c}$$

is the propagation constant for plane TEM waves in the material of the cladding. For large values of r such that $r > a$ and $wr \gg 1$,

$$K_k(wr) \propto \frac{\exp(-wr)}{(wr)^{1/2}}$$

This means that the fields in the cladding die away exponentially at large radial distances from the core. It is these fields that we have previously called the evanescent wave.

So far we have only quoted solutions for the axial components of **E** and **H**. The transverse components may be derived from these using the following relations:

$$
\left.\begin{aligned}
E_r &= -\frac{\mathrm{j}}{A^2}\left[\beta\frac{\partial E_z}{\partial r} + \frac{\mu_r\mu_0\omega}{r}\frac{\partial H_z}{\partial \phi}\right]; \quad E_\phi = -\frac{\mathrm{j}}{A^2}\left[\frac{\beta}{r}\frac{\partial E_z}{\partial r} - \mu_r\mu_0\omega\frac{\partial H_z}{\partial r}\right] \\
H_r &= -\frac{\mathrm{j}}{A^2}\left[\beta\frac{\partial H_z}{\partial r} - \frac{\varepsilon_r\varepsilon_0\omega}{r}\frac{\partial E_z}{\partial \phi}\right]; \quad H_\phi = -\frac{\mathrm{j}}{A^2}\left[\frac{\beta}{r}\frac{\partial H_z}{\partial r} - \varepsilon_r\varepsilon_0\omega\frac{\partial E_z}{\partial \phi}\right]
\end{aligned}\right\} \tag{5.2.5}
$$

where $A^2 = \left(\dfrac{2\pi n}{\lambda}\right)^2 - \beta^2$ and $n(r) = \sqrt{(\mu_r\varepsilon_r(r))}$.

The electromagnetic field patterns of the various modes have some similarity to those obtained in circular metal waveguides, bearing in mind the less dramatic boundary condition at the core–cladding interface. Whereas at a metal boundary the electric field parallel to the surface must vanish (E_z, E_ϕ both zero), in the fiber the radial fields suffer only a slight discontinuity in crossing the interface between core and cladding, and the tangential field components (E_z, E_ϕ, H_z, H_ϕ) must all be continuous there. The need to satisfy these boundary conditions means that for each value of k, only certain discrete values of u and w are permitted in eqns. (5.2.1) and (5.2.3). We shall identify these as u_{km} and w_{km}. Like k, the subscript, m, is an integer. This means that the propagation constant, β, also takes on discrete values, given by:

$$\beta_{km}^2 = \beta_1^2 - u_{km}^2 = w_{km}^2 - \beta_2^2 \tag{5.2.6}$$

Consider first the solutions to eqns. (5.2.1) and (5.2.3), when $k = 0$. These are:

in the core

$$\psi_z = \psi_1 J_0(u_{0m} r) \tag{5.2.7}$$

in the cladding

$$\psi_z = \psi_2 K_0(w_{0m} r) \tag{5.2.8}$$

These solutions represent modes in which the field patterns have radial symmetry. In the ray model each of these $(0, m)$ modes may be thought to correspond to a set of meridional rays having a prescribed angle of obliqueness to the fiber axis. As with circular metal waveguide, there are two sets of solutions, those for which H_z is zero, designated *transverse magnetic* modes (TM_{0m}), and those for which E_z is zero, designated *transverse electric* modes (TE_{0m}). The transverse electric field patterns in the fiber core for the two $k = 0$, $m = 1$ modes are included in Fig. 5.1. The axial and transverse distributions for both the electric and the magnetic fields associated with the TE_{02} mode are illustrated in Fig. 5.2(a).

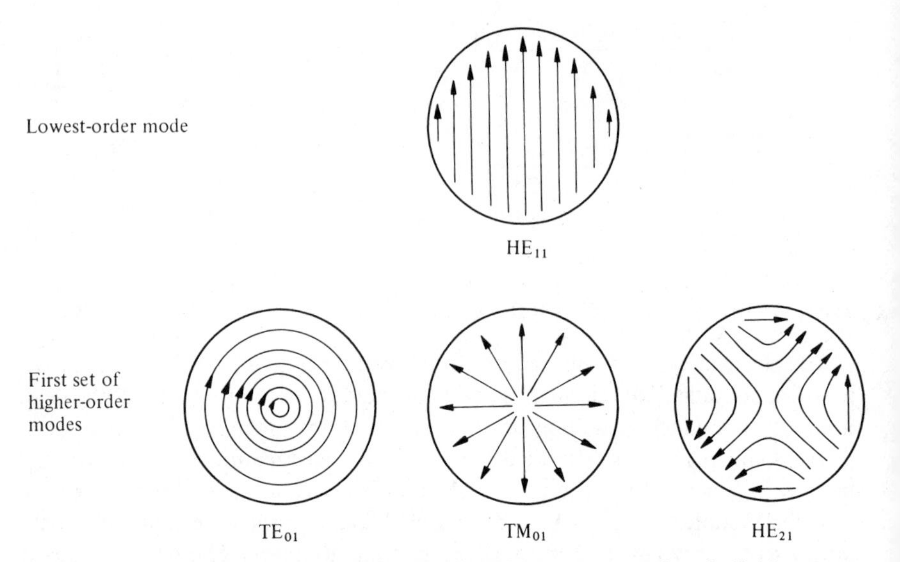

Fig. 5.1 Schematic cross-sections showing the transverse electric field vectors in the fiber core for the four lowest-order modes of an SI fiber.

It is now possible to see the significance of the second mode parameter, m. The Bessel function $J_0(x)$, which describes the field amplitudes in the core, is an oscillatory function. It is plotted in Fig. 5.3. On the other hand the modified Hankel function $K_0(x)$, which describes the fields in the cladding, is a monotonically decaying function. For the $k=0$, $m=1$ mode the magnitude of the parameter u_{01} has to be such that the core–cladding interface, $r=a$, falls within the second half-cycle of $J_0(u_{01}r)$. For $k=0$, $m=2$, $r=a$ must fall within the third half-cycle of $J_0(u_{02}r)$ and so on. The allowed values for u_{0m} are therefore bounded. $u_{01}a$ must be greater than the first nonzero root of the $J_0(x)$ Bessel function and smaller than the second. $u_{02}a$ must be greater than the second root and smaller than the third. And so on. In general, if t_{0m} are the roots of $J_0(x)=0$, then

$$t_{0,m} < u_{0m}a < t_{0,m+1} \tag{5.2.9}$$

Some values of t_{km} are given in Table 5.1, and are marked on Fig. 5.3.

The requirement that the field solutions for the core and cladding should match at $r=a$ does, in fact, put a tighter restriction than that indicated by (5.2.9) on the upper limits to $u_{0m}a$, which we shall discuss shortly.

Let us next consider the solutions when k is a positive integer and not equal to zero. The resulting field patterns no longer have radial symmetry, because of the $\cos k\phi$ term. Now both E_z and H_z may be nonzero, so these modes are

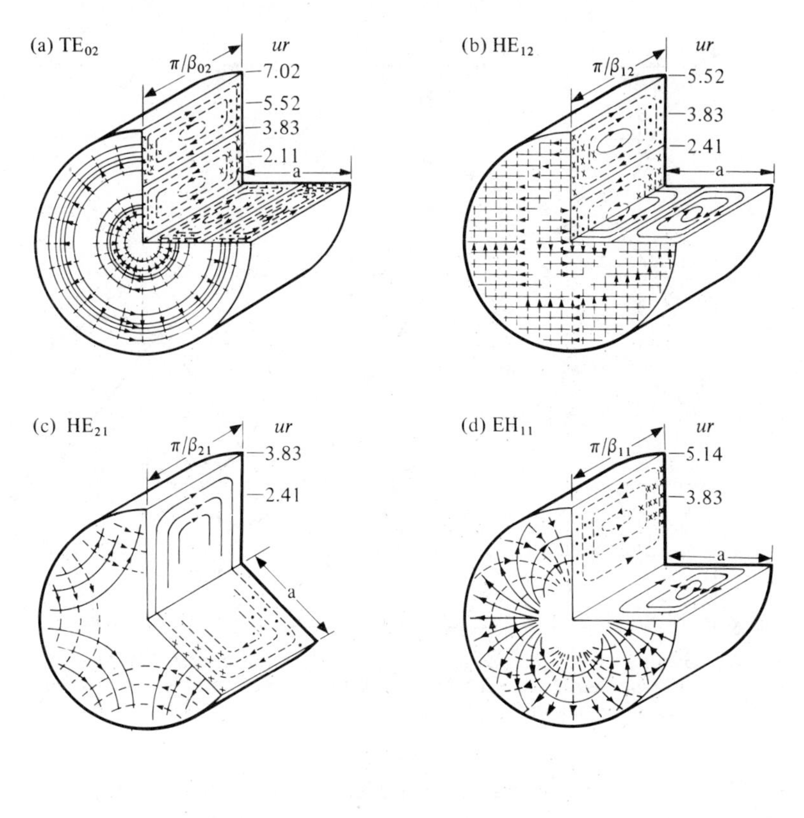

Fig. 5.2 Three-dimensional cut-away representations of the electric and magnetic fields in the core of an SI fiber for some of the low-order modes far from cutoff: (a) TE_{02}; (b) HE_{12}; (c) HE_{21}; (d) EH_{11}. [From E. Snitzer, *J. Opt. Soc. of America* **51**, 491–8 (1961).]

In each case one half-wavelength (π/β_{km}) of the wave in the z-direction is shown.

known as *hybrid* modes. We may associate them with skew rays. For each value of k there are two sets of modes. In one set the axial magnetic field (H_z) makes a greater contribution than E_z to the transverse fields, and these are called HE_{km} modes. In the other set it is the axial electric field (E_z) that makes the greater contribution, and these are known as EH_{km} modes. Again each mode gives rise to a discrete pair of values of u and w (u_{km} and w_{km}) and hence to a particular value of β_{km}. Again the precise values are functions of the optical frequency and have to be determined from the boundary conditions at

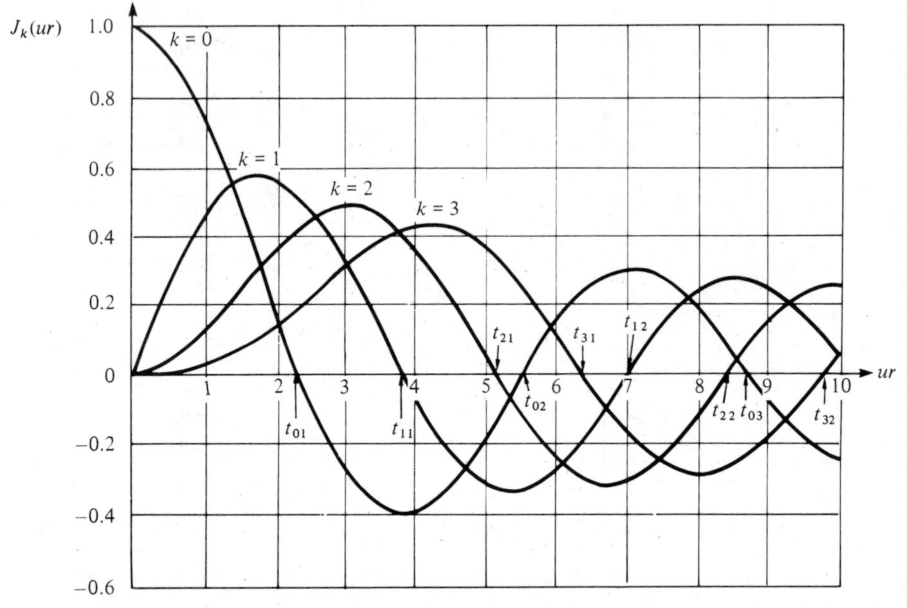

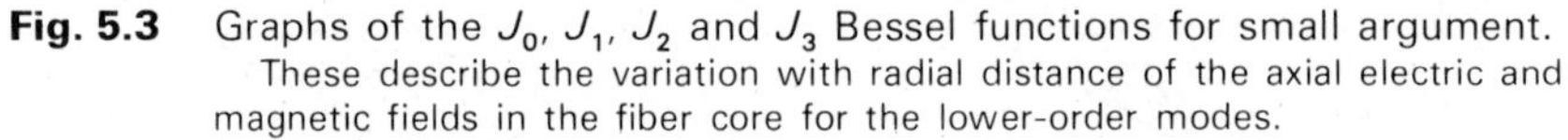

Fig. 5.3 Graphs of the J_0, J_1, J_2 and J_3 Bessel functions for small argument. These describe the variation with radial distance of the axial electric and magnetic fields in the fiber core for the lower-order modes.

$r = a$. The complicated calculations needed to obtain these values have been performed numerically and we shall examine the results in detail later in this section. Their importance lies in the fact that it is the values of u_{km} that, through eqn. (5.2.6), determine the propagation constants, β_{km}, for each mode as a function of the optical frequency. In turn, it is $\beta_{km}(\omega)$ that determine the group and phase velocities of the guided modes and hence the fiber dispersion. The transverse electric field patterns in the fiber core for the HE_{11} and HE_{21} modes are included in Fig. 5.1. Three-dimensional illustrations of the electric and magnetic fields of the HE_{12}, HE_{21} and EH_{11} modes are given in Fig. 5.2(b)–(d).

We will now examine the variation of the propagation constants, β_{km}, with the optical frequency.

Table 5.1 Bessel function roots

$t_{01} = 2.405$	$t_{11} = 3.832$	$t_{21} = 5.136$
$t_{02} = 5.520$	$t_{12} = 7.016$	$t_{22} = 8.417$
$t_{03} = 8.654$	$t_{13} = 10.173$	$t_{23} = 11.620$

At very high frequencies ($\omega \to \infty$), such that the core diameter is large compared with the wavelength of plane waves in the core material, it would seem reasonable to suppose that light launched into the core would propagate approximately as a plane TEM wave having a propagation constant, $\beta \simeq \beta_1$. Under these conditions it is appropriate to return to the idea of rays. We may then associate the higher-order modes with those rays that are more oblique to the fiber axis. The mode propagation parameter, β, defines the apparent phase velocity $(v_p)_z$ of the waves along a line in the axial direction:

$$(v_p)_z = \frac{\omega}{\beta} \tag{5.2.10}$$

Let us associate such a mode with a ray representing a plane TEM wave travelling at an angle Φ to the fiber axis and propagating in the z-direction by multiple reflections as shown in Fig. 5.4. There we can see that while the phase velocity in the direction of the ray is

$$v_p = \lambda_m f = \frac{\omega}{\beta_1} = \frac{c}{n_1} \tag{2.2.5}$$

the apparent phase velocity with which the wavefronts intercept any line parallel to the axis is

$$(v_p)_z = (\lambda_m \sec \Phi) f = \frac{\omega}{\beta_1} \sec \Phi = \frac{\omega}{\beta}$$

$$\therefore \quad \beta = \beta_1 \cos \Phi \tag{5.2.11}$$

We saw in Section 2.1 that the limiting obliqueness, Φ_m, for propagating rays was set by the critical angle, θ_c, such that

$$\sin \theta_c = \cos \Phi_m = n_2/n_1 \tag{2.1.2}$$

Thus, Φ_m determines a minimum value for β given by:

$$\beta_{\min} = \beta_1 \cos \Phi_m = \beta_1 n_2/n_1 = \beta_2 \tag{5.2.12}$$

This means that at any given frequency the lowest order modes (corresponding to axial rays) have a propagation constant near to β_1, whereas the highest-order mode able to propagate (corresponding to the most oblique ray) has a propagation constant, β_2, which is that of plane TEM waves in the cladding.

Figure 5.4 enables us to discuss another important property of the wave solution in terms of the ray model. This is the fact that only a finite number of discrete modes may propagate. In the figure these are represented as plane TEM waves and it is clear that if these are not to interfere destructively after

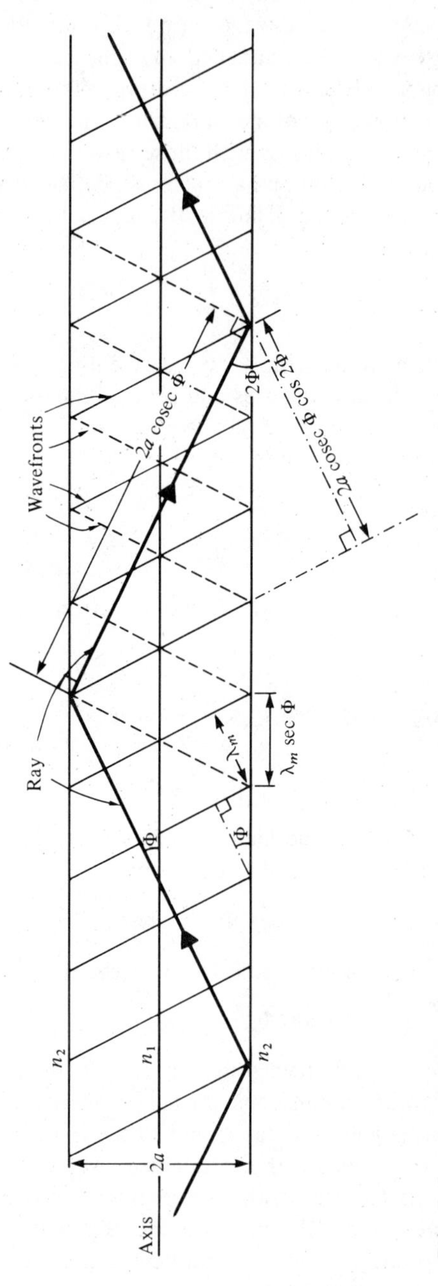

Fig. 5.4 Propagation by multiple reflection in a multimode step-index fiber.

multiple reflections, then the distance $2a$ cosec Φ cos 2Φ must be an integral number of wavelengths:

$$2a \operatorname{cosec} \Phi \cos 2\Phi = i\lambda_m \tag{5.2.13}$$

where i is an integer. Thus only a finite number of discrete propagating angles satisfying eqn. (5.2.13) is permitted.

At low frequencies ($\omega \to 0$) such that the core diameter is small compared with the wavelength of plane waves in either core or cladding material, it would seem reasonable to suppose that the core would have little effect on the wave propagation. This would take place almost entirely in the cladding and with constricting boundary conditions having minimal effect, we should expect unbound and unguided, plane, TEM waves having a propagation constant, $\beta \simeq \beta_2$. There are in fact infinitely many unguided modes and they include, for example, light which enters the core through the core–cladding boundary, passes across the core and leaves from the opposite side.

We might intuitively expect that the propagation constant for the guided modes would lie between the two extreme values, β_1 and β_2, and indeed, the condition

$$\beta_2^2 < \beta_{km}^2 < \beta_1^2 \tag{5.2.14}$$

for the existence of bound, guided-wave solutions is implicit in eqns. (5.2.2) and (5.2.4), if u and w are to be real. It is these guided-wave solutions satisfying (5.2.14) that are of interest to us.

It is convenient at this point to introduce a normalized frequency parameter, V, such that

$$V = \omega/\omega_0$$

where

$$\omega_0 = c/a(n_1^2 - n_2^2)^{1/2} \tag{5.2.15}$$

Thus

$$\begin{aligned} V &= \frac{\omega}{c}(n_1^2 - n_2^2)^{1/2} a = \frac{2\pi a}{\lambda}(n_1^2 - n_2^2)^{1/2} \\ &= (\beta_1^2 - \beta_2^2)^{1/2} a = (u_{km}^2 + w_{km}^2)^{1/2} a \end{aligned} \tag{5.2.16}$$

Equation (5.2.2) may now be re-expressed as

$$\left(\frac{u_{km} a}{V}\right)^2 = (\beta_1^2 - \beta_{km}^2)\frac{a^2}{V^2} = \frac{\beta_1^2 - \beta_{km}^2}{\beta_1^2 - \beta_2^2} \tag{5.2.17}$$

At high frequencies, as $\omega \to \infty$ and $\lambda \to 0$, it is found that the values of u_{km}

approach the following limiting values:

$$\begin{aligned}
&\text{For } k = 0, \quad u_{0m}a \to t_{1,m} \quad &&\text{for the TE}_{0m} \text{ and TM}_{0m} \text{ modes} \\
&\text{For } k > 1, \quad u_{km}a \to t_{k-1,m} \quad &&\text{for the HE}_{km} \text{ modes} \qquad (5.2.18)\\
&\qquad\qquad u_{km}a \to t_{k+1,m} \quad &&\text{for the EH}_{km} \text{ modes}
\end{aligned}$$

In each case $u_{km}a/V \to 0$ as $\omega \to \infty$, so that $\beta_{km} \to \beta_1$.

At low frequencies, as $\omega \to 0$ and $\lambda \to \infty$, we expect that $\beta_{km} \to \beta_2$. Using eqns. (5.2.4) and (5.2.17) we see that this implies $w_{km}a \to 0$ and $u_{km}a \to V$. However, it is a matter of great importance that for all guided-wave modes except one, this limit of $w_{km}a = 0$ and $u_{km}a = V$ is reached at some nonzero cutoff frequency. At optical frequencies below the cutoff frequency the condition (5.2.14) cannot be satisfied and the value of β_{km} for each mode becomes imaginary so that the mode ceases to propagate. The sole exception to this is the HE_{11} mode in which $u_{11}a$ approaches V asymptotically as $\omega \to 0$, and as a result this mode can in principle propagate at all frequencies.

The cutoff conditions, which define this lower limit to the u_{km} values are as follows:

$$\text{For the TE}_{0m}\text{, TM}_{0m} \text{ and EH}_{km} \text{ modes,} \qquad u_{km}a = t_{km} \qquad (5.2.19)$$

$$\text{For the HE}_{1m} \text{ modes} \qquad u_{1m}a = t_{1,m-1} \qquad (5.2.20)$$

For the other HE_{km} modes ($k > 2$) the general cutoff condition is more complicated. It is:

$$(u_{km}a)\frac{J_{k-2}(u_{km}a)}{J_{k-1}(u_{km}a)} = -(k-1)\frac{(n_1^2 - n_2^2)}{n_2^2} \qquad (5.2.21)$$

In the limit of small core–cladding index differences, this becomes simply $J_{k-2}(u_{km}a) = 0$, so that the cut-off condition is

$$u_{km}a = t_{k-2,m} \qquad (5.2.22)$$

The allowed ranges of values of $u_{km}a$ are illustrated schematically in Fig. 5.5. This makes it clear that with a normalized optical frequency, V, lower than $t_{01} = 2.405$, the only mode able to propagate is the HE_{11} mode. Fibers designed to satisfy this condition are known as *single-mode* or *monomode* fibers. They are likely to become extremely important in future developments of high bandwidth optical communication systems and we shall discuss them in more detail in Section 5.5. Meanwhile we should examine the propagation characteristics of the various modes a little more closely.

In Fig. 5.6 we have tried to give a schematic illustration of the way in which the propagation characteristics of the modes vary with frequency. We have

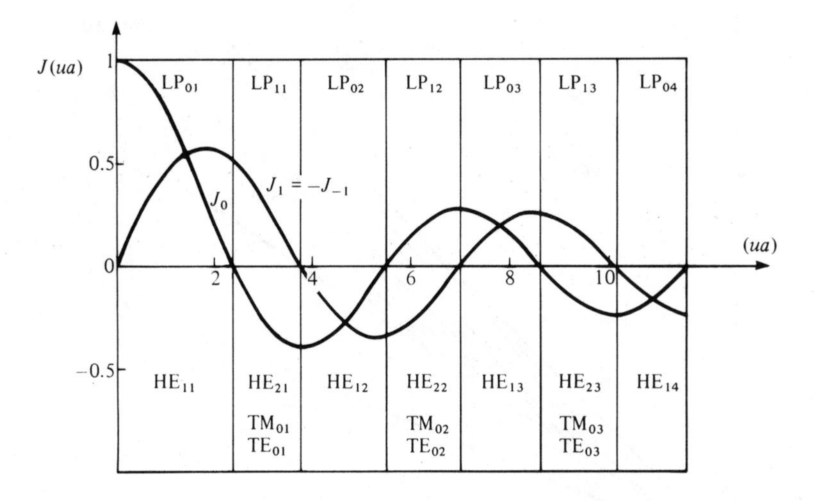

Fig. 5.5 A sketch of the lower-order Bessel functions indicating the range of allowed values of $(u_{km}a)$ for the lower-order modes. [Taken from D. Gloge, Weakly guiding fibers, *Applied Optics* **10**, 2252–8 (1971).]

At high frequencies $(V \to \infty)$ the $(u_{km}a)$ values tend to the right-hand limits of the ranges.

At low frequencies the $(u_{km}a)$ values tend to the left-hand limits, t_{ij}, and at the cutoff frequencies $u_{km}a = V = t_{ij}$.

At any given frequency the values of $(u_{km}a)$ for each of the allowed modes adjust to particular values within the prescribed ranges which permit the boundary conditions to be satisfied. The fields in the cladding $(u_{km}r > u_{km}a)$ die away to zero, the attenuation being most rapid for modes far away from cutoff.

deliberately chosen to plot β against ω, in contrast to the more conventional ω–β diagram, on the grounds that it is clearly the propagation constant, β, that should be regarded as the dependent variable, not the angular frequency, ω. We recall that the phase velocity associated with any particular mode at a given frequency is $v_p = (\beta/\omega)^{-1}$, whereas the group velocity is $v_g = (d\beta/d\omega)^{-1}$. Figure 5.6 illustrates the variation of v_g with frequency for any given mode (waveguide dispersion), and the variation of v_g between the various modes that may propagate at any given frequency (mode dispersion). These characteristics can be further modified to show the additional effects of material dispersion and thereby to demonstrate the inter-relationships between all three causes of dispersion. This we have attempted *very schematically* in Fig. 5.7. There it can be seen that the effect of the material dispersion is to distort the whole pattern of the characteristics. The shortcomings of Fig. 5.6 can be relieved to some

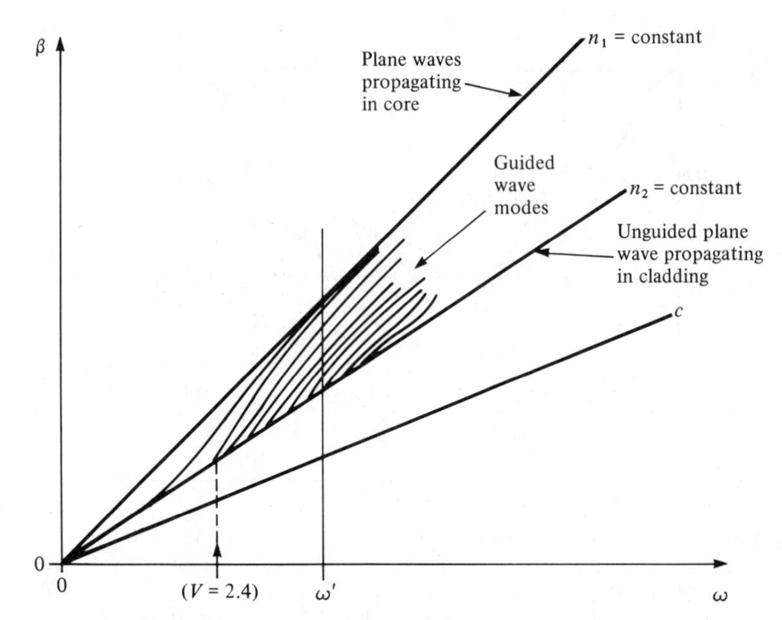

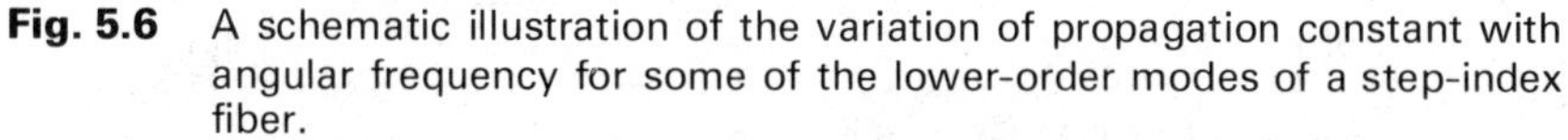

Fig. 5.6 A schematic illustration of the variation of propagation constant with angular frequency for some of the lower-order modes of a step-index fiber.

The diagram is drawn on the assumption that the media are nondispersive, that is, neither n_1 nor n_2 vary with frequency. The whole region between the two lines corresponding to n_1 and n_2 has been greatly exaggerated in an attempt to illustrate the behavior of the modes.

For frequencies such that $V < 2.405$, only one mode can propagate. At higher frequencies (for example ω' marked on the diagram) a number of modes may propagate and each will have a group velocity given by the reciprocal of the slope of the mode characteristic at ω'. It is the variation of $(d\beta/d\omega)^{-1}_{\omega'}$ among the modes that gives rise to intermode dispersion.

extent if the propagation constant, β, is normalized against $2\pi/\lambda$. Figure 5.8 shows some calculated propagation characteristics for the lower-order modes plotted as $\beta\lambda/(2\pi)$ versus V.

5.3 WEAKLY GUIDING SOLUTIONS

It is found that somewhat simpler expressions for the mode field patterns can be obtained in the case that the difference in refractive index between core and cladding is small—that is, $n_1 - n_2 \ll n_1$. Fibers satisfying this condition are referred to as *weakly guiding* fibers. Then the electromagnetic field patterns

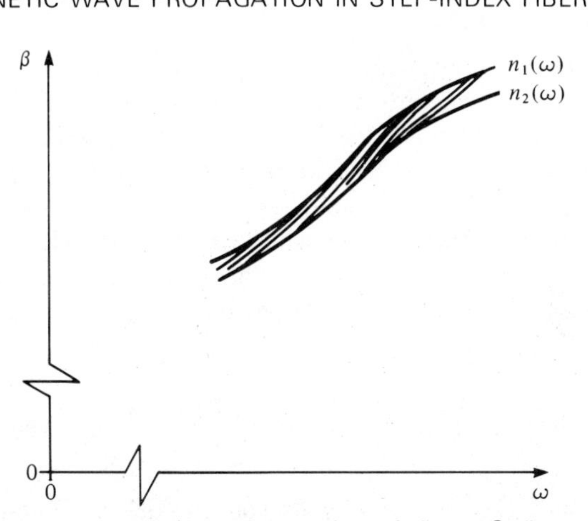

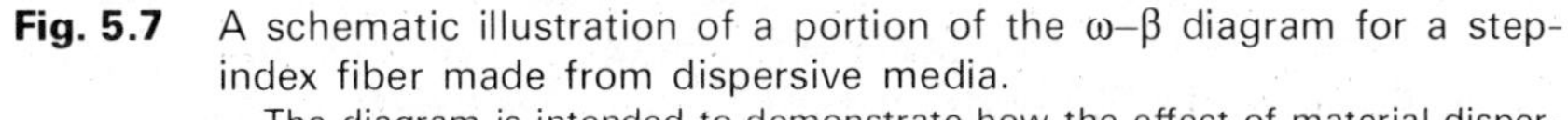

Fig. 5.7 A schematic illustration of a portion of the ω–β diagram for a step-index fiber made from dispersive media.

The diagram is intended to demonstrate how the effect of material dispersion is to distort the whole pattern of the mode propagation characteristics.

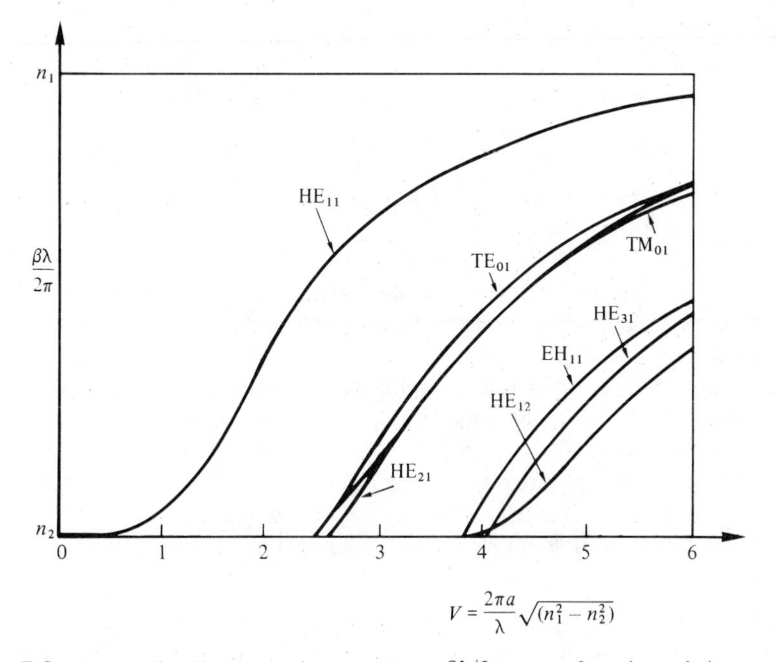

Fig. 5.8 Normalized propagation constant, $\beta\lambda/2\pi$, as a function of the normalized frequency parameter, V, for some of the lowest order modes of a step-index fiber.

and the propagation constants for the mode pairs $HE_{k+1,m}$ and $EH_{k-1,m}$ are very similar, as are those of the three modes in each TM_{0m}, TE_{0m} and HE_{2m} group. This is illustrated in Fig. 5.8 for $(k, m) = (0, 1)$ and $(2, 1)$. The differences within each pair or group reduce to zero as $(n_1 - n_2)/n_1 \rightarrow 0$, that is in the weakly guiding limit. Then, combining such groups of modes together gives rise to sets of linearly polarized waves in which the transverse electric and magnetic fields are each parallel and mutually orthogonal over the whole fiber cross-section. Because of this the fields can best be expressed using Cartesian field components, E_x, E_y, H_x, H_y:

$$\text{within the core,} \quad \begin{bmatrix} E_y \\ H_x \end{bmatrix} = \begin{bmatrix} E_2 \\ H_2 \end{bmatrix} J_{\kappa,m}(u_{\kappa m} r) \cos \kappa\phi \tag{5.3.1}$$

and $E_x = H_y = 0$, where E_2 and H_2 are constant fields, κ and m are integers, x and y are Cartesian coordinates of arbitrary orientation in the plane perpendicular to the fiber axis, and ϕ represents the aximuthal angle.

$$\text{in the cladding,} \quad \begin{bmatrix} E_y \\ H_x \end{bmatrix} = \begin{bmatrix} E_3 \\ H_3 \end{bmatrix} K_{\kappa,m}(w_{\kappa m} r) \cos \kappa\phi \tag{5.3.2}$$

Again $E_x = H_y = 0$, E_3 and H_3 are constant fields, κ and m are integers.

The (κ, m) mode is derived by adding together the $(k - 1)$ and $(k + 1)$ solutions for E_z and H_z. Thus in the core, using eqn. (5.2.1),

$$\psi_z = \psi_1 \{ J_{(k-1)}(ur) \cos (k - 1)\phi + J_{(k+1)}(ur) \cos (k + 1)\phi \} \tag{5.3.3}$$

Use can now be made of the Bessel function recursion formula:

$$J_{(k-1)}(ur) + J_{(k+1)}(ur) = (2k/ur) J_k(ur). \tag{5.3.4}$$

Instead of deriving the polar components (ψ_r and ψ_ϕ) of the transverse fields using eqns. (5.2.5), the Cartesian components ψ_x and ψ_y are obtained by means of the transformations $x^2 + y^2 = r^2$ and $y/x = \tan \phi$. It is then found that combining the two solutions in this way causes the E_x and H_y field components to be almost cancelled out when the change of refractive index at the core–cladding boundary is small. It is also found that the longitudinal field components E_z and H_z are much smaller than the main transverse components E_y and H_x. We are thus left with what are almost plane-polarized, transverse electromagnetic wave solutions. They are known as linearly polarized, $LP_{\kappa m}$, modes.

In general

each LP_{0m} mode is derived from an HE_{1m} mode
each LP_{1m} mode from a TE_{0m}, TM_{0m} and HE_{2m} mode
each $LP_{\kappa m}$ mode ($\kappa \geqslant 2$) from an $HE_{(\kappa+1),m}$ and $EH_{(\kappa-1),m}$ mode.

Thus,

HE_{11}	becomes LP_{01}
TE_{01}, TM_{01} and HE_{21}	become LP_{11}
EH_{11} and HE_{31}	become LP_{21}
HE_{12}	becomes LP_{02}, and so on.

As well as the set of solutions for E_y and H_x with E_x and H_y small, there is an equivalent set of solutions for E_x and H_y with E_y and H_x small. In addition there are equivalent solutions with the field polarities reversed, making each $LP_{\kappa m}$ mode comprise four degenerate solutions. The four field distributions for LP_{11} are shown in Fig. 5.9. Detailed derivations of the LP mode solutions are given in Refs. [5.1]–[5.3].

One of the advantages of the weakly guiding approximation is the relative simplicity of the boundary condition that results. This now takes the form

$$u_{\kappa m}\left[\frac{J_{\kappa-1}(u_{\kappa m}a)}{J_{\kappa}(u_{\kappa m}a)}\right] = -w_{\kappa m}\left[\frac{K_{\kappa-1}(w_{\kappa m}a)}{K_{\kappa}(w_{\kappa m}a)}\right] \tag{5.3.5}$$

It still requires numerical solution in order to determine $u_{\kappa m}$ and $w_{\kappa m}$ and hence $\beta_{\kappa m}(\omega)$ for the allowed guided modes.

The cutoff condition, $w_{\kappa m} = 0$, is seen from eqn. (5.3.5) to require that

$$J_{\kappa-1}(u_{\kappa m}a) = 0 \tag{5.3.6}$$

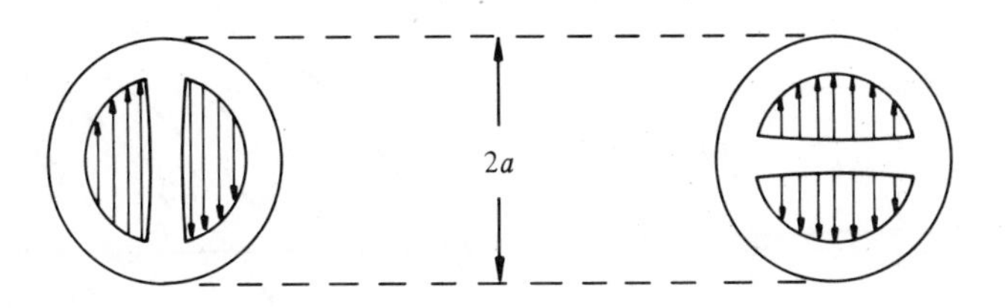

Fig. 5.9 Transverse electric field patterns in the fiber core for the four degenerate LP_{11} modes.

That is, for all except the $\kappa = 0$ modes,

$$u_{\kappa m} a = t_{\kappa-1,m} \tag{5.3.7}$$

This condition is consistent with the cutoff conditions (5.2.19)–(5.2.21), in the weakly guiding limit. When $\kappa = 0$, it is found that, as before, there is no cutoff frequency for the LP_{01} mode, and that for the other LP_{0m} modes the cutoff frequencies are given by

$$u_{0m} a = t_{1,m-1} \tag{5.3.7a}$$

The ranges of the permitted values of $u_{\kappa m} a$ for the $LP_{\kappa m}$ modes were indicated on Fig. 5.5.

We saw in the last section that the propagation constants, $\beta_{\kappa m}$, are confined to the range

$$\beta_2 < \beta_{\kappa m} < \beta_1 \tag{5.2.14}$$

and Fig. 5.8 showed how the normalized propagation constants $(\beta\lambda/2\pi)$ vary from n_2 at cutoff to n_1 at high frequency. A further normalization of the propagation constant enables the characteristics to be calculated and plotted in

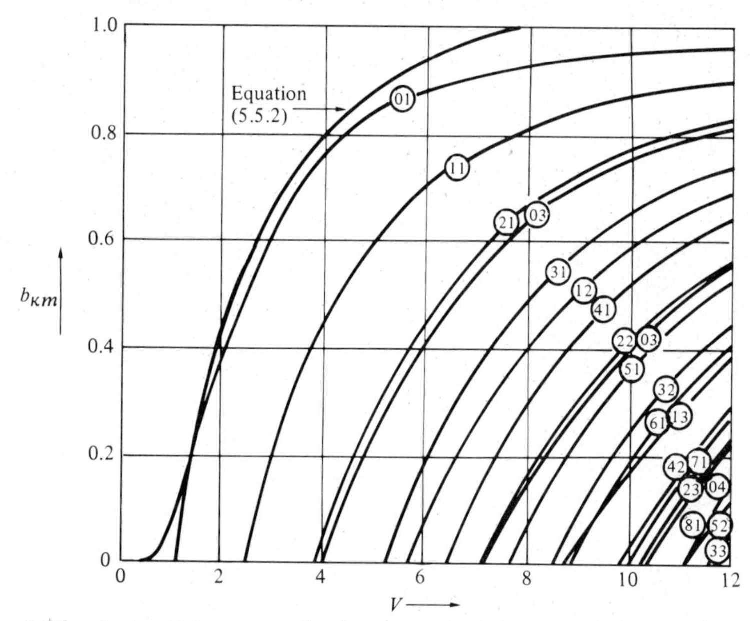

Fig. 5.10 Plots of the normalized propagation parameters $b_{\kappa m}$ vs. the normalized frequency parameter, V, for some of the lowest order modes of a step-index fiber, using the 'weakly guiding fiber' approximation. [Taken from D. Gloge, Weakly guiding fibers, *Appl. Optics* **10**, 2252–8 (1971).]

a way that is quite independent of the fiber parameters, n_1, n_2 and a. We define $b_{\kappa m}$ such that

$$b_{\kappa m} = \frac{(\beta_{\kappa m}\lambda/2\pi)^2 - n_2^2}{n_1^2 - n_2^2} = \frac{\beta_{\kappa m}^2 - \beta_2^2}{\beta_1^2 - \beta_2^2} \tag{5.3.8}$$

$b_{\kappa m}$ is then a new propagation parameter which covers the range of values 0 to 1. The propagation characteristics of the lowest-order modes, calculated using the weakly guiding approximation, are shown in Fig. 5.10 plotted as $b_{\kappa m}$ versus V. We may note in passing that

$$b_{\kappa m} = 1 - \frac{u_{\kappa m}^2 a^2}{V^2} = \frac{w_{\kappa m}^2 a^2}{V^2} \tag{5.3.9}$$

One of the benefits of the mode analysis is that it enables the power density distribution in the fiber for each mode to be derived straightforwardly by evaluating the Poynting vector $\mathbf{E} \times \mathbf{H}$ for the transverse fields over the fiber cross-section. It is then possible to determine the extent to which the electromagnetic wave penetrates the core–cladding interface and spreads into the cladding. The fraction of the total power in each mode that is carried in the core can then be evaluated by integrating over the core cross-section. The results of such calculations in the weakly guiding approximation are shown in Fig. 5.11 for several of the lowest-order LP modes. As expected, most of the power flow is to be found in the fiber core except when modes are near to cutoff.

Most multimode fibers that are used for optical communication propagate many modes. A method for estimating the number of propagating modes will be given shortly. Most of the optical power is carried, then, in the core of the fiber by means of high-order modes that are far from their cutoff frequency. In this case we may simplify the characterization of the modes by making use of the asymptotic expression for Bessel functions of large argument:

$$J_\kappa(x) \simeq (2/\pi x)^{1/2} \cos(x - \kappa\pi/2 - \pi/4) \tag{5.3.10}$$

when $x \gg 1$.

The condition for cutoff is

$$J_{\kappa-1}(u_{\kappa m} a) = 0 \tag{5.3.11}$$

so, for $u_{\kappa m} a \gg 1$, cutoff requires that

$$\cos(u_{\kappa m} a - \kappa\pi/2 + \pi/4) = 0 \tag{5.3.12}$$

that is,

$$u_{\kappa m} a - \kappa\pi/2 + \pi/4 = (m + \tfrac{1}{2})\pi \tag{5.3.13}$$

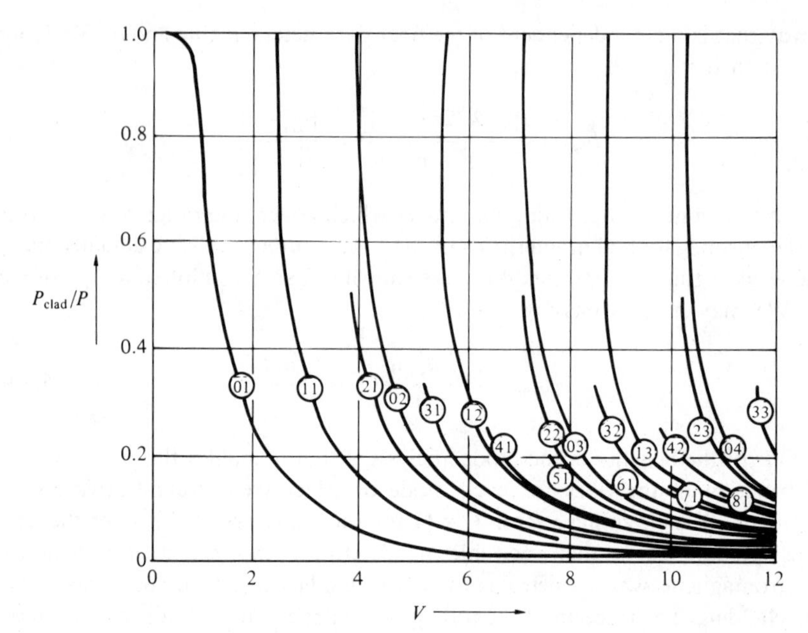

Fig. 5.11 Fraction of the mode power which propagates in the cladding plotted against the normalized frequency parameter, *V*. [Taken from D. Gloge, Weakly guiding fibers, *Appl. Optics* **10**, 2252–8 (1971).]

or

$$u_{\kappa m} a = \frac{\pi}{2} (\kappa + 2m + \tfrac{1}{2}) \tag{5.3.14}$$

Although we have derived eqn. (5.3.14) from the condition for cutoff, the total variation in $u_{\kappa m} a$ over all possible frequencies is limited to something less than π, so the fractional variation in $u_{\kappa m}$ is very small under the conditions of the approximation, namely $(\kappa + 2m + \frac{1}{2}) \gg 1$. This means that all modes with the same value of $(\kappa + 2m)$ are approximately degenerate. It is therefore sensible to identify them by means of a new, integer, mode group number, $q = \kappa + 2m$. Then,

$$u_{\kappa m} a \simeq q\pi/2 \tag{5.3.15}$$

Approximately $(q/2)$ pairs of values of (κ, m) give rise to the same value of q, and each (κ, m) pair corresponds to four modes. Thus there are $2q$ modes associated with each mode group, q. Their propagation constant may be obtained in normalized form either by substituting eqn. (5.3.15) into eqn.

(5.3.9):

$$b_q = b_{\kappa m} = 1 - \frac{u_{\kappa m}^2 a^2}{V^2} = 1 - \frac{\pi^2 q^2}{4V^2} \tag{5.3.16}$$

or explicitly from eqn. (5.2.2):

$$\beta_q^2 = \beta_{\kappa m}^2 = \beta_1^2 - u_{\kappa m}^2 = \beta_1^2 - \frac{\pi^2 q^2}{4a^2} \tag{5.3.17}$$

The largest permitted value of q occurs when $\beta_q = \beta_2$. We shall designate this maximum value of the mode group quantum number by Q. Then,

$$\beta_Q^2 = \beta_2^2 = \beta_1^2 - \frac{\pi^2 Q^2}{4a^2} \tag{5.3.18}$$

That is,

$$Q^2 = (\beta_1^2 - \beta_2^2)\frac{4a^2}{\pi^2} \tag{5.3.19}$$

And, using eqn. (5.2.16),

$$Q = \frac{2V}{\pi} \tag{5.3.20}$$

When V is large, as it must be for this analysis to be valid, the total number of (κ, m) modes that may propagate, M, is given approximately by

$$M = \sum_{q=1}^{Q} (2q) \simeq Q^2 \tag{5.3.21}$$

A slightly better approximation is

$$M \simeq \frac{\pi^2 Q^2}{8} \tag{5.3.22}$$

Using eqn. (5.3.20) we see that this gives

$$M \simeq \frac{\pi^2 Q^2}{8} = \frac{V^2}{2} = \frac{1}{2}\left(\frac{2\pi a}{\lambda}\right)^2 (n_1^2 - n_2^2)$$

$$= 2\pi \frac{A_c}{\lambda^2} (NA)^2 \tag{5.3.23}$$

where A_c is the core area and (NA), as before, is the numerical aperture of the fiber. A relationship of this sort is to be expected because the square of the numerical aperture determines the optical power that can enter the fiber, and the number of propagating modes, if they are all excited equally, determines the total power that can propagate.

To calculate M for a typical, multimode, step-index fiber, let us take $n_1 = 1.46$, $n_2 = 1.45$, and $2a = 50\ \mu\text{m}$. Then $(NA) = 0.17$. At a source wavelength of 0.85 μm, $V = 31.5$, the number of mode groups, $Q = 20$, and the total number of propagating modes given by eqn. (5.3.23) is $M \simeq 500$.

5.4 TIME DISPERSION IN STEP-INDEX FIBERS

Figures 5.6 and 5.7 enable us to visualize the various causes of optical pulse broadening during transmission along a fiber. In Section 2.2.3 we established that the propagation time of an optical signal over a path length, l, is

$$t = \frac{l}{v_g} = l\frac{d\beta}{d\omega} \tag{2.2.36}$$

Then for any one mode, an optical impulse in which the half-power range of optical frequencies is $\Delta\omega$ spreads to a pulse of half-height width, τ, where

$$\tau = \frac{dt}{d\omega}\Delta\omega = l\frac{d^2\beta}{d\omega^2}\Delta\omega \tag{5.4.1}$$

What this indicates is that the waveguide dispersion associated with any particular mode is determined by the second derivative of the propagation characteristic of that mode at the frequency of interest. It is, therefore, represented by the curvature of the mode characteristics shown in Fig. 5.6. Figure 5.7 shows that material dispersion, which arises when $n = c\beta/\omega$ is itself a function of ω, has the effect of distorting the coordinate system of the β versus ω diagram in the region of interest and thus of modifying the propagation characteristics of *all* the modes there. Indeed this effect normally dominates the waveguide dispersion. Larger than either of these effects in a multimode, step-index fiber is that resulting from the range of values of $d\beta_{km}/d\omega$ among the various (k, m) modes propagating at the given frequency. It is this we have called intermode dispersion.

Algebraic representation of these three effects and their inter-relationships is complex, but it can be greatly simplified for weakly guiding fibers by introducing the parameter Δ to represent the fractional change of refractive index at the core–cladding boundary:

$$\Delta = \frac{n_1^2 - n_2^2}{2n_2^2} \tag{5.4.2}$$

For weakly guiding fibers, where $n_1 \simeq n_2 = n$

$$\Delta \simeq \frac{n_1 - n_2}{n} \ll 1 \tag{5.4.3}$$

Rearranging eqn. (5.3.8) gives the propagation constant for the (κ, m) mode in terms of the dimensionless parameter, $b_{\kappa m}$ as

$$\begin{aligned}
\beta_{\kappa m} &= [\beta_2^2 + (\beta_1^2 - \beta_2^2)b_{\kappa m}]^{1/2} \\
&= \frac{2\pi}{\lambda}[n_2^2 + (n_1^2 - n_2^2)b_{\kappa m}]^{1/2} \\
&= \frac{2\pi}{\lambda} n_2(1 + 2\Delta b_{\kappa m})^{1/2} \\
&\simeq \frac{\omega n_2}{c}(1 + \Delta b_{\kappa m} + \cdots)
\end{aligned} \tag{5.4.4}$$

where we have retained only the first term in the binomial expansion, assuming small Δ. Within this approximation the transit time, $t_{\kappa m}$, for the (κ, m) mode over a distance, l, is given by

$$\frac{ct_{\kappa m}}{l} = c\frac{\mathrm{d}\beta_{\kappa m}}{\mathrm{d}\omega} = N_2(1 + \Delta b) + \omega n_2 \frac{\mathrm{d}(\Delta b)}{\mathrm{d}\omega} \tag{5.4.5}$$

In eqn. (5.4.5) we have dropped the (κ, m) subscripts, which are nevertheless implied, and have reintroduced the group index, N_2, where

$$N_2 = \frac{c}{v_g} = \frac{\mathrm{d}(\omega n_2)}{\mathrm{d}\omega} = n_2 + \omega\frac{\mathrm{d}n_2}{\mathrm{d}\omega} \tag{2.2.37}$$

We can see at once that the transit time and hence the total dispersion will depend on three parameters: the material dispersion of the cladding, N_2, the differential material dispersion between the cladding and the core, $\mathrm{d}\Delta/\mathrm{d}\omega$, and the waveguide dispersion, $\mathrm{d}b/\mathrm{d}\omega$. Because the waveguide propagation constant has been expressed in terms of the normalized frequency, V, as in Fig. 5.10, we will take the derivatives of b with respect to V rather than ω. Then

$$\frac{ct}{l} = N_2(1 + \Delta b) + \omega n_2 b \frac{d\Delta}{d\omega} + \omega n_2 \Delta \frac{db}{dV} \frac{dV}{d\omega} \tag{5.4.6}$$

Now, using eqn. (5.2.16), we see that

$$V = \frac{a\omega}{c} (n_1^2 - n_2^2)^{1/2} = \frac{a\omega n_2}{c} (2\Delta)^{1/2} \tag{5.4.7}$$

so that

$$\frac{dV}{d\omega} = \frac{a}{c}\left[(2\Delta)^{1/2} N_2 + \frac{\omega n_2}{(2\Delta)^{1/2}} \frac{d\Delta}{d\omega}\right] = V\left[\frac{N_2}{\omega n_2} + \frac{1}{(2\Delta)} \frac{d\Delta}{d\omega}\right] \tag{5.4.8}$$

It is sometimes convenient to define a normalized differential dispersion parameter, δ, such that

$$\delta = \frac{n_2 \omega}{2N_2 \Delta} \frac{d\Delta}{d\omega} \tag{5.4.9}$$

Then,

$$\frac{dV}{d\omega} = \frac{VN_2}{\omega n_2} (1 + \delta) \tag{5.4.10}$$

and using this expression in eqn. (5.4.6), we obtain

$$\frac{ct}{l} = N_2(1 + \Delta b) + \omega n_2 b \frac{d\Delta}{d\omega} + \Delta N_2 V \frac{db}{dV} (1 + \delta) \tag{5.4.11}$$

Substituting for δ gives

$$\frac{ct}{l} = N_2 \left[1 + \Delta \frac{d(Vb)}{dV}\right] + \omega n_2 \frac{d\Delta}{d\omega}\left(b + \frac{V}{2} \frac{db}{dV}\right) \tag{5.4.12}$$

Equation (5.4.12) may, of course, be obtained directly by substituting eqn. (5.4.8) into the final term of eqn. (5.4.6).

In order to obtain an expression for the pulse spreading, $\tau_{\kappa m}$, of the (κ, m) mode as given by eqn. (5.4.1), we must differentiate again. Doing so term by term gives

$$\frac{c\tau}{l\,\Delta\omega} = c\frac{\mathrm{d}^2\beta}{\mathrm{d}\omega^2}$$

$$= \frac{\mathrm{d}N_2}{\mathrm{d}\omega} + \frac{\mathrm{d}N_2}{\mathrm{d}\omega}\Delta\frac{\mathrm{d}(Vb)}{\mathrm{d}V} + N_2\frac{\mathrm{d}\Delta}{\mathrm{d}\omega}\frac{\mathrm{d}(Vb)}{\mathrm{d}V}$$

$$+ N_2\Delta\frac{\mathrm{d}^2(Vb)}{\mathrm{d}V^2}\left(\frac{VN_2}{\omega n_2} + \frac{V}{2\Delta}\frac{\mathrm{d}\Delta}{\mathrm{d}\omega}\right)$$

$$+ N_2\frac{\mathrm{d}\Delta}{\mathrm{d}\omega}\left(b + \frac{V}{2}\frac{\mathrm{d}b}{\mathrm{d}V}\right) + \omega n_2\frac{\mathrm{d}^2\Delta}{\mathrm{d}\omega^2}\left(b + \frac{V}{2}\frac{\mathrm{d}b}{\mathrm{d}V}\right)$$

$$+ \omega n_2\frac{\mathrm{d}\Delta}{\mathrm{d}\omega}\left(\frac{3}{2}\frac{\mathrm{d}b}{\mathrm{d}V} + \frac{V}{2}\frac{\mathrm{d}^2b}{\mathrm{d}V^2}\right)\left(\frac{VN_2}{\omega n_2} + \frac{V}{2\Delta}\frac{\mathrm{d}\Delta}{\mathrm{d}\omega}\right) \tag{5.4.13}$$

Remembering that $\Delta \ll 1$ and discarding terms involving $(\mathrm{d}\Delta/\mathrm{d}\omega)^2$ and $\mathrm{d}^2\Delta/\mathrm{d}\omega^2$, we can reduce this expression in such a way that the terms describing the three main effects separate out:

$$\frac{c\tau}{l\Delta\omega} = \underbrace{\frac{\mathrm{d}N_2}{\mathrm{d}\omega}\left[1 + \Delta\frac{\mathrm{d}(Vb)}{\mathrm{d}V}\right]}_{\text{material dispersion (cladding)}} + \underbrace{\frac{\Delta N_2^2}{\omega n_2}\frac{V\mathrm{d}^2(Vb)}{\mathrm{d}V^2}}_{\text{waveguide dispersion}} + \underbrace{N_2\frac{\mathrm{d}\Delta}{\mathrm{d}\omega}\frac{\mathrm{d}^2(V^2b)}{\mathrm{d}V^2}}_{\text{differential material dispersion}} \tag{5.4.14}$$

In normal, multimode, step-index fibers all three of the principal terms in eqn. (5.4.14) are dominated by the variation of propagation times among the individual modes. That is, by the variation over the range of (κ, m) modes of the middle term in eqn. (5.4.12), namely $\Delta N_2\,\mathrm{d}(Vb_{\kappa m})/\mathrm{d}V$. The value of $\mathrm{d}(Vb_{\kappa m})/\mathrm{d}V$ approaches unity for all modes far from cutoff, and for the higher-order modes it can be shown to approximate to $(2 - \pi/V)$ at cutoff. Thus for a multimode fiber in which $V \gg 1$, the spread in group delay times among the modes is given approximately by

$$\frac{c\,\Delta t_{\text{mode}}}{l} = \Delta N_2[(2 - \pi/V) - 1]$$

$$= \Delta N_2(1 - \pi/V) \tag{5.4.15}$$

$$\simeq \Delta N_2 \tag{5.4.16}$$

in the limit of large V. With $\Delta \ll 1$ this is similar to the result given by the ray model, eqn. (2.2.36).

This discussion of modes in step-index fibers has involved a great deal of effort for results that may seem to be either trivial or irrelevant. This would be fair comment in the case of the multimode step-index fibers we have considered so far. As we stated at the outset, these can be satisfactorily treated using the ray model. However, the methods we have developed provide a basis for the discussion of single-mode propagation and also for the discussion of multimode propagation in graded-index fibers.

5.5 SINGLE-MODE FIBERS

The intermode dispersion that places such a severe limit on the information capacity of multimode, step-index fibers can be eliminated completely by designing the fiber so that only the HE_{11} mode can propagate. As we showed in Section 5.2, the condition for this is that

$$V = \frac{2\pi a(n_1^2 - n_2^2)^{1/2}}{\lambda} < 2.405 \tag{5.5.1}$$

and it can be achieved by increasing the operating wavelength, by reducing the core diameter, or by reducing the difference in refractive index between core and cladding. Usually the wavelength is decided by other factors and although it is customary to use a small value of Δn, in the region of 0.002, to reduce it any further would cause the fiber to be very sensitive to losses at bends, as we showed in Section 3.1.3. Thus the dimensions of single-mode fiber are typically of the order of those shown in Fig. 2.5(e). If we take $\lambda = 0.85\ \mu m$, $n = 1.46$, $\Delta n = 0.002$, then for single-mode operation, condition (5.5.1) requires that

$$2a < \frac{2.405\lambda}{\pi(n_1^2 - n_2^2)^{1/2}} = 8.5\ \mu m$$

Although the core diameter has to be reduced to around 5 μm to 10 μm, it is usual to maintain the cladding diameter at 60 μm to 100 μm in order to preserve the strength and handling characteristics of the fiber and to reduce its susceptibility to bending and deformation losses.

Excitation of a single-mode fiber requires a source giving a high intensity of radiation from a small area. A semiconductor injection laser is ideal. One of the laser transverse modes of oscillation can be coupled directly into the HE_{11} propagation mode of the fiber. If we were to attempt to couple light from an incoherent source such as an LED into a single-mode fiber, only a tiny fraction of the emitted power would propagate and the source would be hopelessly inefficient.

The attractions of the single-mode system are obvious. Intermode dispersion is completely eliminated. The use of a laser source operating stably in a single transverse and single longitudinal mode means that the spectral width of the light can be extremely small, certainly much less than 1 nm, so material dispersion and waveguide dispersion are minimal also. Even when several longitudinal laser modes are launched during a pulse giving a wavelength spread of 1 or 2 nm, the dispersion of the transmission medium is only likely to limit the information capacity of the system over the very longest of repeater spacings.

In order to estimate the total intramode dispersion, we need to know the value of the parameter $V\mathrm{d}^2(Vb)/\mathrm{d}V^2$ for the fundamental HE_{11} mode, so that the effects of waveguide dispersion can be added to the material dispersion, as in eqn. (5.4.14). The propagation characteristics of this mode were plotted in Figs. 5.8 and 5.10. From them we can derive the values of $V\mathrm{d}^2(Vb)/\mathrm{d}V^2$ shown in Fig. 5.12, plotted as a function of the normalized frequency, V. For values of

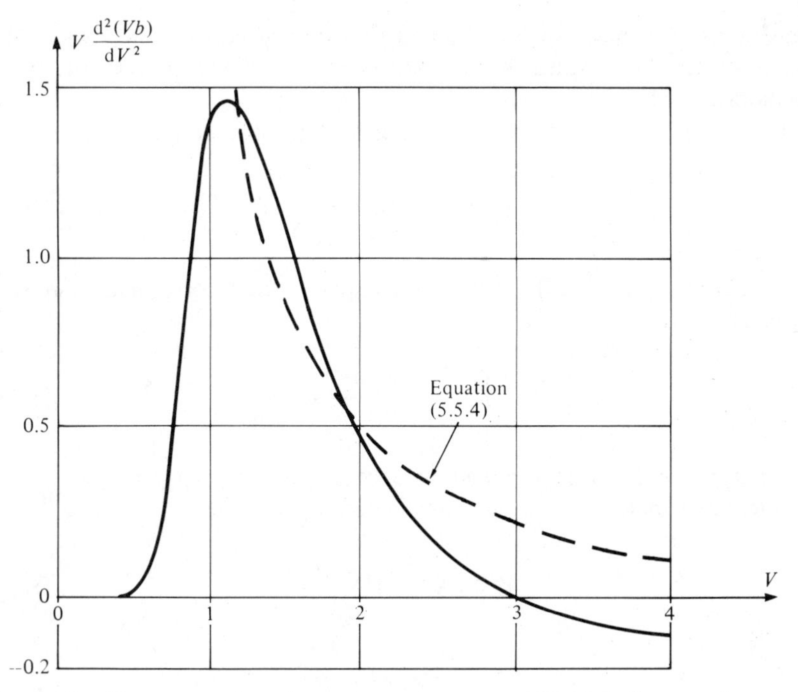

Fig. 5.12 The waveguide dispersion parameter $V\mathrm{d}^2(Vb)/\mathrm{d}V^2$ plotted against the normalized frequency parameter, V, for the HE_{11} mode of a step-index fiber.

V in the range $1.5 < V < 2.4$, a widely used approximation to $b_{11}(V)$ is

$$b = \left(1.1428 - \frac{0.996}{V}\right)^2 \tag{5.5.2}$$

This is shown on Fig. 5.10. If we differentiate it we obtain

$$\frac{\mathrm{d}(Vb)}{\mathrm{d}V} = 1.306 - \frac{0.992}{V^2} \tag{5.5.3}$$

$$V\frac{\mathrm{d}^2(Vb)}{\mathrm{d}V^2} = \frac{1.984}{V^2} \tag{5.5.4}$$

and

$$\frac{\mathrm{d}^2(V^2 b)}{\mathrm{d}V^2} = 2.612 \tag{5.5.5}$$

Equation (5.5.4) has also been plotted as a dashed curve in Fig. 5.12. This demonstrates how sensitive the dispersion is to the precise functional relationship $b(V)$.

In Section 2.2.3 we defined the material dispersion parameter, Y_{m}, to be

$$Y_{\mathrm{m}} = \lambda^2 \frac{\mathrm{d}^2 n}{\mathrm{d}\lambda^2} = \omega \frac{\mathrm{d}N}{\mathrm{d}\omega} \tag{2.2.45}$$

We see from eqn. (5.4.14) that this may require a small correction, so we shall redefine Y_{m} as

$$Y_{\mathrm{m}} = \left[1 + \Delta \frac{\mathrm{d}(Vb)}{\mathrm{d}V}\right] \omega \frac{\mathrm{d}N_2}{\mathrm{d}\omega} \tag{5.5.6}$$

By analogy we may also introduce a waveguide dispersion parameter, Y_{w}, and a differential material dispersion parameter, Y_{d}, as

$$Y_{\mathrm{w}} = \Delta \frac{N_2^2}{n_2} V \frac{\mathrm{d}^2(Vb)}{\mathrm{d}V^2} \tag{5.5.7}$$

and

$$Y_{\mathrm{d}} = N_2 \frac{\mathrm{d}^2(V^2 b)}{\mathrm{d}V^2} \omega \frac{\mathrm{d}\Delta}{\mathrm{d}\omega} \tag{5.5.8}$$

Then, using $\gamma = |\Delta\lambda/\lambda| = |\Delta\omega/\omega|$, eqn. (5.4.14) may be put in the form

$$\frac{c\tau}{l} = |Y_m + Y_w + Y_d|\gamma \tag{5.5.9}$$

In order to establish the relative magnitudes of the three terms in eqn. (5.5.9) for a typical fiber we shall consider as an example a single-mode fiber made with a pure silica cladding and a core doped with germania. We shall examine its behavior at four wavelengths: 0.85 μm, 1.27 μm, 1.35 μm and 1.55 μm, on the assumption that at each wavelength the fiber will be designed so that $\Delta = 0.005$ and $V = 2.0$. We shall also assume that laser sources having a spectral spread of $\gamma = 0.003$ are available at each wavelength. To make $\Delta = 0.005$ will require a germania doping concentration of about 4.5%. With $V = 2.0$, equations (5.5.3)–(5.5.5) give

$$\frac{d(Vb)}{dV} = 1.058, \qquad \frac{Vd^2(Vb)}{dV^2} = 0.496 \qquad \text{and} \qquad \frac{d^2(V^2b)}{dV^2} = 2.612.$$

Although these may be over-estimates, they have been used to calculate the dispersion parameters set out in Table 5.2.

Clearly at a wavelength of 0.85 μm, Y_m dominates the other terms in eqn. (5.5.9). But, as we saw in Section 2.2.3, the value of Y_m decreases to zero and then changes in sign at a wavelength in the region of 1.28 μm. At wavelengths longer than this the waveguide dispersion and the material dispersion become self-cancelling. In practice what this means is that the wavelength of minimum total dispersion in single-mode fibers is shifted to longer wavelengths, in our example to 1.35 μm. The amount of the wavelength increase depends on the magnitude of Y_w and hence on the design of the fiber. Using eqn. (5.5.4) we have

$$Y_w = \Delta \frac{N_2^2}{n_2} \frac{1.984}{V^2} \tag{5.5.10}$$

Substituting

$$V = \left(\frac{2\pi a n_2}{\lambda}\right)(2\Delta)^{1/2} \tag{5.5.11}$$

from eqn. (5.4.7) with (ω/c) replaced by $(2\pi/\lambda)$, this becomes

$$Y_w = \frac{1.984}{2\pi^2} \frac{N_2^2}{n_2^3} \frac{\lambda^2}{(2a)^2} \tag{5.5.12}$$

Thus if we wish to increase the waveguide dispersion, we shall need to decrease

Table 5.2

λ/[μm]	For $V=2$ $(2a)$/[μm]	n_2	N_2	N_2^2/n_2	Y_m	Y_w	Y_d	Y_{tot}	γ	$(\tau/l) =$ $(Y_{tot}\gamma/c)$ [ps/km]	$(Bl) =$ $(l/2\tau)$ [(Gb/s). km]
0.85	3.72	1.453	1.466	1.48	0.0215	0.0037	0.0008	0.026	0.003	260	2
1.27	5.58	1.448	1.462	1.48	0.00015	0.0037	−0.0008	0.003	0.003	30	17
1.35	5.94	1.447	1.462	1.48	−0.0028	0.0037	−0.0010	−0.0001	0.003	1	500
1.55	6.83	1.444	1.463	1.48	−0.0100	0.0037	−0.0016	−0.008	0.003	80	6

the V value of the fiber at the operating wavelength and this requires decreasing the core diameter. In order to bring the total dispersion to zero at 1.55 μm, the wavelength of minimum attenuation, it would be necessary to make $Y_w = 0.0116$. The required value of $V \mathrm{d}^2(Vb)/\mathrm{d}V^2$ is thus $0.0116/(0.005 \times 1.48) = 1.57$. It can be seen from Fig. 5.12 that this is higher than the maximum value of $V\mathrm{d}^2(Vb)/\mathrm{d}V^2$, so the required condition can only be met by increasing Δ as well as by decreasing a. Let us say we shall decrease V to 1.5: then, we must increase Δ to

$$\Delta = \frac{n_2 V^2 Y_w}{1.984 N_2^2} = 0.0089$$

and decrease the core diameter to

$$2a = \left[\frac{1.984 N_2^2 \lambda^2}{2\pi^2 n_2^3 Y_w}\right]^{1/2} = 3.84\ \mu\mathrm{m}.$$

These effects are further illustrated in Fig. 5.13, which confirms that it is quite possible to develop a single-mode fiber in which the dispersion minimum is made to coincide with the attenuation minimum in the region of 1.55 μm. Figure 5.14 presents some published experimental results obtained with such a fiber. After propagating through 20 km of fiber the pulse width has increased from 0.38 ns to 0.40 ns. Time dispersion is better than 10 ps/km and the bandwidth–distance product of the fiber is better than 50 GHz.km. There is an attenuation penalty in making the fiber in this way. Increasing Δ requires high doping levels and is likely to increase the infra-red absorption and Rayleigh scattering limits. Decreasing V is likely to increase microbending and waveguide losses as discussed in Section 6.6. The result is a fiber attenuation similar to that of the best fibers at 1.3 μm.

In the region of the dispersion minimum effects which we have so far neglected determine the lower limit of dispersion that can be achieved. One in particular that has been studied is stress-induced birefringence. As a result of stress in a fiber, light waves having different planes of polarization propagate at different speeds. That is, the refractive index is a function of the plane of polarization. This is expected to set a lower limit to dispersion of about 5 ps/km. A suggestion to minimize the effect is to launch waves with a particular plane of polarization into a fiber having a core of elliptical cross-section. However, a bandwidth–distance product of 100 GHz.km does not at present constitute a serious limitation!

So far we have tacitly assumed that single-mode fibers have a perfect step change in refractive index. In general this is not true. Diffusion of the dopants during fiber production causes the core–cladding transition to spread over a distance of a micron or so, and this is a significant fraction of the core radius.

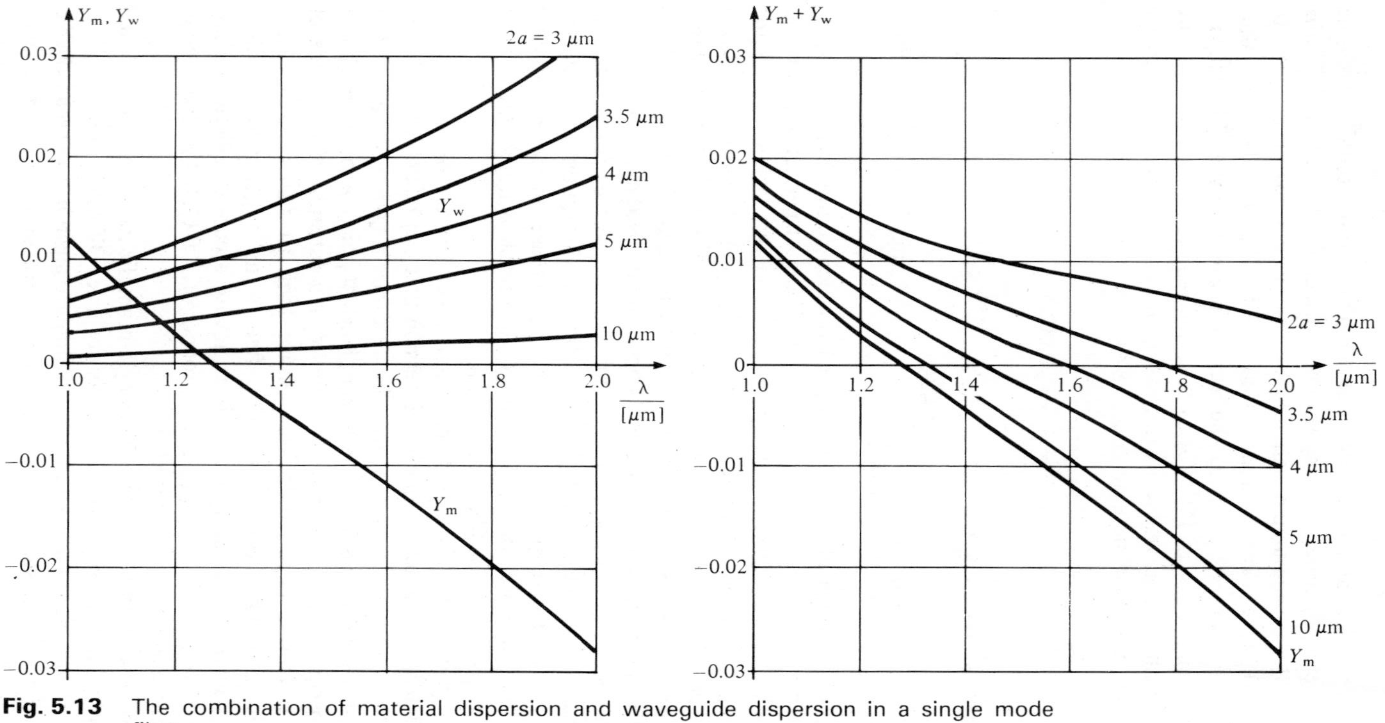

Fig. 5.13 The combination of material dispersion and waveguide dispersion in a single mode fiber.

The graphs show how the dispersion minimum may be shifted to longer wavelength by reducing the core diameter and thus decreasing the normalized frequency, V. The curves for Y_w are derived from eqn. (5.5.12); Y_m is drawn for pure silica cladding with the correction term in eqn. (5.5.6) omitted.

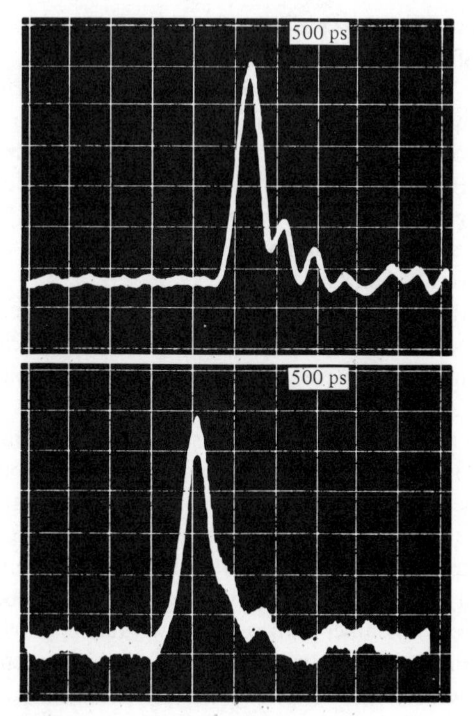

Fig. 5.14 Pulse propagation through 20 km of single-mode fiber in which the dispersion minimum and the attenuation minimum coincide at 1.55 μm. [Results taken from A. Kawana *et al.*, *Ets. Letts.* **16**, 188–9 (28 Feb. 1980).]

The fiber attenuation was approximately 1 dB/km and the total dispersion was about 6 ps/km. The spectral width of the InGaAsP/InP DH laser source used was 4.8 nm. at halfheight.

We have seen in Chapter 4 that the multiple layers deposited in the MCVD process give rise to oscillations in the refractive index, and that evaporation of dopant during preform collapse causes a refractive-index dip on the fiber axis. Both effects occur in single-mode fibers produced by this method and the axial dip in refractive index is serious because the optical power density is greatest on the axis.

All of these perturbations cause the maximum value of V that allows single-mode operation (V_c) to be increased above the value $V_c = 2.405$ given in condition (5.5.1). For the α-profile fibers discussed in Chapter 6, $V_c \simeq 2.405(1 + 2/\alpha)^{1/2}$, where α is defined on p. 152 by eqn. (6.1.8). The perturbations also change the waveguide dispersion parameter plotted in Fig. 5.12. The effect is in general beneficial in that it permits single-mode fibers to be made with larger core diameters. A more controlled way of achieving the same end is to use the refractive-index profile that was shown in Fig. 2.5(f), the so-called W-profile, or depressed cladding, fiber. Then a very low value for $\Delta \simeq (n_1 - n_2)/n_1$ can be used. Although a number of partially guided modes propagate, the evanescent fields penetrate the thin annular ring of low refractive index (n_3) so that all modes except one radiate into the cladding (n_2) and are very lossy. But the cladding fields of the one fully guided mode attenuate rapidly in the region of refractive index n_3 so that the very low Δ does not lead to high bending and microbending losses, but does permit a larger core radius which still satisfies condition (5.5.1).

The obvious disadvantage of single-mode operation comes from the small core diameter and the consequent very close tolerances needed when fibers are joined or terminated. However, the techniques for splicing and connecting fibers which we described in Section 4.3 have been shown to be capable of dealing with single-mode fibers. The necessary lateral tolerances can be relaxed somewhat if the value of V is reduced from around 2 to around 1. At this value some 70% of the total power is carried in the cladding, as can be seen in Fig. 5.11. Indeed, the cross-section occupied by the electromagnetic wave expands as the core diameter is reduced. At $V = 1$ the power density in the fiber is reduced to about $1/e$ of its maximum value at a radius of about $3a$. As we have seen, to make the whole fiber with such a low value of V would render it liable to excessive attenuation because of bending and microbending losses. But it has been suggested that the core diameter might be reduced, in order to reduce V, just in the neighborhood of joints. Whilst lateral alignment is relaxed, angular alignment then becomes more critical, and repairs are precluded from this scheme. Single-mode fibers are less susceptible than multimode fibers to losses caused by small, local lateral displacements of the core, but they are, on the other hand, more wavelength selective: if V increases beyond 2.4, they become multimode; if it reduces below 1.5, beam-spreading increases in the way we have just described. Operation at the dispersion minimum does, of course, require critical control of the source wavelength in any case. None of the difficulties appears intractable, and it seems likely that single-mode fibers will assume increasing importance in those trunk telecommunication applications which require very high bandwidths to be maintained over long transmission paths. Submarine cables are an important example.

PROBLEMS

5.1 Using eqns (5.2.19), (5.2.20) and (5.2.22), and Table 5.1 write down the normalized frequencies at which the twelve lowest order modes of propagation in step-index fibers suffer cutoff.

5.2 Two particular fibers have core and cladding refractive indices of 1.465 and 1.460, respectively. Their core diameters are 50 and 10 μm. In each case calculate the optical wavelengths which correspond to the cutoff frequencies of the lowest-order modes.

5.3 Evaluate the number of modes, M, that will propagate in the two fibers of Problem 5.2 when they are excited by sources of wavelength 1.55 μm and 0.85 μm. Allow for the two-fold degeneracy of each LP_{0m} mode and for the four-fold degeneracy of each $LP_{\kappa m}$ mode when $\kappa \neq 0$. It will be necessary to use tables of the roots of Bessel functions to answer this and Problem 5.4.

5.4 Make a graph of the number of propagating modes versus the normalized frequency V over the range $0 < V < 12.5$ and compare this with the curve of eqn. (5.3.23).

5.5 Compare the normalized frequencies V_c below which propagation is limited to a single mode in fibers having:
(a) a 'step-index' profile ($\alpha = \infty$)
(b) a 'parabolic' profile ($\alpha = 2$)
(c) a 'triangular' profile ($\alpha = 1$).

5.6 A particular fiber has a pure silica cladding and a core doped with a maximum of 13.5% germania, giving $\Delta = 0.0147$. The fiber is required to propagate in a single mode when excited by a laser source of wavelength 1.55 μm. Calculate the maximum permissible core diameter with each of the three profiles of Problem 5.5.

5.7 Confirm the values given in Table 5.2 for the step-index core diameters required to make $V = 2.0$ at each of the four wavelengths 0.85, 1.27, 1.35 and 1.55 μm in a fiber with $\Delta = 0.005$.

5.8 Sketch the variation of the core diameter ($2a$) and the index difference parameter (Δ) with operating wavelength for a single-mode, step-index fiber designed so that $V = 2$ and total dispersion $Y_m + Y_w = 0$. Use Fig. 5.13 to estimate Y_m, neglect any variation of Y_m with Δ and any variation of $n_2 = 1.445$ with λ, and cover the range of wavelengths 1.3–1.7 μm. (Using eqn. (5.5.10) with $N_2^2/n_2 = 1.48$, $\Delta = Y_w/0.74$; using eqn. (5.5.11), $2a = \lambda V/\pi n_2(2\Delta)^{\frac{1}{2}} = 0.31\lambda/\Delta^{\frac{1}{2}} = 0.27\lambda/Y_w^{\frac{1}{2}}$.)

5.9 Show that $\omega \mathrm{d}\Delta/\mathrm{d}\omega = (n_2 N_1 - n_1 N_2) n_1 / n_2^3$.
The core of a step-index fiber is 4 μm in diameter and is doped with 13.5%. GeO_2. The cladding is pure silica, giving $\Delta = 0.0147$. Show that $V = 2.0$ when $\lambda = 1.55$ μm. Compare the relative magnitudes of Y_m, Y_w and Y_d at this wavelength using the following data:

At 1.55 μm, $n_1 = 1.4655$, $N_1 = 1.4823$, $n_2 = 1.4444$, $N_2 = 1.4628$, $Y_{m_2} = 0.0101$.
At $V = 2.0$, $\mathrm{d}(Vb)/\mathrm{d}V = 1.06$, $\mathrm{d}^2(Vb)/\mathrm{d}V^2 = 0.48$.

If the fiber is excited by means of a laser source having an r.m.s. spectral width of 2 nm, calculate the bandwidth of a 10 km length.

REFERENCES

5.1 D. Marcuse, *Theory of Dielectric Optical Waveguides*, Academic Press (1974).
5.2 H.-G. Unger, *Planar Optical Waveguides and Fibres*, O.U.P. (1977).
5.3 M. J. Adams, *An Introduction to Optical Waveguides*, John Wiley (1981).

SUMMARY

Step-index fibers act as dielectric waveguides permitting specific modes of guided electromagnetic wave propagation characterized by the integer mode numbers (k, m). Each has a characteristic propagation constant β_{km}. With $\beta_1 = n_1\omega/c$ and $\beta_2 = n_2\omega/c$ (propagation constants of plane transverse electromagnetic (TEM) waves in the core and cladding respectively) $\beta_2 < \beta_{km} < \beta_1$.

Intermode (multipath) dispersion is the result of the variation of β_{km} among the (k, m) modes at a given optical frequency.

Intramode dispersion, $\tau_{km}(\omega)/l = (d^2\beta_{km}/d\omega^2)\Delta\omega$, is the algebraic sum of material dispersion $(Y_m + Y_d)$ and waveguide (Y_w) dispersion (see eqn. (5.4.14)).

For weakly guiding fibers, $\Delta \simeq (n_1 - n_2)/n_1 \ll 1$, the normalized frequency $V = (\beta_1^2 - \beta_2^2)a$ and the normalized propagation constant, $b_{\kappa m} = (\beta_{\kappa m}^2 - \beta_2^2)/(\beta_1^2 - \beta_2^2)$ enable solutions to be generalized.

The number of propagating modes is $M \simeq V^2/2$ when $M \gg 1$. These fall into $Q = 2V/\pi$ groups, each identified by the mode group number $q = \kappa + 2m$ and containing $2q$ modes. Then $\beta_q^2 \simeq \beta_1^2 - \pi^2 q^2/4a^2$.

For each mode there is a minimum cutoff frequency below which it will not propagate. For $V < 2.4048$ only one mode, HE_{11} (or LP_{01}), will propagate. Fibers made to satisfy this condition are known as single-mode, or monomode, fibers. They require a laser source, but suffer no intermode dispersion. The cutoff frequency is increased by departures from an ideal step-index profile.

Fibers may be designed with sufficient waveguide dispersion to shift the dispersion minimum to 1.5 or 1.6 μm (Fig. 5.13), but such fibers tend to have losses similar to conventional single-mode fibers at 1.3 μm.

6

Electromagnetic Wave Propagation in Graded-index Fibers

6.1 MODES IN GRADED-INDEX FIBERS

6.1.1 Introduction

In Section 2.1.3 we showed on the basis of a ray model that by profiling the variation of refractive index between the core and the cladding of an optical fiber, it was possible to cause rays following quite different trajectories to have similar propagation velocities along the fiber. Thus the effective pulse width during which most of the propagating power is to be found can be very significantly reduced. In this section we shall take the theory of the propagation of light in such inhomogeneous media a stage further.

Ideally we would wish to find solutions to the electromagnetic wave equations in graded-index fibers, and to derive expressions for their dispersive properties. However, any general analysis soon becomes intractable. For a start we note in Appendix 1 that in order to obtain the wave equation (A1.2) we put

$$\operatorname{div} \mathbf{E} = \operatorname{div} (\mathbf{D}/\varepsilon_r \varepsilon_0) = 0$$

In an inhomogeneous medium with ε_r a function of position this is not a valid assumption, because

$$\operatorname{div} (\mathbf{D}/\varepsilon_r \varepsilon_0) = \frac{1}{\varepsilon_0} \nabla \cdot (\mathbf{D}/\varepsilon_r) = \frac{1}{\varepsilon_0} \left[\frac{1}{\varepsilon_r} \nabla \cdot \mathbf{D} + \mathbf{D} \cdot \nabla \left(\frac{1}{\varepsilon_r} \right) \right] \tag{6.1.1}$$

With $\nabla \cdot \mathbf{D} = 0$ and putting $n = \sqrt{\varepsilon_r}$, this term becomes

$$\frac{\mathbf{D}}{\varepsilon_0} \cdot \nabla \left(\frac{1}{\varepsilon_r} \right) = -\frac{\mathbf{E}}{\varepsilon_r} \cdot \nabla \varepsilon_r = -\frac{\mathbf{E}}{n^2} \cdot \nabla (n^2) \tag{6.1.2}$$

and the wave equation takes on the form

$$\nabla^2 \mathbf{E} + \nabla\left[\frac{\mathbf{E}\cdot\nabla(n^2)}{n^2}\right] - \frac{n^2}{c^2}\frac{\partial^2 \mathbf{E}}{\partial t^2} = 0 \qquad (6.1.3)$$

A correction term involving $(1/n^2)\nabla(n^2)$ also appears in the equation for **H**.

In all graded-index fibers of practical interest $\nabla(n^2)$ is small enough to make these new terms negligible so that they have no significant effect on the propagation characteristics of the electromagnetic waves in the fiber. We shall assume this to be the case throughout the rest of the section. We shall, therefore, be seeking solutions to eqn. (A1.2(d)):

$$\nabla^2 \psi - \frac{n^2}{c^2}\frac{\partial^2 \psi}{\partial t^2} = 0 \qquad (6.1.4)$$

where ψ still stands for **E** or **H**, but where n is now a slowly varying function of radial position.

Exact solutions to equation (6.1.4) can be obtained in terms of known functions when the profile takes on the parabolic form

$$n(r) = n_0[1 - 2\Delta(r/a)^2]^{1/2} \quad \text{for } r \leqslant a$$

and

$$n(r) = n_0[1 - 2\Delta]^{1/2} = n_c \quad \text{for } r > a \qquad (6.1.5)$$

Note that n_0 is the refractive index on the fiber axis and that when $\Delta \ll 1$,

$$\Delta = \frac{n_0^2 - n_c^2}{2n_0^2} \simeq \frac{n_0 - n_c}{n_0} \qquad (6.1.6)$$

Rather than consider this special case in isolation we may deal with a general axi-symmetric index profile by using the form

$$n(r) = n_0[1 - 2\Delta f(r)]^{1/2} \quad \text{for } r \leqslant a$$
$$n(r) = n_0[1 - 2\Delta]^{1/2} = n_c \quad \text{for } r > a \qquad (6.1.7)$$

where $f(0) = 0$ and $f(a) = 1$.

One particular function, $f(r)$, that has been widely used, describes the so-called 'α-profile'. This is

$$f(r) = (r/a)^\alpha \qquad (6.1.8)$$

The profile parameter, α, may take on values between unity and infinity. The parabolic index of eqn. (6.1.5) is obtained by putting $\alpha = 2$, and $\alpha = \infty$ may be thought of as a limiting case representing a step-index profile. We should be aware, however, that putting $\alpha = \infty$ into eqns. (6.1.8) and (6.1.7) does lead to a

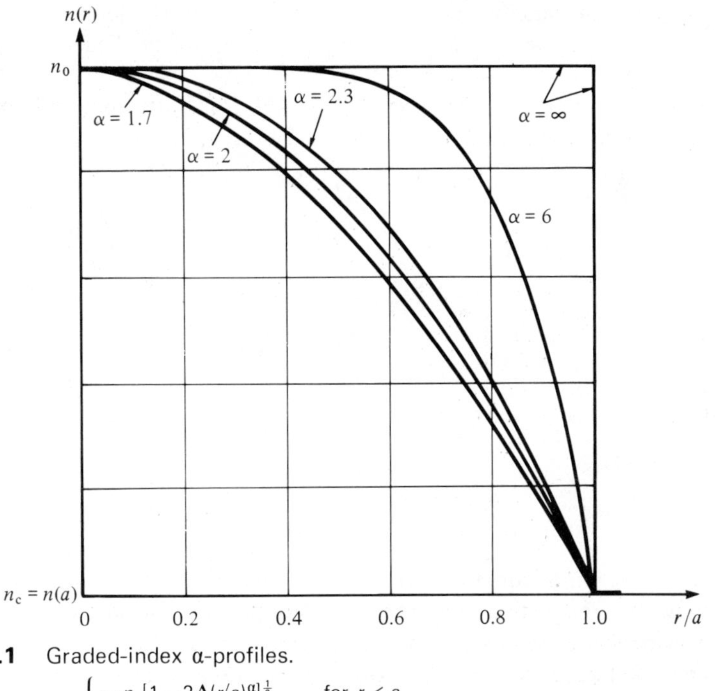

Fig. 6.1 Graded-index α-profiles.

$$n(r) \begin{cases} = n_0[1 - 2\Delta(r/a)^{\alpha}]^{\frac{1}{2}} & \text{for } r < a \\ = n_0[1 - 2\Delta]^{\frac{1}{2}} = n_c & \text{for } r > a \end{cases}$$

Refractive index profiles with $\alpha = 1.7, 2.0, 2.3, 6.0$ and the step-index, $\alpha = \infty$, are shown.

definition of Δ via eqn. (6.1.6) that is slightly different from the definition used in Chapter 5. Comparison of eqn. (6.1.6) with eqn. (5.4.2) will make this clear. In the limit of small Δ the difference becomes negligible. Some α-profiles are plotted in Fig. 6.1.

6.1.2 Approximate Solution

We next note from the discussion in Appendix 1 that the solution of the wave equation, (A1.2) or (6.1.4), demands that the radial distributions of the electric and magnetic fields, ψ_r, satisfy eqn. (A1.7), namely:

$$\frac{d^2\psi_r}{dr^2} + \frac{1}{r}\frac{d\psi_r}{dr} + \left[\frac{\omega^2}{c^2}n^2(r) - \beta^2 - \frac{k^2}{r^2}\right]\psi_r = 0 \tag{6.1.9}$$

In this equation β is the propagation constant which characterizes the periodicity of the fields in the axial direction through the term $e^{j\beta z}$; k is an integer which characterizes the azimuthal periodicity of the fields through the term $e^{jk\phi}$.

The first step in dealing with eqn. (6.1.9) is to eliminate the term in $d\psi_r/dr$. This may be achieved by means of the substitution

$$U = r^{1/2}\psi_r \tag{6.1.10}$$

Then eqn. (6.1.9) becomes

$$\frac{d^2 U}{dr^2} + \left[\frac{\omega^2 n^2(r)}{c^2} - \beta^2 - \frac{(k^2 - \frac{1}{4})}{r^2}\right] U = 0 \tag{6.1.11}$$

We shall write the new term in square brackets as K^2. Thus

$$K^2 = \frac{\omega^2 n^2(r)}{c^2} - \beta^2 - \frac{k^2 - \frac{1}{4}}{r^2} \tag{6.1.12}$$

An examination of eqn. (6.1.12) shows that for light of a particular angular frequency, ω, propagating in a mode having a given propagation constant, β, there will be some regions of the fiber where K is real and some where it is imaginary. Reference to Fig. 6.2 shows that K is real within a range of radii bounded by r_1 and r_2, where

$$\frac{\omega^2 n^2(r)}{c^2} - \beta^2 \geqslant \frac{k^2 - \frac{1}{4}}{r^2} \tag{6.1.13}$$

In Appendix 2 we show that within this region the field distributions vary periodically with radial distance, whereas outside these radii, they decay exponentially. By analogy with the discussion in Section 5.2, we associate such periodic field distributions with guided modes bound to the cylindrical shell bordered by r_1 and r_2, and fringed by decaying evanescent fields inside and outside these radii.

When $k = 0$, there is no lower boundary, r_1, and the periodic nature of the guided mode fields extends to the fiber axis. As before such modes may be thought to correspond to meridional rays; the fields have radial symmetry.

The periodic field distributions between r_1 and r_2 are required to match the decaying fields outside, and in consequence only certain discrete values of the propagation constant, β, satisfy the boundary conditions and give rise to allowed modes of propagation. As before, we shall designate these permitted values of β as β_{km}. The integer k is the azimuthal mode number. The integer m represents the number of half-periods of ψ_r between r_1 and r_2. In Appendix 2

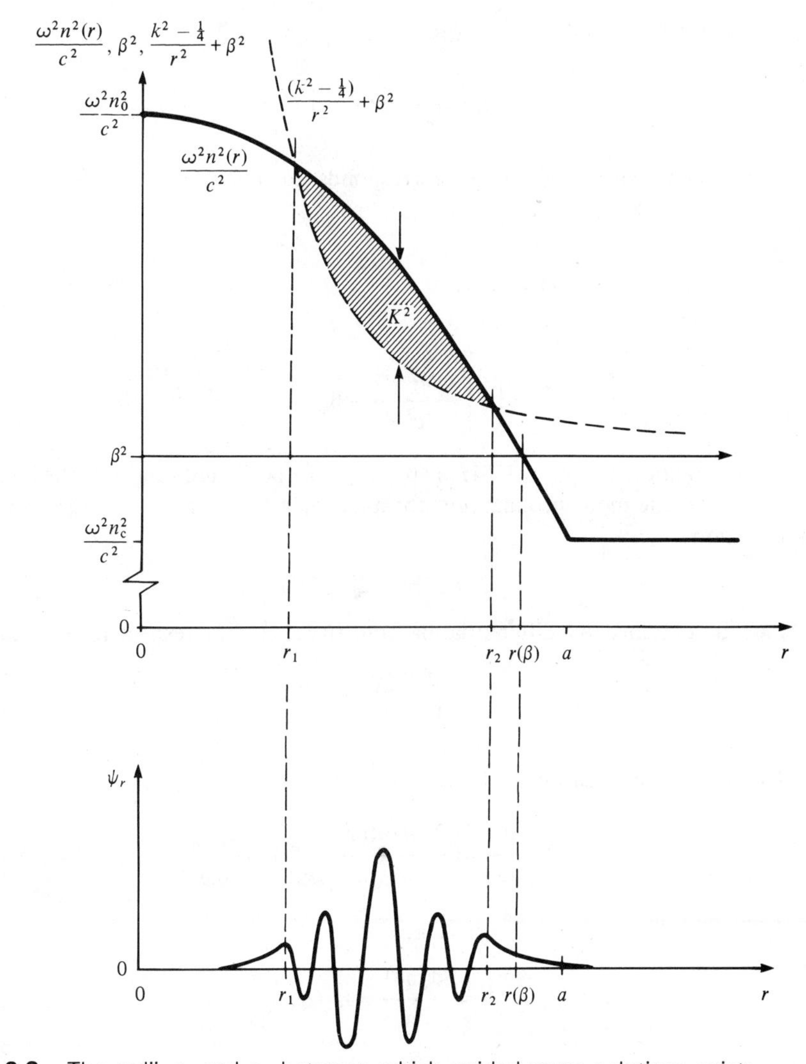

Fig. 6.2 The radii, r_1 and r_2, between which guided wave solutions exist: (a) diagram of $[\omega^2 n^2(r)]/c^2$, β^2 and $[(k^2 - \frac{1}{4})/r^2] + \beta^2$ plotted against r, indicating the region bounded by r_1 and r_2 in which

$$K = \left[\frac{\omega^2 n^2(r)}{c^2} - \beta^2 - \frac{(k^2 - \frac{1}{4})}{r^2}\right]^{\frac{1}{2}}$$

is real;
(b) schematic illustration showing the oscillatory solutions for the guided wave fields between r_1 and r_2 and the decaying evanescent fields outside this region.

we have shown that

$$\psi_r = \frac{1}{r^{1/2}} \exp\{j \int K \, dr\} \tag{A2.14}$$

Thus each half-period of ψ_r corresponds to $\int K \, dr = \pi$, and m is given approximately by

$$m\pi \simeq \int_{r_1}^{r_2} K \, dr \tag{6.1.14}$$

$$\simeq \int_{r_1}^{r_2} \left[\frac{\omega^2 n^2(r)}{c^2} - \beta_{km}^2 - \frac{(k^2 - \frac{1}{4})}{r^2} \right]^{1/2} dr$$

The solution of eqn. (6.1.14) for β_{km} over the permitted ranges of the integers k, m gives the mode-propagation constants as a function of ω. Mode propagation times,

$$t_{km} = l \, d\beta_{km}/d\omega \tag{2.2.40}$$

may be obtained by differentiating eqn. (6.1.14) with respect to ω. Thus,

$$\int_{r_1}^{r_2} \frac{dK}{d\omega} \, dr = 0 \tag{6.1.15}$$

Differentiating equation (6.1.12) gives

$$2K \frac{dK}{d\omega} = \frac{2\omega n}{c^2} \frac{d(\omega n)}{d\omega} - 2\beta_{km} \frac{d\beta_{km}}{d\omega} \tag{6.1.16}$$

Then putting

$$N = \frac{d(\omega n)}{d\omega} = n + \omega \frac{dn}{d\omega} \tag{2.2.27}$$

we obtain

$$\int_{r_1}^{r_2} \frac{dK}{d\omega} \, dr = \int_{r_1}^{r_2} \frac{\omega n N}{K c^2} \, dr - \int_{r_1}^{r_2} \frac{\beta_{km}}{K} \frac{d\beta_{km}}{d\omega} \, dr = 0 \tag{6.1.17}$$

Hence

$$\beta_{km} \frac{d\beta_{km}}{d\omega} \int_{r_1}^{r_2} \frac{dr}{K} = \frac{\omega}{c^2} \int_{r_1}^{r_2} \frac{nN}{K} \, dr$$

and

$$\frac{t_{km}}{l} = \frac{\mathrm{d}\beta_{km}}{\mathrm{d}\omega} = \frac{\int_{r_1}^{r_2} (nN/K)\,\mathrm{d}r}{(c^2/\omega)\beta_{km} \int_{r_1}^{r_2} (1/K)\,\mathrm{d}r} \tag{6.1.18}$$

With

$$n^2 = n_0^2[1 - 2\Delta f(r)] \tag{6.1.7}$$

we may put

$$nN = n^2 + n\omega \frac{\mathrm{d}n}{\mathrm{d}\omega} = n^2 + \frac{\omega}{2} \frac{\mathrm{d}(n^2)}{\mathrm{d}\omega}$$

and provided $f(r)$ is not a function of ω,

$$nN = n_0^2[1 - 2\Delta f(r)] + \omega n \frac{\mathrm{d}n_0}{\mathrm{d}\omega}[1 - 2\Delta f(r)] - \omega n_0^2 f(r) \frac{\mathrm{d}\Delta}{\mathrm{d}\omega}$$

$$= n_0 N_0 \left[1 - 2\Delta f(r) - \frac{n_0}{N_0} \Delta f(r) \frac{\omega}{\Delta} \frac{\mathrm{d}\Delta}{\mathrm{d}\omega}\right]$$

$$= n_0 N_0 [1 - 2\Delta(1 + \delta) f(r)] \tag{6.1.19}$$

where

$$\delta = \frac{n_0}{2N_0} \frac{\omega}{\Delta} \frac{\mathrm{d}\Delta}{\mathrm{d}\omega} \tag{6.1.20}$$

is a dispersion parameter of the fiber which is similar to but not quite the same as that introduced in eqn. (5.4.9).

Equation (6.1.18) then becomes

$$\frac{t_{km}}{l} = \frac{\omega n_0 N_0}{c^2 \beta_{km}} \left\{1 - \frac{2\Delta(1 + \delta) \int_{r_1}^{r_2} (f(r)/K)\,\mathrm{d}r}{\int_{r_1}^{r_2} (1/K)\,\mathrm{d}r}\right\} \tag{6.1.21}$$

Equation (6.1.21) can be evaluated in principle for known values of n_0, N_0, δ, Δ and $f(r)$. Clearly this would not be an easy task, so we shall seek simpler and more explicit expressions for β_{km} and $\mathrm{d}\beta_{km}/\mathrm{d}\omega$, for α-profiles, using a rather more circuitous argument.

6.1.3 Number of Propagating Modes

Referring again to eqn. (6.1.14), we see that the largest value of m arises when $k = 0$ (implying that $r_1 = 0$) and when β takes on its smallest possible value. The largest value of k also occurs when β is a minimum, but when $K^2 = 0$. Because the third term in the expression for K goes to infinity when $r = 0$, $m_{\max}$ cannot be obtained without some difficulty. However, we can more easily derive a value for $k_{\max}$ and for the total number of propagating modes, M.

The values of β must lie between the value of the propagation constant for plane TEM waves in the cladding, β_c, and the value of the propagation constant for plane TEM waves in the material of the fiber axis, β_0. That is,

$$\beta_0 = \frac{\omega n_0}{c} > \beta > \frac{\omega n_c}{c} = \beta_c \tag{6.1.22}$$

To estimate $k_{\max}$, we put $\beta = \beta_c$, $K^2 = 0$ and neglect $\frac{1}{4}$ in comparison with $k^2_{\max}$. Then

$$k_{\max} = r\left[\frac{\omega^2 n^2}{c^2} - \beta_c^2\right]^{1/2} \tag{6.1.23}$$

The total number of propagating modes, M, may be estimated by summing the values of m given by eqn. (6.1.14) with $\beta = \beta_c$ over all possible values of k. When $k_{\max}$ is large, as in a typical multimode, graded-index fiber, the summation may be replaced by an integration over k, and $k^2 - \frac{1}{4}$ may again be approximated by k^2. Each of the modes described by a given pair of values of (k, m) is four-fold degenerate, since two directions of polarization and two orthogonal orientations of the fields are possible for each. Thus, to obtain M, the double integral must be multiplied by a factor of 4, giving

$$M = 4\int_{k=0}^{k_{\max}} m \, \mathrm{d}k$$

$$= \frac{4}{\pi}\int_{k=0}^{k_{\max}}\int_{r=r_1}^{r_2} \left[\frac{\omega^2 n^2(r)}{c^2} - \beta_c^2 - \frac{k^2}{r^2}\right]^{1/2} \mathrm{d}r \, \mathrm{d}k \tag{6.1.24}$$

Changing the order of integration and putting in the full range of possible values of r gives

$$M = \frac{4}{\pi}\int_{r=0}^{a}\int_{k=0}^{k_{\max}} \frac{1}{r}(k^2_{\max} - k^2)^{1/2} \, \mathrm{d}k \, \mathrm{d}r$$

$$=\frac{4}{\pi}\int_{r=0}^{a}\frac{1}{r}\left[\tfrac{1}{2}k_{\max}^2\sin^{-1}(k/k_{\max})+\tfrac{1}{2}k(k_{\max}^2-k^2)^{1/2}\right]_0^{k_{\max}}\mathrm{d}r$$

$$=\frac{4}{\pi}\int_{r=0}^{a}\frac{1}{r}\frac{\pi k_{\max}^2}{4}\,\mathrm{d}r$$

$$=\int_{r=0}^{a} r\left[\frac{\omega^2 n^2(r)}{c^2}-\frac{\omega^2 n_c^2}{c^2}\right]\mathrm{d}r \tag{6.1.25}$$

Substituting the α-profile for $n(r)$ gives

$$M=\int_{r=0}^{a} r\,\frac{2\omega^2 n_0^2}{c^2}\,\Delta[1-(r/a)^{\alpha}]\,\mathrm{d}r$$

$$=\frac{2\omega^2 n_0^2}{c^2}\,\Delta\left[\frac{r^2}{2}-\frac{r^{(\alpha+2)}}{a^{\alpha}(\alpha+2)}\right]_0^a$$

$$=2\beta_0^2\Delta a^2\left(\frac{1}{2}-\frac{1}{\alpha+2}\right)$$

$$=\frac{\alpha}{(\alpha+2)}\,a^2\beta_0^2\Delta \tag{6.1.26}$$

When $\alpha=\infty$, corresponding to a step-index fiber, the number of modes given by eqn. (6.1.26) is

$$M_{\infty}=a^2\beta_0^2\Delta=\frac{a^2\omega^2 n_0^2\Delta}{c^2} \tag{6.1.27}$$

$$=\frac{a^2\omega^2(n_0^2-n_c^2)}{2c^2}=\frac{1}{2}\left(\frac{2\pi a}{\lambda}\right)^2(n_1^2-n_2^2) \tag{5.3.23}$$

where we have substituted for Δ using eqn. (6.1.6) and put $n_0=n_1$ and $n_c=n_2$ in order to show that there is complete agreement with the value for M_{∞} quoted in Section 5.3. Thus

$$M=\frac{\alpha}{(\alpha+2)}M_{\infty} \tag{6.1.28}$$

We note in passing that the normalized frequency, V, may be defined for a graded-index fiber by

$$V = \frac{a\omega}{c}(n_0^2 - n_c^2)^{1/2}$$

by analogy with eqn. (5.2.16), so that

$$V = \frac{a\omega}{c} n_0 (2\Delta)^{1/2} = a\beta_0 (2\Delta)^{1/2} \tag{6.1.29}$$

Thus,

$$M_\infty = \tfrac{1}{2}V^2 \tag{5.3.21}$$

still applies.

When $\alpha = 2$, corresponding to a parabolic-index fiber, only half this number of modes can propagate. The implication is that when a step-index and a parabolic-index fiber having the same core diameter and the same total variation of refractive index are illuminated by a source which excites all modes equally, then the graded-index fiber will propagate only half the total power carried by the step-index fiber. Its effective numerical aperture is reduced by a factor of two.

6.1.4 Variation of the Propagation Constant

As a first step towards determining the dispersion properties of graded-index fibers, we may estimate the number of modes which, for light of a given angular frequency, ω, have a propagation constant, β, greater than some specified value, β_1. We simply need to evaluate eqn. (6.1.24) with β_c replaced by β_1 and with the maximum radius in which guided modes propagate changed from a to $r_2(\beta_1)$. We shall call this number of modes $p(\beta_1)$. Then

$$p(\beta_1) = \int_{r=0}^{r_2(\beta_1)} r \left[\frac{\omega^2 n^2(r)}{c^2} - \beta_1^2 \right] \mathrm{d}r \tag{6.1.30}$$

For α-profile fibers $r_2(\beta_1)$ is given by

$$\frac{\omega^2 n^2(r_2)}{c^2} = \beta_1^2 \tag{6.1.31}$$

$$\therefore \quad \frac{\omega^2 n_0^2}{c^2}[1 - 2\Delta(r_2/a)^\alpha] = \beta_0^2[1 - 2\Delta(r_2/a)^\alpha] = \beta_1^2 \tag{6.1.32}$$

$$\therefore \quad (r_2/a)^\alpha = (1 - \beta_1^2/\beta_0^2)/2\Delta = (\beta_0^2 - \beta_1^2)/2\Delta\beta_0^2 \tag{6.1.33}$$

Using this limit,

$$p(\beta_1) = \int_0^{r_2(\beta_1)} r\left(\frac{\omega^2 n_0^2}{c^2}\,[1 - 2\Delta(r/a)^\alpha] - \beta_1^2\right) dr$$

$$= \int_0^{r_2(\beta_1)} [(\beta_0^2 - \beta_1^2)r - 2\Delta\beta_0^2(r/a)^\alpha r]\, dr$$

$$= \left[(\beta_0^2 - \beta_1^2)\frac{r_2^2}{2} - \frac{2\Delta\beta_0^2 a^2 (r_2/a)^{(\alpha+2)}}{(\alpha + 2)}\right]$$

$$= \left[\frac{(\beta_0^2 - \beta_1^2)}{2} a^2 \frac{(\beta_0^2 - \beta_1^2)^{2/\alpha}}{(2\Delta\beta_0^2)^{2/\alpha}} - \frac{2\Delta\beta_0^2 a^2}{(\alpha + 2)} \frac{(\beta_0^2 - \beta_1^2)^{(\alpha+2)/\alpha}}{(2\Delta\beta_0^2)^{(\alpha+2)/\alpha}}\right]$$

$$= \left[\frac{(\beta_0^2 - \beta_1^2)}{(2\Delta\beta_0^2)}\right]^{(\alpha+2)/\alpha} \left[\frac{a^2 2\Delta\beta_0^2}{2} - \frac{2\Delta\beta_0^2 a^2}{(\alpha + 2)}\right]$$

$$= \frac{\alpha}{(\alpha + 2)} a^2 \Delta\beta_0^2 \left[\frac{\beta_0^2 - \beta_1^2}{2\Delta\beta_0^2}\right]^{(\alpha+2)/\alpha} \qquad (6.1.34)$$

Note that

$$\frac{p(\beta_1)}{M} = \left[\frac{\beta_0^2 - \beta_1^2}{2\Delta\beta_0^2}\right]^{(\alpha+2)/\alpha} \qquad (6.1.35)$$

Equation (6.1.35) expresses the fraction of the total number of propagating modes that have a propagation constant greater than β_1. There is no advantage in retaining the subscript for β_1 in this equation and we shall write $p(\beta)$ to represent the number of modes having a propagation constant greater than β. As we found with step-index fibers in the last section, when we are considering high-order modes ($k \gg 1$, $m \gg 1$) far from cutoff, the permitted fractional variation in propagation constant is small for any given mode compared to the variation between modes. We may, therefore, rearrange eqn. (6.1.35), dropping the β_1 subscript, and use it as an approximate expression for the propagation constant of the pth mode:

$$\beta = \beta_0\left[1 - 2\Delta\left(\frac{p}{M}\right)^{\alpha/(\alpha+2)}\right]^{1/2} \qquad (6.1.36)$$

For step-index fibers, $\alpha = \infty$ and eqn. (6.1.36) becomes

$$\beta = \beta_0[1 - 2\Delta\, p/M]^{1/2} \qquad (6.1.37)$$

We have already obtained an expression for the propagation constants of modes in weakly guiding step-index fibers in terms of the mode group number, q:

$$\beta = \beta_0 \left[1 - \frac{\pi^2}{4a^2}\frac{q^2}{\beta_0^2}\right]^{1/2} \tag{5.3.17}$$

We now express the core propagation constant as β_0. Using eqns. (5.3.20) and (6.1.28), we have

$$V = \frac{\pi Q}{2} = \beta_0 a (2\Delta)^{1/2} \tag{6.1.38}$$

so that eqn. (5.3.17) becomes

$$\beta = \beta_0 [1 - 2\Delta(q/Q)^2]^{1/2} \tag{6.1.39}$$

Making comparison between eqn. (6.1.39) and eqn. (6.1.37) and bearing in mind that we have already established that $M \simeq Q^2$, it is clear that $p \simeq q^2$ for the step-index fiber. The result $(p/M) \simeq (q/Q)^2$ can be shown to hold for all α-profile fibers, so that eqn. (6.1.36) takes the form

$$\beta = \beta_0 [1 - 2\Delta(q/Q)^{2\alpha/(\alpha+2)}]^{1/2} \tag{6.1.40}$$

Then from eqn. (6.1.26) we obtain

$$Q \simeq \left(\frac{\alpha\Delta}{\alpha + 2}\right)^{1/2} a\beta_0 \tag{6.1.41}$$

Equation (6.1.40) gives us an approximate expression for the propagation constant of those modes having the mode group number, $q = k + 2m$. For $q \gg 1$ this group comprises approximately $2q$ modes. We expect all the modes of such a group to propagate with a similar phase velocity: this was shown to be so in the special case of the step-index fiber ($\alpha = \infty$) in Section 5.3 and has been shown in a more rigorous analysis† to be valid also for the more general α-profile.

Equation (6.1.40) will be used in Section 6.3 to derive an expression for the dispersion in multimode α-profile fibers and hence to determine the value of α that will minimize this dispersion. But we see at once from this equation the way in which the various groups of modes are distributed with respect to their propagation constants. Take first the case of the step-index fiber with $\alpha = \infty$

† C. N. Kurtz and W. Streifer, Guided waves in inhomogeneous focusing media. Part II: Asymptotic solution for general weak inhomogeneity, *IEEE Trans. on Microwave Theory and Techniques*, **MTT-17**, 250–3 (1969).

and

$$\beta = \beta_0[1 - 2\Delta(q/Q)^2]^{1/2} \simeq \beta_0[1 - \Delta(q/Q)^2] \qquad (6.1.42)$$

The difference in propagation constant between adjacent mode groups is given by

$$\Delta\beta = \frac{d\beta}{dq} \simeq -2\beta_0 \Delta q/Q^2 \qquad (6.1.43)$$

The separation between the mode groups increases as q increases, but because each group consists of $2q$ modes the average distribution of the modes with respect to β is constant.

Consider next the parabolic distribution with $\alpha = 2$. Now

$$\beta = \beta_0[1 - 2\Delta(q/Q)]^{1/2} \simeq \beta_0[1 - \Delta q/Q] \qquad (6.1.44)$$

and

$$\Delta\beta = \frac{d\beta}{dq} \simeq -\Delta\beta_0/Q \qquad (6.1.45)$$

so that it is now the mode groups which are evenly distributed with respect to β.

6.2 THE EQUIVALENCE OF THE WKB APPROXIMATION AND THE RAY MODEL

In order to obtain useful solutions to the wave equation in Sections 5.3 and 6.1, respectively for multimode step- and graded-index fibers, we have found it necessary to restrict the discussion in three ways. We have been able to consider only higher-order modes at frequencies far from cutoff in weakly guiding fibers. Under these conditions we have found that the solutions approximate locally to linearly polarized, plane, transverse electromagnetic waves. These conditions are just those required for optical propagation to be described by a ray model, and these two apparently very different approaches can be shown to be equivalent.

Points of constant phase in a plane TEM wave form planes perpendicular to the direction of propagation. These planes are referred to as wavefronts, and a 'ray' may be thought of as a trajectory, orthogonal to the wavefronts and travelling in the direction of propagation, as in Fig. 5.4.

The equation for the trajectory of a ray representing waves propagating through a slowly varying, inhomogeneous medium has already been quoted as

$$\frac{\mathrm{d}}{\mathrm{d}s}\left(n\frac{\mathrm{d}\mathbf{r}}{\mathrm{d}s}\right) = \nabla n \tag{2.1.21}$$

Here $\mathbf{r}(s)$ is the general position vector of a point on the ray, s is the distance of this point measured along the ray trajectory and n is the refractive index of the medium. Note that $\mathrm{d}\mathbf{r}/\mathrm{d}s$ is a unit vector tangential to the ray at location s. In this section we shall use this equation to establish the behavior of a general skew ray in a fiber with a graded-index core, in which n has radial symmetry. This makes it convenient to express the general ray position, $\mathbf{r}$, in cylindrical polar coordinates, (r, ϕ, z), based on the fiber axis.

Consider the ray shown in Fig. 6.3(a). This enters the fiber at the point $P(r_0, \phi_0, 0)$ and its initial trajectory makes angles α_0, β_0 and γ_0 with the coordinate axes Ox, Oy, and Oz, respectively.

We show in Section A3.1 of Appendix 3 that the vector equation (2.1.21) may be re-expressed in terms of the following three scalar differential equations:

$$n(r)\frac{\mathrm{d}z}{\mathrm{d}s} = n(r_0)\left(\frac{\mathrm{d}z}{\mathrm{d}s}\right)_{z=0} = E \tag{6.2.1}$$

$$r^2\frac{\mathrm{d}\phi}{\mathrm{d}z} = r_0^2\left(\frac{\mathrm{d}\phi}{\mathrm{d}z}\right)_{z=0} = l \tag{6.2.2}$$

and

$$\frac{\mathrm{d}^2 r}{\mathrm{d}z^2} = \frac{l^2}{r^3} + \frac{n}{E^2}\frac{\mathrm{d}n}{\mathrm{d}r} \tag{6.2.3}$$

where

$$E = n(r_0)\cos\gamma_0 \tag{6.2.4}$$

and

$$l = r_0 \sec\gamma_0 (\cos\beta_0 \cos\phi_0 - \cos\alpha_0 \sin\phi_0) \tag{6.2.5}$$

E and l are constants which define the trajectory of the ray through the fiber. They, and thus the trajectory for a given fiber, depend only on the initial conditions, that is on the position and direction of the ray as it enters the fiber at $z = 0$.

In Section A3.2 of Appendix 3 we show how eqn. (6.2.3) can be integrated

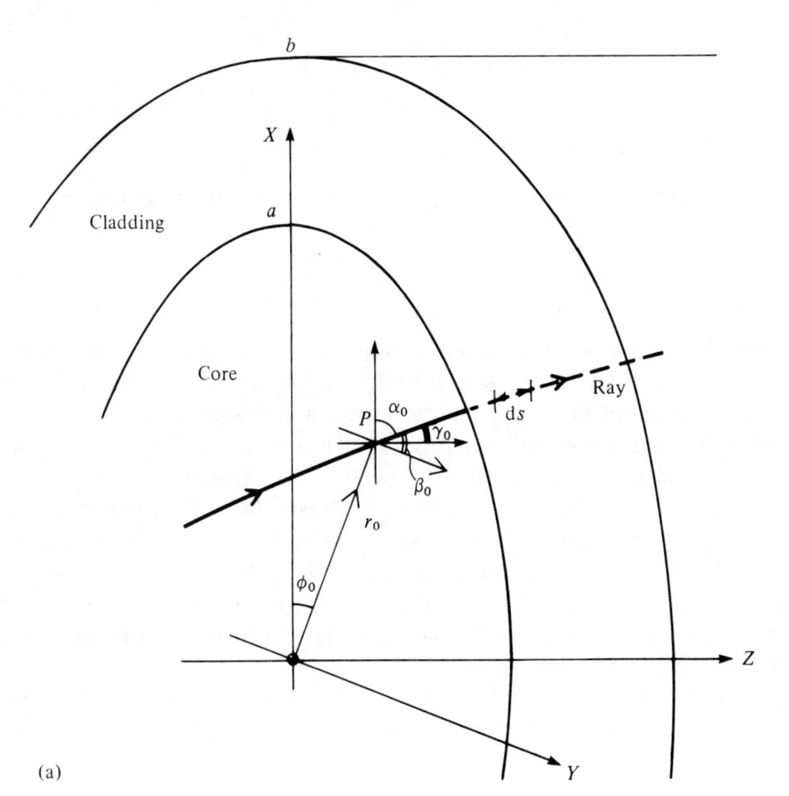

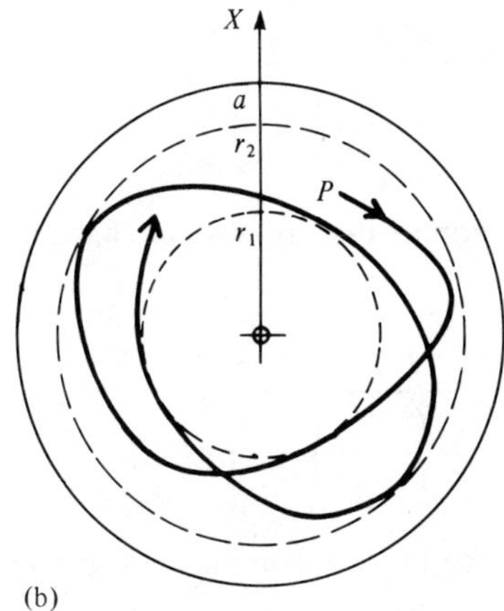

Fig. 6.3 A ray entering a GI fiber at point P and propagating along the core: (a) ray enters fiber at P, coordinates $(r_0, \phi_0, 0)$ with direction cosines $(\alpha_0, \beta_0, \gamma_0)$; (b) transverse section looking along the fiber axis and showing the ray trajectory within the fiber core, bounded by the caustics at r_1 and r_2.

to give

$$\frac{dr}{dz} = \left(\frac{n^2(r)}{E^2} - \frac{l^2}{r^2} - 1 \right)^{1/2} \tag{6.2.6}$$

and we define the radii r_1 and r_2 to be the roots of the equation

$$n^2(r) - \frac{E^2 l^2}{r^2} - E^2 = 0 \tag{6.2.7}$$

A number of things can be deduced immediately from eqns. (6.2.6) and (6.2.7). First, r_1 and r_2 are functions of the refractive index profile, $n(r)$, and the initial launch conditions of the ray, as given by E and l. Secondly, dr/dz is real only within the region bounded by r_1 and r_2. Thirdly, the radii r_1 and r_2 represent turning points at which $dr/dz = 0$ and the ray trajectory has no radial component. Fourthly, ray trajectories for which eqn. (6.2.7) has two real roots are confined to the region between r_1 and r_2; trajectories which do not satisfy this condition will not be guided by the fiber. The cylindrical surfaces of radii r_1 and r_2 bounding the ray trajectories are known in optics as *caustics*. Fifthly, the condition that rays should pass through the fiber axis (the condition for meridional rays) is seen to be $l = 0$.

A further integration of eqn. (6.2.6) shows that the ray trajectory will be periodic in r over an axial distance z_0. In Appendix 3, we show that z_0 is given by

$$z_0 = 2\int_{r_1}^{r_2} \frac{dr}{\left(\frac{n^2(r)}{E^2} - \frac{l^2}{r^2} - 1 \right)^{1/2}} \tag{6.2.8}$$

Over the distance z_0 the azimuthal position of the ray always changes by the same angle, Φ, given by

$$\Phi = 2\int_{r_1}^{r_2} \frac{(l/r^2)\, dr}{\left(\frac{n^2(r)}{E^2} - \frac{l^2}{r^2} - 1 \right)^{1/2}} \tag{6.2.9}$$

Thus the ray paths along the fiber have the form of complicated spiral helices, as shown in Fig. 6.3(b).

It is a fundamental principle of optics, deriving from Fermat's principle of least time, that in any medium all rays which start from a point X and arrive at another point Y travel from X to Y in the same time, independent of the path taken. Applying this to the fiber, we see that if it were possible to find a profile which would make z_0 and Φ constant and equal to each other for all values of E and l, then a fiber with such a profile would be free from multipath dispersion. It is known that no such profile exists. However, the variation in z_0 and Φ over the range of values of E and l of rays guided by the fiber is a measure of the fiber dispersion. Minimizing the total spread in the values of z_0 and Φ minimizes the multipath dispersion.

The dispersion of α-profile fibers which are free from material dispersion has been treated by this method,† giving identical results to those that we shall present in Section 6.3 using a mode analysis. The ray model provides a much simpler and clearer physical picture of the propagation in optical fibers, but it does not permit the intimate relationship between the different causes of dispersion that we shall discuss in Section 6.5 to be brought out clearly.

6.3 INTERMODE DISPERSION IN GRADED-INDEX FIBERS

6.3.1 Neglecting Material Dispersion

In the ray model fiber dispersion is represented by the range of variation in optical path lengths among rays travelling along different trajectories. Using the model of guided wave modes, it is represented by the range of values of $d\beta/d\omega$ for the various permitted modes. Within the approximations we have used for graded-index fibers, the two models are equivalent and are found to lead to the same result for fiber dispersion. Here we shall approach the problem from the wave model. We shall start from eqn. (6.1.40) for the propagation constant, β_q, of the group of modes having the mode group number q. We shall seek expressions for the dispersion in α-profile fibers, and hence obtain the value of α that will minimize this dispersion. We shall assume that all modes suffer equal attenuation. Thus,

$$\beta_q = \beta_0[1 - 2\Delta(q/Q)^{2\alpha/(\alpha+2)}]^{1/2} \tag{6.1.40}$$

where

$$Q = \left[\frac{\alpha\Delta}{(\alpha+2)}\right]^{1/2} \beta_0 a \tag{6.1.41}$$

and

$$\beta_0 = \frac{n_0 \omega}{c} \tag{6.1.22}$$

† M. Eve, Rays and time dispersion in multimode graded core fibers, *Optical and Quantum Electronics*, **8**, 285–93 (1976).

We shall simplify the notation as follows:

Put

$$\beta_q = \beta_0[1 - 2\Delta\xi]^{1/2} \tag{6.3.1}$$

where

$$\xi = (q/Q)^{2\alpha/(\alpha+2)}$$

$$= \left[\frac{(\alpha+2)}{\alpha\Delta}\right]^{\alpha/(\alpha+2)} \left[\frac{cq}{n_0 \omega a}\right]^{2\alpha/(\alpha+2)} \tag{6.3.2}$$

An impulse of light of optical angular frequency, ω, launched into the modes of group q, propagates an axial distance, l, along the fiber in a time, t_q, given by:

$$t_q = l\frac{d\beta_q}{d\omega} \tag{6.3.3}$$

Consider first a fiber made from hypothetical materials which have no material dispersion. As a result, neither n_0 nor Δ is a function of the optical frequency, so we may write

$$\xi = K_1 \omega^{-2\alpha/(\alpha+2)} \tag{6.3.4}$$

where K_1 is a parameter that is independent of the optical frequency. Differentiating eqn. (6.3.1) leads to

$$\frac{t_q}{l} = \frac{d\beta_0}{d\omega}[1 - 2\Delta\xi]^{1/2} + \tfrac{1}{2}\beta_0[1 - 2\Delta\xi]^{-1/2}(-2\Delta)\frac{d\xi}{d\omega} \tag{6.3.5}$$

Now, differentiating eqn. (6.3.4) gives

$$\frac{d\xi}{d\omega} = -\frac{2\alpha}{(\alpha+2)}\frac{\xi}{\omega} \tag{6.3.6}$$

$$\therefore \quad \frac{t_q}{l} = \frac{n_0}{c}[1 - 2\Delta\xi]^{1/2} - \frac{n_0\omega}{c}\Delta[1 - 2\Delta\xi]^{-1/2}\frac{(-2\alpha)}{(\alpha+2)}\frac{\xi}{\omega}$$

and

$$\frac{ct_q}{n_0 l} = [1 - 2\Delta\xi]^{-1/2}\left[1 - 2\Delta\xi + \frac{2\alpha\Delta}{(\alpha+2)}\xi\right]$$

$$\simeq [1 + \Delta\xi + \tfrac{3}{2}(\Delta\xi)^2 + \cdots]\left[1 - \frac{4\Delta\xi}{\alpha+2}\right]$$

$$= 1 + \frac{(\alpha - 2)}{(\alpha + 2)} \Delta\xi + \frac{(3\alpha - 2)}{(\alpha + 2)} \frac{(\Delta\xi)^2}{2} + \cdots \tag{6.3.7}$$

The extreme mode groups are characterized by $q = 2$ and $q = Q$—that is, by $\xi \simeq 0$ and $\xi = 1$—and their propagation times are given by

$$\frac{ct_0}{n_0 l} \simeq 1$$

and (6.3.8)

$$\frac{ct_Q}{n_0 l} = 1 + \frac{(\alpha - 2)}{(\alpha + 2)} \Delta + \frac{(3\alpha - 2)}{(\alpha + 2)} \frac{\Delta^2}{2} + \cdots$$

Except for fibers in which α is in the region of 2, these mode groups are the ones having the fastest and the slowest propagation velocities. Then the mode dispersion is simply the difference between t_0 and t_Q. To first order it is

$$\Delta T = |t_0 - t_Q| = \frac{n_0 l}{c} \frac{|\alpha - 2|}{(\alpha + 2)} \Delta \tag{6.3.9}$$

For $\alpha > 2$, the higher-order mode groups travel more slowly than the lower-order groups and the fiber is undercompensated for mode dispersion. For $\alpha < 2$ the converse is true and the fiber is overcompensated. In fact the dispersion

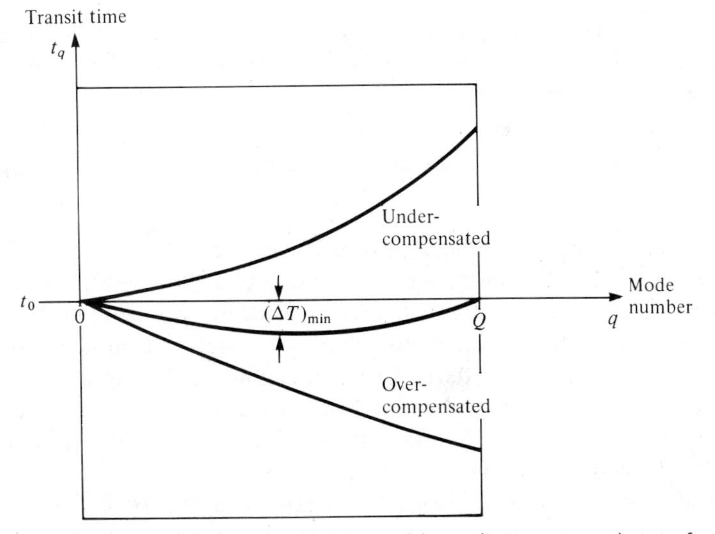

Fig. 6.4 Variation of propagation time, t_q, with mode-group number, q, for over- and undercompensated, α-profile fibers.

minimum occurs at a value of α rather less than 2, when t_0 and t_Q have the same value but the intermediate mode groups travel more quickly. This is illustrated in Fig. 6.4. The condition for minimum dispersion is then

$$t_0 = t_Q \tag{6.3.10}$$

$$\therefore \quad 1 = 1 + \frac{(\alpha - 2)}{(\alpha + 2)}\Delta + \frac{(3\alpha - 2)}{(\alpha + 2)}\frac{\Delta^2}{2}$$

$$\therefore \quad 2\alpha - 4 + (3\alpha - 2)\Delta = 0$$

$$\begin{aligned}\therefore \quad \alpha = \frac{4 + 2\Delta}{2 + 3\Delta} &= (2(1 + \tfrac{1}{2}\Delta)(1 + \tfrac{3}{2}\Delta)^{-1} \\ &\simeq 2(1 + \tfrac{1}{2}\Delta)(1 - \tfrac{3}{2}\Delta + \cdots) \\ &\simeq 2(1 - \Delta)\end{aligned} \tag{6.3.11}$$

With this value of α, substitution of eqn. (6.3.11) into (6.3.7) shows that

$$\begin{aligned}\frac{c}{n_0 l}(t_0 - t_q) &= \frac{2\Delta^2}{(4 - 2\Delta)}\xi - \frac{(4 - 6\Delta)}{(4 - 2\Delta)}\frac{\Delta^2}{2}\xi^2 \\ &\simeq \frac{\Delta^2}{2}\xi(1 - \xi)\end{aligned} \tag{6.3.12}$$

This difference in propagation time between the lowest-order mode and the qth mode group is, of course, zero for $\xi = 0$ and $\xi = 1$ and reaches its maximum value at $\xi = \frac{1}{2}$ when

$$t_0 - t_q = \Delta T_{\text{min}} = \frac{n_0 l \Delta^2}{8c} \tag{6.3.13}$$

Obtaining the optimum value of α can therefore have a considerable and quite remarkable effect on the mode dispersion of a graded-index fiber. Comparison with eqn. (2.1.18) shows that it is reduced to a fraction ($\Delta/8$) of that of a step-index fiber having the same values of n_0 and Δ. Consider a fiber for which $n_0 = 1.5$ and $\Delta = 0.01$. The dispersion of a step-index fiber is approximately $\Delta T/l = n_0\Delta/c = 50$ ns/km. An α-profile, graded-index fiber having the same n_0 and Δ and with $\alpha = 2(1 - \Delta) = 1.98$ would have a theoretical mode dispersion of $\Delta T/l = (n_0\Delta/c)(\Delta/8) = 62.5$ ps/km. The bandwidth–distance product would be approximately 8 GHz.km, an improvement of almost three orders of magnitude. Figure 6.5 shows a plot of the theoretical mode dispersion against α for such a fiber and this makes two

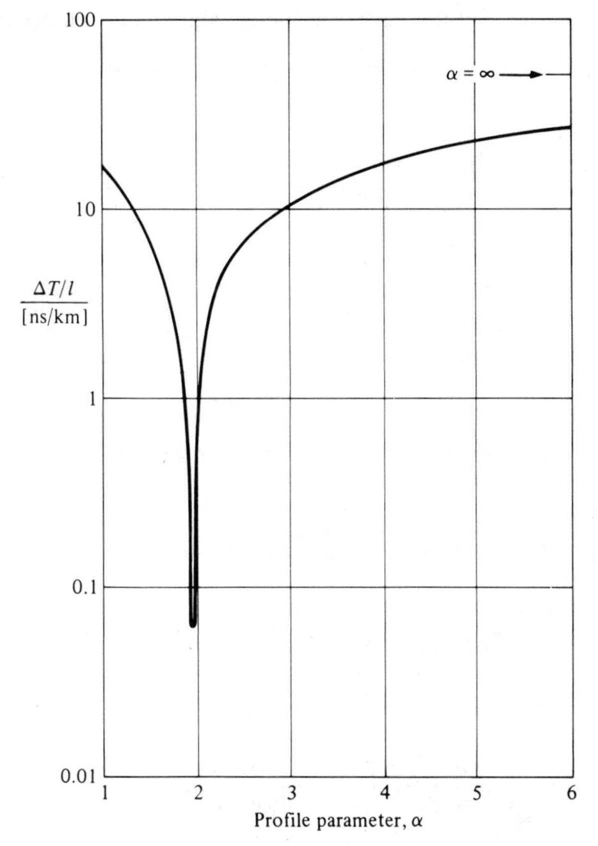

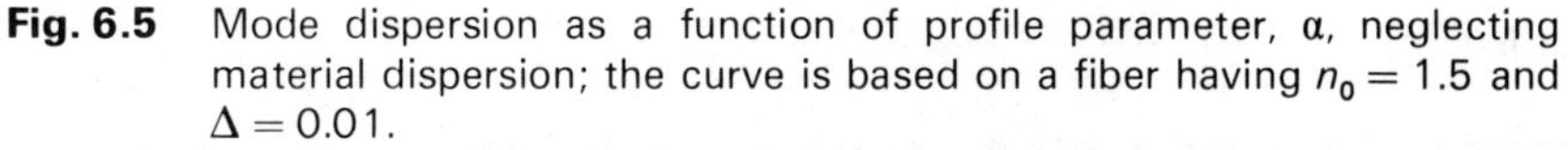
Fig. 6.5 Mode dispersion as a function of profile parameter, α, neglecting material dispersion; the curve is based on a fiber having $n_0 = 1.5$ and $\Delta = 0.01$.

points clear. First, in order to obtain the minimum dispersion the value of α has to be controlled very closely indeed. But second, *any* grading which approximates to a parabolic profile causes a significant reduction in the mode dispersion of a fiber.

6.3.2 Including Material Dispersion

We will now seek the optimum profile when the dispersive properties of the fiber are taken into account. This means that in the expanded form of eqn. (6.1.40) we have to allow n_0 and Δ as well as Q to be functions of the optical frequency. The argument is simply a repeat of the one we have just given for dispersionless media, but the algebra becomes rather more cumbersome. Again

$$\beta_q = \beta_0[1 - 2\Delta\xi]^{1/2} \tag{6.3.1}$$

but

$$\frac{t_q}{l} = \frac{d\beta_q}{d\omega} = \frac{d\beta_0}{d\omega}[1-2\Delta\xi]^{1/2} - \beta_0[1-2\Delta\xi]^{-1/2}\left(\Delta\frac{d\xi}{d\omega} + \xi\frac{d\Delta}{d\omega}\right)$$

$$= \frac{N_0}{c}[1-2\Delta\xi]^{1/2} - \frac{n_0\omega}{c}[1-2\Delta\xi]^{-1/2}\left(\Delta\frac{d\xi}{d\omega} + \xi\frac{d\Delta}{d\omega}\right) \tag{6.3.14}$$

Put

$$\xi = K_2(\Delta n_0^2\omega^2)^{-\alpha/(\alpha+2)} \tag{6.3.15}$$

where K_2 is a parameter that is independent of ω. Then

$$\frac{d\xi}{d\omega} = \left(\frac{-\alpha}{\alpha+2}\right)\frac{\xi}{\Delta n_0^2\omega^2}\frac{d}{d\omega}(\Delta n_0^2\omega^2)$$

$$= \left(\frac{-\alpha}{\alpha+2}\right)\frac{\xi}{\Delta n_0^2\omega^2}\left(\Delta\cdot 2n_0\omega N_0 + n_0^2\omega^2\frac{d\Delta}{d\omega}\right) \tag{6.3.16}$$

using $N_0 = d(n_0\omega)/d\omega$.

$$\therefore \quad \frac{d\xi}{d\omega} = \left(\frac{-\alpha}{\alpha+2}\right)\frac{N_0}{n_0}\frac{\xi}{\omega}2(1+\delta) \tag{6.3.17}$$

where

$$\delta = \frac{n_0}{2N_0}\frac{\omega}{\Delta}\frac{d\Delta}{d\omega} = -\frac{n_0}{2N_0}\frac{\lambda}{\Delta}\frac{d\Delta}{d\lambda} \tag{6.1.20}$$

Equation (6.3.14) then simplifies to

$$\frac{ct_q}{l} = N_0[1-2\Delta\xi]^{1/2} - n_0\omega[1-2\Delta\xi]^{-1/2}\left(-\frac{\alpha\Delta}{(\alpha+2)}\frac{N_0\xi}{n_0\omega}2(1+\delta) + \xi\frac{d\Delta}{d\omega}\right)$$

$$= N_0[1-2\Delta\xi]^{-1/2}\left[(1-2\Delta\xi) + \frac{\alpha\Delta}{(\alpha+2)}2\xi(1+\delta) - \frac{n_0\omega}{N_0}\xi\frac{d\Delta}{d\omega}\right]$$

$$= N_0[1-2\Delta\xi]^{-1/2}\left[1 - 2\Delta\xi\left(1 - \frac{\alpha(1+\delta)}{(\alpha+2)} + \delta\right)\right] \tag{6.3.18}$$

using eqn. (6.1.20). Applying a binomial expansion to the first term in square brackets gives

$$\frac{ct_q}{l} \simeq N_0[1 + \Delta\xi + \tfrac{3}{2}(\Delta\xi)^2 + \cdots]\left[1 - \frac{4(1+\delta)}{(\alpha+2)}\Delta\xi\right]$$

$$\simeq N_0\left[1 + \frac{(\alpha + 2 - 4 - 4\delta)}{(\alpha+2)}\Delta\xi + \frac{(3\alpha + 6 - 8 - 8\delta)}{2(\alpha+2)}(\Delta\xi)^2 + \cdots\right]$$

$$\simeq N_0\left[1 + \frac{(\alpha - 2 - 4\delta)}{(\alpha+2)}\Delta\xi + \frac{(3\alpha - 2 - 8\delta)}{(\alpha+2)}\frac{(\Delta\xi)^2}{2} + \cdots\right] \quad (6.3.19)$$

Again the extreme mode groups are represented by $\xi = 0$ and $\xi = 1$. And for values of α away from the optimum, their mode groups will be the ones having the longest and the shortest propagation times.

Then

$$c\frac{\Delta T}{l} = c\frac{|t_0 - t_Q|}{l} = N_0\Delta\frac{(\alpha - 2 - 4\delta)}{(\alpha+2)} \quad (6.3.20)$$

taking only the first-order terms.

Again, if we neglect third-order terms, the minimum time dispersion arises when $t_0 = t_Q$, that is

$$\frac{(\alpha - 2 - 4\delta)}{(\alpha+2)}\Delta + \frac{(3\alpha - 2 - 8\delta)}{(\alpha+2)}\frac{\Delta^2}{2} = 0$$

$$\therefore \quad \alpha(1 + \tfrac{3}{2}\Delta) = 2(1 + 2\delta + \tfrac{1}{2}\Delta - \delta\Delta)$$

$$\therefore \quad \alpha \simeq 2(1 + 2\delta + \tfrac{1}{2}\Delta)(1 + \tfrac{3}{2}\Delta)^{-1}$$

$$\simeq 2(1 + 2\delta + \tfrac{1}{2}\Delta)(1 - \tfrac{3}{2}\Delta + \cdots)$$

$$\simeq 2(1 + 2\delta - \Delta) = \alpha_{opt} \quad (6.3.21)$$

Then, the variation of group delay with the mode-group parameter, ξ, is

$$\frac{c}{l}(t_0 - t_q) = N_0\left[\frac{2\Delta^2\xi}{(4 + 4\delta - 2\Delta)} - \frac{(4 + 4\delta - 6\Delta)}{(4 + 4\delta - 2\Delta)}\frac{\Delta^2\xi^2}{2}\right]$$

$$\simeq \frac{N_0\Delta^2\xi}{2}(1 - \xi) \quad (6.3.22)$$

where we have neglected the terms in δ and Δ, compared to 1. Equation

(6.3.22) has its maximum value at $\xi = \frac{1}{2}$, when

$$(t_0 - t_q) = \Delta T_{\min} = \frac{N_0 l \Delta^2}{8c} \tag{6.3.23}$$

We see that the presence of material dispersion has two effects on the inter-mode dispersion in α-profile, graded-index fibers:

(a) It leads to the appearance of the group index, N_0, in the equation for minimum mode dispersion, eqn. (6.3.23).
(b) It alters the optimum value of α required to achieve this minimum mode dispersion by an amount that depends on $d\Delta/d\lambda$, through δ in eqn. (6.3.21).

Remember we have not yet included the intra-mode dispersion: that is, the pulse-broadening effects of the material dispersion and waveguide dispersion for each mode. This question will be taken up in the next section.

The mere act of doping the material of the fiber core in order to produce the desired variations of refractive index in general causes unwanted variations with radius of the dispersive properties of the glass. This is why Δ may become a function of wavelength. (Indeed, so may the α-value of the profile itself, which is not something we have allowed for in the analysis.) These effects, which result from the material properties, can be compensated by a control of the refractive-index profile, that is, by controlling α. When α is less than 2, the higher-order modes (the more oblique rays) propagate more quickly than the lower-order modes (the axial rays). This can compensate for a difference in refractive index between core and cladding that increases with increasing wavelength. When α is greater than 2, it is the axial rays that travel more quickly, and this can compensate for a value of Δ that decreases with increasing wavelength. In consequence, the optimum value of α depends on the choice of impurity material used to produce the refractive-index variation, and on the wavelength to be used. Values of $\delta(\lambda)$ for different possible dopants can be estimated from the known dispersion data which we discussed in Section 2.2.2. These can then be used to obtain the optimum values of α at different wavelengths. Results are shown in Fig. 6.6.

The fiber designer has at his disposal a number of possible dopants which, used in combination, may give him a considerable degree of control over the profile of a graded-index fiber and its dispersive properties. A rather more general analysis has shown that it should be possible to reduce the intermode dispersion to a minimum value of

$$\frac{c}{N_0 l} \Delta T_{\min} = \frac{[1 - (1 - 2\Delta)^{1/4}]^2}{[1 + (1 - 2\Delta)^{1/2}]} \simeq \frac{\Delta^2}{8} \quad \text{for } \Delta \ll 1 \tag{6.3.24}$$

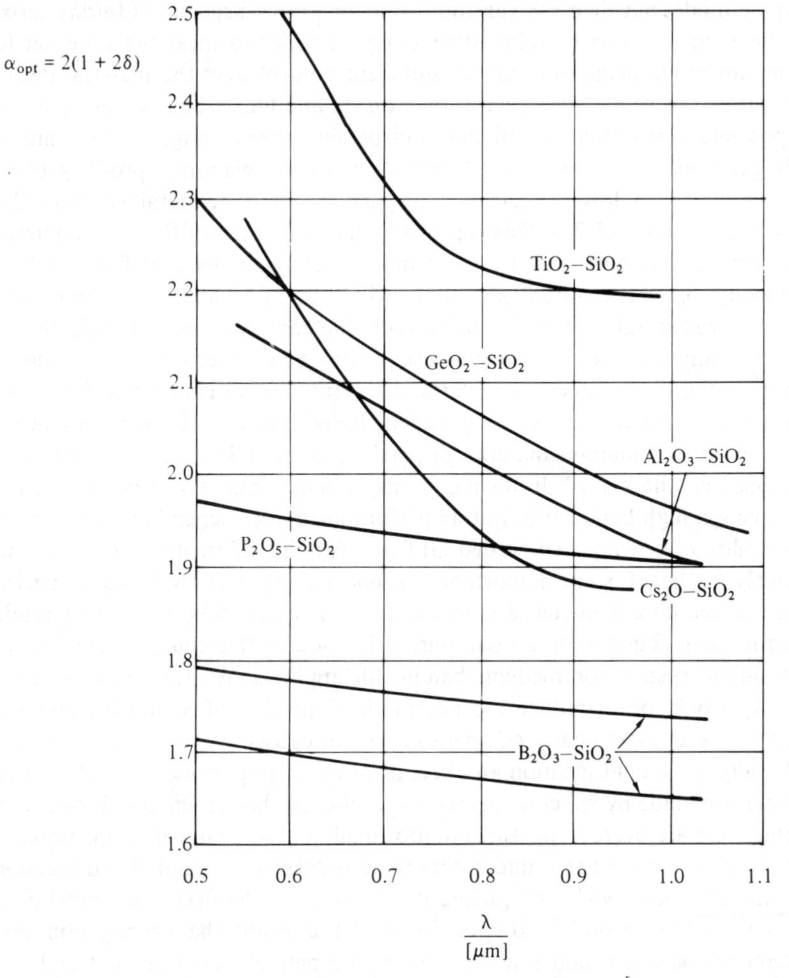

Fig. 6.6 Optimum values of α as a function of wavelength. [Data derived from the interference measurements of H. M. Presby and I. P. Kaminow, *Appl. Optics* **15**, 3029–36 (1976).]

The values of α_{opt} have been calculated from the dispersion parameters $\delta(\lambda)$ using eqn. (6.3.21) with $\Delta = 0$. The cladding material is assumed to be pure silica.

for a very wide range of fiber profiles. The theory applies to all fibers having cylindrical symmetry and made from isotropic materials. When the material dispersion is fixed, then the optimum profile is uniquely determined. Indeed, if the material dispersion is taken as being constant across the core of the fiber as in the theory we have just presented, then the optimum profile is the α-profile with α given by eqn. (6.3.21). Alternatively, an arbitrary choice of index profile may be made, but then the variation of the material dispersion ($dn/d\lambda$) across the fiber core has to be tailored precisely in order to meet the criterion for minimum mode dispersion. Given sufficient control over the material dispersion properties of the glass (that is over $dn/d\lambda$ and higher derivatives) it should be possible to maintain the minimum dispersion over a range of wavelengths.

Suggestions have been made that with still more elaborate profiling of the core index, even lower intermode dispersion could be obtained than that indicated by eqn. (6.3.23). This is probably more a matter of theoretical interest than practical importance. It seems more likely that demand for very high bandwidth optical communication links will be met by single-mode fiber rather than by graded-index fibers requiring such fine control in their manufacture. It would in any case be necessary to use narrow-band laser sources in order to minimize the direct effects of material dispersion. There is one possible exception to this, and that concerns a system based on an LED operating at the wavelength of minimum material dispersion, about 1.3 μm. Such a system is discussed in Chapter 17. It has the merits of being relatively cheap and simple yet giving a high bandwidth. But its performance does depend crucially on the availability of a really well graded GI fiber. A quite different type of fiber that is likely to be of some importance is one having a fairly large numerical aperture and core diameter, and in which the index grading is relatively crude. Used in conjunction with a broad-band LED source this could realize a cheap and simple system for medium bandwidth applications. The purpose of the grading would be to reduce the intermode dispersion of a simple step-index fiber to a level more nearly comparable to that caused by material dispersion.

Finally we should mention an effect of practical importance when long runs of fiber are made by splicing together a number of shorter lengths. When some of the fibers are overcompensated (α too small) and some undercompensated (α too large), it is found that alternating them reduces the cumulative dispersion. Groups of modes which propagate more slowly in the first fiber travel more quickly in the second, and vice versa. As a result the propagation time differences between modes at the end of the pair of fibers are reduced.

6.4 INTRAMODE DISPERSION IN GRADED-INDEX FIBERS

In this section eqn. (6.1.40) will be used as the basis for calculating the time dispersion *within* each mode group in a graded-index fiber. In Section 5.4 we showed that this intramode dispersion was given by

$$\tau = l \frac{d^2 \beta}{d\omega^2} \Delta\omega \tag{5.4.1}$$

where $\Delta\omega$ is the spectral spread of the source. We have defined $\Delta\omega$ as the range of frequencies over which the power spectral density exceeds half its peak value, and τ is then the period during which the received power exceeds half its peak value. τ is sometimes referred to as the full-width-at-half-height (FWHH) of the pulse. We will start by evaluating eqn. (5.4.1) for the qth mode-group propagating in a fiber having no material dispersion, i.e., n_0 and Δ do not vary with ω. This gives us an expression for the pure waveguide dispersion.

In the last section we showed that

$$\beta_q = \beta_0[1 - 2\Delta\xi]^{1/2} \tag{6.3.1}$$

and

$$\therefore \quad \frac{d\beta_q}{d\omega} = \frac{n_0}{c}\left[1 + \frac{(\alpha-2)}{(\alpha+2)}\Delta\xi + \cdots\right] \tag{6.3.7}$$

retaining only the first term in the expansion. Differentiating again and using eqn. (6.3.6) gives

$$\frac{d^2\beta_q}{d\omega^2} = \frac{n_0\Delta}{c}\frac{(\alpha-2)}{(\alpha+2)^2}(-2\alpha)\frac{\xi}{\omega} \tag{6.4.1}$$

Using eqn. (5.4.1) the pulse width for the qth mode group becomes:

$$\tau_q = l\frac{d^2\beta_q}{d\omega^2}\Delta\omega = -\frac{n_0 l\Delta}{\omega c}\frac{2\alpha(\alpha-2)}{(\alpha+2)^2}\xi\Delta\omega \tag{6.4.2}$$

For $q = Q$, $\xi = 1$ and

$$\tau_Q = -\frac{n_0 l\Delta}{\omega c}\frac{2\alpha(\alpha-2)}{(\alpha+2)^2}\Delta\omega \tag{6.4.3}$$

Waveguide dispersion is seen to be a function of α. When α is near to 2 it is significantly reduced along with intermode dispersion.

We will next evaluate eqn. (5.4.1) when n_0 and Δ do vary with ω. In the previous section we obtained

$$\frac{d\beta_q}{d\omega} = \frac{N_0}{c}\left[1 + \frac{(\alpha-2-4\delta)}{(\alpha+2)}\Delta\xi + \cdots\right] \tag{6.3.19}$$

where we have retained only the first term. Differentiating again gives

$$c\frac{d^2\beta_q}{d\omega^2} = \frac{dN_0}{d\omega}\left[1 + \frac{(\alpha-2-4\delta)}{(\alpha+2)}\Delta\xi\right]$$

$$+ \frac{N_0}{(\alpha+2)}\left[(\alpha-2-4\delta)\Delta\frac{d\xi}{d\omega} + (\alpha-2-4\delta)\xi\frac{d\Delta}{d\omega} - 4\Delta\xi\frac{d\delta}{d\omega}\right] \tag{6.4.4}$$

In the second term on the right-hand side we may substitute for $d\xi/d\omega$ using eqn. (6.3.17) and for $d\Delta/d\omega$ using eqn. (6.1.20), and neglect $d\delta/d\omega$. Then

$$c\frac{d^2\beta_q}{d\omega^2} = \frac{dN_0}{d\omega}\left[1 + \frac{(\alpha - 2 - 4\delta)\Delta\xi}{(\alpha + 2)}\right]$$

$$+ \frac{N_0(\alpha - 2 - 4\delta)}{(\alpha + 2)}\left[\frac{\Delta(-\alpha)}{(\alpha + 2)} \frac{N_0 \xi 2(1 + \delta)}{n_0 \omega} + \frac{\xi \cdot 2N_0\Delta\delta}{n_0\omega}\right]$$

$$= \frac{dN_0}{d\omega}\left[1 + \frac{(\alpha - 2 - 4\delta)\Delta\xi}{(\alpha + 2)}\right] - \frac{2(\alpha - 2\delta)(\alpha - 2 - 4\delta)}{(\alpha + 2)^2}\frac{N_0^2\Delta\xi}{n_0\omega} \tag{6.4.5}$$

material dispersion waveguide dispersion

Under almost all circumstances material dispersion dominates. The small second term may be simplified further by putting $n_0 \simeq N_0$ and, neglecting δ in comparison with α, the pulse width of any given mode group characterized by the parameter ξ becomes

$$\frac{c\tau_q}{l} = c\frac{d^2\beta_q}{d\omega^2}\Delta\omega = \left[\omega\frac{dN_0}{d\omega} - \frac{2\alpha(\alpha - 2 - 4\delta)}{(\alpha + 2)^2}\Delta N_0 \xi\right]\frac{\Delta\omega}{\omega} \tag{6.4.6}$$

The waveguide dispersion term is important only in the region of the material dispersion minimum when $dN_0/d\omega = 0$, and it then has the effect of shifting the wavelength of minimum total dispersion by an amount that varies with mode group. Even this effect is negligible when the α value is chosen so as to minimize intermode dispersion, that is, $\alpha \simeq 2(1 + 2\delta) \simeq 2$. And if this is not done, intermode dispersion is dominant anyway.

6.5 TOTAL DISPERSION IN GRADED-INDEX FIBERS

We start by emphasizing again that in basing this analysis of dispersion in graded-index fibers on eqn. (6.1.40), we are restricting its validity to those high-order modes in multimode fibers that are far from cutoff. We are tacitly assuming that the majority of the optical power carried by the fiber propagates in such modes. In this section we shall seek the r.m.s. variation of propagation times averaged over these modes. We shall assume that the light comes from an optical source of spectral half-width $\Delta\omega$, or r.m.s. width σ_ω, and is equally distributed among the propagating modes.

In order to estimate the total dispersion it is necessary to combine the inter-mode and intramode effects. It is clear from the discussion of the previous two sections that these are independent and uncorrelated. For this reason they can most conveniently be combined by adding together the mean square pulse widths that each would produce independently in the absence of the other. As was discussed in Section 2.4, the total r.m.s. pulse width, σ, is then given by

$$\sigma = (\sigma_1^2 + \sigma_2^2)^{1/2} \tag{2.4.10}$$

where σ_1 is the r.m.s. pulse width resulting from intermode dispersion alone and σ_2 is the r.m.s. pulse width resulting from intramode dispersion alone.

It was shown in the last section that intramode dispersion involves a material dispersion term that is the same for all modes, and other terms, principally a waveguide term, that vary from mode to mode. The latter terms involve the mode parameter ξ, as in eqn. (6.4.6). Except possibly in the vicinity of the material dispersion minimum, the mode-independent term dominates all others, and to simplify the algebra we shall retain only this term. In eqn. (6.4.6) the intramode dispersion was expressed as the 'full-width at half-height' spread, τ_q, of each mode group, q, using the FWHH spectral line-width, $\Delta\omega$. For the purposes of this section it is more convenient to use instead the r.m.s. pulse width, σ_2, which can be expressed directly in terms of the r.m.s. spectral line width, σ_ω. Thus eqn. (6.4.6) reduces to eqn. (2.4.12) when we retain only the first term:

$$\sigma_2 = \frac{l}{c}\left|\omega \frac{dN_0}{d\omega}\right|\frac{\sigma_\omega}{\omega} = \frac{l}{c}\left|\lambda^2 \frac{d^2 n}{d\lambda^2}\right|\frac{\sigma_\lambda}{\lambda} = \frac{l}{c}|Y_m|\gamma_\sigma \tag{6.5.1}$$

Next it is necessary to determine σ_1, the r.m.s. pulse width resulting from intermode dispersion when σ_ω is very small. To do this we need to extend the work of Section 6.3 where we examined the total spread in propagation times. To determine σ_1 we need to evaluate eqn. (2.4.3) on the assumption that the optical power is launched onto the fiber in the form of an impulse of unit energy distributed equally among the propagating modes of the fiber. The received power will then arrive in a sequence of weighted impulses as the mode groups arrive after their respective transit times. Each of the Q mode groups is identified by the mode group parameter, q, and consists of $2q$ independent modes. Thus the energy arriving at time t_q with mode group q is proportional to q. The total number of individual modes is

$$M = \sum_{q=0}^{Q} 2q = Q(Q+1) \simeq Q^2 \tag{6.5.2}$$

Since we are using the results of the WKB analysis, we have to assume that Q

is large and that the results may be inaccurate for the lowest and highest order modes, that is, when q is small and when it approaches Q. With unit input energy equally distributed among the modes, the energy in each is $1/Q^2$, and that in each mode group is $2q/Q^2$.

We shall examine the distribution of propagation times for the mode groups for some representative values of the index-grading parameter, α. From these we can obtain the impulse responses of the fibers and hence deduce values for σ_1. The analysis will be based on eqn. (6.3.19):

$$t_q = \frac{N_0 l}{c}\left[1 + \frac{(\alpha - 2 - 4\delta)}{(\alpha + 2)}\Delta\left(\frac{q}{Q}\right)^{2\alpha/(\alpha+2)} + \frac{(3\alpha - 2 - 8\delta)}{(\alpha + 2)}\frac{\Delta^2}{2}\left(\frac{q}{Q}\right)^{4\alpha/(\alpha+2)}\right] \tag{6.3.19}$$

It is more convenient to normalize these propagation times against that of the lowest-order mode, $t_0 = N_0 l/c$. We can also use t_0 as the origin for the mode arrival times. A normalized propagation time, T_q, is thus defined as

$$T_q = \frac{ct_q}{N_0 l} - 1 = \frac{(\alpha - 2 - 4\delta)}{(\alpha + 2)}\Delta\left(\frac{q}{Q}\right)^{2\alpha/(\alpha+2)} + \frac{(3\alpha - 2 - 8\delta)}{(\alpha + 2)}\frac{\Delta^2}{2}\left(\frac{q}{Q}\right)^{4\alpha/(\alpha+2)} \tag{6.5.3}$$

$$= Aq^{2\alpha/(\alpha+2)} + Bq^{4\alpha/(\alpha+2)} \tag{6.5.4}$$

where

$$A = \frac{(\alpha - 2 - 4\delta)\Delta}{(\alpha + 2)Q^{2\alpha/(\alpha+2)}} \quad \text{and} \quad B = \frac{(3\alpha - 2 - 8\delta)\Delta^2}{2(\alpha + 2)Q^{4\alpha/(\alpha+2)}} \tag{6.5.5}$$

Now the time interval between pulses is given by $\mathrm{d}T_q/\mathrm{d}q$, and so the rate at which the mode groups arrive is $1/(\mathrm{d}T_q/\mathrm{d}q)$. With each mode group carrying an energy of $2q/Q^2$, the average power received at any time, T, is given by:

$$P(T) = \frac{2q/Q^2}{\mathrm{d}T_q/\mathrm{d}q} \tag{6.5.6}$$

This is the smoothed-out impulse response of the fiber. It is what is observed when the receiver acts as a low-pass filter unable to distinguish the separate impulses caused by the arrival of the individual mode groups.

A general expression for $\sigma_1(\alpha)$ can be obtained by substituting eqns. (6.5.6) and (6.5.3) into eqn. (2.4.3):

$$\left(\frac{c\sigma_1}{N_0 l}\right)^2 = \int_0^{T_Q} T^2 P(T)\,\mathrm{d}T - \left[\int_0^{T_Q} TP(T)\,\mathrm{d}T\right]^2 \tag{6.5.7}$$

$$= \int_0^Q T^2 P(T) \frac{\mathrm{d}T}{\mathrm{d}q}\,\mathrm{d}q - \left[\int_0^Q TP(T)\frac{\mathrm{d}T}{\mathrm{d}q}\,\mathrm{d}q\right]^2$$

$$= \int_0^Q T^2 \frac{2q}{Q^2}\,\mathrm{d}q - \left[\int_0^Q T\frac{2q}{Q^2}\,\mathrm{d}q\right]^2 \tag{6.5.8}$$

If the expression for T given in eqn. (6.5.3) is now substituted, then after some straightforward manipulation σ_1 may be expressed as

$$\left(\frac{c\sigma_1}{\Delta N_0 l}\right)^2 = \frac{(\alpha-2-4\delta)^2\alpha^2}{4(\alpha+1)^2(\alpha+2)(3\alpha+2)} + \frac{(\alpha-2-4\delta)(3\alpha-2-8\delta)\alpha^2\Delta}{2(\alpha+1)(\alpha+2)(2\alpha+1)(3\alpha+2)}$$

$$+ \frac{(3\alpha-2-8\delta)^2\alpha^2\Delta^2}{(\alpha+2)(3\alpha+2)^2(5\alpha+2)} \tag{6.5.9}$$

Rather than just substituting α-values blindly into this equation, we will first examine the impulse responses obtained for some specific examples and then deduce values of σ_1 from these.

Consider first a step-index fiber, for which $\alpha = \infty$. Then, only the first term in eqn. (6.5.3) is significant and

$$T_q = \frac{\Delta q^2}{Q^2} \tag{6.5.10}$$

Thus the power is received during a period $T_Q - T_0 = \Delta$. This is in normalized time and corresponds to a real-time interval of $(\Delta N_0 l/c)$. During this period the average received power is constant. Using eqn. (6.5.6),

$$P(T) = \frac{2q/Q^2}{2\Delta q/Q^2} = \frac{1}{\Delta} \tag{6.5.11}$$

In Fig. 6.7(a) we have illustrated both the distribution of T_q and the smoothed-out impulse response for a step-index fiber with $Q = 20$. In Section 2.4 we showed that the r.m.s. width of a pulse of constant amplitude was $(1/2\sqrt{3})$ times its duration, eqn. (2.4.13). Thus the normalized r.m.s. width of the impulse response of a step-index fiber is

$$\frac{c\sigma_1}{N_0 l} = \frac{\Delta}{2\sqrt{3}} = 0.289\Delta \tag{6.5.12}$$

Putting $\alpha = \infty$ into the first term of eqn. (6.5.9) yields the same result.

Let us take another specific example, this time of a slightly graded fiber with $\alpha = 6$. Then, again taking only the first term in eqn. (6.5.3) and neglecting δ in

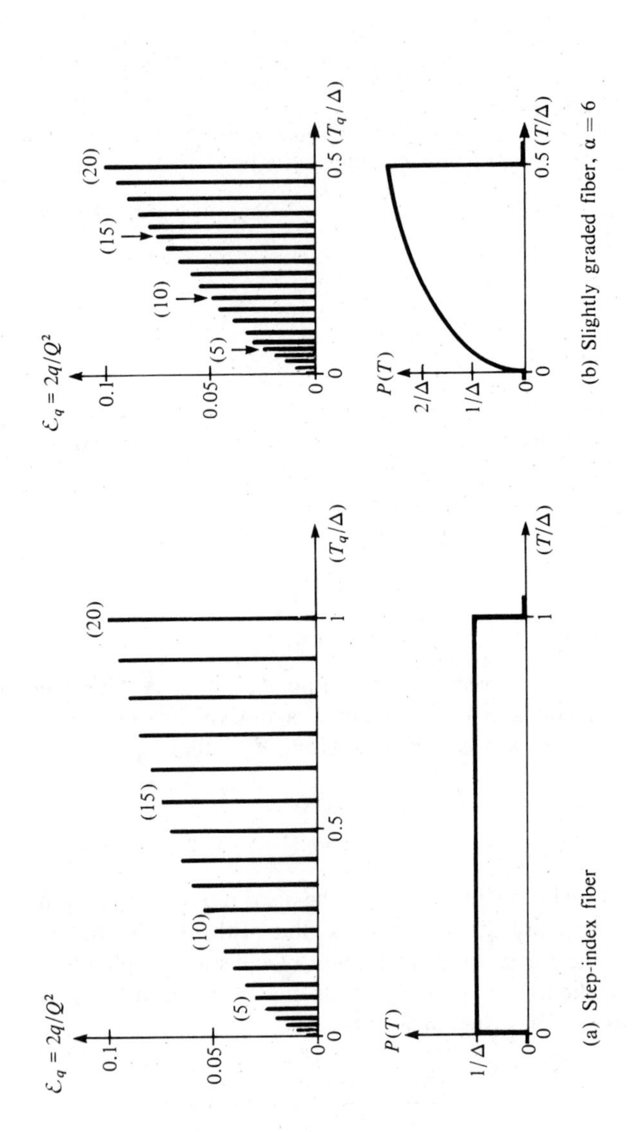

(a) Step-index fiber

(b) Slightly graded fiber, $\alpha = 6$

continued

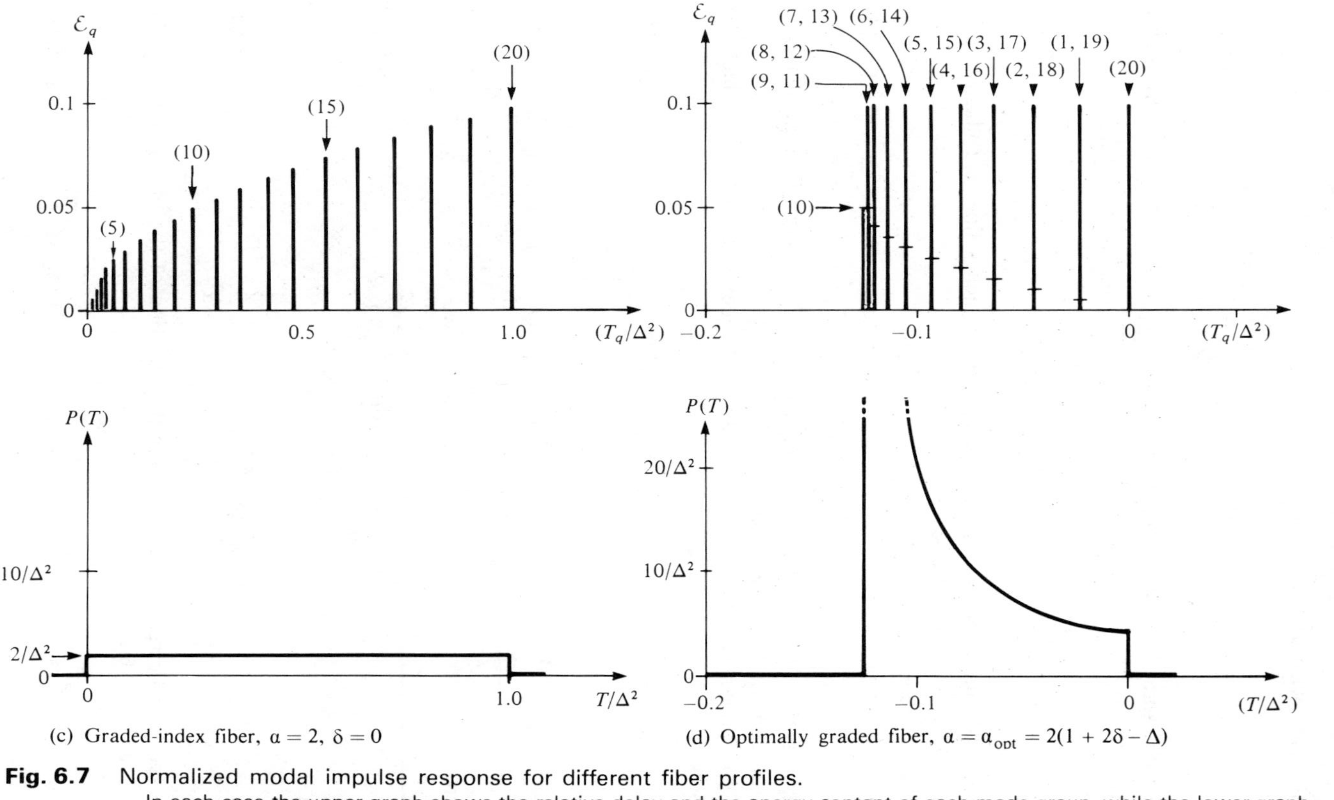

(c) Graded-index fiber, $\alpha = 2$, $\delta = 0$

(d) Optimally graded fiber, $\alpha = \alpha_{\text{opt}} = 2(1 + 2\delta - \Delta)$

Fig. 6.7 Normalized modal impulse response for different fiber profiles.

In each case the upper graph shows the relative delay and the energy content of each mode group, while the lower graph shows the smoothed out average power received. The curves are drawn on the assumption that a total of 20 mode groups are launched.

comparison with unity,

$$T_q = Aq^{3/2} = \frac{\Delta q^{3/2}}{2Q^{3/2}} \tag{6.5.13}$$

The total spread of the impulse response is $T_Q - T_0 = \Delta/2$ and the power distribution is

$$P(T) = \frac{4}{3}\left(\frac{2}{\Delta}\right)^{4/3} T^{1/3} \tag{6.5.14}$$

The modal and smoothed impulse responses of such a fiber are illustrated in Fig. 6.7(b). The r.m.s. pulse width is

$$\frac{c\sigma_1}{N_0 l} = \frac{3}{7\sqrt{10}}\Delta = 0.14\Delta \tag{6.5.15}$$

in agreement with the first term in eqn. (6.5.9). This is smaller than that of a step-index fiber by a factor of rather more than 2.

As a further example, consider a parabolic-index fiber, $\alpha = 2$, in which the dispersion factor δ is zero. Then

$$T_q = Aq + Bq^2 \tag{6.5.16}$$

where

$$A = \frac{\delta\Delta}{2Q} = 0, \qquad B = \frac{1}{2}\left(\frac{\Delta}{Q}\right)^2 \qquad \text{and} \qquad T_Q - T_0 = \frac{\Delta^2}{2}$$

Then

$$\frac{dT_q}{dq} = A + 2Bq = \left(\frac{\Delta}{Q}\right)^2 q \tag{6.5.17}$$

and

$$P(T) = \frac{2q}{Q^2}\frac{1}{(\Delta/Q)^2 q} = \frac{2}{\Delta^2} \tag{6.5.18}$$

The received power is constant over the duration of the pulse, $\Delta^2/2$, so the r.m.s. width is given by

$$\frac{c\sigma_1}{N_0 l} = \frac{1}{2\sqrt{3}}\frac{\Delta^2}{2} = \frac{\Delta^2}{4\sqrt{3}} = 0.144\Delta^2 \tag{6.5.19}$$

This response is shown in Fig. 6.7(c). With $\Delta \ll 1$ a very significant reduction in r.m.s. pulse width has been achieved. When $\delta \neq 0$ a result similar to eqn. (6.5.19) is obtained for $\alpha = 2(1 + 2\delta)$, as can easily be demonstrated by substituting into eqn. (6.5.9).

Finally let us deal with the α-value that was shown in Section 5.4 to minimize the total pulse length at $\Delta^2/8$, namely, $\alpha = 2(1 + 2\delta - \Delta)$. Again we assume $\delta \ll 1$ and $\Delta \ll 1$. Then

$$T_q = -\frac{\Delta^2}{2Q} q + \frac{\Delta^2}{2Q^2} q^2 \tag{6.5.20}$$

so that T_Q and T_0 are both zero and $T_{Q/2} = -\Delta^2/8$. Thus

$$\frac{\mathrm{d}T_q}{\mathrm{d}q} = -\frac{\Delta^2}{2Q} + \frac{\Delta^2 q}{Q^2} \tag{6.5.21}$$

and

$$P(T) = \frac{2q/Q^2}{|\Delta^2 q/Q^2 - \Delta^2/2Q|} = \frac{2}{\Delta^2}\frac{1}{|1 - Q/2q|} \tag{6.5.22}$$

We now need to solve eqn. (6.5.20) for q and substitute back into eqn. (6.5.22). Put $y = -Q/2q$ so that eqn. (6.5.22) becomes

$$P(T) = \frac{2}{\Delta^2}\frac{1}{|1 + y|} \tag{6.5.23}$$

where, using eqn. (6.5.20),

$$y^2 - \frac{\Delta^2}{4T_q} y - \frac{\Delta^2}{8T_q} = 0 \tag{6.5.24}$$

$$\therefore \quad y = \frac{\Delta^2}{8T_q}\left[1 \pm \left(1 + \frac{8T_q}{\Delta^2}\right)^{1/2}\right] \tag{6.5.25}$$

There are two solutions for each y and hence for each q. The total power is given by the arithmetic sum of the two solutions. Put

$$S = \Delta^2/8T_q \qquad \text{and} \qquad R^2 = 1 + \frac{1}{S} = \frac{(1 + S)}{S}$$

to simplify the algebra. Then summing the two solutions gives

$$\frac{\Delta^2}{2} P(T) = \frac{1}{1 + S(1 - R)} - \frac{1}{1 + S(1 + R)} = \frac{2RS}{(1 + S)^2 - R^2 S^2}$$

$$= \frac{2RS}{1 + S} = \frac{2}{R}$$

$$\therefore \quad P(T) = \frac{4}{\Delta^2} \left(1 + \frac{8T_q}{\Delta^2}\right)^{-1/2} \tag{6.5.26}$$

The modal and smoothed-out impulse responses for this case are shown in Fig. 6.7(d). Evaluation of eqn. (2.4.3) for this function gives

$$\frac{c\sigma_1}{N_0 l} = \frac{\Delta^2}{12\sqrt{5}} = 0.037\Delta^2 \tag{6.5.27}$$

Substituting $\alpha = 2 + 4\delta - 2\Delta$ directly into eqn. (6.5.9) and assuming $\delta \ll 1$, $\Delta \ll 1$, yields the same result.

It should be noted that the value of $\alpha = \alpha_{opt} = 2(1 + 2\delta - \Delta)$ was chosen to minimize the total pulse spreading. But it does not produce the smallest theoretical value for σ_1. It can be shown that the minimum value for σ_1 as given by eqn. (6.5.9) is obtained by choosing a value for α close to $2(1 + 2\delta - 6\Delta/5)$. Then σ_1 becomes

$$\left(\frac{c\sigma_1}{N_0 l}\right)_{min} = \frac{\Delta^2}{20\sqrt{3}} = 0.029\Delta^2 \tag{6.5.28}$$

Such distinctions are theoretical niceties rather than matters of practical significance. What is important for the fiber manufacturer is to know how much the mode dispersion will increase when a given tolerance is allowed on the α-value. We can gain some feel for this from the analysis just presented. More sophisticated theories have been published in which the effects of statistical variations in α and the effects of profile perturbations like those shown in Fig. 4.7 on the mode propagation parameters have been calculated. We can only reiterate that to achieve a level of mode dispersion close to the theoretical minimum would require the correct α-value to be maintained with an accuracy of the order of Δ—that is, to ± 0.01. However, even a roughly graded profile does produce a significant reduction in pulse width.

We set out in this section to determine the total effect when mode dispersion and material dispersion are present together. In Fig. 6.8 we have plotted total dispersion as a function of α when varying amounts of material dispersion are present. The total dispersion is obtained by putting eqns. (6.5.9) and (6.5.1)

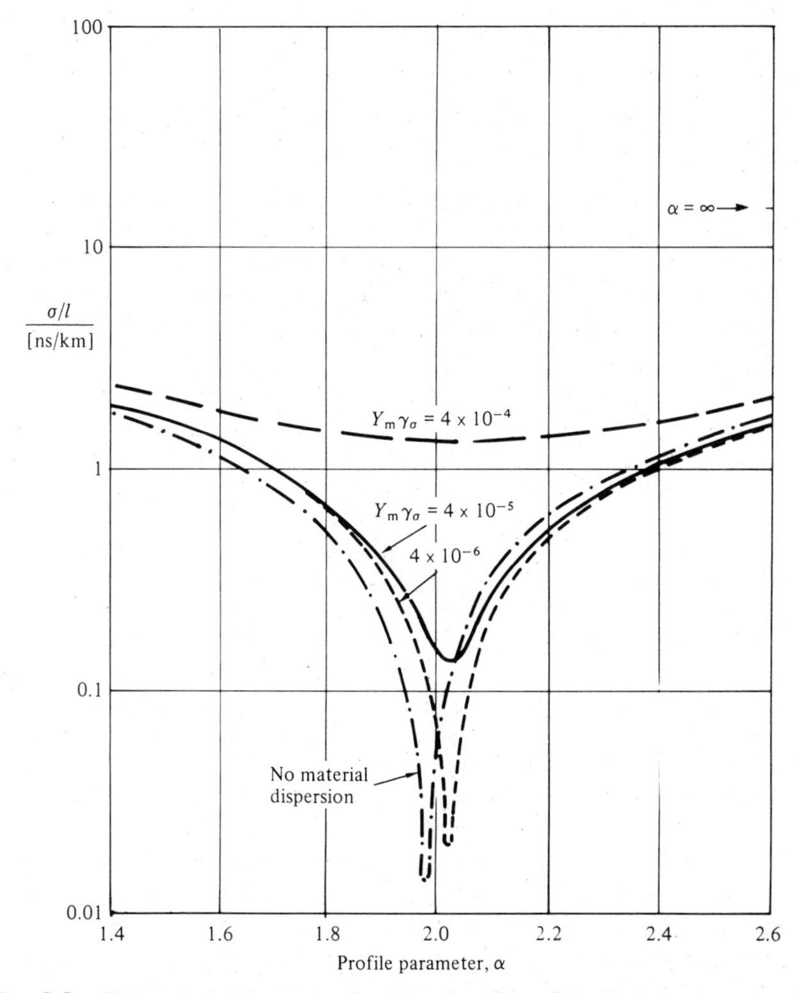

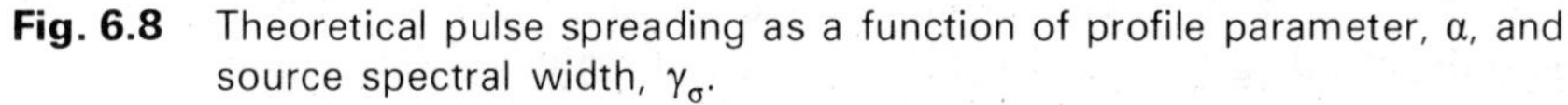

Fig. 6.8 Theoretical pulse spreading as a function of profile parameter, α, and source spectral width, γ_σ.

The calculations are based loosely on a fiber having a germania-doped silica core and a silica cladding, and at a source wavelength of 0.85 μm. We have taken $\Delta = 0.01$, $N_0 = 1.5$, $\delta = 0.01$, $Y_m = 0.02$ for the fiber characteristics and $\gamma_\sigma = 0.02$, 0.002, 0.0002 as characteristic of LED, multimode laser and single-mode laser sources, respectively. The curve for a dispersionless fiber, $Y_m = 0$, $\delta = 0$, is also shown, for comparison.

into eqn. (2.4.10). For α-values not too close to two this becomes

$$\left(\frac{c\sigma}{l}\right) = \frac{c}{l}(\sigma_1^2 + \sigma_2^2)^{1/2} \tag{2.4.10}$$

$$= \left[\frac{(\alpha - 2 - 4\delta)^2 \alpha^2 \Delta^2 N_0^2}{4(\alpha + 1)^2(\alpha + 2)(3\alpha + 2)} + (Y_m \gamma_\sigma)^2\right]^{1/2} \tag{6.5.29}$$

For α-values in the region of 2 the full expression for σ_1 given in eqn. (6.5.9) has to be used. In such well-graded fibers the material dispersion can easily dominate the mode dispersion unless a long wavelength source or one having a narrow spectral width is used. This is clearly brought out in the figure, as is the small shift in the optimum α-value caused by the differential dispersion parameter, δ.

6.6 MODE COUPLING

In Section 4.4 we gave a largely qualitative discussion of some of the phenomena that arise because of defects and nonuniformities in the dielectric waveguide, and we highlighted the problem of characterizing fiber properties such as attenuation and dispersion when they vary along the fiber length. In this section we shall carry some of these matters further in the light of the modal description of optical propagation in fibers. This is a subject that has lent itself to a great deal of detailed and complex mathematical analysis, beyond the scope of the present text. However, the underlying ideas are quite simple and a number of important results can be obtained very easily.

First we will deal with *differential mode attenuation.* It can be seen in Fig. 5.11 that for those modes near to cutoff a higher proportion of their power propagates in the cladding than is the case for the lower-order modes. It is also usually the case that, either by intention or by circumstance, the outer layers of the cladding material are more lossy than the material of the core and inner cladding layers. As a result, those modes having fields which extend further into the cladding are more highly attenuated than those which are confined to the core and its immediate boundary. When a full spectrum of modes is launched onto a multimode fiber from a diffuse source, then the total attenuation at the launch end of the fiber is high, until these high-order modes are removed and an equilibrium mode distribution propagates giving a uniform rate of attenuation. This may take up to 1 km to achieve.

Limiting the number of modes launched onto the fiber, sensible though this is, may not eliminate the effect because the mis-match caused by any splice or connector that is less than perfectly aligned disturbs the equilibrium mode distribution and may launch a number of the more rapidly attenuated modes into the next section of the fiber. Thus the losses caused by a connector or splice may not be limited to those that occur just at the joint itself.

It should be mentioned that there is a significant number of modes which, while they are not fully guided, do nevertheless propagate considerable distances along the fiber. These are known as *leaky modes*. For these,

$$\beta_c^2 > \beta^2 > \beta_c^2 - k^2/a^2 \tag{6.6.1}$$

as shown in Fig. 6.9. They do not satisfy the condition

$$\beta^2 > \beta_c^2 \tag{5.2.14}$$

and so do not constitute fully guided modes. But neither do they couple directly to radiating modes in the cladding: there is a region between radii r_2 ($<a$) and r_3 ($>a$) in which evanescent fields exist, and this inhibits coupling between the 'guided' part of the mode and the freely propagating region and so causes these modes to be partially trapped within the core. The effect of leaky modes is discussed in terms of the ray model in Appendix A3.4.

Unwanted high-order and leaky modes can easily be removed by bending a short length of fiber into a tight curve. This has the effect of locally lowering the value of V, the normalized optical frequency, and as a result a number of higher-order modes that would be guided or only slowly leaky in a straight length of the fiber find themselves below cutoff, so they propagate out into the cladding. This is sometimes referred to as *mode stripping*.

Mode coupling is a quite distinct phenomenon that may also lead to non-uniform fiber characteristics. It is the principal cause of the microbending and waveguide losses referred to in Section 3.1.3. In that section we discussed the way in which non-uniformities and irregularities in the optical waveguide affect the propagation characteristics. The variations may be the consequences of material composition changes, or of changes in the core radius or simply crookedness in the lie of the fiber axis brought about by external stress. Figure 6.10 illustrates the way in which small bends in the fiber (microbends) may be brought about by laying the fiber on an uneven surface. Another cause may be associated with the helical lay of a fiber in a multifiber cable. All such non-uniformities are assumed here to be very small in magnitude. Gross irregularities give rise to large local losses. When the variations take place over distances that are short compared to the optical wavelength, they contribute to the Rayleigh scattering discussed in Section 3.1.3. Then a small fraction of the power propagating in all the guided modes may be scattered into unguided radiation. Irregularities that extend over distances large compared to the optical wavelength are unable to cause this, but what they can cause is an

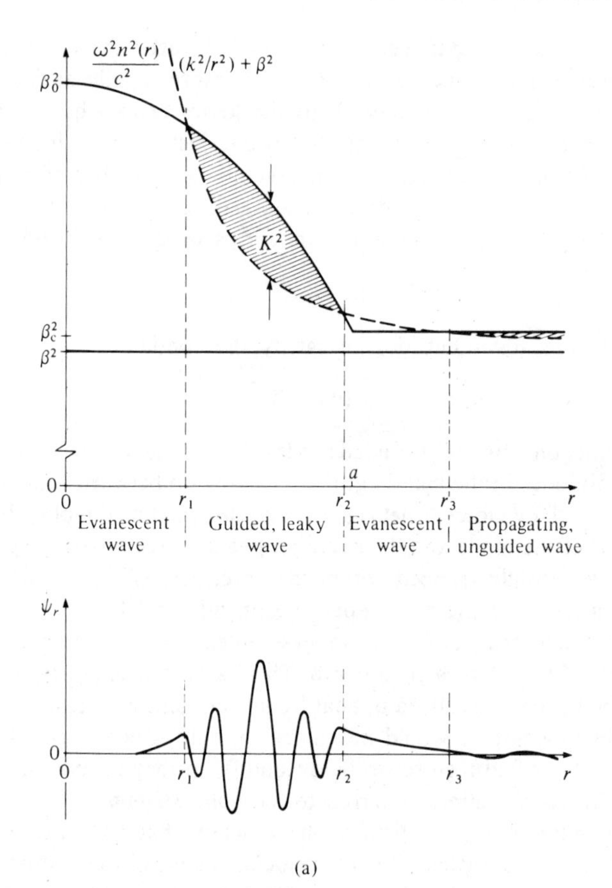

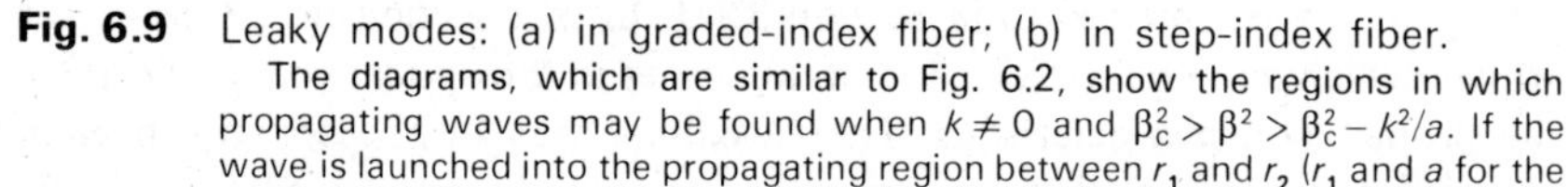

Fig. 6.9 Leaky modes: (a) in graded-index fiber; (b) in step-index fiber.
The diagrams, which are similar to Fig. 6.2, show the regions in which propagating waves may be found when $k \neq 0$ and $\beta_c^2 > \beta^2 > \beta_c^2 - k^2/a$. If the wave is launched into the propagating region between r_1 and r_2 (r_1 and a for the

exchange of energy between adjacent modes. In the case of modes near to cutoff this may lead to energy being transferred from a high-order guided mode to a leaky or an unguided mode, which is subsequently radiated out of the fiber. When one considers that 1 km of fiber corresponds to something of the order of 10^9 optical wavelengths, it becomes clear that even a very small perturbation of the waveguide, if it extends over a significant fraction of the fiber length, will be sufficient to disturb the propagation characteristics.

Energy is most readily exchanged between modes having field patterns that overlap to a considerable degree, and disturbances are most effective at stimulating this when they have a periodicity equal to the beat period between the two modes. We can express the latter requirement as follows. Let the

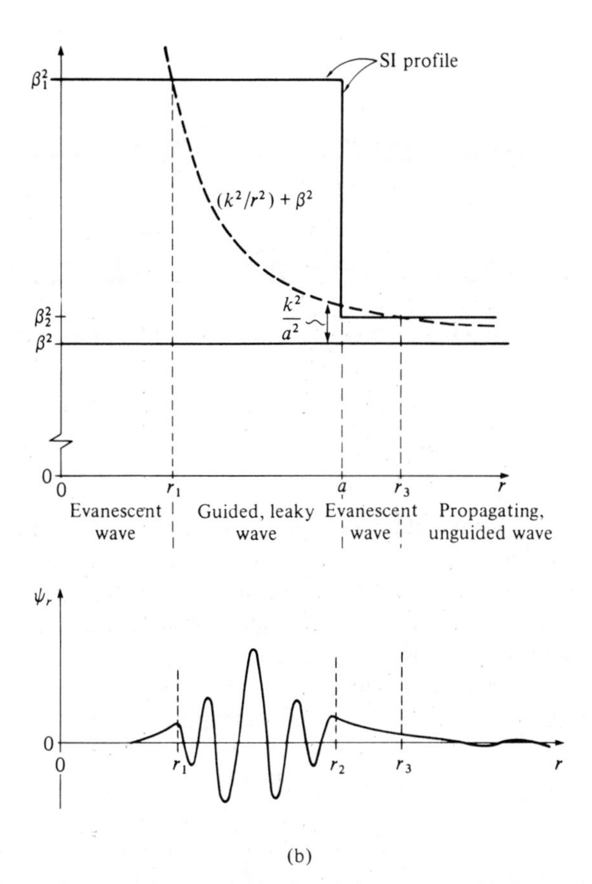

(b)

SI fiber) it is unable to couple directly into an unguided wave in the cladding because of the evanescent region between r_2 and r_3. It may therefore propagate within the core for a considerable distance before finally radiating away.

waveguide perturbation be periodic with a wavelength, Λ, and let the propagation constants of the two modes in question be β_i and β_j. Then coupling will be strongest when

$$\Lambda = \frac{2\pi}{|\beta_i - \beta_j|} = \frac{2\pi}{\Delta\beta} \tag{6.6.2}$$

Coupling between modes having widely differing propagation constants demands a periodic perturbation of short wavelength.

In general, perturbations show a continuous spectrum of length scales and lead to the coupling of a wide range of modes. However, it should be possible

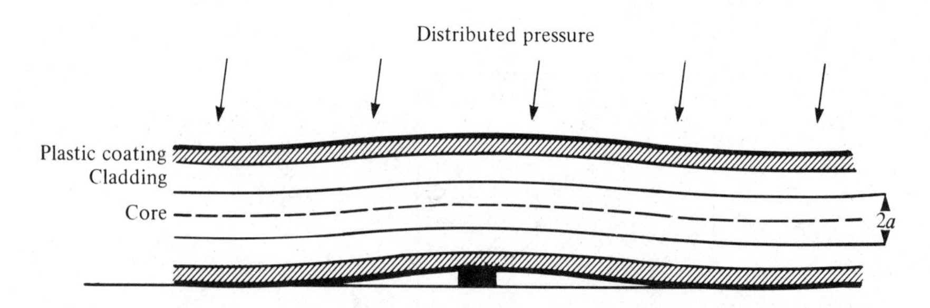

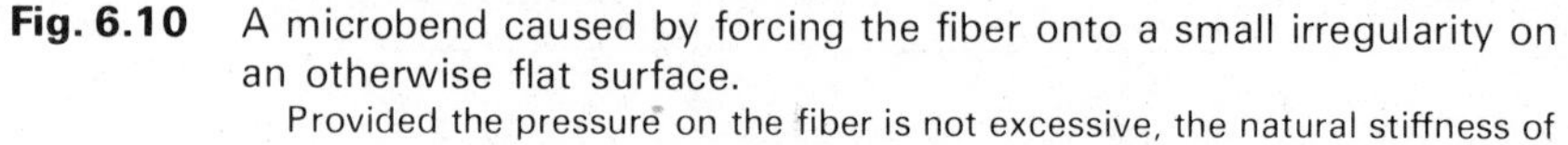

Fig. 6.10 A microbend caused by forcing the fiber onto a small irregularity on an otherwise flat surface.

Provided the pressure on the fiber is not excessive, the natural stiffness of the fiber and the compliance of the plastic coating will smooth the transition over an axial distance of several millimeters.

by careful control of manufacturing processes and cable design to eliminate very rapid variations and thus to prevent coupling between modes with more widely spaced β-values. In particular, we shall normally want to prevent the high-order modes near to cutoff from coupling into unguided modes, because this will increase attenuation.

We have established that the modes tend to propagate in mode groups with all the modes in each group having closely similar propagation constants. The modes in any one group may well interchange energy among themselves without the fiber performance being affected. But coupling between modes from different groups would be more serious. We established that the propagation constant, β_q, for the modes in group q was given by

$$\beta_q = \beta_0 \left[1 - 2\Delta\left(\frac{q}{Q}\right)^{2\alpha/(\alpha+2)}\right]^{1/2} \tag{6.1.40}$$

where β_0 is the propagation constant for plane TEM waves in the material of the fiber axis, α is the refractive index profile parameter, and Δ is the fractional index difference between core and cladding. Q is the total number of guided mode groups. For a core radius, a,

$$Q = \left(\frac{\alpha\Delta}{\alpha + 2}\right)^{1/2} a\beta_0 \tag{6.1.41}$$

Now, the spacing between adjacent mode groups is

$$\Delta\beta = \beta_q - \beta_{q-1} = \frac{d\beta_q}{dq}$$

$$= -\frac{2\alpha\Delta}{(\alpha+2)}\frac{\beta_0}{Q}\left(\frac{q}{Q}\right)^{(\alpha-2)/(\alpha+2)}\left[1-2\Delta\left(\frac{q}{Q}\right)^{2\alpha/(\alpha+2)}\right]^{-1/2}$$

$$\simeq (-)\frac{2}{a}\left(\frac{\alpha\Delta}{\alpha+2}\right)^{1/2}\left(\frac{q}{Q}\right)^{(\alpha-2)/(\alpha+2)} \tag{6.6.3}$$

The term in square brackets can be neglected for $\Delta \ll 1$, and the minus sign has no significance and will be dropped.

For step-index fiber, $\alpha = \infty$

$$\therefore \quad \Delta\beta = \frac{2}{a}\Delta^{1/2}\left(\frac{q}{Q}\right) \tag{6.6.4}$$

Thus $\Delta\beta$ varies from $(2\Delta^{1/2})/(aQ)$ for the lowest-order mode groups to $(2\Delta^{1/2})/a$ for the mode groups nearest to cutoff. As an example, consider a fiber in which $a = 25$ μm, $\Delta = 0.01$ and $Q = 20$. Then, $\Delta\beta$ varies from 400 m^{-1} to 8000 m^{-1} and $\Lambda = 2\pi/\Delta\beta$ varies from 16 mm to 0.8 mm. Quite gentle perturbations give rise to coupling between the lower-order modes, but, if distortion on a millimeter scale can be avoided, then coupling between higher-order modes with its associated microbending and waveguide losses will be eliminated.

For graded-index fiber with $\alpha = 2$,

$$\Delta\beta = \frac{(2\Delta)^{1/2}}{a}\left[1-2\Delta\left(\frac{q}{Q}\right)\right]^{-1/2}$$

$$\simeq \frac{(2\Delta)^{1/2}}{a} \tag{6.6.5}$$

Putting in the same values for a and Δ gives $\Delta\beta = 5700$ m^{-1} and $\Lambda = 1.1$ mm. We should expect the onset of coupling between all mode groups to require a perturbation periodicity in the region of 1 mm. A sharp increase in attenuation would be expected at the same time.

With single-mode fibers microbending and waveguide losses occur if there is coupling between the guided LP_{01} mode of propagation constant, β_{01}, and the unguided modes of propagation constant, β_2. To prevent this it is necessary to keep the minimum wavelength, Λ_m, of any perturbations sufficiently long that

$$\beta_{01} - \beta_2 > 2\pi/\Lambda_m \tag{6.6.6}$$

We previously defined the normalized propagation constant, b_{01}, as

$$b_{01} = \frac{\beta_{01}^2 - \beta_2^2}{\beta_1^2 - \beta_2^2} \tag{5.3.8}$$

which enables us to write

$$\beta_{01} - \beta_2 = \frac{2\pi n_2 \Delta}{\lambda} b_{01} \tag{6.6.7}$$

using the definition of Δ given in eqn. (5.4.3) for small Δ. Then condition (6.6.6) becomes

$$b_{01} > \frac{\lambda}{\Lambda_{\mathrm{m}} n_2 \Delta} \tag{6.6.8}$$

If we take $\Lambda_{\mathrm{m}} = 1$ mm, $n_2 = 1.5$, $\Delta = 0.005$ and $\lambda = 0.85$ μm, then (6.6.8) requires $b_{01} > 0.1$. Reference to Fig. 5.10 shows that it is then necessary to have $V > 1.2$ in order to avoid excessive microbending losses.

In multimode fibers, strong coupling between the modes can be induced very easily by sandwiching the fiber between rough surfaces, such as sandpaper, a few centimeters in length. This ensures that all the modes in the fiber are uniformly excited, and it is known as *mode scrambling*. On the other hand, microbending losses can be almost completely eliminated by good cable design. Jacketing the fiber in a soft cushioning material and encasing the whole in a fairly rigid outer tube enables the stiffness of the fiber itself to smooth out imposed irregularities and thereby to increase Λ to the order of centimeters.

Although full mode coupling is likely to cause an increase in attenuation, it can at the same time have the beneficial effect of decreasing dispersion. Consider an optical pulse launched into a particular, pure mode of a multimode optical fiber. As it propagates, mode coupling causes some of the optical power to excite other modes having different propagation velocities. As a result the pulse spreads out. Now think of the light launched onto the fiber so that it excites all the guided modes equally. Some of the optical power which starts in a particular mode again couples into adjacent modes as it propagates. Many such transitions take place before the light finally reaches the detector at the end of the fiber. As a result, it propagates in many different modes during its journey and thus travels at an overall speed that is an average of the modal group velocities. The same argument applies to all of the optical power and in consequence the full pulse spreading is not observed; no light spends all its time in the fastest mode and none in the slowest mode. Both the total and the r.m.s. pulse lengths are reduced in comparison with those of an otherwise similar fiber having no mode coupling.

With mode coupling it is found that the pulse spreading is initially proportional to the propagation distance, l, but that after a distance, l_c, an equilibrium distribution of energy among the modes is established and maintained, and the pulse width then increases with the square root of distance.

Thus, for $0 < l < l_c$ the pulse width increases as

$$\sigma(l) = \sigma_0 l \tag{6.6.9}$$

where σ_0 is given by eqn. (6.5.29). For $l > l_c$,

$$\sigma(l) = \sigma_0 (l_c l)^{1/2} \tag{6.6.10}$$

The distance l_c decreases with an increase in the coupling strength. Equations (6.6.9) and (6.6.10) should be regarded as asymptotic solutions with $\sigma(l)$ making a smooth transition between them.

It is impractical to couple all the guided modes in this way without at the same time coupling guided modes to unguided ones. So the considerable benefit of the reduced pulse spreading is won at the expense of an increase in attenuation. The additional attenuation, α_c dB/km, increases as the coupling strength increases. Thus α_c and l_c are related by

$$\alpha_c l_c = K = \text{constant} \tag{6.6.11}$$

where $K = 0.1$ dB for step-index fibers and increases towards 0.5 dB as the fiber profile becomes parabolic. It is thus possible to trade off dispersion with attenuation.

The analysis we have given is found to be reasonably valid for step-index and poorly graded fibers, but effects in near ideal graded-index fibers are more complex. It should perhaps be said that now that quite low dispersion fibers can be made by both double-crucible and vapor-deposition methods, it is unlikely that the technique of deliberately introducing mode coupling in order to reduce dispersion will retain much practical importance.

PROBLEMS

6.1 An α-profile, graded-index fiber has a core diameter of 60 μm and an α-value of 2.0. It is excited by a source of wavelength 0.85 μm, where it has an axial refractive index of 1.460 and a cladding refractive index of 1.450.

(a) Calculate values for $\beta_0 a$, Δ and V.
(b) Estimate the total number of propagating modes and mode-groups.
(c) Calculate the difference in the propagation constants of adjacent mode-groups.
(d) Estimate the effective numerical aperture of the fiber.

6.2 Show that the differential material dispersion coefficient δ is given by $n_c(n_c N_0 - n_0 N_c)/(n_0^2 - n_c^2)N_0$.

In the fiber of Problem 6.1, $N_0 = 1.474$ and $N_c = 1.463$ at 0.85 μm:

(a) Calculate the optimum value of α required to minimize multimode dispersion.

(b) Calculate the improvement in the r.m.s. multimode dispersion (ns/km) that could be obtained with the optimum profile, compared with $\alpha = 2$.
(c) Sketch the impulse response in each case.

6.3 Derive eqn. (6.5.9) from eqns (6.5.3) and (6.5.8).

6.4 A graded-index fiber having a germania-doped silica core of 60 μm diameter and a pure silica cladding is designed to work at the wavelength of minimum material dispersion, that is, at 1.3 μm, where $n_0 = 1.460$, $N_0 = 1.474$, $n_c = 1.447$, $N_c = 1.462$. Using eqn. (6.5.9) calculate values for σ_1/l in the vicinity of the multimode dispersion minimum in order to determine the maximum permitted tolerance on the grading parameter α that will still ensure that σ_1/l does not exceed 0.25 ns/km. (A short computer program will assist this calculation.)

6.5 An inexpensive, graded-index, multimode fiber is under consideration for a CATV distribution network. The core diameter is 150 μm and $\Delta = 0.02$. The proposed optical source has an r.m.s. spectral spread $\gamma_\sigma = 0.04$ about 0.83 μm, at which wavelength $N_0 = 1.50$ and the material dispersion parameter $Y_m = 0.030$. Neglecting δ in eqn. (6.5.29), estimate the degree of index-profiling (i.e. the maximum permitted value of α) required to reduce the multimode dispersion to the level of the material dispersion.

6.6 Estimate the approximate bandwidth–distance products of the various fibers discussed in Problems 6.2, 6.4 and 6.5.

SUMMARY

Approximate solutions to the equations for guided electromagnetic wave propagation in graded-index fibers may be obtained for the range of α-profiles:

$$\begin{aligned} n(r) &= n_0[1 - 2\Delta(r/a)^\alpha]^{1/2} && r < a \\ &= n_0[1 - 2\Delta]^{1/2} = n_c && r > a \end{aligned}$$

with $1 \leqslant \alpha \leqslant \infty$. These solutions assume $(a/n)(\mathrm{d}n/\mathrm{d}r) \ll 1$ and $V = (\beta_0^2 - \beta_c^2)^{1/2} a \gg 1$ and may be shown to be equivalent to the ray analysis.

Note that $\alpha = \infty$ corresponds to a step-index and $\alpha = 2$ is a parabolic index (see Fig. 6.1).

The number of propagating modes is

$$M = \frac{\alpha}{(\alpha + 2)} a^2 \beta_0^2 \Delta = \frac{\alpha}{(\alpha + 2)} M_\infty$$

and they again form Q degenerate groups with propagation constants given by

$$\beta_q = \beta_0[1 - 2\Delta(q/Q)^{2\alpha/(\alpha+2)}]^{1/2}$$

Except when α is in the region of 2, the total spread in propagation times is

$$\Delta T \simeq \frac{N_0 l}{c} \frac{(\alpha - 2 - 4\delta)}{(\alpha + 2)} \Delta$$

where

$$\delta = -\frac{n_0}{2N_0} \frac{\lambda}{\Delta} \frac{d\Delta}{d\lambda}$$

The minimum value of ΔT occurs when $\alpha = 2(1 + 2\delta - \Delta)$ and thus depends on the dispersive properties of the dopant through δ. Its value is

$$(\Delta T)_{\text{min}} \simeq \frac{N_0 l}{8c} \Delta^2$$

For $\Delta \simeq 0.01$ this represents a reduction of about 10^3 in intermode dispersion.

Root mean square pulse widths (σ_1) may be calculated by taking into account the energy and arrival time of each mode group (see Fig. 6.7). Minimization of σ_1 requires a value of $\alpha \simeq 2(1 + 2\delta - 6\Delta/5)$. Then

$$(\sigma_1)_{\text{min}} = \frac{0.029 N_0 l}{c} \Delta^2$$

Total dispersion, $(\sigma/l) = [(\sigma_1/l)^2 + (\sigma_2/l)^2]^{1/2}$, is shown as a function of α on Fig. 6.8 for varying levels of material dispersion, $(\sigma_2/l) = Y_{\text{m}} \gamma_\sigma / c$.

Fiber irregularities having a periodicity of about 1 mm or less cause strong coupling between mode groups. This averages out the overall propagation time and causes σ_1 to increase as $l^{1/2}$. It also causes coupling to unguided modes and so increases attenuation.

7

Basic Semiconductor Properties

7.1 INTRODUCTION

Whereas Chapters 2–6 have dealt with the optical fiber as a transmission medium, Chapters 7–13 will be concerned with the generation and detection of the optical signals.

The properties required of an optical source for optical fiber communication are easily stated. It must have a high radiance† over a narrow band of wavelengths in the range 0.8–1.7 μm. The output should be easily modulated. The emissive area should be no greater than that of the fiber core. The angular distribution of the radiation should as far as possible match the acceptance cone of the fiber. These requirements are quite different in several ways from the properties sought in conventional optical sources where it is normally the total radiant power† and often the total luminous† (i.e. visible) output that matters. Efficiency, cost and reliability are always important and should not be overlooked, neither should the efficient coupling of the output radiation into the fiber, the stability of the output power—recognizing that it may be subject both to long-term drift and to high-frequency fluctuations—nor the likely life of the source. Semiconductor sources generating light at a p–n junction by the process known as injection luminescence fulfill these requirements better than any others.

At the receiver end of the system the detector used to convert the received optical power back into an electrical signal also requires the highest possible conversion efficiency. Its response time must match the optical modulation rate and its active area should match that of the fiber core. Two important further requirements are that it should not introduce a significant, excess random noise into the received signal and that the self-capacitance that it presents to the input of the receiver amplifier should be minimal. In all respects semiconductor junction detectors are much the most appropriate.

Essentially two types of semiconductor optical source may be used: the light-emitting diode (LED) and the laser diode. The radiation from an LED is

† A brief introduction to radiometric and photometric quantities and their definitions is given in Appendix 4.

incoherent and is emitted over a wide range of angles. It covers a broad spectrum of wavelengths, with, typically, $\gamma = |\Delta\lambda/\lambda| \simeq 0.03$. To a good approximation the emitted power is proportional to the diode current, although there is a tendency for the output to saturate at high power levels as the semiconductor temperature rises. The modulation rate is limited to about 100 MHz. Above this, efficiency is sacrificed.

The behavior of the laser diode is more complicated. It is constructed so that above a threshold current the light generation mechanism changes. At low currents radiation is produced in the same way as in the LED, by spontaneous emission. But above the threshold the generation becomes dominated by stimulated emission. As a result there is a change in the character of the emitted radiation: it becomes more directional, more coherent and its spectrum takes the form of one or more very narrow lines. Both the threshold current and the spectrum are sensitive to temperature and thus may vary with the ambient conditions and between or during pulses of high power. The benefits to set against this increased complexity are that a higher power can be coupled onto a fiber and that significantly higher modulation frequencies may be used.

There are also two types of semiconductor photodetector to consider: the p–i–n photodiode and the avalanche photodiode (APD). Both will convert the incident radiation back to an electric current, provided that the wavelength is shorter than some threshold value. The current generated is directly proportional to the received power. Both types are normally biassed, but the APD uses a high bias voltage (100–300 V) in order to produce internal multiplication of the initially generated current. This gain is strongly temperature dependent. The APD is thus a more complex detector to make and to use, but it does offer increased sensitivity to very low optical powers, and so in most cases it allows an extension of the repeater spacing in an optical communication link.

For reasons that will be discussed in detail in Chapters 8 and 12 the ranges of optical wavelengths over which these semiconductor sources and detectors are most efficient are determined by the semiconductor band-gap energy. They are, therefore, specific to the semiconductors used. As we shall see, it is best for the band-gap energy of the detector material to be slightly less than that of the source material. The first generation of optical fiber communication systems has been exclusively based on gallium arsenide (GaAs) sources and silicon (Si) detectors, both of which were already highly developed and readily available. However, it should be said that a great deal of further development was needed in order to make optical sources suitable for communication purposes. This particular pair of materials is ideal for operation at wavelengths in the region of 0.8–0.9 μm, but to take advantage of the improved fiber propagation characteristics at longer wavelengths requires the use of different semiconductors for both source and detector. Germanium (Ge) detectors are already well developed for wavelengths out to 1.7 μm, but some quite new semiconductor materials have been developed for use as sources and detectors at wavelengths longer than 1 μm.

In order to appreciate the design criteria applicable to semiconductor optical sources and detectors and to understand those properties which may limit their performance we shall need to delve into the theory of semiconductors and semiconductor junctions and to examine the detailed construction of certain representative devices. In the remainder of this chapter we give a brief outline of the basic semiconductor properties relevant to the devices used for optical generation and detection. In Chapter 8 we discuss the physical origin of the optical emission mechanism. In Chapter 9 we describe the structure of a type of semiconductor diode that gives a high output efficiency and has been essential for the room-temperature operation of continuously running semiconductor lasers. This is the heterostructure. Laser action in semiconductors is the subject of Chapters 10 and 11, and finally in Chapters 12 and 13 we deal with p–i–n and APD detectors.

7.2 INTRINSIC AND IMPURITY SEMICONDUCTORS

7.2.1 Intrinsic Semiconductors

The emission of radiation as a result of the recombination of excess electrons and holes created at a forward-biassed p–n junction is known as *injection luminescence.* In the converse process light which enters a piece of semiconductor may give rise to the creation of an electron–hole pair in a quantum interaction in which a single photon is absorbed. If there is an electric field present, as there is at a p–n junction, the electron and hole will separate and a change in current may be detected in any electrical circuit of which the semiconductor is an element. In this section we shall give a brief review of the properties of some of the semiconductor materials that may be used for generating or detecting light by means of these mechanisms.

Materials that are intrinsic semiconductors are characterized by an electrical conductivity that is appreciably lower than that of metals and rises rapidly with temperature. These properties may be understood in terms of the familiar diagram showing the permitted electron energy levels within the material. The levels fall into bands as shown in Fig. 7.1(a). The vacuum level, $\mathcal{E}_0$, represents the energy of an electron at rest outside the surface of the semiconductor. The highest band of allowed levels inside the material, the conduction band, extends from the vacuum level down to an energy, $\mathcal{E}_c$, and is normally empty at low temperatures. The depth of this band, the energy difference, $\chi = \mathcal{E}_0 - \mathcal{E}_c$, is known as the *electron affinity* of the material. The conduction band is separated by the energy gap, $\mathcal{E}_g$, from the next highest band, the valence band, which is normally completely filled.

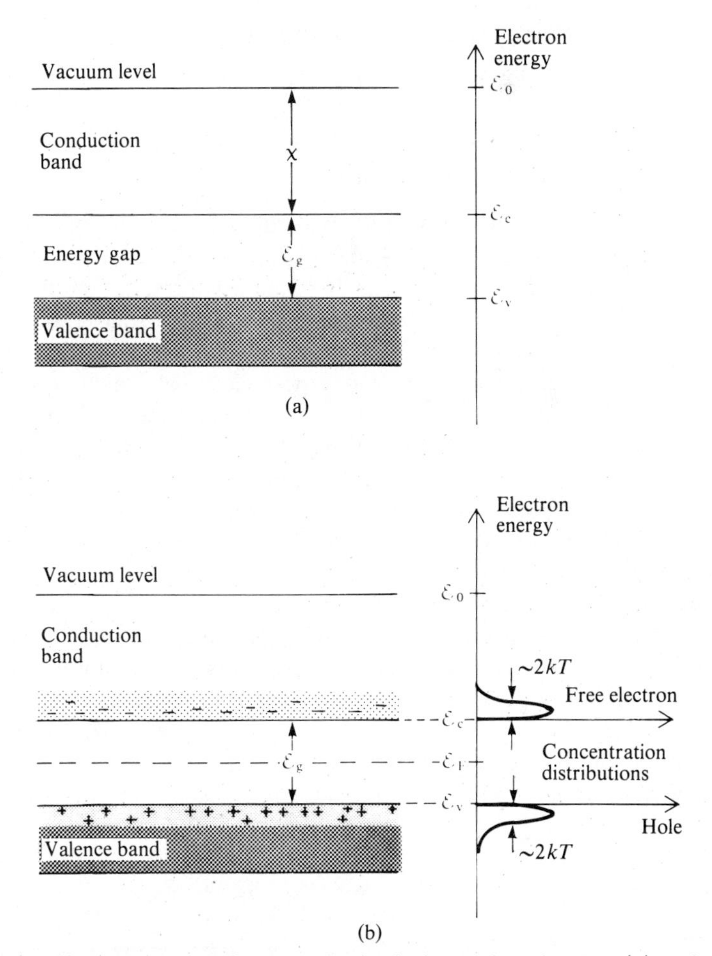

Fig. 7.1 Electron energy diagrams for intrinsic semiconductors: (a) an intrinsic semiconductor at low temperatures behaves as an insulator; (b) an intrinsic semiconductor such that $10 < \mathcal{E}_g/kT < 100$, with the result that a significant concentration of electrons is thermally excited across the forbidden energy gap, $\mathcal{E}_g$. This sets up the electron and hole concentration distributions shown to the right.

As the temperature is raised, some electrons are thermally excited across the band gap, giving rise to an appreciable concentration, n, of free electrons in the conduction band and leaving behind an equal concentration, p, of vacancies or *holes* in the valence band. This is shown schematically in Fig. 7.1(b). Both the free electrons and the holes are mobile within the material so both may contribute to the electrical conductivity. Their concentration is known as the intrinsic carrier concentration, n_i, and is given by

$$n = p = n_i = K \exp(-\mathcal{E}_g/2kT) \tag{7.2.1}$$

where

$$K = 2(2\pi kT/h^2)^{3/2}(m_e m_h)^{3/4} \tag{7.2.2}$$

is a constant that is characteristic of the material and varies as the $\frac{3}{2}$ power of the temperature, T. In eqns. (7.2.1) and (7.2.2) k is Boltzmann's constant, 1.38×10^{-23} J/K; h is Planck's constant, 6.626×10^{-34} J.s; m_e and m_h are the effective masses of the electrons and holes, which are often smaller, by a factor of ten or more, than the free-space electron rest mass of 9.11×10^{-31} kg.

To a first approximation we should expect electrical conductivity to be proportional to the carrier concentration, so the electrical conductivity of a band-gap material ought to vary exponentially with the ratio $(\mathcal{E}_g/kT)$. When $(\mathcal{E}_g/kT)$ exceeds about 100, the electrical conductivity is so low that the material would be regarded as an insulator. When $(\mathcal{E}_g/kT)$ is less than about 10, the material behaves as a semimetal.

In Fig. 7.1(b) we have indicated a reference energy, $\mathcal{E}_F$, that is characteristic of the material. This is the ***Fermi energy***. An electron energy level at $\mathcal{E}_F$, should one exist there, would have a 50% probability of being occupied. The energy difference, $\phi = \mathcal{E}_0 - \mathcal{E}_F$, is known as the *work function*. Energy levels more than a few kT above $\mathcal{E}_F$ have a very low probability of being occupied; those more than a few kT below $\mathcal{E}_F$ are very unlikely to be empty. The probability, $F(\mathcal{E})$, that an allowed electronic state at an energy, $\mathcal{E}$, be occupied is given by the Fermi function:

$$F(\mathcal{E}) = \frac{1}{[1 + \exp(\mathcal{E} - \mathcal{E}_F)/kT]} \tag{7.2.3}$$

When $(\mathcal{E} - \mathcal{E}_F) \gg kT$,

$$F(\mathcal{E}) \simeq \exp\left[-\frac{(\mathcal{E} - \mathcal{E}_F)}{kT}\right] \tag{7.2.4}$$

When $(\mathcal{E}_F - \mathcal{E}) \gg kT$,

$$1 - F(\mathcal{E}) \simeq \exp\left[-\frac{(\mathcal{E}_F - \mathcal{E})}{kT}\right] \tag{7.2.5}$$

In the valence band $[1 - F(\mathcal{E})]$ represents the probability that a level at energy, $\mathcal{E}$, is unoccupied, that is, that there is a hole present at that energy.

The elements germanium and silicon from group IV(b) of the periodic table, for which $\mathcal{E}_g = 0.7$ eV and 1.1 eV respectively, are well-known as room-temperature semiconductors. So too are several binary compounds made from elements from groups III(b) and V(b) of the periodic table, in particular aluminum, gallium and indium from group III and phosphorus, arsenic and antimony from group V. Two common examples are gallium arsenide, GaAs, and indium phosphide, InP. To aid this discussion a periodic table of the elements is reproduced as Table 7.1.

The interatomic bonds in all these semiconducting materials are predominantly covalent and the materials crystallize in a tetrahedral structure similar to those shown in Fig. 7.2. The group III and group V elements are almost totally miscible in the solid state, so that more complex ternary and quaternary solid solutions can be formed. Examples include $Ga(As_{1-y}P_y)$, $(Ga_{1-x}Al_x)As$, $(In_{1-x}Ga_x)(As_{1-y}P_y)$ and $(Ga_{1-x}Al_x)(As_ySb_{1-y})$. The parameters x and $1-x$, and y and $1-y$ simply represent the proportions of the pairs of group III and group V elements, respectively, and cover the whole range of values from 0 to 1. We have adopted the convention of using x and $1-x$ for the group III elements, and y and $1-y$ for the group V elements. Furthermore, x and y are the proportions of the material with the lower atomic weight.

Each of these semiconductors has its own characteristic properties. These include the band-gap energy, $\mathcal{E}_g$, the electron and hole mobilities, μ_e and μ_h, and the crystal lattice spacing, a_0. These and other properties vary continuously with the parameters x and y for the mixed compounds. For this reason the study of the ternary and quaternary systems, seeking semiconductors with the best combination of properties, has played an important part in the development of sources and detectors for longer wavelengths. Of the many possibilities, the quaternary system InGaAsP has proved to be the most suitable for LEDs and lasers, while the ternary InGaAs has been used to make detectors which compare favorably with germanium. With so many available combinations these preferences are of course something which may change with further work.

Some particular properties of the group IV and the III–V binary compound semiconductors are set out in Table 7.2. Much of this information, together with references, is to be found in Part B of Ref. [7.1]. The variations of $\mathcal{E}_g$ and a_0 with x and y are displayed diagrammatically for the ternary compounds in Fig. 7.3 and for two of the quaternary systems, InGaAsP and GaAlAsSb, in Fig. 7.4.

The range of semiconductor materials can be further extended by the inclusion of elements from groups II, IV and VI of the periodic table, whilst maintaining the correct total number of valence electrons. It has not so far been found possible to make useful communication sources or detectors from such materials, so we shall not consider them here.

Table 7.1 The periodic table of the elements

Group O	Group I a	Group I b	Group II a	Group II b	Group III a	Group III b	Group IV a	Group IV b
2 He Helium	3 Li Lithium	1 H Hydrogen		4 Be Beryllium		5 B Boron		6 C Carbon
10 Ne Neon	11 Na Sodium			12 Mg Magnesium		13 Al Aluminum		14 Si Silicon
18 A Argon	19 K Potassium	29 Cu Copper	20 Ca Calcium	30 Zn Zinc	21 Sc Scandium	31 Ga Gallium	22 Ti Titanium	32 Ge Germanium
36 Kr Krypton	37 Rb Rubidium	47 Ag Silver	38 Sr Strontium	48 Cd Cadmium	39 Y Yttrium	49 In Indium	40 Zr Zirconium	50 Sn Tin
54 Xe Xenon	55 Cs Cesium	79 Au Gold	56 Ba Barium	80 Hg Mercury	57 La – 71 Lu Lanthanum Lutecium (Rare Earths)	81 Tl Thallium	72 Hf Hafnium	82 Pb Lead
86 Rn Radon	(87 Fr) Francium		88 Ra Radium		89 Ac Actinium		90 Th Thorium	

Table 7.2 Room temperature† properties of semiconductors

Material	Band gap Energy [eV]	Band gap Type§	Cutoff wavelength [μm]	Lattice spacing a_0/[nm]	Effective masses $\frac{m_e}{m_{e0}}$	Effective masses $\frac{m_h}{m_{e0}}$	Intrinsic carrier concentration n_i/[m^{-3}]
Ge	0.66	I	1.88	0.5657	0.22	0.3	2×10^{19}
Si	1.11	I	1.15	0.5431	0.97	0.5	2×10^{16}
AlP	2.45	I	0.52	0.5451	—	0.70	$\sim 10^{5}$
AlAs	2.16	I	0.57	0.5661	0.15	0.79	$\sim 10^{7}$
AlSb	1.58	I	0.75	0.6135	0.12	0.98	$\sim 10^{11}$
GaP	2.26	I	0.55	0.5451	0.82	0.60	$\sim 10^{6}$
GaAs	1.42	D	0.87	0.5653	0.07	0.48	$\sim 10^{13}$
GaSb	0.73	D	1.70	0.6096	0.04	0.44	$\sim 10^{19}$
InP	1.35	D	0.92	0.5870	0.08	0.64	$\sim 10^{14}$
InAs	0.36	D	3.5	0.6058	0.02	0.40	$\sim 10^{21}$
InSb	0.17	D	7.3	0.6479	0.014	0.40	$\sim 10^{22}$

† Except, of course, the melting points.

Group V a	Group V b	Group VI a	Group VI b	Group VII a	Group VII b	Group VIII		
	7 N Nitrogen		8 O Oxygen		9 F Fluorine			
	15 P Phosphorus		16 S Sulfur		17 Cl Chlorine			
23 V Vanadium	33 As Arsenic	24 Cr Chromium	34 Se Selenium	25 Mn Manganese	35 Br Bromine	26 Fe Iron	27 Co Cobalt	28 Ni Nickel
41 Nb Niobium	51 Sb Antimony	42 Mo Molybdenum	52 Te Tellurium	(43 Tc) Technetium	53 I Iodine	44 Ru Ruthenium	45 Rh Rhodium	46 Pd Palladium
73 Ta Tantalum	83 Bi Bismuth	74 W Tungsten	84 Po Polonium	75 Re Rhenium	(85 At) Astatine	76 Os Osmium	77 Ir Iridium (Noble metals)	78 Pt Platinum
(91 Pa) Protactinium		92 U Uranium						

Electron affinity χ/[eV]	Carrier mobilities μ_e [m²/V.s]	μ_h [m²/V.s]	Relative permittivity ε_r	Refractive index μ	Thermal conductivity [W/(m.K.)]	Melting point [K]
4.13	0.39	0.19	16	4.0	64	1210
4.01	0.15	0.06	11.8	3.5	145	1693
—	—	—	9	3.0	90	2803
(2.62)	0.03	—	10.1	3.2	91	2013
3.64	0.02	0.04	14.4	(3.7)	57	1333
(4.0)	0.017	0.01	11.1	3.45	77	1740
4.05	0.85	0.04	13.1	3.7	44	1511
4.03	0.40	0.14	15.7	3.8	33	983
(4.4)	0.4	0.015	12.4	3.45	68	1335
4.54	3.2	0.05	14.6	3.5	27	1215
4.59	7.8	0.4	17.7	4.0	17	798

§ 'D' indicates, a direct band-gap material. 'I' indicates an indirect band-gap material (see Section 8.3).

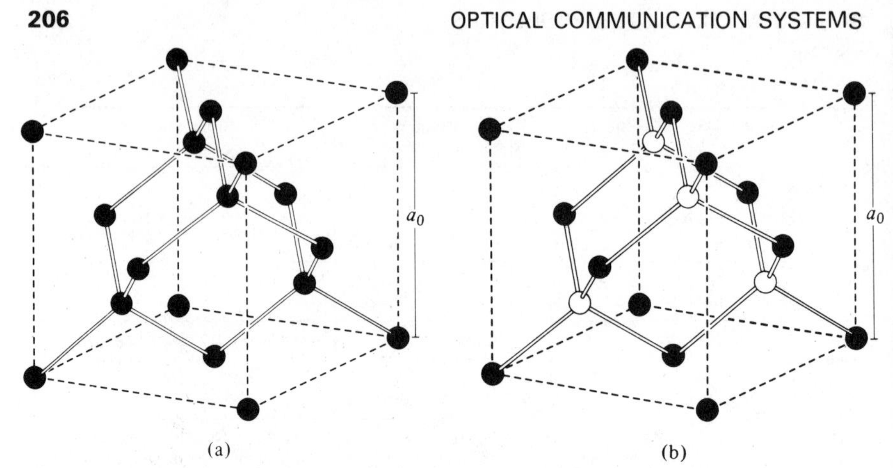

Fig. 7.2 Tetrahedral crystal structures of group IV elements and III–V compound semiconductors: (a) crystal structure of diamond, typical of C, Si, Ge; (b) crystal structure of zincblende (ZnS) typical of GaAs and other III–V compounds. [From C. Kittel, *Introduction to Solid-State Physics*, 4th edn, John Wiley (1971).]

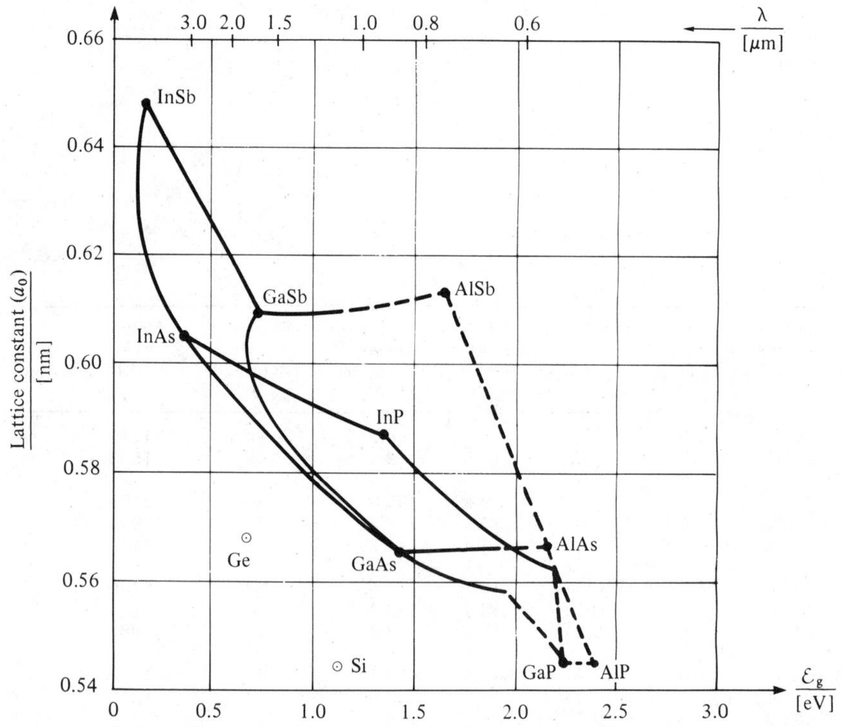

Fig. 7.3 Diagram showing the variation of lattice parameter, a_0, with band-gap energy, $\mathcal{E}_g$, as the composition of the III–V ternary compounds is varied. The dotted lines indicate composition ranges which give rise to indirect band-gap material.

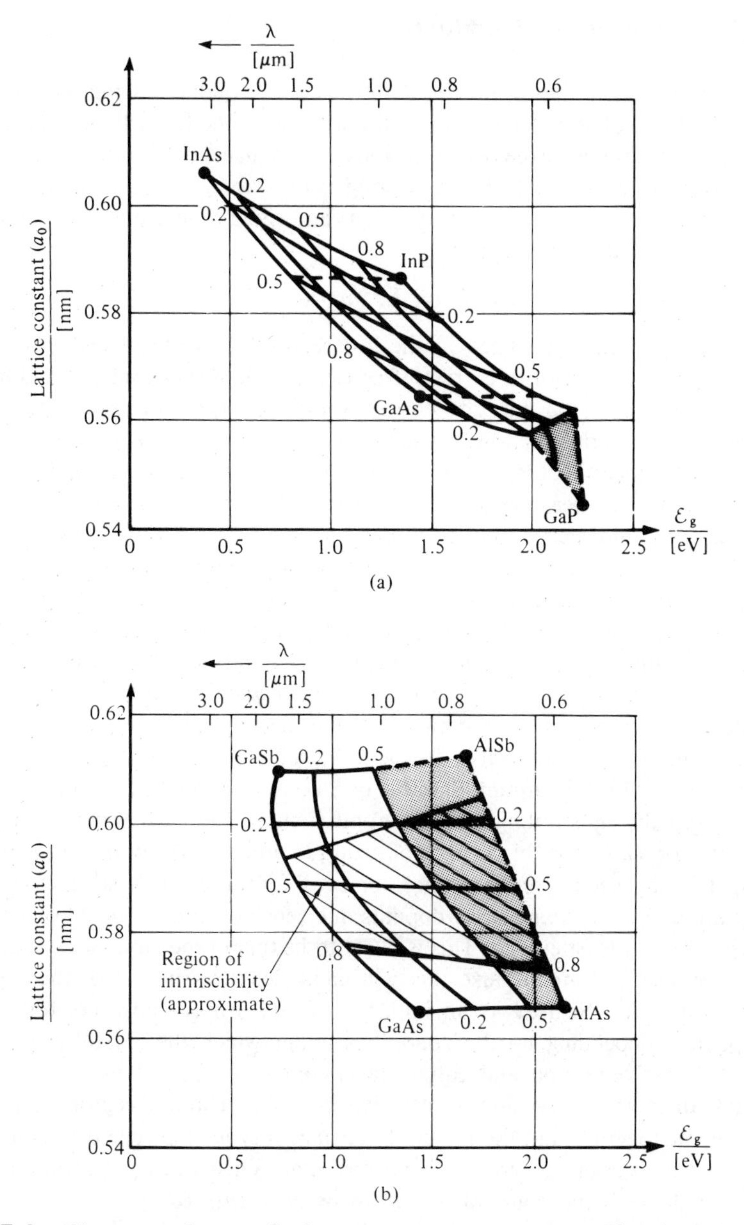

Fig. 7.4 Diagrams of a_0 vs $\mathcal{E}_g$ for two quaternary systems. The shaded areas indicate composition ranges which give rise to indirect band-gap material: (a) $(In_{1-x}Ga_x)(As_{1-y}P_y)$—the dashed lines indicate the ranges of band-gap energies that can be obtained with quaternary material that is lattice-matched to substrates of GaAs and InP; (b) $(Ga_{1-x}Al_x)(As_ySb_{1-y})$—the striped region cannot be obtained because of the immiscibility of the components.

7.2.2 Impurity Semiconductors

The electrical properties of nearly all intrinsic semiconductors can be modified by the addition of small quantities of impurity elements. In particular they may be doped so that an excess of electrons is produced (n-type) or so that an excess of holes is created (p-type). As long as the doping concentrations are not excessive, the product of the electron and hole concentrations remains independent of the doping level:

$$np = n_i^2 = K^2 \exp(-\mathcal{E}_g/kT) \tag{7.2.6}$$

using the symbols introduced in the last section. This means that in doped material there are majority carriers (free electrons in n-type material and holes in p-type material) and minority carriers (holes in n-type and free electrons in p-type). The material is known as an *impurity semiconductor*, or an impurity-controlled semiconductor, when the doping concentration is so high that it, rather than the temperature, is the main factor determining the total number of free carriers and hence the electrical conductivity. Electron energy level diagrams for n-type and p-type impurity semiconductors are shown in Fig. 7.5.

The group IV semiconductors, Si and Ge, can be made n-type by introducing small concentrations of the group V elements, P or As, as *donor*, substitutional impurities. They can be made p-type with small concentrations of group III elements, B or Ga, which act as *acceptor* impurities.

The III–V compound semiconductors can be made n-type by substituting some atoms from group VI of the periodic table (e.g. Se, Te) onto group V lattice sites, or by substituting Si or Ge or Sn onto group III sites. They may be made p-type by substituting divalent atoms such as Zn or Cd in place of group III atoms or Si, Ge or Sn in place of group V atoms. Clearly the atoms from group IV may become either donors or acceptors in a III–V semiconductor. They are known as *amphoteric* dopants and tend to occupy whichever lattice site requires the least energy. This is likely to be that of the atom nearest in size to the impurity atom because this will cause least disruption of the crystal lattice. However, Si used as an impurity in GaAs may substitute for either component, depending on the conditions under which the crystal is formed. Thus successive n-type and p-type layers may be created using the same dopant. In most semiconductor materials both donor and acceptor impurities are present together and the material becomes n-type or p-type depending on which is present in the greater concentration. When the concentrations are closely balanced, the material is said to be *compensated*.

It can be seen in Fig. 7.5 that the n-type impurities give rise to localized energy levels just below $\mathcal{E}_c$. Because this is usually also well above $\mathcal{E}_F$, almost all of the donors are ionized. However, a consequence of the enhanced free-electron population is that the position of the Fermi energy is raised within the

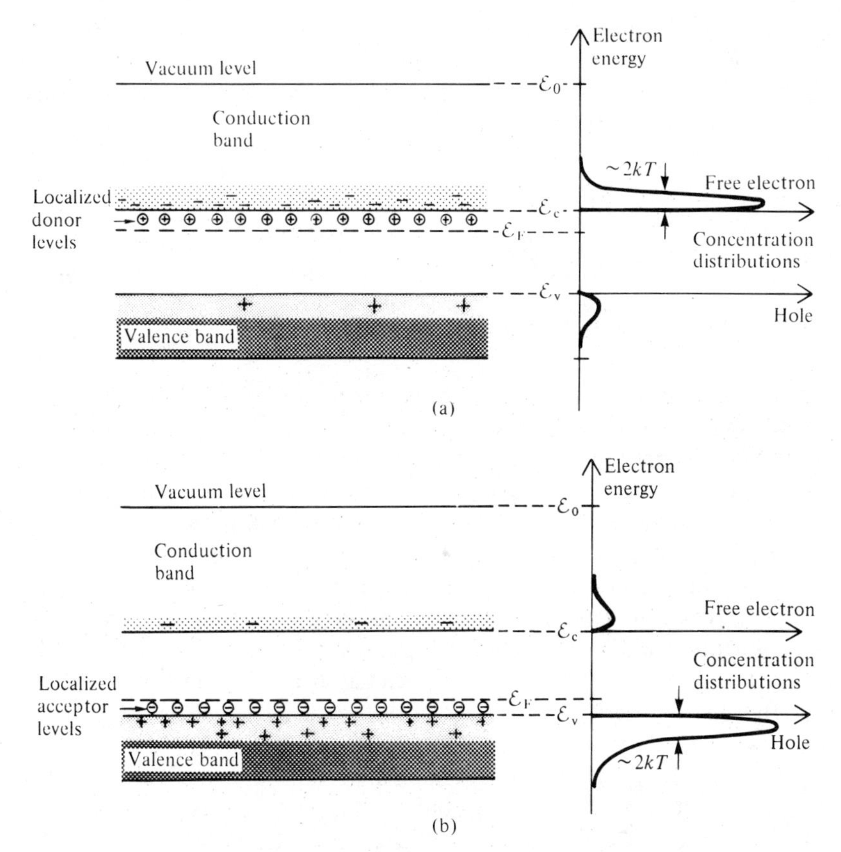

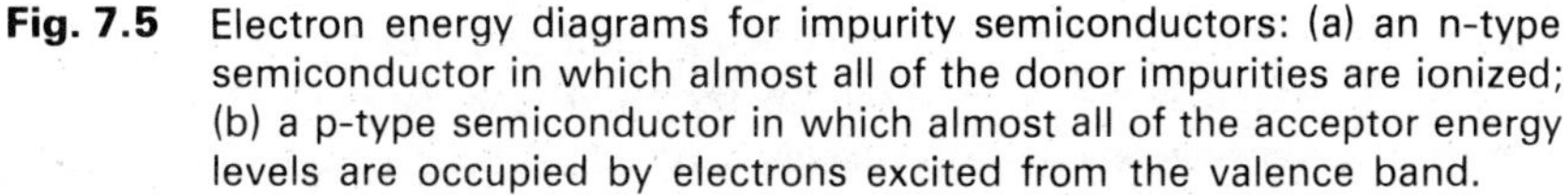

Fig. 7.5 Electron energy diagrams for impurity semiconductors: (a) an n-type semiconductor in which almost all of the donor impurities are ionized; (b) a p-type semiconductor in which almost all of the acceptor energy levels are occupied by electrons excited from the valence band.

band gap. Conversely the p-type impurities give rise to localized energy levels in the band gap near to $\mathcal{E}_v$. Being well below $\mathcal{E}_F$ these acceptor levels are normally occupied by electrons excited out of the valence band. The increased hole population has the effect of lowering $\mathcal{E}_F$.

In what has been said so far it has been tacitly assumed that the doping concentrations remain small, not greater than about 10^{24} m^{-3} in Si and Ge and several orders of magnitude less than this in III–V semiconductors. When such high doping levels are reached, three important effects arise. The first is that the impurity levels interact with one another and as a result the range of energy levels they occupy spreads and may start to merge with the band edge. The second is that the band edge itself is perturbed and a *band tail* is formed. The band gap is therefore made narrower. The third is that the Fermi energy moves up into the conduction band in n-type material, or down into the valence band in p-type. The semiconductor is then said to be ***degenerate***. These effects will be illustrated in more detail in Section 8.2.

7.3 THE p–n JUNCTION

7.3.1 The p–n Junction in Equilibrium

It is normally possible to form an abrupt junction in the bulk semiconductor between one region doped n-type and another that is p-type. In equilibrium, that is with no applied voltages or thermal gradients, the Fermi level is uniform throughout the material. There is then a detailed balance between the electron concentrations at all energies in all parts. As a result of this, the electron energy levels across an n^+p junction are as shown in Fig. 7.6. The use of the designation n^+ is simply to indicate that the n-type region is rather more heavily doped than the p-type region. As a result its Fermi energy is positioned quite close to the conduction band edge. The internal potential difference, V_D, set up across the junction is known as the diffusion potential. It has the effect of obstructing the diffusion of majority carriers out of their respective regions, which would otherwise result from the abrupt change in concentration. Thus the flows of electrons out of and into the n-type region are brought into equilibrium and the same applies to the flows of holes out of and into the p-type region.

We may deduce from eqns. (7.2.4) and (7.2.5) that the majority carrier concentrations will decrease exponentially as the internal potential varies from its value in the bulk material. As a result the region around the junction where the potential is varying is relatively depleted of carriers. Indeed it is normally referred to as the ***depletion region*** or the ***depletion layer***. The variation of the concentration of free electrons and holes across a p–n junction in equilibrium is shown in Fig. 7.7. The following notation has been used:

n_{n0} is the equilibrium concentration of majority electrons in the n-type material.

p_{n0} is the equilibrium concentration of minority holes in the n-type material.

p_{p0} is the equilibrium concentration of majority holes in the p-type material.

n_{p0} is the equilibrium concentration of minority electrons in the p-type material.

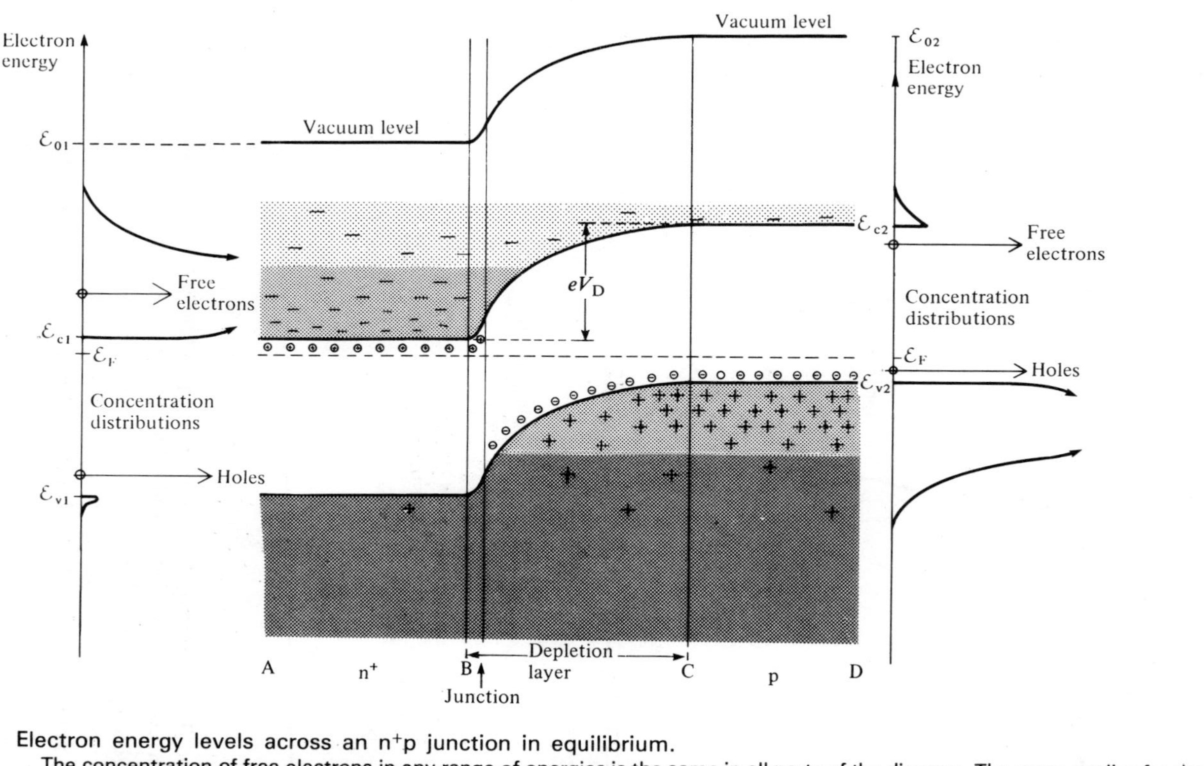

Fig. 7.6 Electron energy levels across an n^+p junction in equilibrium.
The concentration of free electrons in any range of energies is the same in all parts of the diagram. The same applies for the holes. The diagrams on the left and right hand edges indicate schematically the free-electron and hole concentration distributions in the bulk semiconductor regions AB and CD. Note the change in the vacuum level, $\mathcal{E}_0$, outside the surfaces of the n-type and the p-type material. This implies that there is a contact potential difference across the junction.

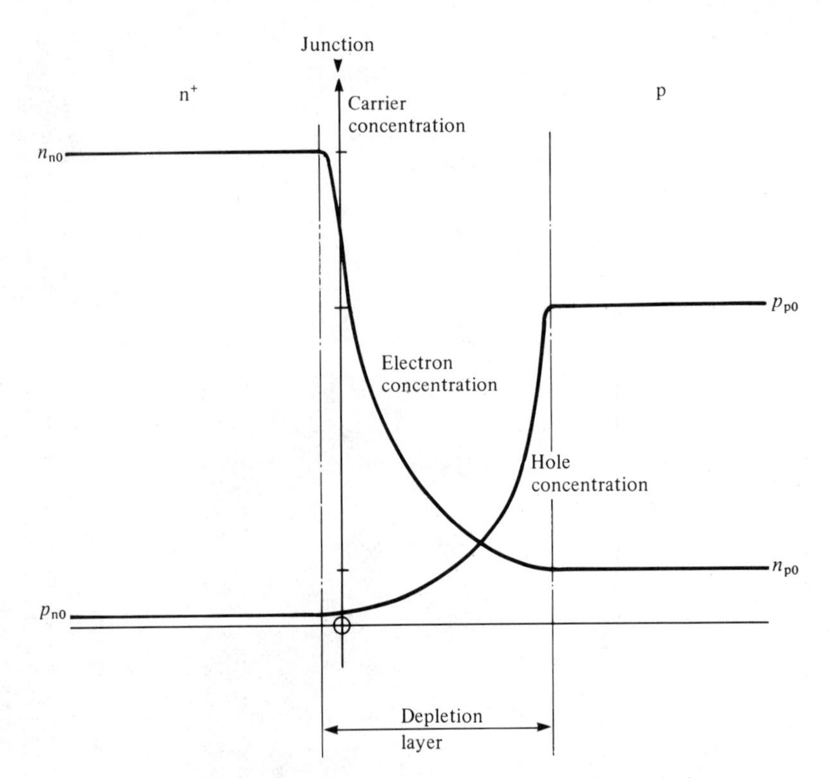

Fig. 7.7 Schematic illustration of the carrier concentration variations across an n^+p junction in equilibrium.

Thus

$$n_{n0}p_{n0} = n_{p0}p_{p0} = n_i^2 \tag{7.3.1}$$

We may note at this point that in an n-type impurity semiconductor, the free-electron concentration, n_{n0}, will effectively be just the net concentration of donor impurities, n_D. These are assumed to be all ionized and to swamp the hole concentration. Similarly in a p-type semiconductor we may assume that p_{p0} is simply n_A, the net concentration of acceptor impurities. Then using eqn. (7.3.1) we may put

$$n_{p0} = n_i^2/p_{p0} \simeq n_i^2/n_A \tag{7.3.2}$$

and

$$p_{n0} = n_i^2/n_{n0} \simeq n_i^2/n_D \tag{7.3.3}$$

7.3.2 The Biassed p–n Junction

An external potential difference applied to the n-type and p-type regions, across the junction, will upset the equilibrium. Almost all of the applied voltage will appear across the junction where it will have the effect of raising or lowering the potential barrier, depending on its polarity. When the potential of the p-type region is made more positive than that of the n-type region by a voltage V, then the barrier is lowered. As a result, the flux of the majority carriers across the junction increases by the factor $\exp(eV/kT)$, but the counter flows of minority carriers remain unchanged. We should thus expect the current/voltage characteristic of the junction to take on the form

$$I_1 = I_{s1}[\exp(eV/kT) - 1] \quad (7.3.4)$$

where I_{s1}, the saturation current, represents the balancing currents that flow in each direction across the junction in the absence of any external voltage. The saturation current, I_{s1}, is proportional to the area of the junction and to n_i^2. It thus depends exponentially on $(\mathcal{E}_g/kT)$.

Junctions produced in most semiconductor materials are found to show deviations from eqn. (7.3.4). These mainly result from two effects. The first is the generation and recombination of carriers either at the surface or within the depletion layer. This can give rise to an additional current following an I–V characteristic of the form

$$I_2 = I_{s2}[\exp(eV/2kT) - 1] \quad (7.3.5)$$

As I_{s2} is proportional to n_i, I_2 tends to be more important in wide band-gap semiconductors where it tends to dominate the total current $I = I_1 + I_2$ at low values. The second effect is the result of ohmic losses within the bulk semiconductor and at the contacts. This causes a further voltage, proportional to the total current, to appear at the device terminals. It tends to dominate the junction characteristics at high currents. Most of these matters are fully discussed in many texts on semiconductor devices, and readers are referred in particular to [7.2] for more detailed theory on this and on all aspects of semiconductor device behavior. It is a subject we shall return to in Section 9.1.

It should be noted that eqns. (7.3.4) and (7.3.5) apply for negative as well as positive values of V. With $V < 0$, $I \rightarrow I_s = I_{s1} + I_{s2}$. For reasons that will become clear, optical detectors in communication systems are normally p–n junctions operated with $V < 0$, that is in reverse bias. Further consideration of this mode of operation is deferred to Section 7.6.

What matters to us now is that with $V > 0$, when the junction is forward biassed, excess carriers are injected into the bulk semiconductor on either side of the depletion layer, that is, into regions where they are the minority carrier.

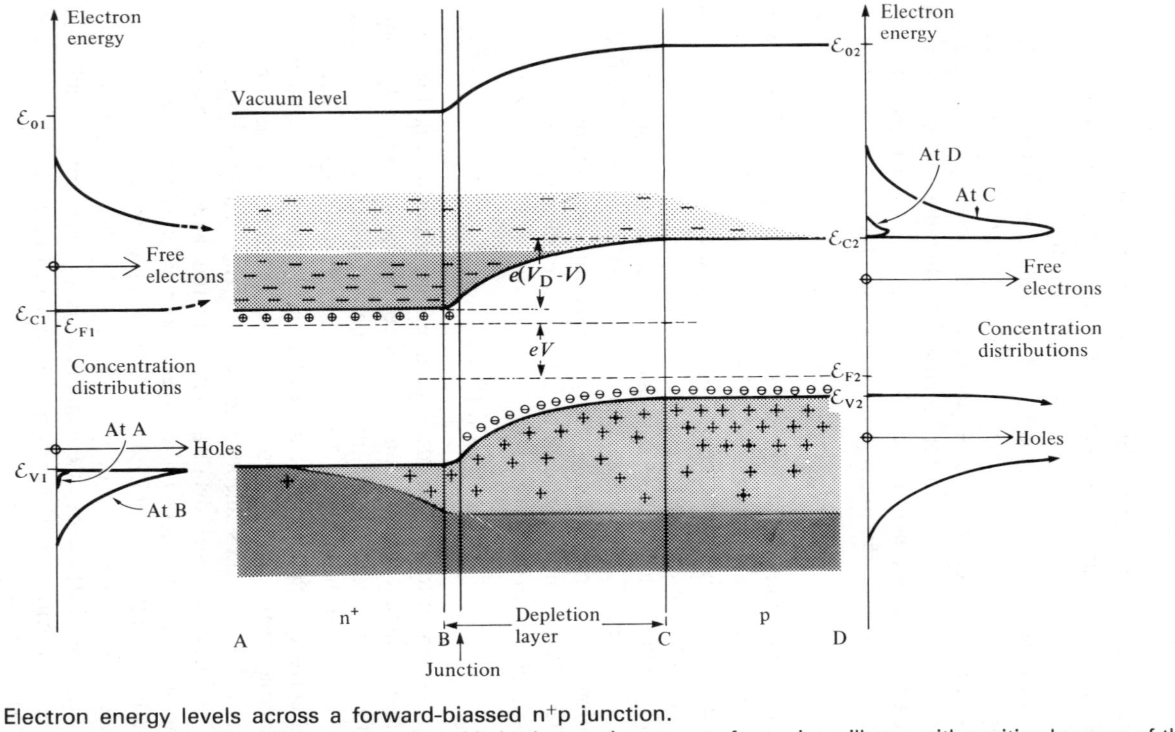

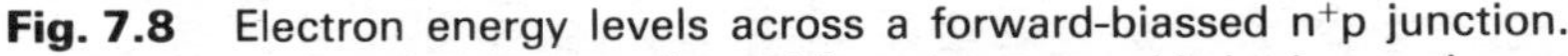

Fig. 7.8 Electron energy levels across a forward-biassed n^+p junction.

Here the concentrations of free-electrons and holes in any given range of energies will vary with position because of the carrier diffusion that takes place outside the depletion layer. For electrons in the conduction band the distribution remains in quasi-equilibrium from A to C. Diffusion between C and D causes the electron concentration to fall until the equilibrium concentration distribution characteristic of the bulk material is reached. Holes remain in quasi-equilibrium from D to B. Diffusion between B and A causes the hole concentration to fall until the equilibrium concentration, p_{n0} is reached.

Because the electric field in the bulk semiconductor remains small, only sufficient to support the counterflow of majority carriers, the excess minorities flow away from either side of the depletion layer by diffusion. The minority carrier concentration gradient that supports this flow represents a balance between the injection of carriers over the depletion layer, which gives rise to the excess concentration, and the net carrier recombination, which is proportional to the excess concentration. It is in the recombination process that radiation may be created, thereby generating injection luminescence. It is, therefore, the regions just outside the depletion layer that we need to consider most carefully in order to understand the behavior of optical sources.

The electron energy levels on either side of a forward-biassed n^+p junction are shown in Fig. 7.8. The effect of forward bias on the carrier concentrations

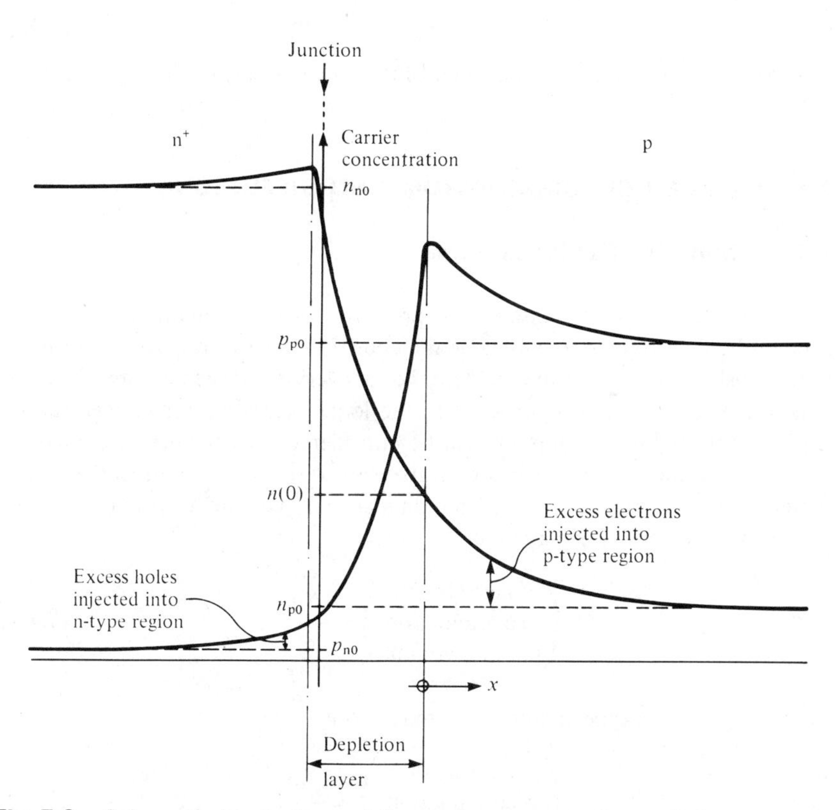

Fig. 7.9 Schematic illustrations of the carrier concentration variation across a forward-biassed n^+p junction. [This is a very schematic sketch since the variation of carrier concentration under realistic bias conditions is much too large to portray accurately on a linear scale.]

across the junction is shown in Fig. 7.9 which should be compared with Fig. 7.7. It embodies the results of the analysis that follows.

In nonequilibrium conditions we will continue to use n and p for the local, instantaneous values of the free-electron and hole concentrations. Because it should always be clear whether we are dealing with n-type or p-type material further subscripts are felt to be unnecessary. The excess electron concentration in the p-type region outside the depletion layer may be written as $\Delta n = n - n_{p0}$. Because, by definition, the region is free of space charge, there is an equal excess concentration of the majority holes. Thus

$$\Delta n = n - n_{p0} = p - p_{p0} \tag{7.3.6}$$

Similarly, in the n-type region we may have an excess concentration of minority holes given by

$$\Delta p = p - p_{n0} = n - n_{n0} \tag{7.3.7}$$

We next have to consider the behavior of these excess carriers in greater detail.

7.4 CARRIER RECOMBINATION AND DIFFUSION

7.4.1 Minority Carrier Lifetime

In this section we shall examine the processes of recombination and carrier flow in the field-free regions outside the depletion layer of a biassed p–n junction. We shall assume for the moment that at every point the net rate of recombination of carriers is proportional to the local excess carrier concentration. The validity of this assumption will be considered further when we come to deal with the different processes that give rise to recombination. In the p-type material the net rate of recombination per unit volume is proportional to $(n - n_{p0}) = \Delta n$, and we may write

$$\begin{pmatrix} \text{net rate of} \\ \text{recombination} \\ \text{per unit volume} \end{pmatrix} = \frac{\Delta n}{\tau_p} \tag{7.4.1}$$

Similarly in the n-type material we may write

$$\begin{pmatrix} \text{net rate of} \\ \text{recombination} \\ \text{per unit volume} \end{pmatrix} = \frac{\Delta p}{\tau_n} \tag{7.4.2}$$

It is easy to show that the constants of proportionality, τ_p and τ_n, are the mean lifetimes of the excess minority carriers in the p-type and n-type material respectively: consider a piece of p-type semiconductor in which a uniform excess of carriers has previously been produced by some external agency. Then eqn. (7.4.1) applies uniformly throughout the material, and the net rate of recombination is everywhere equal to the rate of reduction of carrier concentration. Thus

$$\begin{pmatrix} \text{net rate of} \\ \text{recombination} \\ \text{per unit volume} \end{pmatrix} = -\frac{\mathrm{d}n}{\mathrm{d}t} = -\frac{\mathrm{d}(\Delta n)}{\mathrm{d}t} = \frac{\Delta n}{\tau_\mathrm{p}} \qquad (7.4.3)$$

and, solving for $\Delta n = \Delta n(0)$ at $t = 0$, we have

$$\Delta n(t) = \Delta n(0) \exp(-t/\tau_\mathrm{p}) \qquad (7.4.4)$$

The mean lifetime of the excess minority carriers is then given by

$$\langle t \rangle = \frac{\int_0^\infty t\, \Delta n(0) \exp(-t/\tau_\mathrm{p})\, \mathrm{d}t}{\int_0^\infty \Delta n(0) \exp(-t/\tau_\mathrm{p})\, \mathrm{d}t} = \tau_\mathrm{p} \qquad (7.4.5)$$

An identical argument may be applied to the holes in the n-type material, whose mean lifetime is τ_n.

7.4.2 The Diffusion Length

Let us return to the forward-biassed junction and examine a small cross-section of the electron flow into the p-type region. Consider the carriers passing through an element of area, δA, perpendicular to the flow and of thickness, δx, situated a distance, x, from the depletion-layer edge. This is illustrated in Fig. 7.10. The net rate, $(\mathrm{d}N/\mathrm{d}t)$, at which electrons accumulate in the volume, $\delta A\, \delta x$, is given by

$$\frac{\mathrm{d}N}{\mathrm{d}t} = D_\mathrm{e}\left(\frac{\mathrm{d}n}{\mathrm{d}x}\right)_x \delta A - D_\mathrm{e}\left(\frac{\mathrm{d}n}{\mathrm{d}x}\right)_{x+\delta x} \delta A$$

$$= -D_\mathrm{e}\left(\frac{\mathrm{d}^2 n}{\mathrm{d}x^2}\right)_x \delta x\, \delta A \qquad (7.4.6)$$

The electron diffusion coefficient, D_e, has units of m^2/s and is related to the

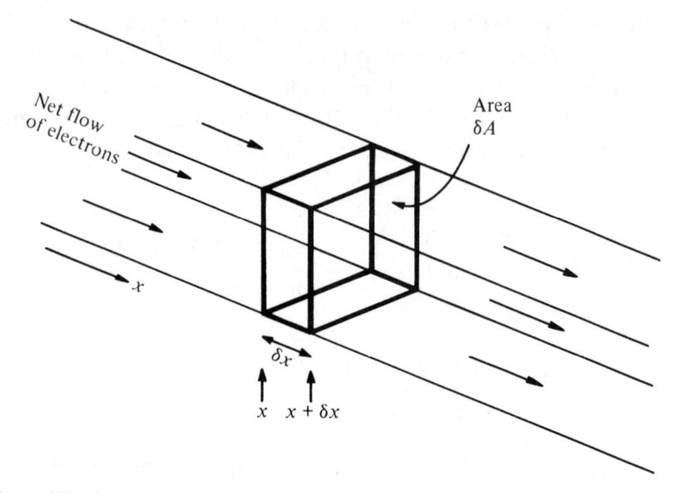

Fig. 7.10 Diffusion of electrons.
With an electron concentration $n(x)$ the flux of electrons per unit area is given by $D_e(dn/dx)_x$. The net rate of recombination of electrons and holes within the volume $\delta A\,\delta x$ is balanced by the net influx of electrons, that is by the change in $D_e\,(dn/dx)$ between x and $x + \delta x$.

electron mobility, μ_e, by the Einstein relation

$$D_e = \mu_e \frac{kT}{e} \tag{7.4.7}$$

An expression similar to eqn. (7.4.6) could be written for the holes entering the n-type region. This would involve the diffusion coefficient for holes, D_h, with

$$D_h = \mu_h \frac{kT}{e} \tag{7.4.8}$$

Under steady-state conditions the net ingress of carriers into the volume $\delta A\ \delta x$ is balanced by the net rate of recombination of the excess carriers within the volume. From eqn. (7.4.1) this is simply

$$-\frac{dN}{dt} = \frac{(n - n_{p0})}{\tau_p}\,\delta A\ \delta x \tag{7.4.9}$$

Combining eqns. (7.4.6) and (7.4.9), we obtain

$$D_e \frac{d^2 n}{dx^2} = \frac{(n - n_{p0})}{\tau_p} \tag{7.4.10}$$

That is,

$$\frac{\mathrm{d}^2(\Delta n)}{\mathrm{d}x^2} = \frac{\Delta n}{D_e \tau_p} \tag{7.4.11}$$

Subject to the boundary conditions $\Delta n = \Delta n(0)$ at $x = 0$ and $\Delta n = 0$ at $x = \infty$, this integrates to give

$$\begin{aligned} \Delta n(x) &= \Delta n(0) \exp\left(-x/\sqrt{(D_e \tau_p)}\right) \\ &= \Delta n(0) \exp(-x/L_p) \end{aligned} \tag{7.4.12}$$

where

$$L_p = (D_e \tau_p)^{1/2} \tag{7.4.13}$$

is known as the minority carrier diffusion length in the p-type material. An analogous parameter, L_n, defined by

$$L_n = (D_h \tau_n)^{1/2} \tag{7.4.14}$$

may be obtained for the n-type region, and an expression similar to eqn. (7.4.12) relates the decay of the excess injected hole concentration there.

7.5 INJECTION EFFICIENCY

It is in the regions within one or two diffusion lengths on either side of the depletion layer that most net carrier recombination takes place, and it is from these regions that the recombination radiation is emitted. In order to ensure that this radiation is predominantly generated from one side of the junction rather than the other, it is normal for semiconductors used to make optical sources to be asymmetrically doped. Usually the n-type material is much more heavily doped than the p-type material so that an n^+p junction is formed. It may be that $n_D \sim 10^{24}$ m^{-3} whereas $n_A \sim 10^{22}$ m^{-3}. Then the forward-biassed current is carried across the junction mainly by electrons injected into the lightly doped p-type region, and it is from there that most of the recombination radiation is emitted. For this case we may define an injection efficiency, η_{inj}, as the ratio of the electron current across the junction to the total current flowing. We will show next how the value of η_{inj} is determined by the characteristics of the n and the p regions.

The current density of electrons, J_e, entering the p-type region is given by

$$J_e = -eD_e \left(\frac{\mathrm{d}n}{\mathrm{d}x}\right)_{x=0} = \frac{eD_e}{L_p} \Delta n(0) = \frac{eD_e}{L_p} [n(0) - n_{p0}] \tag{7.5.1}$$

where we have obtained $(dn/dx)_{x=0}$ by differentiating eqn. (7.4.12). Now the concentration $n(0)$ at $x = 0$ is related to the electron concentration in the bulk of the n-type region on the other side of the potential barrier by the Boltzmann equation. In equilibrium the height of the potential barrier is simply the diffusion potential, V_D, and

$$n(0) = n_{p0} = n_{n0} \exp(-eV_D/kT) \tag{7.5.2}$$

When the potential barrier is lowered by the applied forward bias voltage, V, this becomes

$$\begin{aligned} n(0) &= n_{n0} \exp[-e(V_D - V)/kT] \\ &= n_{p0} \exp(eV/kT) \end{aligned} \tag{7.5.3}$$

Substitution into eqn. (7.5.1) gives

$$J_e = \frac{eD_e}{L_p} n_{p0}[\exp(eV/kT) - 1] \tag{7.5.4}$$

Exactly similar arguments in respect of the holes entering the n-type region lead to a corresponding expression for their current density:

$$J_h = \frac{eD_h}{L_n} p_{n0}[\exp(eV/kT) - 1] \tag{7.5.5}$$

Provided that any carrier generation or recombination taking place within the depletion layer, or at the semiconductor surface around the junction periphery (I_2 in eqn. (7.3.5)), may be neglected, then the total current density crossing the junction is simply

$$J = J_e + J_h \tag{7.5.6}$$

and the injection efficiency is

$$\eta_{inj} = \frac{J_e}{J} = \frac{J_e}{J_e + J_h} = \frac{1}{1 + J_h/J_e} \tag{7.5.7}$$

$$= \frac{1}{1 + \dfrac{D_h p_{n0}}{L_n} \dfrac{L_p}{D_e n_{p0}}} = \frac{1}{1 + \dfrac{D_h L_p n_A}{D_e L_n n_D}} \tag{7.5.8}$$

using eqns. (7.3.2) and (7.3.3). For η_{inj} to approach unity it is clearly necessary for the ratio n_D/n_A to be as large as possible.

The theory we have just given verifies eqn. (7.3.4) for the diffusion current flowing across an ideal, abrupt junction:

$$I_1 = I_{s1}[\exp(eV/kT) - 1] \tag{7.3.4}$$

Substituting for J_e and J_h in eqn. (7.5.6) and multiplying by the junction area, A, we obtain

$$I_1 = JA = (J_e + J_h)A$$

$$= eA\left(\frac{D_e n_{p0}}{L_p} + \frac{D_h p_{n0}}{L_n}\right)[\exp(eV/kT) - 1] \qquad (7.5.9)$$

Thus

$$I_{s1} = eA\left(\frac{D_e n_{p0}}{L_p} + \frac{D_h p_{n0}}{L_n}\right)$$

$$= eA\left(\frac{D_e}{L_p}\frac{1}{n_A} + \frac{D_h}{L_n}\frac{1}{n_D}\right)n_i^2$$

$$= eA\left(\frac{D_e}{L_p}\frac{1}{n_A} + \frac{D_h}{L_n}\frac{1}{n_D}\right)K^2 \exp(-\mathcal{E}_g/kT) \qquad (7.5.10)$$

$$= eA\left[\left(\frac{D_e}{\tau_p}\right)^{1/2}\frac{1}{n_A} + \left(\frac{D_h}{\tau_n}\right)^{1/2}\frac{1}{n_D}\right]K^2 \exp(-\mathcal{E}_g/kT) \qquad (7.5.10a)$$

where we have again made use of eqns. (7.3.2) and (7.3.3) and have also introduced eqn. (7.2.6) in order to bring out the importance of temperature.

7.6 THE DEPLETION LAYER

Finally we ought to examine the depletion layer itself. Let us assume that the concentrations of free electrons and holes are negligible throughout the depletion region. Then, the ionized donors and acceptors set up the static charge distribution shown in Fig. 7.11(a). It is this that supports the barrier potential, $(V_D - V)$, where V_D is the diffusion potential and V the applied voltage. If the depletion layer extends a distance, l_p, into the p-type region and a distance, l_n, into the n-type region, then the charge per unit area stored within it is

$$\frac{|Q|}{A} = en_A l_p = en_D l_n \qquad (7.6.1)$$

with $n_D \gg n_A$, $l_p \gg l_n$.

Poisson's law will apply throughout the region. Let us again set the x-coordinate perpendicular to the junction and with the lightly doped p-type

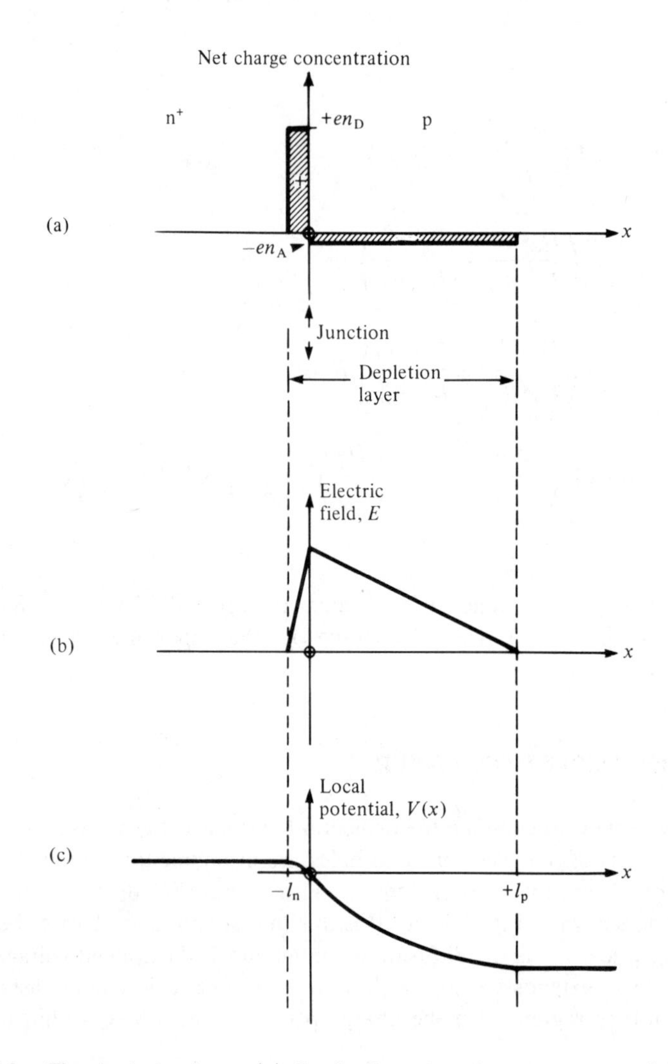

Fig. 7.11 The depletion layer: (a) distribution of static space-charge; (b) electric field; (c) potential variation.

region on the side $x > 0$, but now we shall put the origin of the x-coordinate at the junction itself. The electric field, E, is zero for all $x < -l_n$ and for all $x > l_p$. It is maximum at $x = 0$. The local potential, $V(x)$, is now defined to be zero at $x = 0$ since this simplifies the algebra. Then, applying Poisson's law in the region $x > 0$, we have

$$\frac{d^2 V(x)}{dx^2} = \frac{en_A}{\varepsilon_r \varepsilon_0} \tag{7.6.2}$$

where $\varepsilon_r \varepsilon_0$ is the permittivity of the material. Integrating gives

$$\frac{dV(x)}{dx} = \frac{en_A(x - l_p)}{\varepsilon_r \varepsilon_0} = -E \tag{7.6.3}$$

Integrating again gives

$$V(x) = \frac{en_A}{\varepsilon_r \varepsilon_0}\left(\frac{x^2}{2} - l_p x\right) \tag{7.6.4}$$

In performing each integration we have inserted the appropriate boundary conditions.

These expressions, together with the corresponding ones for the n-type region, are plotted in Fig. 7.11(b) and (c). The total change of potential is

$$(V_D - V) = V(-l_n) - V(l_p)$$

$$= \frac{en_D l_n^2}{2\varepsilon_r \varepsilon_0} + \frac{en_A l_p^2}{2\varepsilon_r \varepsilon_0} \tag{7.6.5}$$

$$= \frac{(Q/A)^2}{2e\varepsilon_r \varepsilon_0}\left[\frac{1}{n_D} + \frac{1}{n_A}\right] \tag{7.6.6}$$

The greater part of the potential change takes place in the lightly doped p-type region, so, neglecting the first term on the right-hand-side, eqn. (7.6.5) may be rearranged to give the depletion-layer thickness as

$$l_D \simeq l_p = \left[\frac{2\varepsilon_r \varepsilon_0 (V_D - V)}{en_A}\right]^{1/2} \tag{7.6.7}$$

Note also that

$$\frac{|Q|}{A} \simeq [2en_A \varepsilon_r \varepsilon_0 (V_D - V)]^{1/2} \tag{7.6.8}$$

Although, so far, we have concentrated on the forward-biassed junction, it is important to be clear that eqns. (7.6.7) and (7.6.8) for the depletion-layer thickness and charge apply equally well for the ideal, reverse-biassed junction. The depletion-layer thickness decreases under forward bias as does the charge stored. With reverse bias, $V < 0$, both increase, and at large values both become proportional to the square root of the reverse bias voltage. A feature of eqn. (7.6.7) that is important in the design of detectors is the inverse variation of the depletion-layer thickness with the square root of the doping concentration. We will defer further consideration of this until Chapter 12.

7.7 AN EQUIVALENT CIRCUIT FOR THE p–n JUNCTION

The response of a p–n junction to changes of applied voltage can be modelled quite well by the equivalent circuit of Fig. 7.12. The components R and L represent the resistance and the inductance of the bulk semiconductor, the contacts and the leads to the device terminals. We shall neglect them in the immediate discussion of the junction. The two capacitances, C_J and C_D, and the resistance, r_D, are all nonlinear components whose values depend on the bias conditions.

The junction capacitance, C_J, dominates the reverse-bias behavior but is

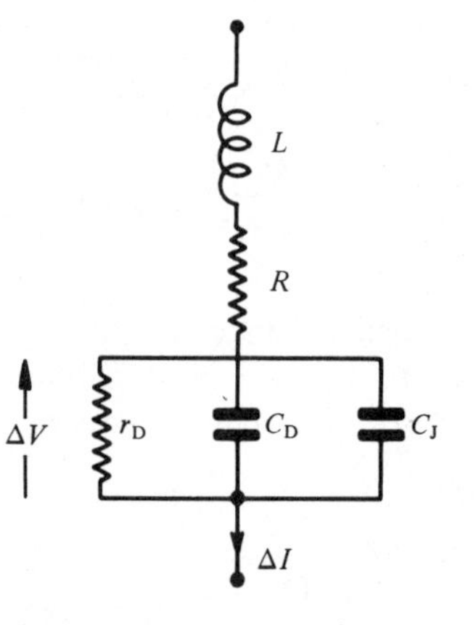

Fig. 7.12 Equivalent circuit for the p–n junction.

often negligible under forward-bias conditions. It is a true capacitance resulting from the charge, Q, stored in the depletion layer, but because of the distributed nature of this charge it is a nonlinear function of the applied voltage, V. If a small change, ΔV, in V brings about a small change, ΔQ, in the depletion-layer charge, then we may define C_J as

$$C_J = \left|\frac{\Delta Q}{\Delta V}\right| \tag{7.7.1}$$

Differentiation of eqn. (7.6.8) shows that

$$\frac{C_J}{A} = \left[\frac{en_A \varepsilon_r \varepsilon_0}{2(V_D - V)}\right]^{1/2} \tag{7.7.2}$$

It is sometimes convenient to express C_J in terms of the capacitance, C_0, of the unbiassed junction,

$$\frac{C_0}{A} = \left(\frac{en_A \varepsilon_r \varepsilon_0}{2V_D}\right)^{1/2} \tag{7.7.3}$$

and

$$C_J = \frac{C_0}{(1 - V/V_D)^{1/2}} \tag{7.7.4}$$

If we substitute the following typical values into eqn. (7.7.2): $\varepsilon_r = 12$ and $n_A = 10^{23}$ m^{-3}, then with $e = 1.6 \times 10^{-19}$ C and $\varepsilon_0 = 8.85 \times 10^{-12}$ F/m,

$$\frac{C_J/A}{[\mathrm{F/m^2}]} = \frac{C_J/A}{[\mathrm{pF/\mu m^2}]} \simeq \frac{10^{-3}[\mathrm{V}^{1/2}]}{(V_D - V)^{1/2}} \tag{7.7.5}$$

Consider a junction having an area of 10^4 μm^2, and $V_D \simeq 1$ V. Unbiassed, $C_0 \simeq 10$ pF. With strong forward bias C_J would be about an order of magnitude greater; with 100 V reverse bias it would be about an order of magnitude less.

The diffusion capacitance, C_D, is not a true capacitance. It is always shunted by the resistance, r_D, and together they represent the additional free carriers that have to be stored in the regions outside the depletion layer in order to support a current, I, across the junction. In practice C_D and r_D can only be defined satisfactorily, and are only significant, when the junction is strongly forward-biassed so that $\exp(eV/kT) \gg 1$. Each depends both on the bias level and on the frequency of any modulation. Analysis of the effects of modulating the applied voltage is deferred until the next chapter so that the optical response may be dealt with at the same time. Here we will just quote the results. The junction is assumed to be forward-biassed by a steady voltage, V_0, so that a steady current, I_0, given by eqn. (7.3.4) flows. Superimposed on this voltage is a small sinusoidal voltage fluctuation, $\hat{V}_1 \cos \omega t$, which we represent by the phasor, $V_1(\omega)$. This produces a current fluctuation, $\hat{I}_1 \cos(\omega t + \phi)$, which may be represented by the phasor

$$I_1(\omega) = Y_D(\omega)V_1(\omega) \tag{7.7.6}$$

where

$$Y_D(\omega) = \frac{1}{r_D} + j\omega C_D \tag{7.7.7}$$

is the admittance of the junction. In Section 8.6 we shall show that at low frequencies such that $\omega\tau_p \ll 1$,

$$r_D \simeq \frac{kT}{eI_0} \tag{7.7.8}$$

and that for the ideal, asymmetric junction in which the junction current is mainly carried by the electrons,

$$C_D \simeq \frac{eI_0\tau_p}{2kT} \tag{7.7.9}$$

Thus,

$$r_D C_D \simeq \tau_p/2 \tag{7.7.10}$$

An n^+p junction at room temperature carrying a forward current of 10 mA and having a value of τ_p of 10 ns may, therefore, be represented by $r_D \simeq 2.6\ \Omega$ and $C_D \simeq 2$ nF. In comparison C_J would normally be negligible.

PROBLEMS

7.1 Calculate the room temperature values of $\mathcal{E}_g/kT$ for the semiconductors listed in Table 7.2. Indicate which of these materials might be expected to show semiconducting properties at 77 K and which at 400 K.

7.2 Calculate room temperature values of the diffusion coefficients of electrons and holes in the semiconductors listed in Table 7.2.

7.3 Of the two components I_1 and I_2 of the total diode current I expressed in equations (7.3.4) and (7.3.5), explain which you would expect to be dominant
(a) at high rather than low currents;
(b) at high rather than low temperatures;
(c) in wide rather than narrow band-gap semiconductors.

7.4 The data shown in Fig. 9.11(a) on p. 274 have been fitted to eqns (7.3.4) and (7.3.5). From the graphs estimate values for I_{s1} and I_{s2} and for the effective ohmic resistances of the diodes.

7.5 Derive an expression for the injection efficiency across a p–n homojunction formed in a direct band-gap semiconductor in which radiative recombination dominates other recombination processes. Hence calculate the injection efficiency in a p⁺n GaAs diode in which $n_A = 10^{24}\ \text{m}^{-3}$ and $n_D = 10^{21}\ \text{m}^{-3}$. Assume eqn. (8.4.13), p. 242, and its converse for n-type material, putting $r = 10^{-16}\ \text{m}^3/\text{s}$.

REFERENCES

7.1 H. C. Casey, Jr. and M. B. Panish, *Heterostructure Lasers: Part A. Fundamental Principles* (1978); *Part B. Materials and Operating Characteristics*, Academic Press (1979).
7.2 S. M. Sze, *Physics of Semiconductor Devices*, 2nd ed, Wiley (1981).

SUMMARY

The sources and detectors used in optical fiber systems are exclusively semiconductor junction devices: light-emitting diodes, injection lasers, p–i–n and avalanche photodiodes.

A key semiconductor parameter is the band-gap energy. This determines the intrinsic carrier concentration:

$$n_i = 2(2\pi kT/h^2)^{3/2}(m_e m_h)^{3/4} \exp(-\mathcal{E}_g/2kT).$$

Values of $\mathcal{E}_g$, n_i and other parameters of Si, Ge, and the binary III–V compound semiconductors are set out in Table 7.2.

Band-gap energy and lattice spacing vary continuously with the composition of ternary and quaternary III–V compounds. Using quaternary and some ternary materials it is possible to grow as a single crystal semiconductors having different band-gap energies. Some examples are shown in Figs. 7.3 and 7.4.

GaAlAs/GaAs sources, used with Si detectors, are well established for wavelengths around 0.85 μm. For longer wavelengths InGaAsP/InP sources and Ge or InGaAs/InP detectors may be used over the range 1.0–1.7 μm.

Semiconductor p–n$^+$ junction properties may be summarized as follows:

(a) Current $I = I_{s1}[\exp(eV/kT) - 1] + I_{s2}[\exp(eV/2kT) - 1]$

The first term comprises diffusion currents and is proportional to n_i^2. The second term comprises generation/recombination currents and is proportional to n_i.

(b) Depletion layer thickness is $l_D \simeq [2\varepsilon_r \varepsilon_0 (V_D - V)/en_A]^{1/2}$.

(c) The equivalent circuit of a reverse-biassed junction is dominated by the depletion-layer capacitance: $C_J = C_0(1 - V/V_D)^{-1/2}$, where $C_0 \simeq (\varepsilon_r \varepsilon_0 en_A/2V_D)^{1/2}$.

(d) The forward-biassed junction may be represented at low frequencies by the admittance

$$Y_D = 1/r_D + j\omega C_D = (eI_0/kT)(1 + j\omega\tau_p/2).$$

8

Injection Luminescence

8.1 RECOMBINATION PROCESSES

The recombination of electrons and holes in a semiconductor may be brought about by many independent, competing, parallel processes. It is sometimes convenient to distinguish between direct band-to-band transitions on the one hand, and transitions which involve intermediate steps on the other. More important for us is the different distinction that can be made between the radiative and the nonradiative recombination processes. In a nonradiative transition most of the energy of recombination is released into the crystal as thermal energy. In a radiative process most of the energy of recombination is released as a single quantum of radiation. When recombination takes place in several stages it is possible for more than one quantum of longer wavelength radiation to be emitted.

Several different kinds of recombination process are illustrated in Fig. 8.1. Of these, the one which principally concerns us is the direct, radiative, band-to-band transition shown as (a) in the figure. The main competing nonradiative transitions are those involving trapping levels lying deep in the band gap, shown as (c) and (d) in Fig. 8.1. Such trapping levels may be caused by impurity atoms such as gold in silicon, or by dislocations or other crystal lattice defects, and they occur in large numbers at the semiconductor surface. The perturbation of the band structure at a free surface causes a much higher proportion of the trapping levels there to act as effective recombination centers. The proximity of the surface, or of a macroscopic flaw in the material, or of an interface where there is a discontinuity in the crystal structure, thus has a considerable effect on the recombination processes.

8.2 THE SPECTRUM OF RECOMBINATION RADIATION

When the recombination energy is released into a single photon of energy, $\mathcal{E}_{ph}$, the frequency, f, and the free-space wavelength, λ, of the radiation are

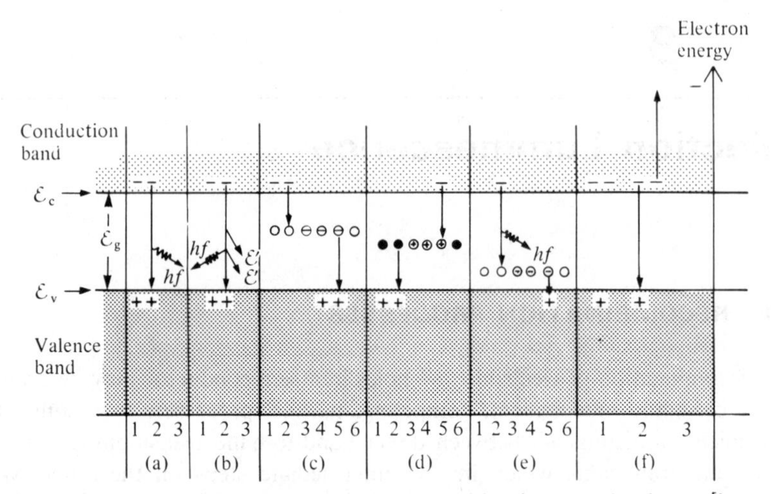

Fig. 8.1 Some possible electron–hole recombination mechanisms [in each case the sequence of situations and events should be followed from left to right through the different stages involved]: (a) direct, band-to-band, radiative transition; (b) radiative, band-to-band transition involving one or more phonons of energy $\mathcal{E}'$, see Section 8.2; (c) transitions, probably nonradiative, involving a deep-lying acceptor trap; (d) transitions, probably nonradiative, involving a deep-lying donor level; (e) transitions, radiative and nonradiative, involving a shallow acceptor level; (f) a nonradiative, 'Auger' recombination transition.

Free electron	–	Filled donor level	●
Hole	+	Empty donor level	⊕
Empty acceptor level	O	Photon emission	hf
Filled acceptor level	⊖	Phonon emission	$\mathcal{E}'$

determined by Planck's law:

$$\mathcal{E}_{ph} = hf = hc/\lambda \tag{8.2.1}$$

$$\therefore \quad \lambda = \frac{hc}{\mathcal{E}_{ph}} = \frac{1.24[\mu m]}{\mathcal{E}_{ph}/[eV]} \tag{8.2.2}$$

In order to determine the distribution with wavelength or photon energy of the radiated recombination energy, we need to proceed as follows. The probability that an electron of energy $\mathcal{E}_2$ will recombine with a hole at energy $\mathcal{E}_1$ is proportional to the concentration of electrons at $\mathcal{E}_2$, $n(\mathcal{E}_2)$, and to the concentration of holes at $\mathcal{E}_1$, $p(\mathcal{E}_1)$. The probability of a photon of energy, $\mathcal{E}_{ph}$, being radiated may be obtained by integrating the product $n(\mathcal{E}_2)p(\mathcal{E}_1)$ over all values of $\mathcal{E}_1$ (or $\mathcal{E}_2$) subject to the constraint that $\mathcal{E}_2 - \mathcal{E}_1 = \mathcal{E}_{ph}$. To calculate this distribution we need to know the carrier concentration distributions in the valence and conduction bands, $p(\mathcal{E}_1)$ and $n(\mathcal{E}_2)$. An indication of these was given in Fig. 7.8 for the regions close to a forward-biassed p–n junction. According to the simple model used in that diagram the distributions rise very sharply at the band edge, they then change relatively slightly over an energy range of about kT, and finally they decrease exponentially as $\exp(-\mathcal{E}/kT)$, the Boltzmann relation. The two questions we need to answer are: how do these distributions come about; and how can we represent them in order to deduce the spectral composition of the recombination radiation?

The concentration distribution of electrons at an energy, $\mathcal{E}_2$, in the conduction band is the product of two terms. The first is the distribution of allowed energy states per unit volume per unit range of energy at energy, $\mathcal{E}_2$, in the conduction band, which is written as $S_c(\mathcal{E}_2)$. The second is the probability that these states are occupied, which was introduced in Section 7.2.1 as the Fermi function, $F(\mathcal{E}_2)$. Thus

$$n(\mathcal{E}_2) = S_c(\mathcal{E}_2)F(\mathcal{E}_2) \tag{8.2.3}$$

Similarly the concentration distribution of holes at an energy, $\mathcal{E}_1$, in the valence band can be written as

$$p(\mathcal{E}_1) = S_v(\mathcal{E}_1)[1 - F(\mathcal{E}_1)] \tag{8.2.4}$$

where $S_v(\mathcal{E}_1)$ is the distribution of allowed energy states per unit volume per unit range of energies at $\mathcal{E}_1$.

A very simple model for the behavior of carriers in a semiconductor allows theoretical expressions for $S_c(\mathcal{E}_2)$ and $S_v(\mathcal{E}_1)$ to be derived. In this model the carriers are assumed to move around inside the material like free charged particles. The only allowance made for what are really very complex dynamic conditions is that the motion of the carriers is characterized by an effective mass, m_e for the electrons and m_h for the holes, these being different from the normal electron rest mass of $m_{e0} = 9.11 \times 10^{-31}$ kg. The effective masses are characteristic of the material and have been listed with the other semiconductor properties in Table 7.2. According to this model the density of states functions will be given by

$$S_c(\mathcal{E}_2) = 4\pi(2m_e/h^2)^{3/2}(\mathcal{E}_2 - \mathcal{E}_c)^{1/2} \tag{8.2.5}$$

and

$$S_v(\mathcal{E}_1) = 4\pi(2m_h/h^2)^{3/2}(\mathcal{E}_v - \mathcal{E}_1)^{1/2} \tag{8.2.6}$$

These functions are plotted in Fig. 8.2(a) for a material in which $(m_e/m_{e0}) = 0.07$ and $(m_h/m_{e0}) = 0.5$, which are the measured values for carriers in GaAs.

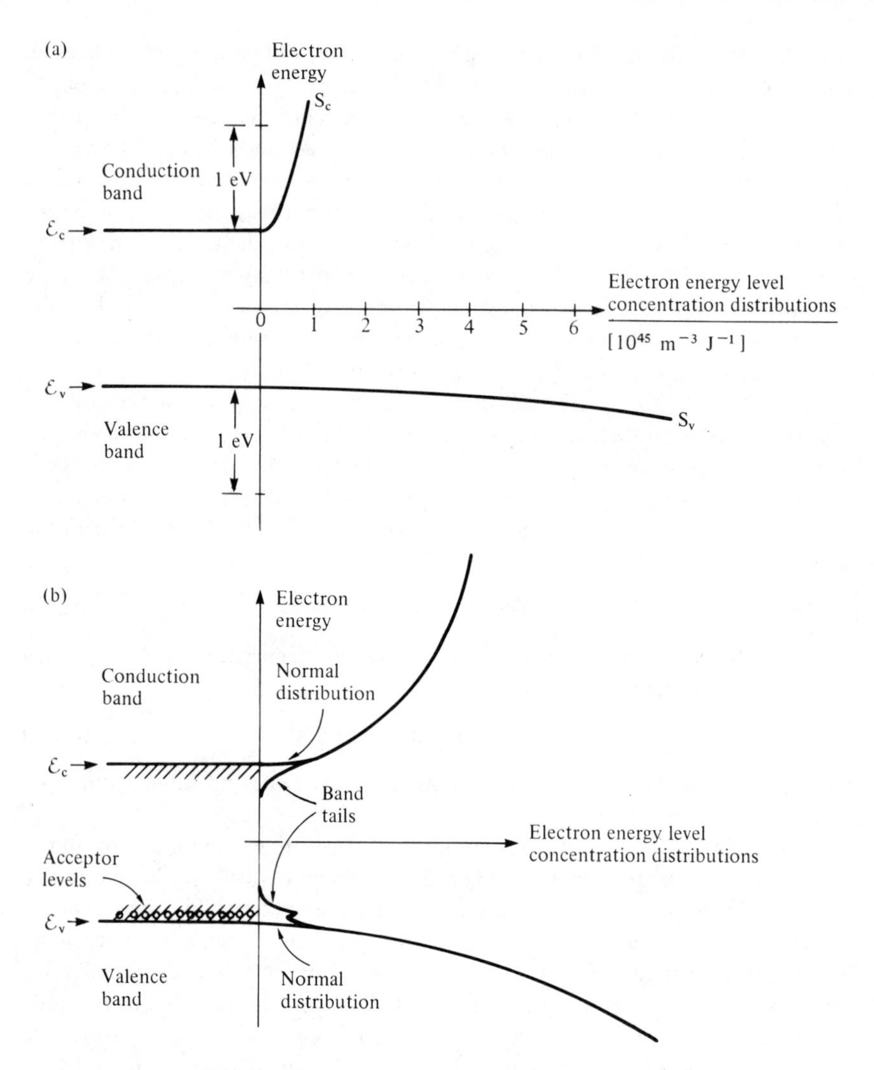

Fig. 8.2 Carrier concentration distribution functions: (a) for pure GaAs; (b) showing schematically the effect of band-tailing with a moderately high acceptor impurity concentration ($\sim 10^{23}$ m^{-3}).

We mentioned in Section 7.2.2 that high concentrations of impurities would distort the density of states functions, S_c and S_v, at the band edges. Typically 'band-tails' are formed, as shown in Fig. 8.2(b), with the result that the band gap is effectively made narrower. Analytical expressions for the density of states in this case become a great deal more complicated.

In order to obtain some idea of the spectral distribution of recombination radiation we shall simplify the expressions for the carrier distribution functions still further. We shall assume that the concentration of free electrons falls exponentially with increasing energy from the conduction band edge at $\mathcal{E}_c$. Similarly, the concentration of holes will be assumed to decrease exponentially with the energy difference from the top of the valence band at $\mathcal{E}_v$. Thus we put

$$n(\mathcal{E}_2) \simeq A \exp[-(\mathcal{E}_2 - \mathcal{E}_c)/kT] \tag{8.2.7}$$

$$p(\mathcal{E}_1) \simeq B \exp[-(\mathcal{E}_v - \mathcal{E}_1)/kT] \tag{8.2.8}$$

where A and B are constants related to the total concentrations of free electrons and holes. These distributions are shown in Fig. 8.3. Now, the spectral distribution of the radiated power density as a function of photon energy is given by

$$P(\mathcal{E}_{ph}) \propto \int n(\mathcal{E}_2) p(\mathcal{E}_1)\, d\mathcal{E}_2 \tag{8.2.9}$$

subject to the constraint

$$\mathcal{E}_2 - \mathcal{E}_1 = \mathcal{E}_{ph} \tag{8.2.10}$$

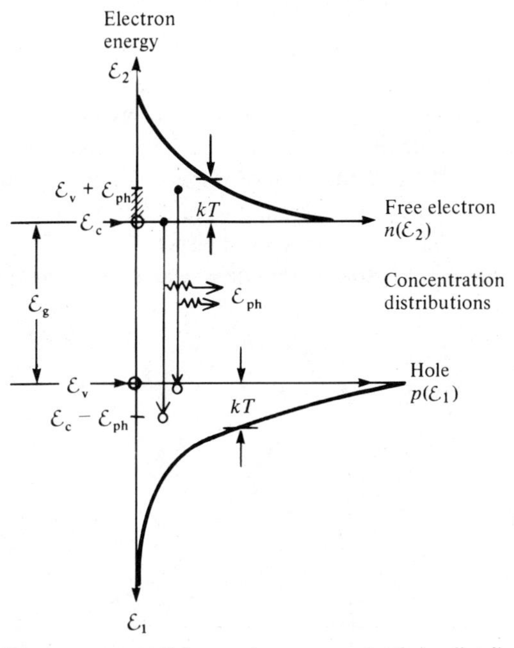

Fig. 8.3 Simplified, exponential, carrier concentration distribution functions assumed in eqns (8.2.7) and (8.2.8).

so that,

$$P(\mathcal{E}_{ph}) = K\int_{\mathcal{E}_2=\mathcal{E}_c}^{\mathcal{E}_v+\mathcal{E}_{ph}} \exp[-(\mathcal{E}_2-\mathcal{E}_c)/kT]\exp[-(\mathcal{E}_v-\mathcal{E}_1)/kT]\,d\mathcal{E}_2 \tag{8.2.11}$$

We have chosen to take the integration over the allowable range of energies in the conduction band. This is shown in Fig. 8.3, also. If $\mathcal{E}_2$ were to exceed $(\mathcal{E}_v + \mathcal{E}_{ph})$, then $\mathcal{E}_1$ would be in the band gap, so $p(\mathcal{E}_1)$ would be zero. The constants A and B have been subsumed in a new constant, K, which includes the transition probability, that is it involves the radiative lifetime, τ_{rr}, of the electrons. Then

$$\begin{aligned}P(\mathcal{E}_{ph}) &= K\int_{\mathcal{E}_2=\mathcal{E}_c}^{\mathcal{E}_v+\mathcal{E}_{ph}} \exp[(\mathcal{E}_c-\mathcal{E}_v)/kT]\exp[-(\mathcal{E}_2-\mathcal{E}_1)/kT]\,d\mathcal{E}_2\\ &= K\exp(\mathcal{E}_g/kT)\exp(-\mathcal{E}_{ph}/kT)\int_{\mathcal{E}_c}^{\mathcal{E}_v+\mathcal{E}_{ph}} d\mathcal{E}_2\\ &= K(\mathcal{E}_{ph}-\mathcal{E}_g)\exp[-(\mathcal{E}_{ph}-\mathcal{E}_g)/kT]\end{aligned} \tag{8.2.12}$$

On the basis of this analysis we should expect the spectral distribution of the emitted recombination radiation to have the form given in eqn. (8.2.12) and plotted in Fig. 8.4. The observed spectrum is always much more symmetrical than this, as in the examples given in Fig. 8.5. There are a number of reasons. First, the use of high donor and acceptor impurity concentrations is normal in LEDs and lasers, and it causes the band-edge distortion shown in Fig. 8.2(b). Secondly, the transition may involve an interaction with the crystal lattice as well as the emission of a photon. In that case a small fraction of the recombina-

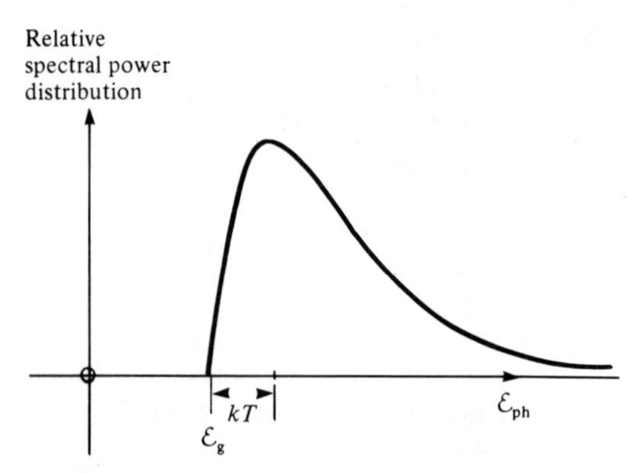

Fig. 8.4 Spectral distribution according to eqn. (8.2.12).

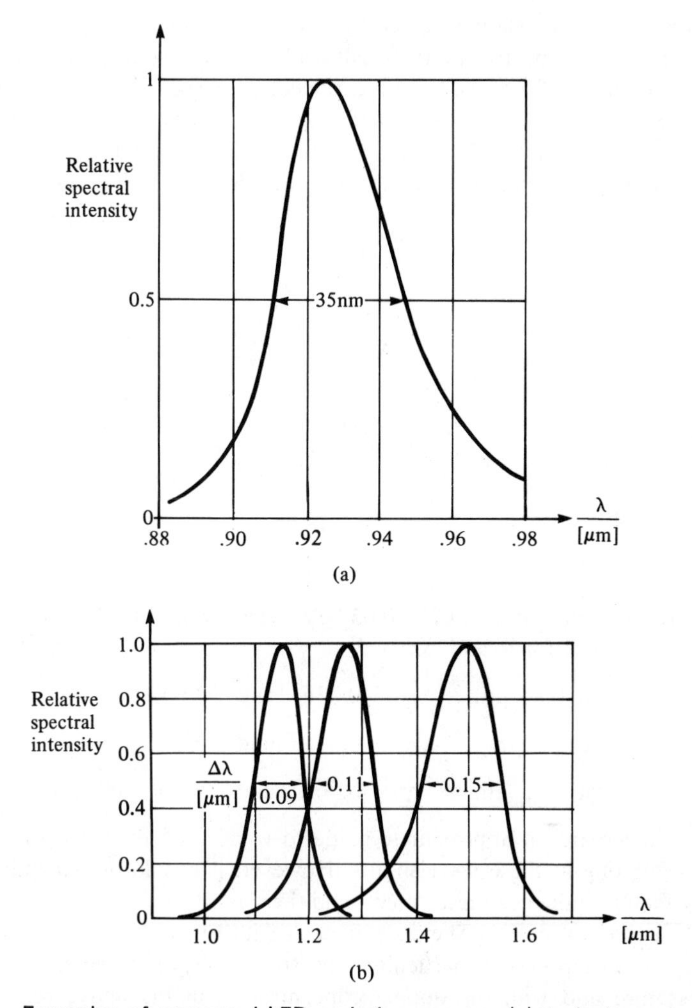

Fig. 8.5 Examples of measured LED emission spectra: (a) typical spectrum of a Si-doped GaAs diode; (b) spectra from a set of InGaAsP LEDs with three different active layer compositions. In these devices $\Delta \mathcal{E}_{ph} \simeq 3.3kT$. [Taken from O. Wada *et al.*, Performance and reliability of high radiance InGaAsP/InP DH LEDs operating in the 1.15–1.5 μm wavelength region, *IEEE Jnl. of Quantum Ets.*, **QE-18**, 368–74 (1982); © 1982 IEEE.]

tion energy (as much as 0.05 eV) may be carried away as the quantized energy associated with the lattice vibration (the optical or acoustic phonon) that is excited at the same time. Thirdly, the transition may involve, as an intermediate step, one of the impurity levels near to the band edge. As a result, rather less than the band-gap energy is released in the main radiative transition.

From Fig. 8.4 we should expect that the peak emission intensity would be at $\mathcal{E}_g + kT$, and the spectral width at half peak intensity would be $2.4kT$. Because of the additional effects we have mentioned, the peak intensity is often shifted to rather longer wavelengths. Usually, the spectral half-width is found to lie between $1.5kT$ and $3.5kT$. The corresponding spread in terms of the wavelength of the emitted radiation may be derived as follows.

$$\lambda = \frac{hc}{\mathcal{E}_{ph}} \tag{8.2.2}$$

$$\therefore \quad \Delta\lambda = -\frac{hc}{\mathcal{E}_{ph}^2}\Delta\mathcal{E}_{ph} \tag{8.2.13}$$

$$\therefore \quad \gamma = \left|\frac{\Delta\lambda}{\lambda}\right| = \frac{\Delta\mathcal{E}_{ph}}{\mathcal{E}_{ph}} \simeq \frac{2kT}{\mathcal{E}_{ph}} \tag{8.2.14}$$

$$\simeq \frac{2kT\lambda}{1.24[\mu\text{m.eV}]} \tag{8.2.15}$$

At room temperature $2kT = 0.052$ eV, which would lead us to predict the following values for γ and $\Delta\lambda$:

at 0.85 μm, $\gamma = 0.036$ and $\Delta\lambda = 30$ nm

at 1.3 μm, $\gamma = 0.055$ and $\Delta\lambda = 70$ nm

at 1.55 μm, $\gamma = 0.065$ and $\Delta\lambda = 100$ nm

These values are an approximate guide to what is observed in practice from LEDs emitting at these wavelengths. However, it should be remembered that many factors influence these values, and there is considerable variation among different types of device. The discrepancies with Fig. 8.5 are within the normal range of variation. In particular the spectral spread increases with the temperature and with the doping concentration in the active region of the device.

8.3 DIRECT AND INDIRECT BAND-GAP SEMICONDUCTORS

A parameter that needs to be made as large as possible in an optical source is the *internal quantum efficiency*, η_{int}. This is defined as the ratio of the number of photons generated to the number of carriers crossing the junction. Clearly its value depends on the relative probabilities of the radiative and nonradiative recombination processes. These in turn depend on the structure of the junction, the level of impurities in the semiconductor, and most of all on the type of semiconductor.

Semiconductors such as silicon, germanium and gallium phosphide have what is known as an *indirect band gap.* In simple terms this means that an electron in an energy state near to the bottom of the conduction band has a momentum in the crystal that is quite different from that of an electron in an energy state near to the top of the valence band. Figure 8.6(a) is an attempt to illustrate this. It shows that a direct band-to-band transition can take place only if some means can be found to absorb the surplus momentum.† One way would be for a high-energy lattice vibration (a phonon) to be launched during recombination. Whilst this might carry away the excess momentum, it would also take up a significant proportion of the recombination energy, up to 0.05 eV. But more important than this, the need to initiate two events simultaneously (producing a phonon as well as a photon) greatly reduces the probability of this type of recombination transition. As a result nonradiative processes, in particular those that make use of trapping levels near to the center of the band gap, tend to dominate recombination in indirect band-gap semiconductors, and the internal quantum efficiency is very low.

Other semiconductor materials may have a *direct band gap* of the type illustrated by Fig. 8.6(b). In this case an electron with the lowest allowed energy in the conduction band does have a crystal momentum very similar to that of the electrons at the highest allowed valence-band energies. There is, then, a high probability of direct, band-to-band transitions, and the internal quantum efficiency for photon generation is also relatively high.

The nature of the band gap of the III–V binary compounds was indicated in Table 7.2. Some of the ternary and quaternary compounds may change from direct to indirect band-gap materials as their composition is varied. This was indicated on Figs. 7.3 and 7.4. There is a general tendency for the narrower band gaps to be direct and the wider ones to be indirect. This is fortunate for the development of near-infra-red sources for optical fiber communications, but represents a serious problem in the production of efficient, visible LEDs. For the latter, the indirect band-gap material, gallium phosphide, is often used, heavily doped with impurities such as N and ZnO. These give rise to trapping levels near to the band edges which promote radiative transitions like (e) in Fig. 8.1. The photon energy of the emitted light is not much less than $\mathcal{E}_g$. Even so, the optical conversion efficiency is too low to make a good source for optical communication. We shall, therefore, restrict our discussion to materials having a direct band gap.

† Readers familiar with solid-state physics theory will recognize what is here called 'momentum' to be the electron wavenumber.

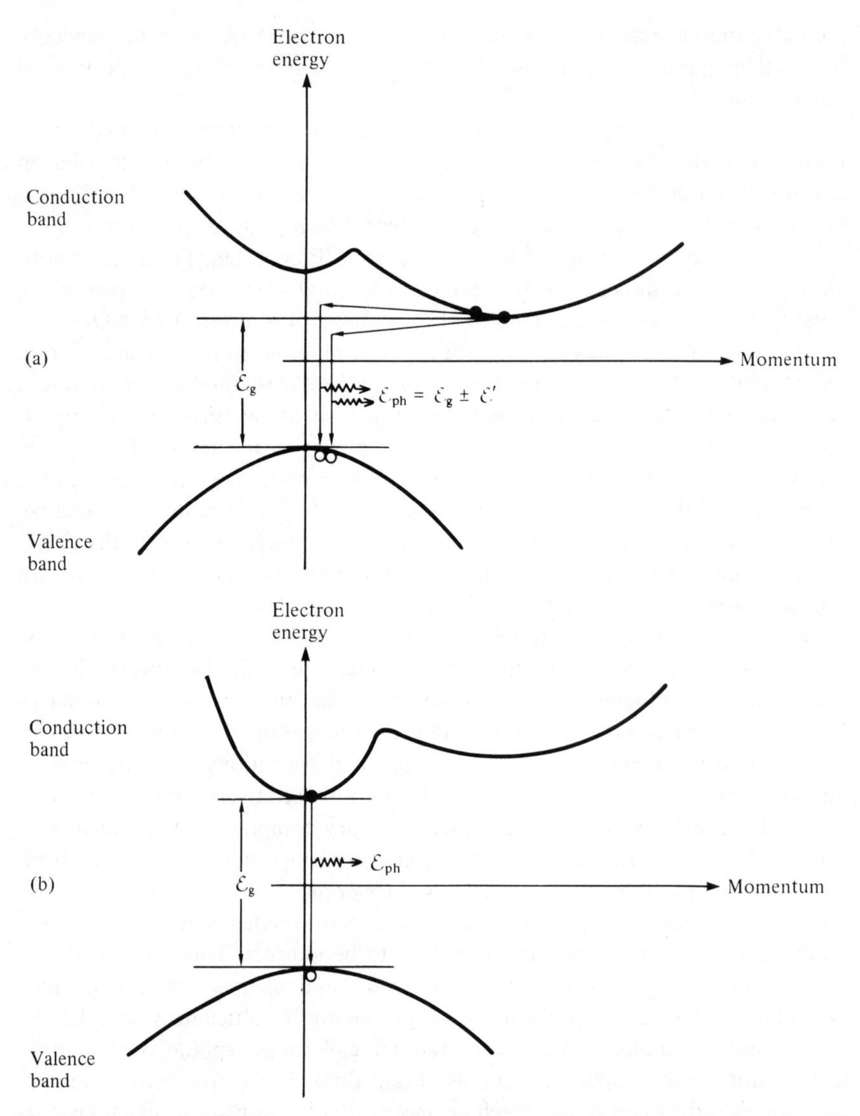

Fig. 8.6 Schematic energy–momentum diagrams for direct and indirect band-gap semiconductors: (a) indirect band-gap material—recombination may only take place if the change of momentum can be absorbed. This either requires trapping levels or, as shown, the participation of an optical or acoustic phonon whose energy, $\mathcal{E}'$, will either add to or subtract from the band-gap energy; (b) direct band-gap material—there is now nothing to prevent direct band-to-band transitions from the lowest conduction band levels. Because of the exponential fall of free electron concentration with energy, it is only these lowest levels that have a finite probability of being occupied.

Gallium arsenide is a direct band-gap material and it has been used for many years as the basis of a number of different types of semiconductor device: transferred electron devices, field effect transistors, LEDs and lasers. Of all the III–V semiconductors it is the one of which there is the greatest technological experience and expertise. It is thus natural that it should have been the first choice for sources for optical communication. From eqn. (8.2.2) we should expect the wavelength of peak emission to be in the region of

$$\frac{\lambda}{[\mu\mathrm{m}]} \simeq \frac{1.24}{\mathcal{E}_g/[\mathrm{eV}]} \simeq \frac{1.24}{1.42} = 0.87 \tag{8.3.1}$$

Because of the effect of high doping concentrations on the band edge and the recombination processes, the peak wavelength is often slightly longer than this and may be as high as 0.92 μm, as can be seen in Fig. 8.5(a).

The ternary system $Ga_{1-x}Al_xAs$ has a direct band gap over the composition range $0 < x < 0.37$, giving a range of values for $\mathcal{E}_g$ of 1.42 eV to 1.92 eV. The corresponding wavelength range is 0.87 μm to 0.65 μm. One of the great advantages of this material derives from the very small difference between the lattice parameters of GaAs and AlAs, 0.5653 nm and 0.5661 nm, respectively, which enables single crystal layers of GaAlAs to be grown on a substrate of GaAs. In consequence GaAlAs sources have dominated the early phases of optical fiber transmission.

Sources for radiation of longer wavelengths, in particular, 1.3 μm and 1.55 μm, require the use of semiconductors with narrower band gaps, around 0.95 eV and 0.8 eV, respectively. Possible III–V compound semiconductors may be identified on Fig. 7.3. Most interest has focussed on the two quaternary systems shown in Fig. 7.4, and specific examples of devices made from these materials will be discussed in Chapter 11. Reference to part (a) of Fig. 7.4 shows that GaAs may be used as a substrate for a range of lattice-matched compounds in the InGaAsP system having energy gaps between 1.42 eV (GaAs) and 2.0 eV ($In_{0.5}Ga_{0.5}P$). Devices made from these materials might be expected to cover the range of wavelengths from 0.87 μm to 0.62 μm. Potentially more important is the range of lattice-matched compounds in this system that can be grown on a substrate of InP. Their band gaps vary from 1.35 eV (InP) down to 0.74 eV ($In_{0.53}Ga_{0.47}As$) and indicate a possible wavelength range of 0.92 μm to 1.67 μm. Some emission spectra from such compounds were given in Fig. 8.5(b). The binary material GaSb may be considered as a substrate for a range of lattice-matched compounds in the system GaAlAsSb shown in Fig. 7.4(b) and also for a range in the system InGaAsSb. The energy gaps vary between 0.73 eV and 1.0 eV (direct–indirect transition) for GaAlAsSb and thus cover the wavelength range 1.24 μm to 1.7 μm. The other system would permit wavelengths longer than 1.7 μm to be used, if required.

8.4 THE INTERNAL QUANTUM EFFICIENCY

We shall next consider the relative magnitude of the minority carrier lifetime in direct and indirect band-gap semiconductors, and examine its effect on the optical generation efficiency and on other important parameters. The discussion will concentrate on excess electrons injected into a p-type region, but exactly analogous arguments may be made for holes injected into n-type material.

In a p-type region the rate of recombination of excess electrons per unit volume is given by

$$-\frac{\mathrm{d}n}{\mathrm{d}t} = \frac{(n - n_{\mathrm{p0}})}{\tau_{\mathrm{p}}} \tag{7.4.3}$$

We wish to distinguish between the radiative and the nonradiative recombination processes and therefore put

$$-\frac{\mathrm{d}n}{\mathrm{d}t} = -\left(\frac{\mathrm{d}n}{\mathrm{d}t}\right)_{\mathrm{rr}} - \left(\frac{\mathrm{d}n}{\mathrm{d}t}\right)_{\mathrm{nr}} \tag{8.4.1}$$

with

$$-\left(\frac{\mathrm{d}n}{\mathrm{d}t}\right)_{\mathrm{rr}} = \frac{(n - n_{\mathrm{p0}})}{\tau_{\mathrm{rr}}} \tag{8.4.2}$$

representing the rate of loss of carriers by radiative recombination, and

$$-\left(\frac{\mathrm{d}n}{\mathrm{d}t}\right)_{\mathrm{nr}} = \frac{(n - n_{\mathrm{p0}})}{\tau_{\mathrm{nr}}} \tag{8.4.3}$$

representing the rate of loss of carriers by nonradiative transitions. The parameters τ_{rr} and τ_{nr} may be thought of as the minority carrier lifetimes that would result if only the radiative, or the nonradiative, transitions were effective. Note that

$$\frac{1}{\tau_{\mathrm{p}}} = \frac{1}{\tau_{\mathrm{rr}}} + \frac{1}{\tau_{\mathrm{nr}}}$$

The internal quantum efficiency in bulk material is given by

$$\eta_{\mathrm{int}} = \frac{(\mathrm{d}n/\mathrm{d}t)_{\mathrm{rr}}}{(\mathrm{d}n/\mathrm{d}t)} = \frac{1/\tau_{\mathrm{rr}}}{1/\tau_{\mathrm{rr}} + 1/\tau_{\mathrm{nr}}} = \frac{1}{1 + \tau_{\mathrm{rr}}/\tau_{\mathrm{nr}}} \tag{8.4.4}$$

So for an efficient source it is essential that $\tau_{\mathrm{rr}}/\tau_{\mathrm{nr}}$ be as small as possible.

The internal quantum efficiency may be defined more generally as the ratio of the rate of photon generation within the semiconductor, $\dot{N}_{ph}$, to the rate, $\dot{N}_c$, at which carriers are injected across the junction. This definition permits effects other than the bulk recombination processes to be taken into account, as we shall see in later sections. When the thickness of the bulk semiconductor on either side of the junction is large compared to L_p and L_n, both definitions are the same. This is because all the injected carriers will eventually recombine in the semiconductor and a fraction η_{int} will recombine radiatively. Thus $\dot{N}_{ph} = \eta_{int}\dot{N}_c$. An injection current $I = \dot{N}_c e$ therefore generates an optical power Φ_s in the semiconductor which is given by

$$\Phi_s = \dot{N}_{ph}\mathcal{E}_{ph} = \eta_{int}\dot{N}_c\mathcal{E}_{ph} = \eta_{int}(I/e)\mathcal{E}_{ph} \tag{8.4.5}$$

where $\mathcal{E}_{ph} = hc/\lambda$ is the photon energy.

Impurity atoms, grain boundaries, dislocations and other crystal defects all provide trapping mechanisms which promote nonradiative decay. As a result, η_{nr} is reduced in inverse proportion to the concentration, N_{tr}, of these traps. In an efficient optical source, N_{tr} must be maintained as small as possible. In general such trapping levels are located near to the middle of the band gap, and when they dominate the nonradiative carrier lifetime, an order-of-magnitude estimate for τ_{nr} may be obtained from

$$\frac{\tau_{nr}}{[\mathrm{s}]} \simeq \frac{10^{14}[\mathrm{m}^3]}{N_{tr}} \tag{8.4.6}$$

Thus a trap concentration of as little as $10^{21}\ \mathrm{m}^{-3}$ is sufficient to reduce τ_{nr} to 100 ns.

The rate at which carriers recombine by direct band-to-band transitions is proportional to the local electron concentration and to the local hole concentration. Thus

$$\frac{\begin{pmatrix}\text{Rate of radiative}\\ \text{recombination of}\\ \text{carriers}\end{pmatrix}}{[\text{number/m}^3/\text{s}]} = rnp \tag{8.4.7}$$

where $r/[\mathrm{m}^3/\mathrm{s}]$ is a recombination constant characteristic of the material. In equilibrium this rate of recombination just balances the rate of thermal generation of carriers, G, which is also characteristic of the material and is an exponential function of $(\mathcal{E}_g/kT)$. Thus

$$rn_{p0}p_{p0} = rn_i^2 = G \tag{8.4.8}$$

When there is an excess of carriers over the equilibrium concentrations, the *net* rate of radiative recombination is given by

$$-\left(\frac{\mathrm{d}n}{\mathrm{d}t}\right)_{\mathrm{rr}} = rnp - G = r(n_{\mathrm{p0}} + \Delta n)(p_{\mathrm{p0}} + \Delta p) - rn_{\mathrm{p0}}p_{\mathrm{p0}} \tag{8.4.9}$$

Now, $\Delta n = \Delta p$, so that

$$-\left(\frac{\mathrm{d}n}{\mathrm{d}t}\right)_{\mathrm{rr}} = r(n_{\mathrm{p0}} + p_{\mathrm{p0}} + \Delta n)\,\Delta n \tag{8.4.10}$$

By definition,

$$\tau_{\mathrm{rr}} = \frac{\Delta n}{-(\mathrm{d}n/\mathrm{d}t)_{\mathrm{rr}}} \tag{8.4.2}$$

$$\therefore \quad \tau_{\mathrm{rr}} = \frac{1}{r(n_{\mathrm{p0}} + p_{\mathrm{p0}} + \Delta n)} \tag{8.4.11}$$

Up to now we have assumed that τ_{p} and thus τ_{rr} were constant. Indeed it was on that basis that we were able to derive eqn. (7.4.12) and the results of Section 7.5. Equation (8.4.11) shows that this will only be true under conditions of moderate carrier injection such that $\Delta n \ll p_{\mathrm{p0}}$.

According to eqn. (8.4.11) the value of τ_{rr} would be greatest in intrinsic material when $n_{\mathrm{p0}} = p_{\mathrm{p0}} = n_{\mathrm{i}} \gg \Delta n$. Then

$$(\tau_{\mathrm{rr}})_{\mathrm{intrinsic}} = \frac{1}{2rn_{\mathrm{i}}} \tag{8.4.12}$$

but this is not a case of any interest to us. In p-type material such that $n_{\mathrm{p0}} \ll p_{\mathrm{p0}}$ and $\Delta n \ll p_{\mathrm{p0}}$

$$\tau_{\mathrm{rr}} = \frac{1}{rp_{\mathrm{p0}}} = \frac{1}{rn_{\mathrm{A}}} \tag{8.4.13}$$

Provided that these conditions are satisfied, τ_{rr} is indeed independent of the excess carrier concentration. However, with high injection levels we may well have $\Delta n \gg p_{\mathrm{p0}}$, in which case

$$(\tau_{\mathrm{rr}})_{\mathrm{high\ injection}} = \frac{1}{r\,\Delta n} \tag{8.4.14}$$

In this case τ_{rr} becomes a function of n and may therefore vary with time and with position, and the analysis becomes a great deal more complicated.

In an indirect band-gap material an order-of-magnitude value for r would be something in the region of 10^{-21} m^3/s. In a direct band-gap material r would be more like 10^{-16} m^3/s. These values enable us to bring out dramatically, with the aid of a simple example, the considerable effect that the nature of the band gap has on the internal quantum efficiency. Consider conditions of low carrier injection into p-type material so that eqn. (8.4.6) determines the nonradiative lifetime and eqn. (8.4.13) determines the radiative lifetime of the carriers. Assume that $\tau_{nr} \simeq 100$ ns and that $p_{p0} = n_A = 10^{23}$ m^{-3}. Then in an indirect band-gap material such as silicon, $\tau_{rr} \simeq 10$ ms and

$$\eta_{int} = \frac{1}{(1 + \tau_{rr}/\tau_{nr})} \simeq \frac{1}{(1 + 10^{-2}/10^{-7})} \simeq 10^{-5} \qquad (8.4.15)$$

On the other hand, in a direct band-gap material such as gallium arsenide, $\tau_{rr} \simeq 100$ ns, and

$$\eta_{int} \simeq \frac{1}{1 + 10^{-7}/10^{-7}} = 0.5 \qquad (8.4.16)$$

Values as high as this are obtained in LEDs made from direct band-gap materials.

A knowledge of the carrier lifetime enables an estimate of the diffusion length, $L_p = \sqrt{(D_e \tau_p)}$, to be made. The diffusion coefficient, D_e, increases with temperature but may decrease at high doping concentrations. For gallium arsenide at room temperature and with an impurity level of about 10^{23} m^{-3}, D_e is about 10^{-2} m^2/s. Using the same values of 100 ns for both τ_{rr} and τ_{nr}, $\tau_p = 50$ ns and

$$L_p = \sqrt{(D_e \tau_p)} = 22 \ \mu m \qquad (8.4.17)$$

The discontinuity to the crystal lattice caused by the semiconductor surface gives rise to a large number of local energy states throughout the band gap. Any other major discontinuity in the crystal structure will have the same effect. This applies particularly to the heterojunctions, which we shall discuss in Chapter 9, where there is a change in material within the crystal. At the interface there is likely to be a high concentration of energy levels which act as recombination centers and form a sink for excess carriers. Recombination via these states is mainly nonradiative, so the presence of a surface or interface within one or two diffusion lengths of the junction may seriously diminish the internal quantum efficiency of a device. As with recombination in the bulk material, we may postulate the net rate of carrier recombination at the surface to be proportional to the excess carrier concentration, Δn, there. The net recombination has to be supplied by a flux of carriers which disappear by recombination at the surface. In the simple case shown in Fig. 8.7 an interface perpendicular to the carrier flow causes a change, ΔJ, in the carrier current

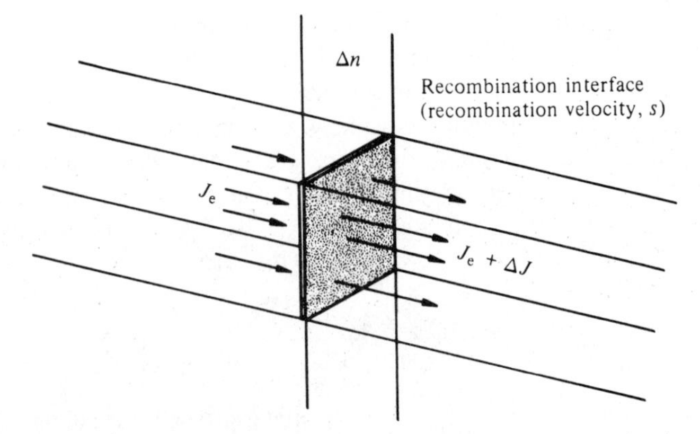

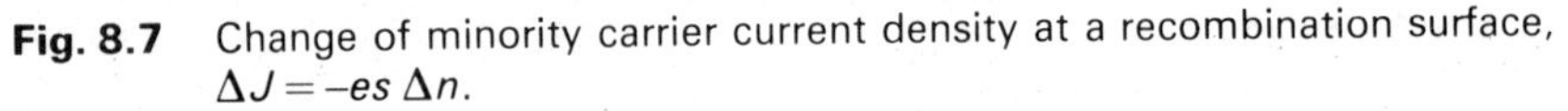
Fig. 8.7 Change of minority carrier current density at a recombination surface, $\Delta J = -es\,\Delta n$.

density given by

$$\Delta J = -es\ \Delta n \tag{8.4.18}$$

where s, which has the dimensions of speed, is usually referred to as the surface recombination velocity. Values for s vary widely. They range from 1 to 10^3 m/s for a free semiconductor–air surface, depending on its physical and chemical condition. Values as low as 0.01 to 0.1 m/s occur at silicon surfaces well passivated with thermally grown oxide.

8.5 THE EXTERNAL QUANTUM EFFICIENCY

Obtaining a high internal quantum efficiency is not of itself sufficient to guarantee a successful semiconductor optical source. Most of the recombination radiation is generated within one or two diffusion lengths of the junction and it is radiated in all directions. It is of course essential that as much as possible should be able to exit effectively from the semiconductor material. The ratio of the number of photons finally emitted to the number of carriers crossing the junction is known as the external quantum efficiency, η_{ext}.

Four main effects cause η_{ext} to be smaller than η_{int}. In the first place only light emitted in the direction of the semiconductor–air surface is useful. Secondly, it is only light reaching the emitting surface at an angle of incidence less than the critical angle, θ_c, that can be transmitted. Thirdly, even some of this light is reflected at the semiconductor–air surface. Fourthly, there is absorption between the point of generation and the emitting surface. The four problems are illustrated in Fig. 8.8.

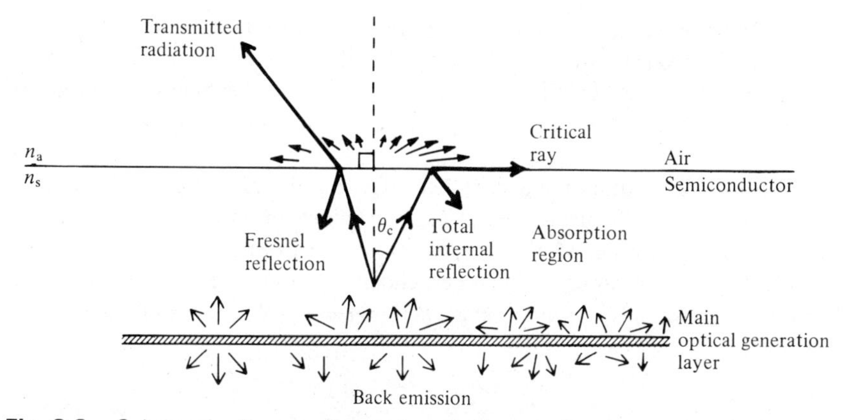

Fig. 8.8 Schematic diagram illustrating the four main causes of optical loss in an LED: back emission, limited acceptance cone of the surface, Fresnel reflection and absorption.

In Fig. 8.8 the light-emitting region is represented as a thin layer radiating in all directions. This behaves as a double-sided diffuse (Lambertian) emitter, and we may use the method given in Section 2.1 to calculate the fraction of the total radiant power that will be emitted such that its angle to the surface normal is less than θ_c. Assume that the power radiated per unit solid angle from the whole emissive area in the direction normal to the emitting surface is I_{s0}. Then the power per unit solid angle radiated in directions making an angle, θ, to the surface normal is $I_{s0} \cos \theta$, and the total radiant flux from both sides of the generation layer is

$$\Phi_s = 2\int_{\theta=0}^{\pi/2} I_{s0} \cos \theta \, 2\pi \sin \theta \, d\theta = 2\pi I_{s0} \tag{8.5.1}$$

The fraction, f', of this total power that is able to escape from the semiconductor–air surface is

$$f' = \frac{1}{2\pi I_{s0}} \int_0^{\theta_c} I_{s0} \cos \theta \, 2\pi \sin \theta \, d\theta = \tfrac{1}{2} \sin^2 \theta_c = \frac{n_a^2}{2n_s^2} \tag{8.5.2}$$

where n_a is the refractive index of the surrounding medium and n_s is the refractive index of the semiconductor. In the case of a gallium arsenide LED emitting into air, $n_a = 1$ and $n_s = 3.7$, so that $\theta_c = 16°$ and $f' = 0.036$.

Note that if the emissive region is regarded as a roughly spherical volume which radiates equally in all directions, then $I(\theta) = I_{s0} =$ constant, and a similar calculation yields a value for f' that is smaller by a factor of two.

In either case, a possibility we have neglected is that some of the radiation emitted back into the bulk semiconductor may be reflected and so may contribute to the emitted radiation. This technique is used to enhance the output of some types of light-emitting diode.|

Even those rays within the acceptance cone ($\theta < \theta_c$) suffer some reflection at the semiconductor–air surface because of the change in refractive index. This is known as Fresnel reflection and is discussed, for example, in Section 1.5 of Ref. [2.1]. Of the radiation incident perpendicularly onto the surface, a fraction R is reflected and only the remaining fraction, $t = 1 - R$, is transmitted, where t is given by

$$t = \frac{4n_a n_s}{(n_a + n_s)^2} = \frac{4n_s}{(1 + n_s)^2} \tag{8.5.3}$$

when $n_a = 1$.

In the case of more oblique rays the transmitted fraction varies little, until it becomes zero at the critical angle, θ_c. For the GaAs–air surface, $t = 0.67$, and

$$f't \simeq \frac{2}{n_s(1 + n_s)^2} = 0.024 \tag{8.5.4}$$

The transmission factor, t, may be increased by 'blooming'. A layer of transparent material, a quarter-wavelength thick, is applied to the semiconductor surface. Its refractive index should lie between that of the semiconductor and that of the surrounding medium. Ideally it should be $(n_a n_s)^{1/2}$, in which case $t \to 1$. Blooming has little effect on the critical angle.

The critical angle loss, f', is clearly much the more serious. There is, however, a fundamental limit to the power that can be coupled from a source of specified radiance and area into any given fiber. This is discussed further in Section 8.6 and Appendix 5, and practical examples of LED–fiber coupling arrangements are shown in Sections 8.6 and 9.3.

The final loss mechanism, self-absorption, cannot be so easily quantified. It is simply the reverse of the process of radiative band-to-band recombination. Within the semiconductor, radiation with a photon energy greater than the band gap ($hf > \mathcal{E}_g$) may interact with a valence electron and as a result excite it into the conduction band. In this way an electron–hole pair is formed and a photon is absorbed from the radiant power. Because this process forms the basis of the semiconductor photodetectors to be treated in Chapter 12, further discussion will be deferred until then. Here we will merely note that when the emitted radiation results from band-to-band recombination, the photon energy of necessity exceeds the band-gap energy and is likely to suffer self-absorption. It is thus essential to keep the distance from the generation region to the emitting surface as short as possible. The danger then is that the surface with its high concentration of trapping levels will come within one or two diffusion lengths of the junction and cause the nonradiative lifetime and hence the internal quantum efficiency to be much reduced. Clearly a compromise has to be reached.

8.6 LED DESIGN FOR OPTICAL COMMUNICATION

The structure of a typical, conventional light-emitting diode is shown in Fig. 8.9. For a visible source GaAsP or GaP doped with N or ZnO would be used.

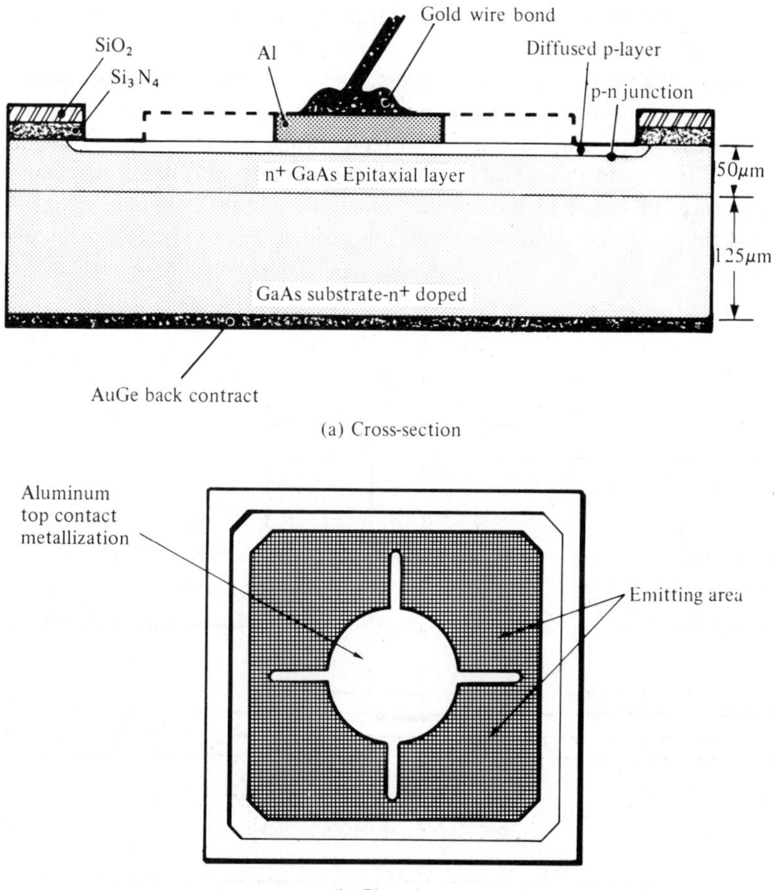

Fig. 8.9 Design of a typical LED. [From *Hewlett-Packard Optoelectronics Applications Manual*, McGraw-Hill. © 1977 Hewlett-Packard Co.; with permission of McGraw-Hill Book Co.]

An alternative design that provides the small emissive area and high radiance needed for optical communication is the Burrus-type, surface-emitting LED. Its essential features are illustrated in Fig. 8.10. Comparison with the more conventional LED of the previous figure shows that the emitted radiation is brought out through what would normally be the 'back' of the diode, that is, through the substrate. This has been etched out in order to minimize the distance between the active layer and the emitting surface. An insulating oxide layer separates what is now the rear contact from the semiconductor except in the light-generating region. In this way the flow of current is concentrated into a laterally well defined active layer.

In the Burrus design the proximity of the active layer to the heat sink means that the thermal impedance is small and that high current densities may be used without causing an excessive rise of temperature. Excessive active-layer temperatures give rise to three effects: the wavelength distribution of the emitted radiation changes with temperature; the internal quantum efficiency falls because there is an increase with temperature in the rate of nonradiative recombination; and the life of the LED falls rapidly with increasing junction temperature. In Fig. 8.11 the output radiation from an LED is seen to fall as the junction temperature increases during a single pulse. The 50% decrease in

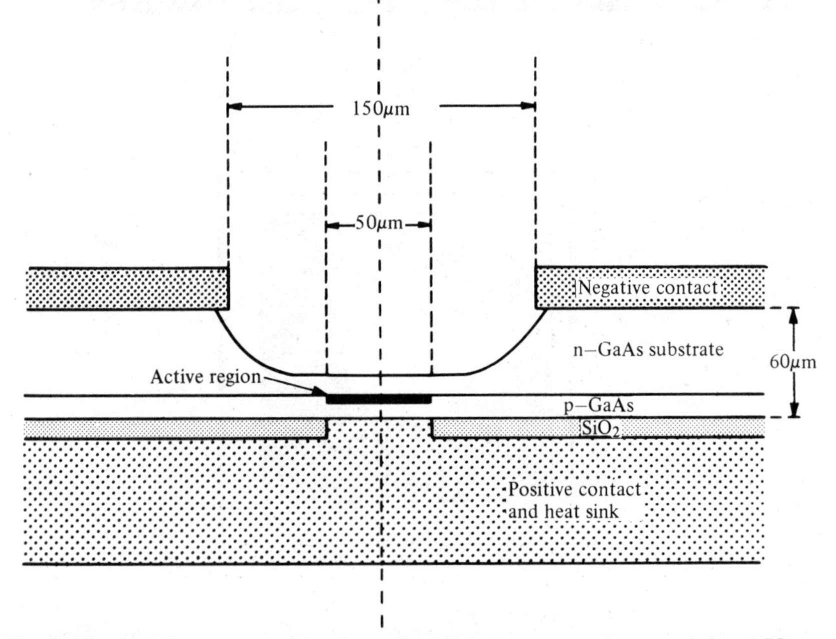

Fig. 8.10 Axial cross-section through a Burrus-type surface-emitting LED.

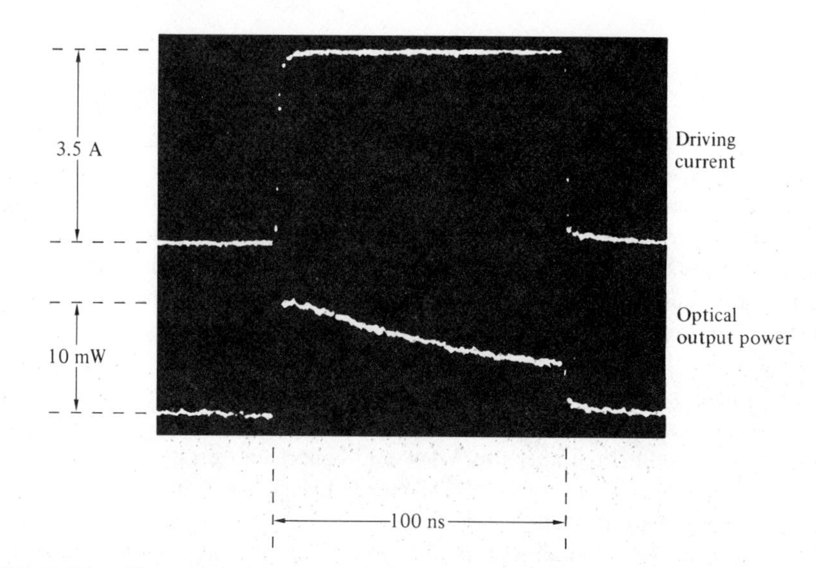

Fig. 8.11 Illustration of the fall in optical output as a result of the rise in junction temperature during a 100 ns pulse. [Redrawn from an oscilloscope photograph from R. W. Dawson and C. A. Burrus, Pulse behavior of high-radiance small-area electroluminescent diodes, *Appl. Optics* **10**, 2367–9 (1971).]

Time resolved wavelength measurements indicated a temperature rise from about 30°C at the start of the pulse to about 90°C at the end.

output power corresponds to a temperature rise from room temperature to about 90–100°C. In general, with GaAs and GaAlAs devices the peak junction temperature should be kept below about 50–100°C.

It may be helpful at this point to give an order-of-magnitude calculation of the likely temperature rise in a Burrus-type diode using a one-dimensional heat-flow model. For GaAs the room-temperature thermal conductivity, η, is 44 W/(m.K), and gets smaller as the temperature rises. Let us take the distance, Δx, from the junction to the heat sink to be 3 μm and the forward voltage drop to be $V = 1.5$ V. This is assumed to occur mainly across the junction. We will neglect the optical power radiated away (!) and estimate the current density J required to produce a 50°C rise in junction temperature (ΔT). This is given by

$$JV = \eta \frac{\Delta T}{\Delta x} \tag{8.6.1}$$

$$\therefore \quad J = \frac{44 \times 50}{1.5 \times 3 \times 10^{-6}} = 4.9 \times 10^{8} = 490 \text{ A/mm}^2$$

With a contact diameter of 50 μm this corresponds to a diode current of about 1 A.

Let us now try to estimate, on the basis of the theory we have developed so far, the optical power that we might expect to be launched onto an optical fiber from a Burrus-type GaAs LED across an intervening, planar air gap. The diameter of the light-generating region is again assumed to be 50 μm. Because the active layer is close to the semiconductor surface and because the critical angle is small, the diameter of the region of the surface which emits the radiation is not much bigger. If the air gap is small and if the fiber core diameter is greater than the emissive area, then we may assume that almost all the emitted radiation is incident onto the face of the fiber core. The question is, what fraction of this lies within the acceptance cone of the fiber? In general LEDs emit as diffuse (Lambertian) sources, and we showed in Section 2.1.2 that a step-index fiber of numerical aperture NA would collect and propagate a fraction $(NA)^2 = (n_1^2 - n_2^2)$ of the total light given off by such a source.

By analogy with the definitions of η_{int} and η_{ext}, we may define a quantum efficiency, η_f, for the overall source–fiber coupling as the ratio of the number of photons usefully launched onto the fiber to the number of carriers crossing the LED junction. Then

$$\eta_f = \eta_{ext}(NA)^2 = \eta_{ext}(n_1^2 - n_2^2) \tag{8.6.2}$$

and if we neglect self-absorption in the semiconductor,

$$\eta_f = \eta_{int} f't(n_1^2 - n_2^2) \tag{8.6.3}$$

For a fiber with $NA = 0.17$ and a GaAs source with $f't = 0.024$, as in eqn. (8.5.4) and with $\eta_{int} = 0.5$, $\eta_f = 0.00035$.

We will make an estimate of the power launched onto the fiber when our diode is forward-biassed with a current of 100 mA (50 A/mm^2). We assume that this corresponds to a voltage drop of 1.5 V giving an electrical power consumption of 150 mW. Let $\mathcal{E}_{ph} = 1.4$ eV. The total optical power produced within the semiconductor is $\Phi_s = \eta_{int}(I/e)\mathcal{E}_{ph} = 70$ mW. The total optical power emitted into the air is $\Phi_a = \eta_{ext}(I/e)\mathcal{E}_{ph} = 1.7$ mW. For a diffuse source this corresponds to a normal radiant intensity of $I_{a0} = \Phi_a/\pi = 0.53$ mW/sr and to a radiance of $L_a = I_{a0}/A_s = 2.7 \times 10^5 = 0.27$ W/mm^2/sr. The power launched onto the fiber is $\Phi_T = \eta_f(I/e)\mathcal{E}_{ph} = 49$ μW.

With the LED and fiber specified, the only way in which this appallingly inefficient coupling can be improved is by minimizing the Fresnel reflection losses. Perhaps the most satisfactory way of doing this is the one illustrated in Fig. 8.12(a). The LED is attached to the fiber using a transparent bonding cement having a refractive index, n_a, similar to that of the fiber. In addition the LED surface is 'bloomed' using a film of dielectric material such as alumina ($n = 1.76$) or silicon monoxide ($n = 1.9$) or silicon nitride ($n \simeq 2.0$). If in

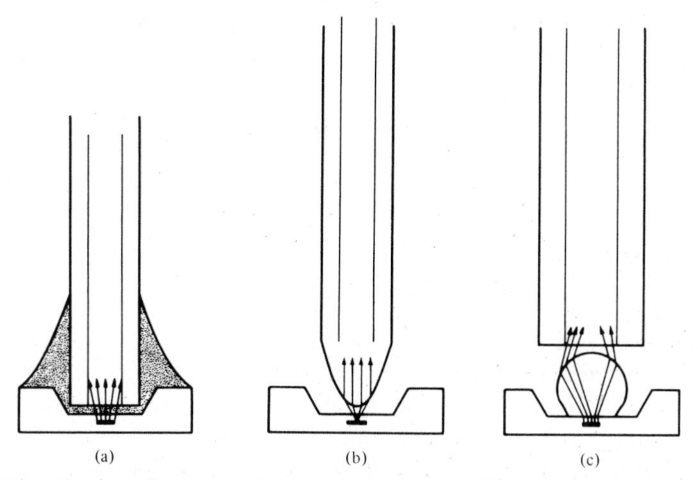

Fig. 8.12 Source–fiber coupling arrangements: (a) the use of an index-matching epoxy; (b) the fiber end is tapered and rounded to form a lens which will collimate divergent rays entering the fiber core; (c) a spherical lens is formed on the LED surface. This will help to collimate the emitted radiation.

Section 2.1.2 we had retained the refractive index, n_a, of the surrounding medium explicitly in the analysis that led to eqn. (2.1.13), we would have found that the fraction of the incident radiation captured and propagated by the fiber was $(NA)^2/n_a^2 = (n_1^2 - n_2^2)/n_a^2$. Using this result and putting $t = 1$, the overall fiber-coupling efficiency, η_f, now becomes

$$\eta_f = \eta_{int} f' \frac{(n_1^2 - n_2^2)}{n_a^2} = \eta_{int} \frac{(n_1^2 - n_2^2)}{2n_s^2} \tag{8.6.4}$$

substituting from eqn. (8.5.2). In our numerical example this gives $\eta_f = 0.00053$ and $\Phi_T = 74\ \mu W$, an improvement of about 50%.

We show in Appendix 5 that lensed couplers like those shown in Fig. 8.12(b) and (c) can improve the coupling efficiency only if the fiber core diameter is increased or the emissive area of the LED is reduced. In Fig. 8.12(b) the spherical end to the fiber can be made quite easily by controlled melting. The radius of curvature can be adjusted by tapering the fiber as shown, or by forming a sphere of larger diameter than the fiber. Any bonding cement should have a low refractive index. In Fig. 8.12(c) the best results can be obtained using a truncated sphere of high refractive index (1.9 or 2.0), and using a cement of refractive index similar to that of the fiber to bond the LED, lens and fiber together. There will have to be a further bonding layer between lens and LED. In both cases Fresnel losses should be minimized by 'blooming'.

With any of these systems the maximum power that can be launched onto the fiber from a diffuse source is shown in Appendix 5 to be limited to

$$\Phi_T \leqslant \eta_{int}\left(\frac{JA_s}{e}\right)\mathcal{E}_{ph}\left(\frac{A_c}{A_s}\right)\frac{(n_1^2 - n_2^2)}{2n_s^2} \tag{8.6.5}$$

where A_s and A_c are the areas of the source and the fiber core, respectively. For a general source of radiance L_s,

$$\Phi_T \leqslant \pi L_s A_c(n_1^2 - n_2^2)/n_s^2 \tag{8.6.6}$$

Although the coupling efficiency may be increased by the use of a smaller diffuse source and a lens coupler, the total power launched will not increase unless the injection current density J can be allowed to increase. If the same total current ($I = JA_s$) can be maintained using the smaller source, then the full benefit of the gain (A_c/A_s) may be realized. But if J is fixed there is no advantage. This serves to emphasize the need for sources of small emissive area and high radiance such as the edge-emitting LED and injection laser which are discussed in Chapters 9–11. These benefit from the highly directional characteristics of the emitted light.

8.7 BEHAVIOR AT HIGH FREQUENCY

In this final section of Chapter 8 we wish to examine the optical output of a luminescent diode when it is electrically modulated at high frequency. We shall assume that a small sinusoidal modulation is superimposed on the normal forward-biassed state. Once again we shall limit the detailed discussion to the case of a long-base, n^+p diode in which electron injection dominates the carrier flow and the excess electron concentration falls to zero far from the junction. In practice the modulation behavior of an LED is likely to be influenced by the proximity of the emitting surface, but while this does affect the detailed form of the carrier distribution functions, the general form of the frequency response remains the same. In Section 7.5 we obtained an equivalent circuit for the LED. Here we shall be concerned to determine the values for r_D and C_D in that circuit, so that the terminal characteristics of the diode at high frequencies are known. We shall also wish to relate the optical output power to the electrical input as a function of frequency.

The steady-state solution for the diffusion region was obtained in Section 7.4.2. Under dynamic conditions the concentration of carriers within the volume element, $\delta A\,\delta x$, in Fig. 7.10 is time-varying. As a result, the condition for carrier conservation, previously expressed in eqn. (7.4.10), now becomes

$$D_e \frac{\partial^2 n(x,t)}{\partial x^2} = \frac{[n(x,t) - n_{p0}]}{\tau_p} + \frac{\partial n(x,t)}{\partial t} \tag{8.7.1}$$

As before, we assume that the electron concentration at the edge of the depletion layer in the p-type region ($x = 0$) is controlled by the concentration of majority electrons in the n-type region and the potential change across the depletion layer. Let this be $V_0 + V_1 \cos \omega t$; then, following eqn. (7.5.3),

$$\begin{aligned} n(0, t) &= n_{p0} \exp [e(V_0 + V_1 \cos \omega t)/kT] \\ &= n_{p0} \exp (eV_0/kT) \exp [(eV_1 \cos \omega t)/kT] \end{aligned} \tag{8.7.2}$$

In this section ω represents the angular frequency of modulation applied to the optical source—not, as elsewhere, to the optical frequency. Similarly, f here refers to the modulation frequency. This should not cause confusion, so the notation has not been complicated by adding subscripts in order to make this distinction. We shall represent the time-varying part of the voltage as the real part of $V_1 \exp j\omega t$, and assume $eV_1 \ll kT$. Then,

$$n(0, t) \simeq n_{p0} \exp (eV_0/kT)\left[1 + \frac{eV_1}{kT} \exp (j\omega t)\right] \tag{8.7.3}$$

which we can divide into the steady-state part

$$n(0) = n_{p0} \exp (eV_0/kT) \tag{8.7.4}$$

and the time-varying part

$$\begin{aligned} n_1(0, t) &= n(0)\frac{eV_1}{kT} \exp (j\omega t) \\ &= n_1(0) \exp (j\omega t) \end{aligned} \tag{8.7.5}$$

where

$$n_1(0) = n_{p0} \frac{eV_1}{kT} \exp\left(\frac{eV_0}{kT}\right) \tag{8.7.6}$$

The solutions to eqn. (8.7.1) will similarly separate into a steady-state solution, $n(x)$, and a time-varying component, $n_1(x, t)$:

$$n(x, t) = n(x) + n_1(x, t) \tag{8.7.7}$$

It is more convenient to work in terms of the excess carrier concentration,

$\Delta n(x, t)$:

$$\begin{aligned}\Delta n(x, t) &= n(x, t) - n_{p0} \\ &= n(x) - n_{p0} + n_1(x, t) \\ &= \Delta n(x) + n_1(x, t)\end{aligned} \tag{8.7.8}$$

because eqn. (8.7.1) then becomes

$$D_e \frac{\partial^2 \Delta n(x, t)}{\partial x^2} = \frac{\Delta n(x, t)}{\tau_p} + \frac{\partial \Delta n(x, t)}{\partial t} \tag{8.7.9}$$

Both parts of eqn. (8.7.8) must satisfy eqn. (8.7.9). Thus the steady-state part remains unchanged:

$$D_e \frac{d^2 \Delta n(x)}{dx^2} = \frac{\Delta n(x)}{\tau_p} \tag{7.4.11}$$

giving

$$\Delta n(x) = \Delta n(0) \exp(-x/\sqrt{(D_e \tau_p)}) \tag{7.4.12}$$

But the time-varying part is

$$D_e \frac{\partial^2 n_1(x, t)}{\partial x^2} = \frac{n_1(x, t)}{\tau_p} + \frac{\partial n_1(x, t)}{\partial t} \tag{8.7.10}$$

Because of the boundary condition at $x = 0$, eqn. (8.7.5), we shall expect solutions of the form

$$n_1(x, t) = n_1(x) \exp(j\omega t) \tag{8.7.11}$$

Substituting eqn. (8.7.11) into eqn. (8.7.10) gives

$$D_e \frac{d^2 n_1(x)}{dx^2} = \frac{n_1(x)}{\tau_p} + j\omega n_1(x) \tag{8.7.12}$$

$$\therefore \quad \frac{d^2 n_1(x)}{dx^2} = \frac{n_1(x)}{L_p^{*2}} \tag{8.7.13}$$

where

$$L_p^{*2} = D_e \tau_p / (1 + j\omega\tau_p) \tag{8.7.14}$$

And so,

$$n_1(x) = n_1(0) \exp(-x/L_p^*) \tag{8.7.15}$$

We therefore have an effective, time-varying diffusion length that is a complex function of frequency. Without enquiring too deeply into the subtleties of this, we shall use eqn. (8.7.15) to calculate the time-varying components of current density, J_1, and optical output power density, P_1, that derive from V_1.

The current density is derived from the gradient of the carrier concentration at $x = 0$:

$$J_1 = -eD_e\left(\frac{dn_1}{dx}\right)_{x=0} = \frac{eD_e n_1(0)}{L_p^*}$$

$$= \frac{eD_e n_{p0}}{L_p^*}\frac{eV_1}{kT}\exp(eV_0/kT) \tag{8.7.16}$$

$$\simeq J_0(1 + j\omega\tau_p)^{1/2}\frac{eV_1}{kT} \tag{8.7.17}$$

where we have used eqn. (8.7.14), and following eqn. (7.5.4) we have put

$$J_0 = \frac{eD_e}{L_p}\Delta n(0) = \frac{eD_e n_{p0}}{L_p}[\exp(eV_0/kT) - 1] \simeq \frac{eD_e n_{p0}}{L_p}\exp(eV_0/kT) \tag{8.7.18}$$

for the steady current density. The admittance of the forward-biassed diode is by definition

$$Y_D = \frac{1}{r_D} + j\omega C_D = \frac{J_1 A}{V_1} = \frac{I_1}{V_1} \tag{8.7.19}$$

where A is the junction area, and I_1 the a.c. component of the current. Therefore, using eqn. (8.7.17),

$$Y_D = \frac{eJ_0 A}{kT}(1 + j\omega\tau_p)^{1/2} \tag{8.7.20}$$

Put $I_0 = J_0 A$, then, squaring both sides, we have

$$Y_D^2 = \frac{1}{r_D^2} + j\frac{2\omega C_D}{r_D} - \omega^2 C_D^2 = \left(\frac{eI_0}{kT}\right)^2(1 + j\omega\tau_p) \tag{8.7.21}$$

Equating the real parts gives

$$\frac{1}{r_D^2} - \omega^2 C_D^2 = \left(\frac{eI_0}{kT}\right)^2 \tag{8.7.22}$$

so that for $\omega r_D C_D \ll 1$,

$$r_D \simeq \frac{kT}{eI_0} \tag{8.7.23}$$

Equating the imaginary parts gives

$$\frac{2C_D}{r_D} = \tau_p \left(\frac{eI_0}{kT}\right)^2 \tag{8.7.24}$$

so that

$$C_D \simeq \frac{\tau_p}{2} \frac{eI_0}{kT} \simeq \frac{\tau_p}{2r_D} \tag{8.7.25}$$

Notice that the validity of eqns. (8.7.23) and (8.7.25) depends on the assumption that $\omega r_D C_D \simeq \omega\tau_p/2 \ll 1$. At high frequencies when $\omega\tau_p \gg 1$, a little manipulation of eqns. (8.7.22) and (8.7.24) produces the results

$$\begin{aligned} r_D &\simeq \frac{kT}{eI_0}(\omega\tau_p/2)^{-1/2} \\ \omega C_D &\simeq \frac{eI_0}{kT}(\omega\tau_p/2)^{1/2} \end{aligned} \tag{8.7.26}$$

Thus $C_D \propto 1/\omega^{1/2}$ and at high frequencies, as C_D falls, the junction behavior again becomes dominated by the depletion layer capacitance, C_J.

The total optical power density is given by the total number of radiative recombinations per second per unit area multiplied by the mean photon energy. Thus

$$P = \int_0^\infty -\overline{\mathcal{E}_{ph}}(dn/dt)_{rr}\, dx \tag{8.7.27}$$

We may separate P into a steady-state component, P_0, and a time-varying component, P_1. Using the definition of η_{int} given in Section 8.4,

$$\left(\frac{dn}{dt}\right)_{rr} = \eta_{int}\left(\frac{dn}{dt}\right) = -\eta_{int}\frac{\Delta n(x,t)}{\tau_p} \tag{8.4.4}$$

thus,

$$P = P_0 + P_1 = \overline{\mathcal{E}_{ph}}\eta_{int}\int_0^\infty \frac{\Delta n(x,t)}{\tau_p}\, dx \tag{8.7.28}$$

Using eqn. (8.7.8) we may write

$$P_0 = \frac{\overline{\mathcal{E}_{ph}}\eta_{int}}{\tau_p}\int_0^\infty \Delta n(x)\,dx \tag{8.7.29}$$

and

$$P_1 = \frac{\overline{\mathcal{E}_{ph}}\eta_{int}}{\tau_p}\int_0^\infty n_1(x, t)\,dx \tag{8.7.30}$$

Substituting eqn. (7.4.12) into eqn. (8.7.29) and integrating, the d.c. component becomes

$$P_0 = \frac{\overline{\mathcal{E}_{ph}}\eta_{int}}{\tau_p}\Delta n(0)L_p \tag{8.7.31}$$

Thus using eqn. (8.7.18),

$$P_0 = J_0\frac{\overline{\mathcal{E}_{ph}}\eta_{int}}{\tau_p}\frac{L_p^2}{eD_e} = \frac{\eta_{int}\overline{\mathcal{E}_{ph}}}{e}J_0 \tag{8.7.32}$$

On the other hand integrating eqn. (8.7.15) to obtain the a.c. component, we arrive at

$$P_1 = \frac{\eta_{int}\overline{\mathcal{E}_{ph}}}{\tau_p}n_1(0)L_p^* \tag{8.7.33}$$

We now use eqn. (8.7.16) to eliminate $n_1(0)$, giving

$$P_1 = J_1\frac{\eta_{int}\overline{\mathcal{E}_{ph}}}{eD_e\tau_p}L_p^{*2} = \frac{\eta_{int}\overline{\mathcal{E}_{ph}}}{e}\frac{J_1}{(1 + j\omega\tau_p)} \tag{8.7.34}$$

Comparing eqns. (8.7.32) and (8.7.34) we see that the a.c. frequency response can be expressed as

$$\left|\frac{P_1}{J_1}\right| = \frac{P_0}{J_0}\frac{1}{(1 + \omega^2\tau_p^2)^{1/2}} \tag{8.7.35}$$

Optical power frequency responses obtained experimentally are often found to follow eqn. (8.7.35) quite closely, as shown in Fig. 8.13. From such a curve τ_p can be estimated to an accuracy of about 1 ns or better. It is important to appreciate that in the equivalent circuit of Fig. 7.12, I_1 is the current flowing in r_D and C_D but does not include any further current flowing in C_J. The driving circuit may cause I_1 itself to vary with frequency and the fall in optical power

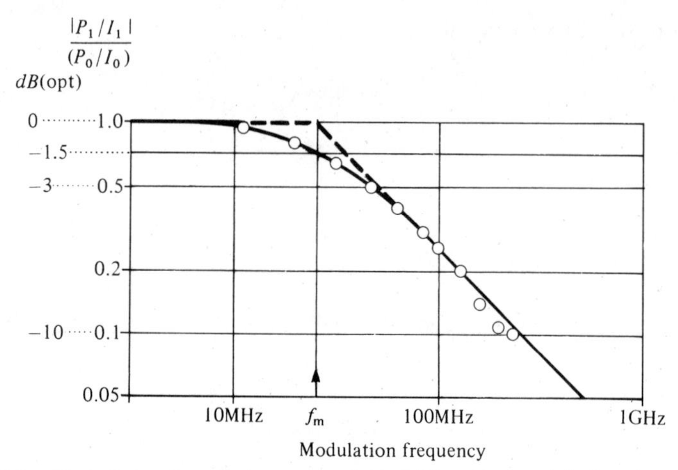

Fig. 8.13 Variation of output power of an LED with modulating frequency. The solid curve is a plot of equation (8.7.35) with $f_m = 1/2\pi\tau_p = 25$ MHz, corresponding to $\tau_p = 6.4$ ns. The circles represent experimental points.

at frequencies of $1/2\pi\tau_p$ and above will be superimposed on this circuit response.

It should be noted that there is always a linear relationship between I_1 and $n_1(0, t)$ and P_1 that does not depend on the small-signal assumption that produced eqn. (8.7.3). The only part of this analysis that depends on the assumption of small signals is the relationship between I_1 and V_1. If we cannot assume that $eV_1 \ll kT$, but do assume that

$$n(0, t) = n(0) + n_1(0) \cos \omega t \tag{8.7.36}$$

then

$$V_1(t) = \frac{kT}{e} \ln\left[1 + \frac{n_1(0)}{n(0)} \cos \omega t\right] \tag{8.7.37}$$

But I_1 and P_1 are still related by eqn. (8.7.35). However, the boundary between the depletion layer and the diffusion region is very difficult to model correctly, and the assumptions made in this analysis have to be treated with great caution when applied to large-signal, transient conditions.

The implications of eqn. (8.7.35) for the modulated power that can be produced in an optical source at high modulation frequencies are very important. At frequencies above $f_m = 1/2\pi\tau_p$, the source efficiency falls as shown in Fig. 8.13. Decreasing τ_p will increase this upper cutoff frequency, but decreasing τ_p by decreasing the lifetime for nonradiative recombination, τ_{nr}, will not increase the optical power obtainable at high frequencies; it will merely decrease the power at low frequencies by reducing the internal quantum efficiency. This follows from eqn. (8.4.5), since

$$f_\mathrm{m} = \frac{1}{2\pi\tau_\mathrm{p}} = \frac{1}{2\pi\eta_\mathrm{int}\tau_\mathrm{rr}} \tag{8.7.38}$$

The effect of reducing τ_{nr} but not τ_{rr} is illustrated in Fig. 8.14. It is thus essential to minimize the radiative lifetime τ_{rr}. This maximizes both the quantum efficiency at low modulation frequencies and the high-frequency cutoff. Reference to eqns. (8.4.13) and (8.4.14) shows that a high doping concentration and a high injection level minimize τ_{rr} and thus maximize the $\eta_{int} f_m$ product. A limit is reached when the doping level becomes so large that the nonradiative lifetime is decreased as well and so the quantum efficiency falls. In gallium arsenide this occurs at an impurity level in the region of 10^{24} m^{-3}, at which $\tau_{rr} \simeq 10$ ns. Thus modulation frequencies much above about 20 MHz can be obtained only at the expense of quantum efficiency. Modulation bandwidths about four times higher than this have been obtained without sacrificing efficiency with InGaAsP/InP LEDs of a type to be discussed in Section 9.3. These can be made so that they emit at wavelengths around either 1.3 μm or 1.5 μm, as was shown in Fig. 8.5(b).

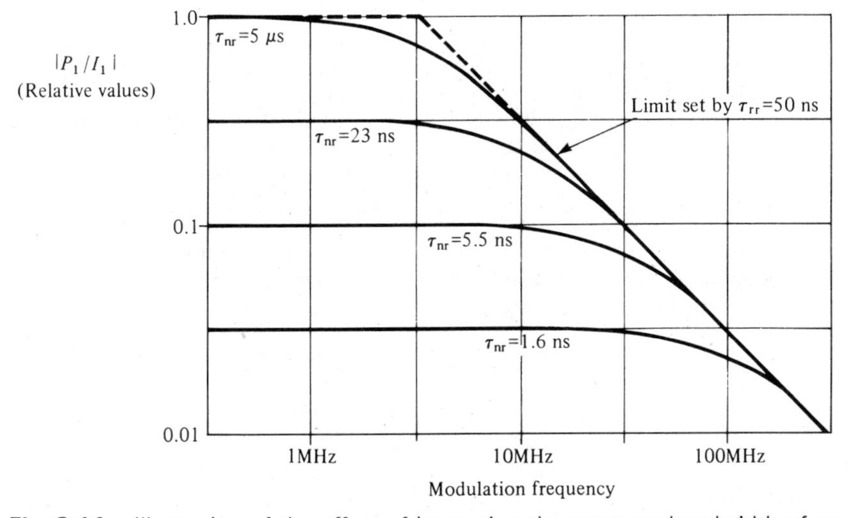

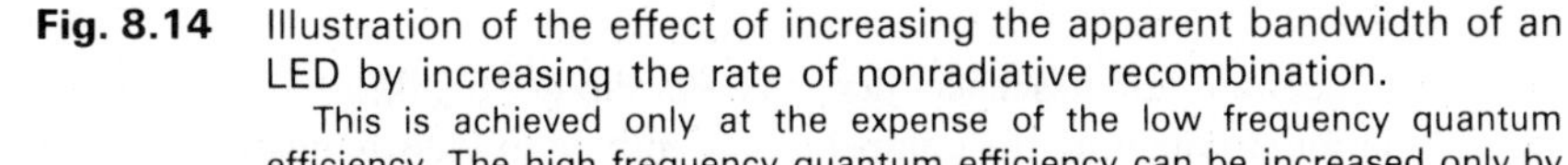

Fig. 8.14 Illustration of the effect of increasing the apparent bandwidth of an LED by increasing the rate of nonradiative recombination.

This is achieved only at the expense of the low frequency quantum efficiency. The high frequency quantum efficiency can be increased only by reducing τ_{rr}.

PROBLEMS

8.1 Assuming density of states functions given by eqns (8.2.5) and (8.2.6), and distributions of electrons in the conduction band and holes in the valence band that are approximated by the Boltzmann distribution, eqns (7.2.4) and (7.2.5), show that $n(\mathcal{E})$ and $p(\mathcal{E})$ reach maximum values at energies $(kT/2)$ from the band edges. Show by inspection that the range of energies over which $n(\mathcal{E})$ and $p(\mathcal{E})$ exceed $(1/\sqrt{2})$ of their maximum values is approximately 1.1 kT.

8.2 Show that the spectral distribution of spontaneous emission given in eqn. (8.2.12) follows from eqn. (10.2.12) on p. 305, when the density of states functions $S_c(\mathcal{E}_2)$ and $S_v(\mathcal{E}_1)$ are assumed to be constant, and the Fermi functions, F_2 and $(1 - F_1)$, are approximated by exponentials.

8.3 Show that the theoretical spectral power distribution given in eqn. (8.2.12) would have its peak value at the photon energy $\mathcal{E}_g + kT$, and would have a halfwidth of 2.4 kT.

8.4 On the assumption that the LED spectral characteristics shown in Fig. 8.5 refer to 300 K, calculate the spectral halfwidths in each case as a multiple of kT.

8.5 Modify eqn. (2.1.13) so that it expresses the fraction of the radiation generated by a plane, isotropic source in a medium of refractive index n_s that will be propagated by a fiber of core index n_1 and cladding index n_2.

8.6 Using the result of Problem 8.5, calculate the overall coupling efficiency for light entering the fiber from an LED source when the refractive indices of the semiconductor, core and cladding are 3.7, 1.460 and 1.455 respectively. The fiber core diameter may be assumed to exceed the diameter of the emissive region and the fiber may be assumed to be bonded directly onto the semiconductor surface with a bonding cement having the same refractive index as the fiber core. Would the use of a domed or dished, rather than a plane, semiconductor surface change the coupling efficiency? And if so, by how much?

8.7 Light from a small source having an angular distribution of radiant intensity given by $I(\theta) = I_0 \cos^a \theta$, where $a > 1$ and θ is the angle to the surface normal, is coupled into a fiber. Calculate the increased coupling efficiency compared with that obtained with a Lambertian source in which $a = 1$.

8.8 Derive eqns (8.7.26) for the high frequency condition $\omega\tau_p \gg 1$.

SUMMARY

Efficient injection luminescence depends on direct, band-to-band radiative recombination dominating the competing recombination mechanisms shown in Fig. 8.1. This requires a direct band-gap material.

Internal quantum efficiency is

$$\eta_{\text{int}} = \frac{1}{1 + \tau_{\text{rr}}/\tau_{\text{nr}}} \quad \text{where} \quad \tau_{\text{rr}} = \frac{1}{r(n_{\text{p0}} + p_{\text{p0}} + \Delta n)}$$

When nonradiative recombination is dominated by bulk trapping, $\tau_{\text{nr}} \simeq (10^{14}/N_{\text{tr}})[\text{m}^{-3}\ \text{s}]$.

The optical wavelength is $\lambda \simeq (1.24/\mathcal{E}_g)[\mu\text{m.eV}]$ with a spread corresponding typically to a range of photon energy of $1.5kT$–$3.5kT$.

Losses due to back emission, absorption, Fresnel reflection and total internal reflection at the semiconductor surface further reduce the available power: $f't = 2/n_s(1 + n_s)^2$. Index matching and 'blooming' can reduce the Fresnel loss. The external quantum efficiency is $\eta_{\text{ext}} = f't\eta_{\text{int}}$.

Optical output power is initially proportional to drive current but then tends to saturate. One cause is a decrease of optical emission with increasing temperature.

The Burrus high-radiance, small area design is shown in Fig. 8.10.

When source diameter is less than fiber core diameter, lens couplers can increase launched power, giving a theoretical maximum coupling efficiency of $\eta_f = \eta_{\text{int}}(n_1^2 - n_2^2)/2n_s^2$.

Frequency response is typically

$$\frac{|P_1/J_1|}{P_0/J_0} = \frac{1}{(1 + \omega^2\tau_p^2)^{1/2}} \quad \text{where} \quad \frac{1}{\tau_p} = \frac{1}{\tau_{\text{rr}}} + \frac{1}{\tau_{\text{nr}}}$$

Reducing τ_p by reducing τ_{nr} increases the frequency response at the expense of reduced quantum efficiency.

9

The Use of Heterostructures

9.1 HETEROJUNCTIONS

9.1.1 Types of Heterojunction

The p–n junctions discussed in Chapters 7 and 8 would have been formed by introducing relatively small concentrations of impurity atoms into what is basically the same semiconductor material. They might be called homojunctions. We have seen, however, in Figs. 7.3 and 7.4 that it is possible to produce quite different semiconductor materials which nevertheless crystallize in the same way and have the same, or almost the same, lattice constant. They may, therefore, be grown together as a single crystal. At the boundary between them there is a change in the band-gap energy, in the electron affinity, in the permittivity, and in all other dependent properties. Such junctions between semiconductors whose lattices match, but which are otherwise different, are known as *heterojunctions.*

At a heterojunction each of the semiconductors may be doped n-type or p-type. There are thus four possible combinations. We shall adopt the convention of indicating the wider band-gap material by an upper-case N or P, and the narrower band-gap material by a lower-case n or p. The four heterojunctions are thus n–N, p–P, n–P and P–n and each may be expected to show a different behavior. Further differences may arise as a result of the relative magnitudes of the electron affinities of the two materials. The electronic behavior to be expected at a heterojunction will be illustrated using the two materials whose electron energy levels in isolated samples are shown in Fig. 9.1. These are based loosely on GaAs and GaAlAs in that the material with the narrower band gap has the larger electron affinity, and the change in $\mathcal{E}_v$ is small because As is common to both semiconductors.

For a plane heterojunction, as with the homojunction discussed in Section 7.6, the electrostatic potential, $V(x)$, is constrained by the one-dimensional Poisson equation,

$$\frac{d^2 V}{dx^2} = -\frac{\rho(x)}{\varepsilon_r \varepsilon_0} \tag{9.1.1}$$

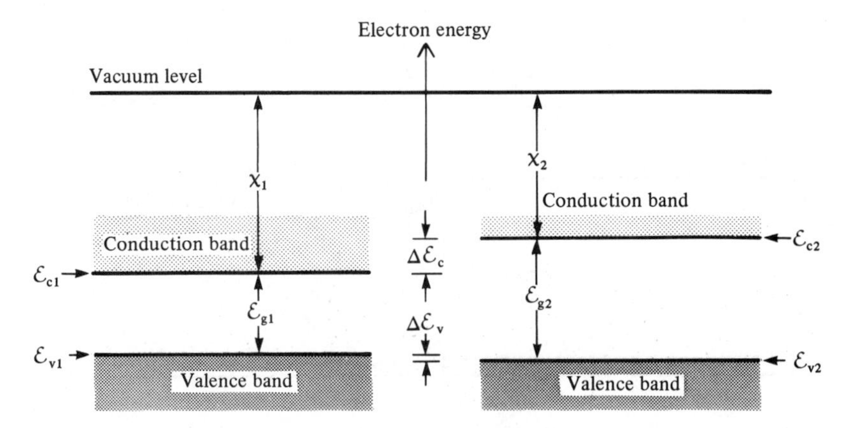

Fig. 9.1 Electron energy levels for two isolated semiconductors in equilibrium.

where ρ is the local net charge concentration, $\varepsilon_r \varepsilon_0$ is the material permittivity, and x is the position coordinate perpendicular to the junction. A change in potential gradient indicates the presence of uncompensated charge. Outside the semiconductor surface the vacuum level $\mathcal{E}_0$ remains continuous at the junction† but inside the material there are step changes of $\Delta\mathcal{E}_c$ and $\Delta\mathcal{E}_v$ in the levels of the band edges. These accommodate the changed conditions inside the semiconductor. In the specific example of a heterojunction between GaAs and $Ga_{0.7}Al_{0.3}As$, $\Delta\mathcal{E}_c \simeq 0.32$ eV and $\Delta\mathcal{E}_v \simeq 0.05$ eV.

In Fig. 9.2 we show the equilibrium energy level diagram for an ideal, abrupt junction between the two materials of Fig. 9.1, when both are doped n-type. The change in potential caused by the junction leads to the formation of a depletion layer for electrons in region 2 and an accumulation layer for electrons in region 1. The effect on the minority holes is negligible. The potential barrier for electrons that occurs at the junction might be expected to give rise to non-ohmic conduction characteristics at low bias voltages. Some very abrupt and lightly doped n–N heterojunctions between GaAs and $Ga_{0.7}Al_{0.3}As$ have been found to act as rectifying junctions as the I–V characteristic of Fig. 9.3 shows. However, n–N heterojunctions in the GaAlAs/GaAs system have generally been found to give ohmic characteristics as in Fig. 9.4, curve (a). Non-ohmic behavior was occasionally observed as in curve (b) in the figure, but such junctions were found to be unstable. The explanation of ohmic behavior may lie in the fact that the junction in practice is not abrupt but is graded over a distance that may be comparable to the thickness of the accumulation and depletion layers. In that case the potential 'spikes' would be

† This statement is not quite true as it is possible for there to be a dipole layer formed at the interface. See, for example, W. R. Frensley and H. Kroemer, Theory of the energy-band lineup at an abrupt semiconductor heterojunction, *Phys. Rev.*, **B16**, 2642–2652 (1977).

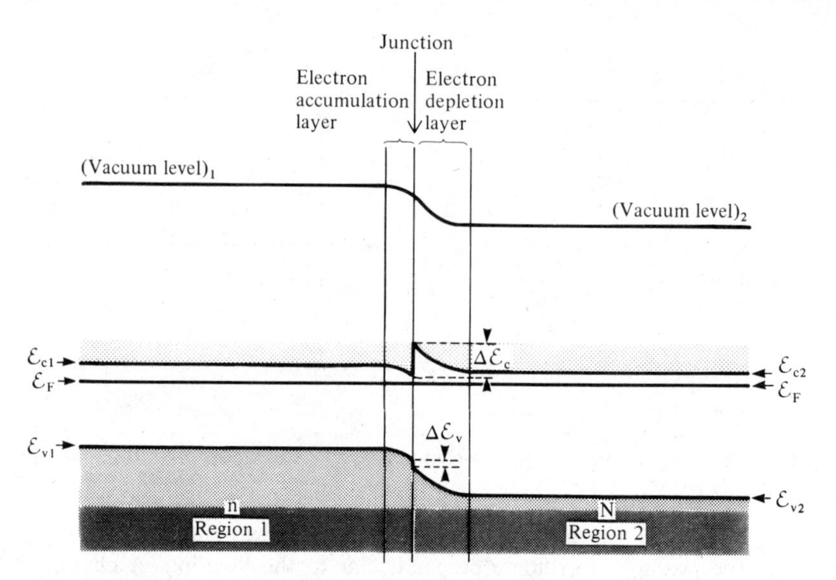

Fig. 9.2 Electron energy levels across an n–N heterojunction in equilibrium.

submerged into the more general variation, and electrons and holes would be able to flow freely across the junction in either direction in response to small applied bias voltages. It has to be said, however, that the detailed structure of the heterojunction is very complicated and not fully understood. A lattice mismatch of as little as 0.1% is sufficient to produce considerable strain at the

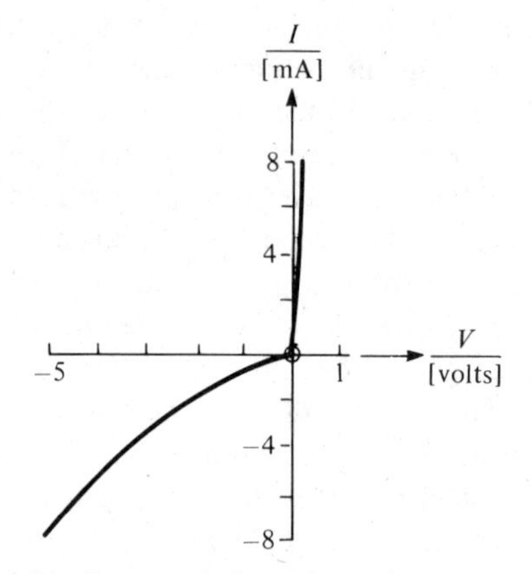

Fig. 9.3 A typical *I–V* characteristic obtained from an n–N, GaAs–GaAlAs heterojunction with the voltage applied to the GaAs layer. [Taken from A. Chandra and L. F. Eastman, Rectification at n–n GaAs:(Ga, Al)As Heterojunctions, *Ets. Letts.* **15**, 90–1, (1 Feb. 1979).]

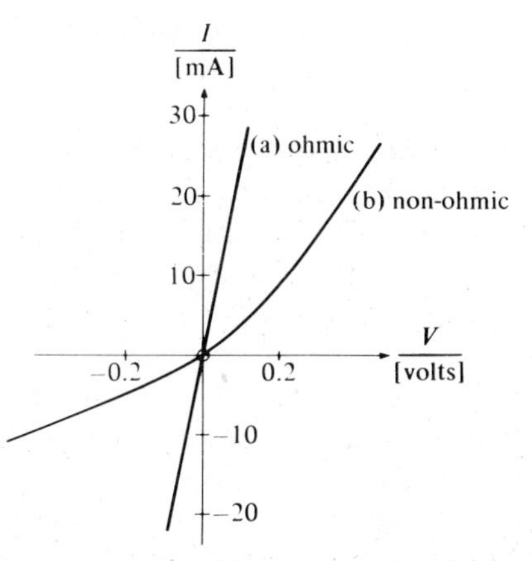

Fig. 9.4 *I–V* characteristics of n–N heterojunctions formed between GaAs and GaAlAs. [Taken from J. F. Womac and R. H. Rediker, The graded-gap $Al_xGa_{1-x}As$–GaAs heterojunction, *J. Appl. Phys.* **43**, 4129–33 (1972).]

junction and give rise to a high concentration of trapping levels. It has been suggested that the recombination velocity, s, at a heterojunction may be approximately related to the fractional change in lattice constant $(\Delta a/a_0)$ by

$$\frac{s}{[\mathrm{m/s}]} \simeq (2 \times 10^5)\frac{\Delta a}{a_0} \tag{9.1.2}$$

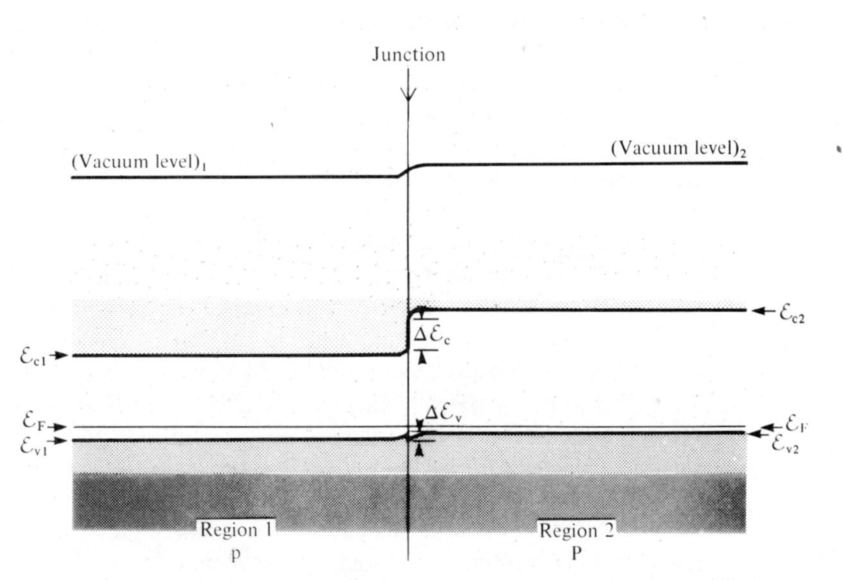

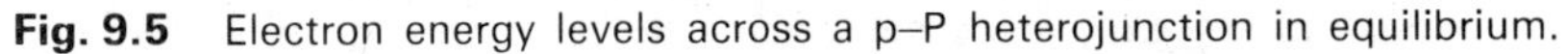

Fig. 9.5 Electron energy levels across a p–P heterojunction in equilibrium.

These trapping levels may cause much larger but very localized potential variations. The need to minimize the disruption of the crystal at the junction represents a severe technological problem.

Figure 9.5 shows the electron energy levels at an ideal, abrupt heterojunction formed between the two semiconductors of Fig. 9.2 when both are doped p-type and the junction is in equilibrium. By virtue of the particular parameter values only the slightest depletion/accumulation layer is formed. Once again the energy-level steps $\Delta\mathcal{E}_c$ and $\Delta\mathcal{E}_v$ serve to equalize the carrier flows in either direction across the junction. With a small applied bias voltage there is no obstruction to the net flow of carriers and ohmic behavior is to be expected.

When an n–P junction is formed between these materials the electron energy levels in equilibrium take on the form shown in Fig. 9.6(a). The effect of a small forward bias voltage is shown in Fig. 9.6(b). In Fig. 9.7 (a) and (b) the corresponding energy level diagrams for a p–N junction in equilibrium and in the forward-biassed state are shown. In these diagrams we have retained the convention of putting the n-type material on the left-hand side and the p-type material on the right. The narrow band-gap material is identified by the subscript 1 and the wider-gap material by 2. In each case a normal diode characteristic is to be expected.

9.1.2 Useful Properties of Heterojunctions

It is possible to identify five properties of heterojunctions that make them especially useful in the fabrication of bright, high efficiency LEDs and semiconductor lasers. These will be summarized here and more detailed discussion and analysis will be given subsequently.

(a) *High Injection Efficiency*
Reference to Figs. 9.6 and 9.7 shows that the majority carriers attempting to leave the narrow band-gap material (material 1) are obstructed by a higher potential barrier than would be the case in a homojunction. This reduces the proportion of the current across the junction carried by the minorities injected into material 2. With a band-gap difference of more than a few times kT this effect is likely to be far more significant than the effect of relative doping concentrations described in eqn. (7.5.8).

(b) *Confinement of the Minority Carriers in a Double Heterostructure*
In the next section we shall discuss a structure that has become very important in the development of optical sources. Two heterojunctions may be used to sandwich a layer of material with narrow band gap between material of wider band gap. A very schematic

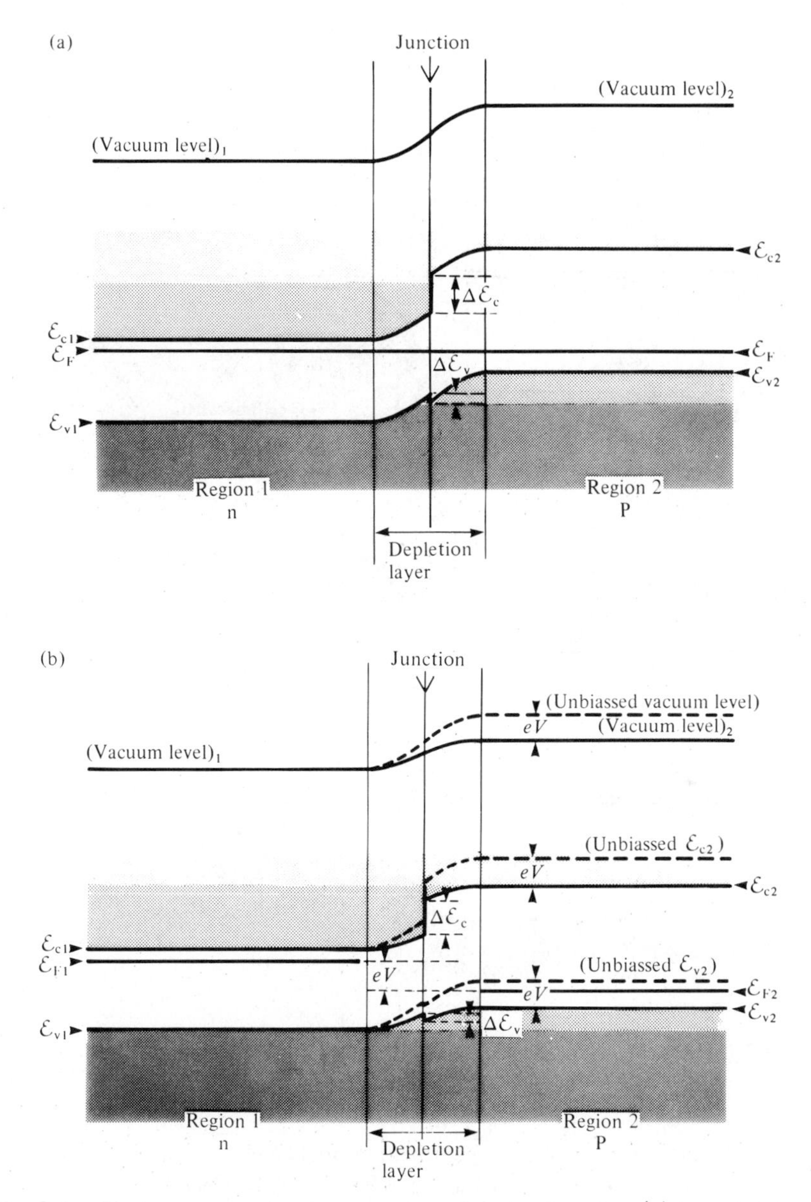

Fig. 9.6 Electron energy levels across an n–P heterojunction: (a) in equilibrium; (b) under forward bias.

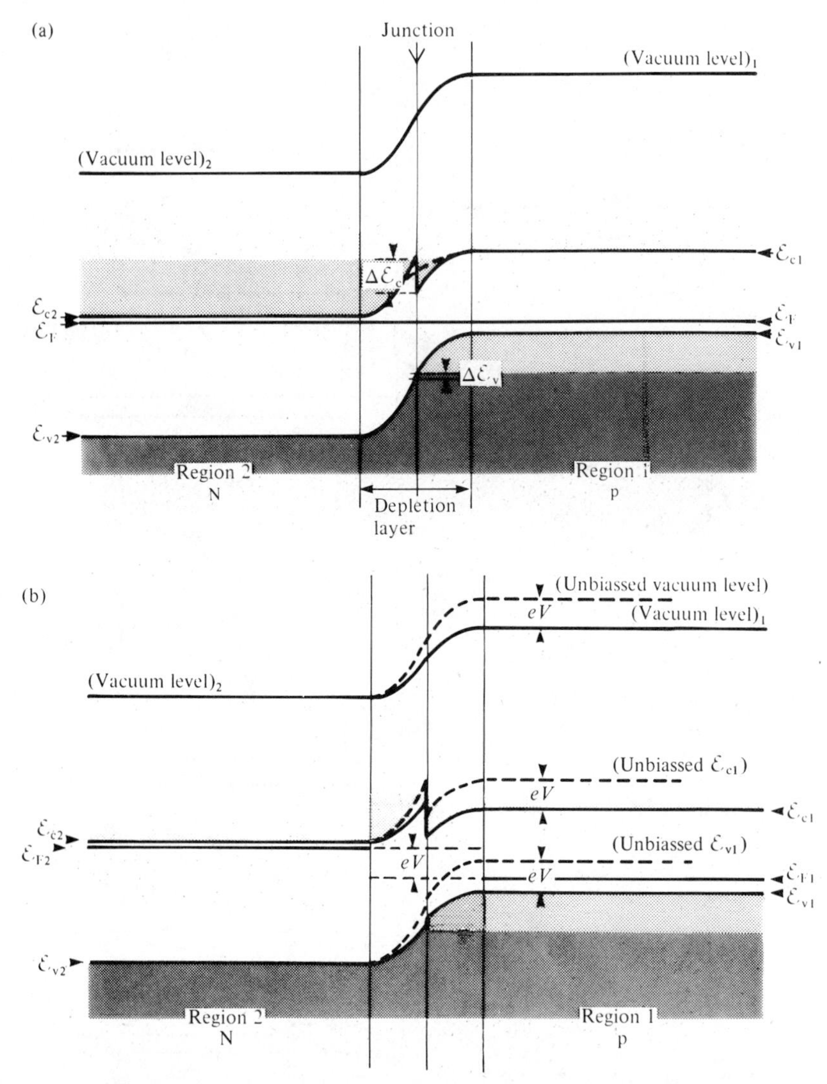

Fig. 9.7 Electron energy levels across an N–p heterojunction: (a) in equilibrium; (b) under forward bias.

diagram of the arrangement that might be used is given in Fig. 9.8. It is known as a double heterostructure. The electron energy levels across an example of such a structure are shown in Fig. 9.9. The minority holes injected under forward bias from region 2 into region 1 are obstructed by the wider band gap at the second junc-

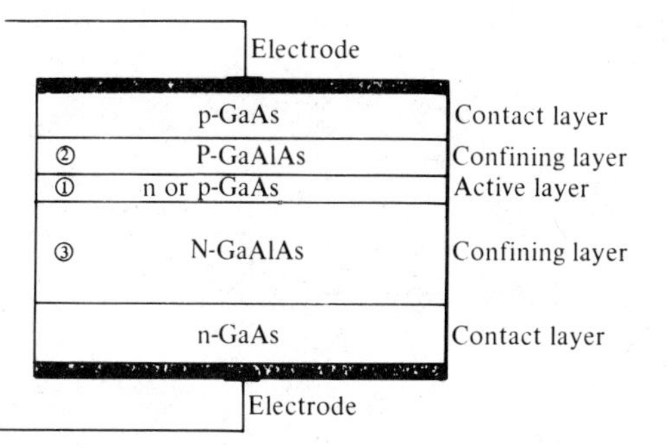

Fig. 9.8 Schematic illustration of a double heterostructure.

tion, J2, with the result that a higher and more uniform minority carrier concentration and a higher rate of recombination can be established within region 1 for a given carrier current. This is illustrated schematically in Fig. 9.10.

(c) *Improvement of Ohmic Contacts*
The use of heterojunctions enables a transition to a narrower band-gap material to be made and this eases the fabrication of good, low resistance, ohmic contacts at the device terminals. This is one of the reasons for the use of the five-layer structure shown in Fig. 9.8.

(d) *Transparency of the Wide Band Gap Material*
Recombination radiation generated in band-to-band transitions in the narrow band gap material cannot promote carrier excitation across the band gap of the wider band gap material. As a result layers 2 and 3 in Figs. 9.8 and 9.9 are much more transparent to the recombination radiation of material 1 than material 1 itself. This effect is made use of in the design of both surface-emitting and edge-emitting LEDs as discussed in Section 9.3.

(e) *Optical Guidance*
Because in general the refractive indexes of the two materials which make a heterojunction are different ($n = \sqrt{\varepsilon_r}$), oblique rays incident on the junction may suffer total internal reflection. In the double heterostructure of Fig. 9.8, if the refractive index of material 1 is higher than that of materials 2 and 3, then recombination radiation generated in 1 may be guided along the layer by multiple reflections at the junctions just as in a clad fiber. This effect, too, is illustrated in Fig. 9.10 and is particularly important in the operation of edge-emitting LEDs and double-heterostructure lasers. It so happens that in nearly every case among the III–V compound semiconductors the wider band-gap material has the lower refractive index. This may be verified by reference to Table 7.2.

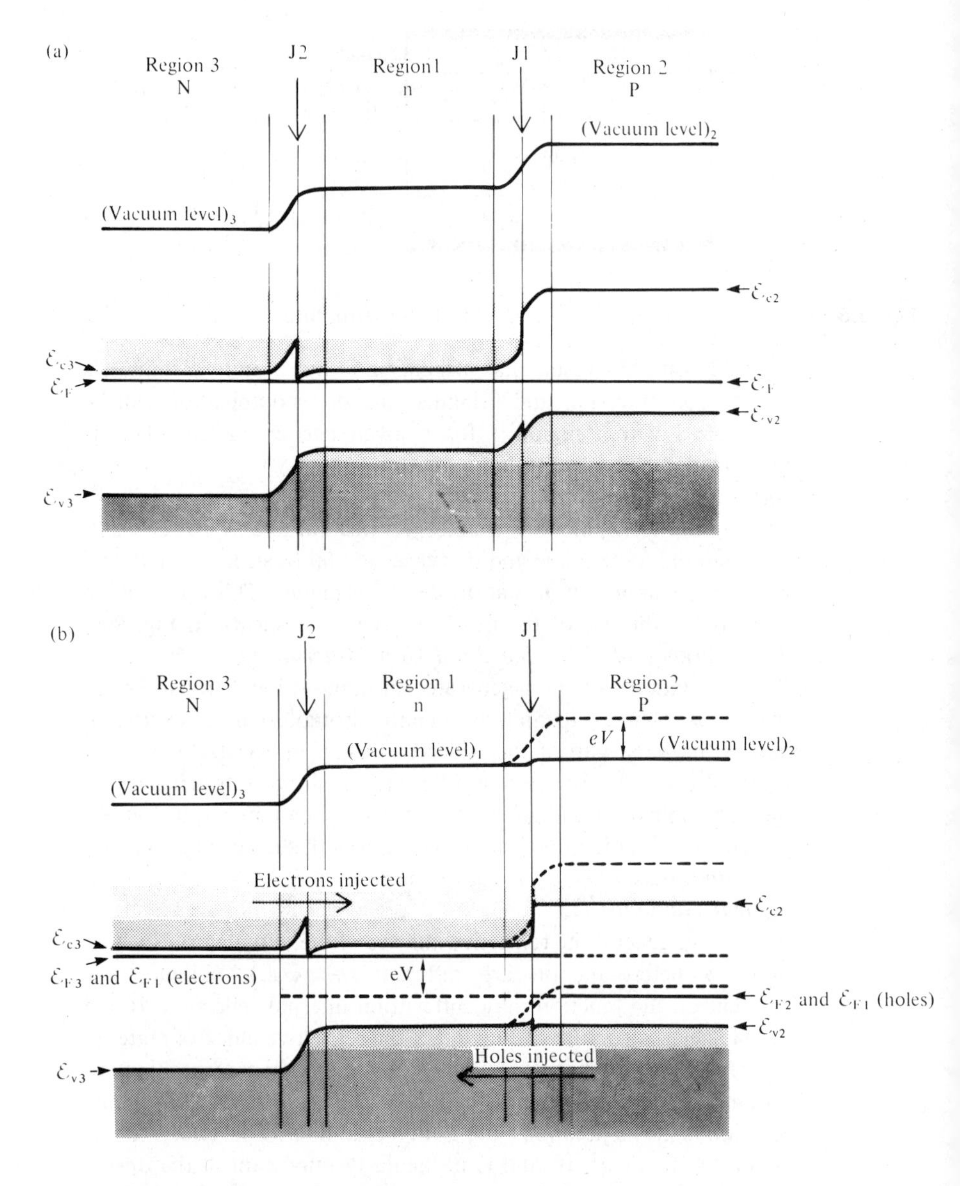

Fig. 9.9 Electron energy levels across an N–n–P double heterostructure: (a) in equilibrium; (b) forward biassed.

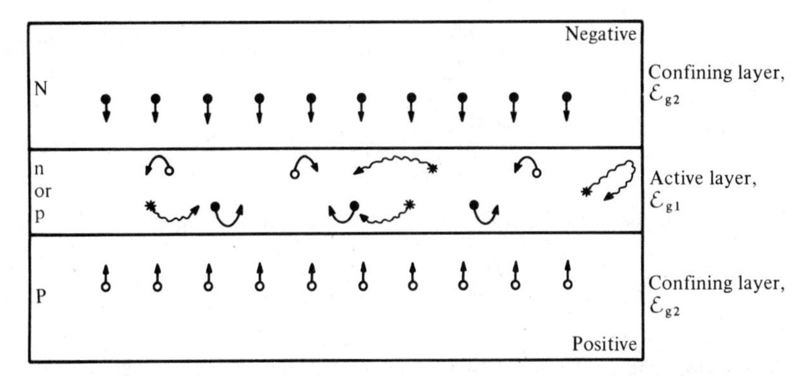

Fig. 9.10 Schematic illustration of optical and carrier confinement in a double heterostructure: electrons ●; holes ○; radiative recombination $\mathcal{E}_{g2} > \mathcal{E}_{g1}$ and $\varepsilon_{r1} > \varepsilon_{r2}$.

9.1.3 Injection Efficiency

We will now derive an expression for the injection efficiency at the p–N heterojunction of Fig. 9.7 on the assumption that any recombination current I_2 is small compared with the diffusion current I_1. This condition will be discussed further in Section 9.1.4. Region 1 is the p-type, narrow band-gap material with an acceptor concentration, n_A, and equilibrium carrier concentrations of $p_{p0} \simeq n_A$ and $n_{p0} \simeq n_{i1}^2/n_A$. Region 2 is the N-type, wide band-gap material having a donor concentration, n_D, and carrier concentrations of $n_{N0} \simeq n_D$ and $p_{N0} \simeq n_{i2}^2/n_D$. Following eqn. (7.2.1) the intrinsic carrier densities may be represented as:

$$n_{i1}^2 = K_1^2 \exp(-\mathcal{E}_{g1}/kT) \tag{9.1.3}$$

$$n_{i2}^2 = K_2^2 \exp(-\mathcal{E}_{g2}/kT) \tag{9.1.4}$$

where K_1 and K_2 are constants characteristic of the materials.

With a forward-bias voltage, V, applied to the junction, the carrier concentration at the edge of the depletion layer in region 1 is again given by eqn. (7.5.3):

$$n(0) = n_{p0} \exp(eV/kT) \tag{7.5.3}$$

Provided that the width of region 1 is much greater than the diffusion length, L_p, the electron current density is given by eqn. (7.5.1):

$$J_e = \frac{eD_e}{L_p}[n(0) - n_{p0}] \tag{7.5.1}$$

$$= \frac{eD_e n_{p0}}{L_p} [\exp (eV/kT) - 1] \tag{7.5.4}$$

$$= \frac{eD_e}{L_p} \frac{n_{i1}^2}{n_A} [\exp (eV/kT) - 1] \tag{9.1.5}$$

By similar arguments the hole current density entering region 2 is given by

$$J_h = \frac{eD_h}{L_N} \frac{n_{i2}^2}{n_D} [\exp (eV/kT) - 1] \tag{9.1.6}$$

Thus, making use of eqns. (9.1.3) and (9.1.4),

$$\frac{J_e}{J_h} = \frac{D_e}{D_h} \frac{L_N}{L_p} \frac{n_D}{n_A} \frac{K_1^2}{K_2^2} \exp [(\mathcal{E}_{g2} - \mathcal{E}_{g1})/kT] \tag{9.1.7}$$

The injection efficiency itself may be obtained by substituting eqn. (9.1.7) into eqn. (7.5.7):

$$\eta_{inj} = \frac{1}{(1 + J_h/J_e)} \tag{7.5.7}$$

As an example, consider two materials having an energy-gap difference of 0.37 eV. This corresponds to the system $GaAs/Ga_{0.7}Al_{0.3}As$. At room temperature the value of the exponential factor in eqn. (9.1.7) is 1.5×10^6 and easily dominates all the other terms in the expression for J_e/J_h. It is thus possible to maintain a high injection efficiency at such a heterojunction even though the ratios n_D/n_A and L_N/L_p may for other reasons be made small.

It is left as an exercise for the reader to show that, in the case of the forward-biassed n–P heterojunction of Fig. 9.6(b), a similar factor enhances the proportion of holes injected into the narrow band-gap semiconductor.

9.1.4 Heterojunction Characteristics

From this discussion it can be seen that most n–N and p–P heterojunctions behave as ohmic resistive elements but that n–P and p–N heterojunctions would be expected to follow a current–voltage characteristic similar to that of a p–n homojunction:

$$I = I_1 + I_2 = I_{s1} [\exp (eV/kT) - 1] + I_{s2} [\exp (eV/2kT) - 1] \tag{9.1.8}$$

The first term on the right-hand side of eqn. (9.1.8) refers to diffusion current I_1 crossing the junction. It is this current that we are principally concerned with because a proportion η_{int} of it provides the recombination radiation we wish to use. The second term, I_2, represents current that will recombine at recombination centers either within the depletion layer or at the semiconductor surface. As the recombination mechanisms involved here are nonradiative, this current serves no useful purpose but rather tends to short out the diffusion current. Indeed for GaAs/GaAlAs heterojunctions I_2 dominates the low current part of the characteristic as can be seen in Fig. 9.11(a). In this case I_2 was almost entirely due to surface recombination and the current was found to vary in direct proportion with the junction perimeter at the semiconductor surface rather than with junction area. I_2 can be increased dramatically by local surface damage. It is important to appreciate that if I_2 is significant at high current levels, it will cause the effective values of the internal quantum efficiency η_{int} and the injection efficiency η_{inj} to be reduced.

Often experimental results have been fitted to an expression of the form

$$I = I_s[\exp(eV/akT) - 1] \tag{9.1.9}$$

where a is a constant which varies from junction to junction and which may change with temperature. It usually lies within the range 1.0 to 3.0. Some typical sets of characteristics are shown in Fig. 9.11(b).

With reverse bias these rectifying junctions support the applied voltage across a wider depletion layer, whose width and self-capacitance may be calculated in the same way as those of a homojunction. The calculations are made slightly more complicated by the change of material properties at the junction, but are otherwise similar to those which produced eqns. (7.6.7) and (7.7.2).

9.2 THE DOUBLE HETEROSTRUCTURE

9.2.1 Carrier Confinement

The double heterostructure of Fig. 9.8 is invariably used for optical sources for communication. The narrow band-gap material is sometimes doped n-type and sometimes p-type. Figure 9.9 showed the electron energy levels across an N–n–P double heterostructure under equilibrium and forward-biassed conditions. Figure 9.12 shows similar diagrams for an N–p–P structure. The three regions are identified by the subscripts 1, 2 and 3 as indicated on the diagrams.

In paragraph (b) of Section 9.1.2 it was claimed that the presence of the second heterojunction helped to confine within region 1 those excess minority carriers that are injected over the forward-biassed p–n junction. Here we will derive an analytical expression for the effect that has on the optical power

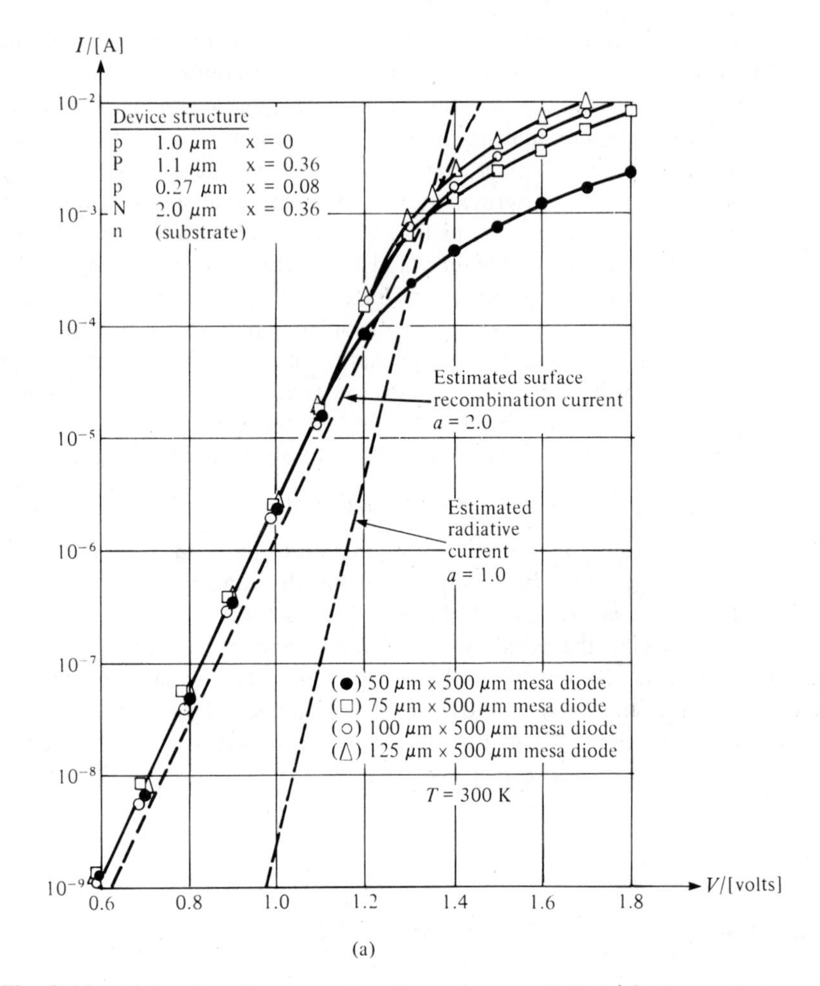

Fig. 9.11 Heterojunction current–voltage characteristics: (a) observed currents in rectangular double-heterostructure mesa diodes. In these devices the current was independent of the width of the active region except for the effect of series resistance at high currents. The conclusion is that I_2 was generated at the junction periphery. [Taken from C. H. Henry *et al.*, The effect of surface recombination on current in $Al_xGa_{1-x}As$ heterojunctions, *J. Appl. Phys.* **49**, 3530–42 (1978).]

generated in the ideal N–p–P structure of Fig. 9.12. It will be left to the reader to show that a corresponding result can be obtained for the holes crossing the N–n–P structure of Fig. 9.9.

We take the origin of the x-coordinate to be at the edge of the depletion layer

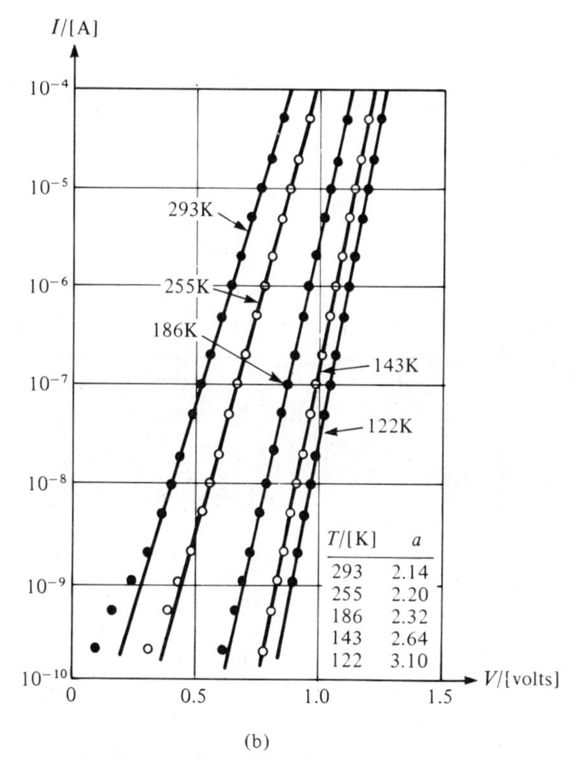

(b)

Fig. 9.11 (b) Currents measured in an n^+P GaAs/GaAlAs heterojunction of area 0.322 mm^2. Recombination current, I_2, dominates, and other effects cause $a > 2$ at low temperature. [Taken from J. F. Womac and R. H. Rediker, The graded-gap $Al_xGa_{1-x}As$–GaAs heterojunction, *J. Appl. Phys.* **43**, 4129–33 (1972).]

of junction J2 in region 3. We assume that the electron concentration distribution is maintained in approximate thermal equilibrium from region 1, through region 2, up to the point $x = 0$ in region 3. From there electrons flow on into region 3 by diffusion. On the basis of the arguments given in Section 7.5, especially eqns (7.5.3) and (7.5.4), we may write

$$n(0) = n_{p3} \exp(eV/kT) \tag{9.2.1}$$

and

$$J_{e2} = \frac{eD_{e3} n_{p3}}{L_{p3}} [\exp(eV/kT) - 1] \tag{9.2.2}$$

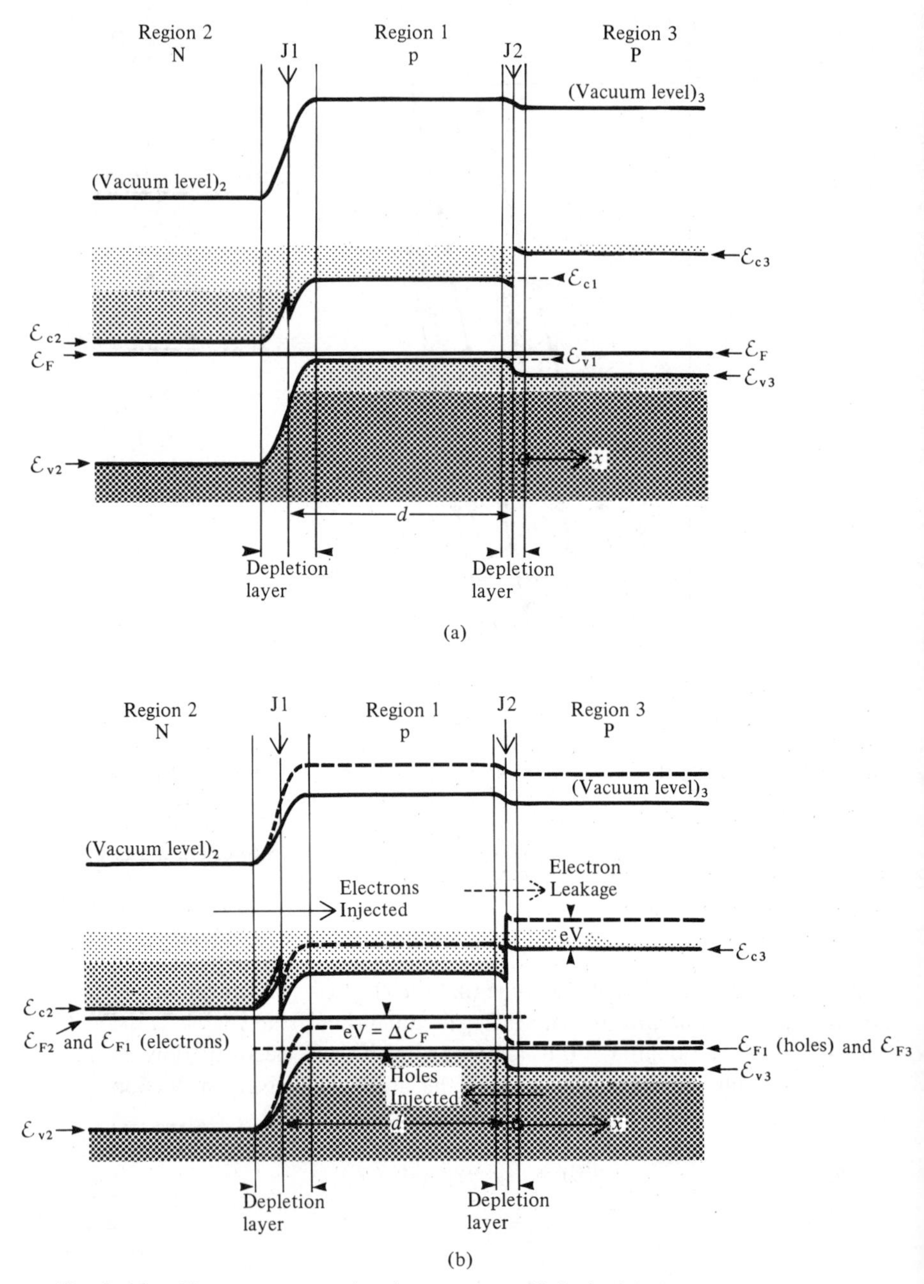

Fig. 9.12 Electron energy levels across an NpP double heterostructure: (a) in equilibrium; (b) forward-biassed.

where

n_{p3} is the equilibrium electron concentration in region 3
D_{e3} is the electron diffusion coefficient in region 3
$L_{p3} = (D_{e3}\tau_{p3})^{1/2}$ is the diffusion length in region 3
V is the forward bias voltage
J_{e2} is the electron current density crossing junction J2.

Within region 1, which is of width, d, the electron concentration is

$$n_1 = n_{p1} \exp (eV/kT) \tag{9.2.3}$$

where n_{p1} is the equilibrium electron concentration in region 1. We are assuming that $d \ll L_{p1} = (D_{e1}\tau_{p1})^{1/2}$, the diffusion length in region 1, so that there is no appreciable variation in the electron concentration over the region. This is something we have already assumed in writing eqn. (9.2.1) and is an assumption we must check later for self-consistency.

The semiconductor of region 1 is characterized by a bulk-material recombination time constant τ_{p1} which is made up of a radiative recombination time constant, τ_{rr1}, and a time-constant for the nonradiative processes, τ_{nr1}. By comparison with eqn. (8.4.4),

$$\frac{1}{\tau_{p1}} = \frac{1}{\tau_{rr1}} + \frac{1}{\tau_{nr1}} \tag{9.2.4}$$

In addition each of the heterojunctions will give rise to a high concentration of local recombination centers that promote largely nonradiative recombination. We may characterize each heterojunction by an interface recombination velocity, s, which we shall assume for simplicity to be the same for each. The net rate of recombination per unit cross-sectional area over the whole of the active region is then given by

$$\begin{pmatrix}\text{Rate of recombination}\\ \text{per unit area}\end{pmatrix} = \frac{n_1 d}{\tau_{rr1}} + \frac{n_1 d}{\tau_{nr1}} + 2n_1 s = \frac{n_1 d}{\tau} \tag{9.2.5}$$

where τ is an effective recombination time constant for the double heterostructure. Then τ is given by

$$\frac{1}{\tau} = \frac{1}{\tau_{rr1}} + \frac{1}{\tau_{nr1}} + \frac{2s}{d} \tag{9.2.6}$$

and the overall internal quantum efficiency is

$$\eta_{int} = \frac{\tau}{\tau_{rr1}} \tag{9.2.7}$$

Because the essential features of the double heterostructure depend on the close proximity of these two heterojunction interfaces, good devices demand as small an interface recombination velocity as possible and thus depend on a very close match of the lattice parameters at the junctions. Measurements indicate that values of s of the order of 10 m/s can be obtained with GaAs/GaAlAs.

It is convenient here, as it was in Section 8.4 when we discussed surface recombination, to express the rate of recombination in terms of the associated carrier current densities. The total electron current density, J_{e1}, injected into region 1 over junction J1 represents almost all the current flowing, if the hole current over J1 can be assumed to be very small ($\eta_{inj} \simeq 1$). We may separate J_{e1} into two components:

$$J_{e1} = J_r + J_{e2} \tag{9.2.8}$$

where J_{e2} is given by eqn. (9.2.2) and represents the 'leakage' current that escapes over the potential barrier at J2 and J_r is the active-layer recombination current density:

$$J_r = \frac{en_1 d}{\tau} \tag{9.2.9}$$

The optical power, P, generated in region 1 per unit cross-sectional area is given by

$$P = \frac{n_1 d}{\tau_{rr1}} \overline{\mathcal{E}_{ph}} = \eta_{int} \frac{n_1 d}{\tau} \overline{\mathcal{E}_{ph}}$$

$$= \eta_{int} \frac{J_r}{e} \overline{\mathcal{E}_{ph}} \tag{9.2.10}$$

where $\overline{\mathcal{E}_{ph}}$ is the mean photon energy. It is important to maximize the ratio of J_r to J_{e2}, which with $(eV/kT) \gg 1$ is given by

$$\frac{J_r}{J_{e2}} = \frac{n_{p1} d}{\tau} \frac{L_{p3}}{D_{e3} n_{p3}} \tag{9.2.11}$$

where we have used eqns. (9.2.9), (9.2.3) and (9.2.2). By analogy with eqns. (7.2.1) and (7.3.2) we may write

$$n_{p1} = \frac{n_{i1}^2}{n_{A1}} = \frac{K_1^2}{n_{A1}} \exp(-\mathcal{E}_{g1}/kT) \tag{9.2.12}$$

and

$$n_{p3} = \frac{n_{i3}^2}{n_{A3}} = \frac{K_3^2}{n_{A3}} \exp(-\mathcal{E}_{g3}/kT) \tag{9.2.13}$$

so that

$$\frac{J_r}{J_{e2}} = \frac{L_{p3}d}{D_{e3}\tau_{p1}} \frac{K_1^2}{K_3^2} \frac{n_{A3}}{n_{A1}} \exp[(\mathcal{E}_{g3} - \mathcal{E}_{g1})/kT] \tag{9.2.14}$$

With $(\mathcal{E}_{g3} - \mathcal{E}_{g1})$ more than about $10kT$, the recombination current J_r is large compared to the leakage current J_{e2} by virtue of the exponential term, independent of the other parameters. Most of the current density crossing the double heterostructure then supplies the recombination processes in the active layer.

The importance of the second heterojunction, J2, is that now most of the optical power, which would normally have been generated over a depth of one or two diffusion lengths, is produced from the thin layer of region 1 whose depth may be made much less than a diffusion length. Not only is the source of radiation much better defined, but the optical power generated per unit volume is much greater as well. This can most easily be shown by comparing the carrier concentration n_1 obtained using the double-heterostructure with the concentration n_1' that would result if junction J2 were eliminated and region 1 extended for many diffusion lengths. With the double heterostructure we have seen that

$$J \simeq J_{e1} \simeq J_r \tag{9.2.15}$$

Then, using eqn. (9.2.9),

$$\frac{en_1}{J} \simeq \frac{\tau}{d} \tag{9.2.16}$$

With a single heterojunction the initial concentration of injected carriers, n_1', according to the discussion of Section 7.5, is given by

$$J \simeq \frac{eD_{e1}n_1'}{L_{p1}} \tag{9.2.17}$$

so that

$$\frac{en_1'}{J} \simeq \frac{L_{p1}}{D_{e1}} \tag{9.2.18}$$

Then, for the same current density,

$$\frac{n_1}{n_1'} \simeq \frac{\tau}{d}\frac{D_{e1}}{L_{p1}} = \frac{L_{p1}}{d} \gg 1 \tag{9.2.19}$$

Provided that $d \ll L_{p1}$, a higher carrier concentration is obtained for a given current density.

Because the electron current diffuses across region 1, there is a small change in electron concentration between J1 and J2. If we call this δn, then

$$J \simeq \frac{en_1 d}{\tau} \simeq \frac{eD_{e1}\,\delta n}{d} \tag{9.2.20}$$

so that

$$\frac{\delta n}{n_1} \simeq \frac{d^2}{L_{p1}^2} \ll 1 \tag{9.2.21}$$

as we assumed at the beginning of the analysis.

In Section 8.4 it was suggested that p-type GaAs with an acceptor concentration of 10^{23} m^{-3} might be expected to have a value of τ_p of about 50 ns. At room temperature L_p would then be about 20 μm. If the central layer of a double heterostructure, the narrow band-gap region, is made no more than 1 μm wide, we can reasonably expect the assumed condition $d \ll L_p$ to prevail. Note that with $s = 10$ m/s the value of $(d/2s)$ is 50 ns also. Thus the presence of the heterojunctions has a significant but not disastrous effect on the internal quantum efficiency. This will be examined further in the next section.

9.2.2 Modulation Bandwidth

If we adapt the discussion of Section 8.7 to the double heterostructure diode, we shall be led to expect that the efficiency of a modulated DH optical source will fall at modulation frequencies higher than $f_m = (1/2\pi\tau)$, where

$$\frac{1}{\tau} = \frac{1}{\tau_{rr1}} + \frac{1}{\tau_{nr1}} + \frac{2s}{d} \tag{9.2.6}$$

We should also expect the internal quantum efficiency at low modulation frequencies to be

$$\eta_{int} = \frac{\tau}{\tau_{rr1}} \tag{9.2.7}$$

According to eqn. (8.4.13), τ_{rr1} should be inversely proportional to the doping concentration in the active layer, n_{A1}, and for a direct band-gap material should be given approximately by

$$\frac{\tau_{rr1}}{[s]} = \frac{1}{rn_{A1}} \simeq \frac{10^{16}}{n_{A1}} [m^{-3}] \tag{9.2.22}$$

Thus,

$$n_{A1} = 10^{23} \text{ m}^{-3} \text{ gives } \tau_{rr1} \simeq 100 \text{ ns}$$
$$n_{A1} = 10^{24} \text{ m}^{-3} \text{ gives } \tau_{rr1} \simeq 10 \text{ ns}$$
$$n_{A1} = 10^{25} \text{ m}^{-3} \text{ gives } \tau_{rr1} \simeq 1 \text{ ns}$$

The great advantage of the double heterostructure is that it enables a high injection efficiency to be maintained even when such very high doping levels are used. The problem is that both of the nonradiative terms in eqn. (9.2.6) are adversely affected by high doping levels. With n_{A1} above about 10^{24} m^{-3}, lattice imperfections cause τ_{nr1} to fall rapidly, and in any case the lower total lifetime reduces L_{p1} and thus necessitates a smaller active layer thickness, d.

Another way of enhancing the modulation bandwidth is to establish conditions of 'high injection'. As we saw in Section 8.4, when the injected electron concentration becomes large compared with the equilibrium hole concentration in a p-type active layer, τ_{rr} is given by eqn. (8.4.14). In the notation of the N–p–P double heterostructure, this becomes

$$(\tau_{rr1})_{\text{high injection}} = \frac{1}{rn_1} \tag{9.2.23}$$

and is a function of time when n_1 is time-varying.

The condition for high injection is

$$\Delta n \simeq n_1 \gg p_{p0} \simeq n_{A1} \tag{9.2.24}$$

That is, using eqn. (9.2.6)

$$J > J_{Cr} = \frac{en_{A1}d}{\tau} = \frac{en_{A1}d}{\eta_{int}\tau_{rr1}} = \frac{ern_{A1}^2 d}{\eta_{int}} \tag{9.2.25}$$

With $r \simeq 10^{-16}$ m^3/s for a direct band-gap material, $n_A = 10^{24}$ m^{-3}, and $\eta_{int} \simeq 1$, this implies a critical current density of

$$J_{Cr} = 1.6 \times 10^{-19} \times 10^{-16} \times 10^{48} d = 1.6 \times 10^{13} d$$

For $d = 1\ \mu\text{m}$,

$$J_{Cr} = 1.6 \times 10^7\ \text{A/m}^2 = 16\ \text{A/mm}^2$$

Assume that above this injection level,

$$\tau \simeq \tau_{rr1} = \frac{1}{rn_1} \tag{9.2.26}$$

so that

$$J \simeq \frac{en_1 d}{\tau_{rr1}} \simeq \frac{ed}{r\tau_{rr1}^2} \tag{9.2.27}$$

$$\therefore \quad \tau_{rr1} \simeq \left(\frac{ed}{rJ}\right)^{1/2} \tag{9.2.28}$$

Some experimental confirmation of this result is shown in Fig. 9.13.

These rather complicated interrelationships may be summarized as follows:

(a) To obtain a high modulation bandwidth without sacrificing internal quantum efficiency, η_{int}, it is necessary to minimize the radiative recombination time constant (τ_{rr1}).

(b) This may be achieved in the first instance by using a high doping concentration (n_{A1}) in the active layer (region 1).

(c) The limit to this is set by the disruption of the lattice that arises when n_{A1} exceeds about $10^{24}\ \text{m}^{-3}$, because this enhances nonradiative recombination mechanisms and η_{int} is reduced.

(d) Higher modulation bandwidth may be obtained under high injection conditions, $J > J_{Cr}$, so that $n_1 \gg n_{A1}$. Then $\tau_{rr1} \propto (d/J)^{1/2}$ so that the bandwidth increases as the square root of the current density and inversely as the square root of the active layer thickness. However, nonlinear effects are then to be expected unless the modulation current is a small fraction of the d.c. current.

9.3 DOUBLE HETEROSTRUCTURE (DH) LEDs

The benefits of the double heterostructure can be readily incorporated into the Burrus design of LED. An example is shown in Fig. 9.14 where it will be noticed that the etched well in the substrate is taken back as far as the n–GaAlAs layer. An etchant which selectively attacks GaAs but not GaAlAs ensures this. As a result there is negligible self-absorption of the emitted radiation between the active layer and the emission surface. One disadvantage suf-

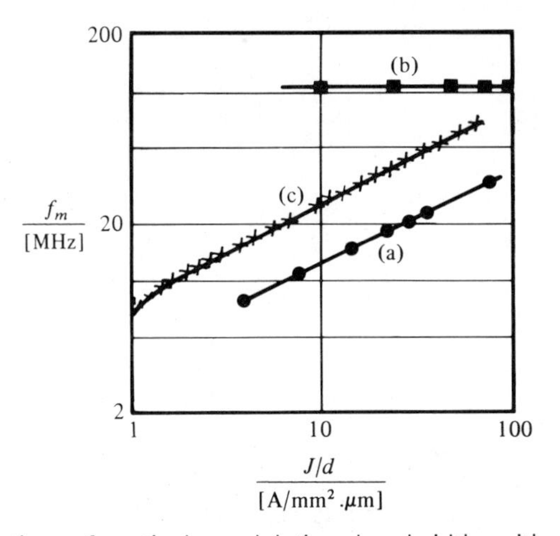

Fig. 9.13 Variation of optical modulation bandwidth with injection level. Measurements (a) and (b) are taken from T. P. Lee and A. G. Dentai, Power and modulation bandwidth of GaAs–AlGaAs high-radiance LEDs for optical communication systems, *IEEE Jnl. of Qu. Ets.* **QE-14**, 150–9 (1978). In (a) the active-layer doping concentration was $2 \times 10^{23} m^{-3}$, in (b) it was $1.5 \times 10^{25} m^{-3}$. Measurements (c) are taken from Itsuo Umebu *et al.*, InGaAsP/InP DH LEDs for fibre-optical communications, *Ets. Letts.* **14**, 499–500 (1978). They refer to a DH LED with a 1 μm thick, undoped active layer of $In_{0.75}Ga_{0.25}As_{0.44}P_{0.56}$ giving emission at 1.24 μm.

In both cases the cutoff frequency, f_m, was defined from the $-1.5\ dB_{opt}$ point on modulated power vs. frequency characteristic for a small modulation (~10%) about the bias current. According to eqns (8.7.38) and (9.2.28),

$$f_m = \frac{1}{2\pi\eta_{int}\tau_{rr1}} = \frac{1}{2\pi\eta_{int}}\left(\frac{rJ}{ed}\right)^{\frac{1}{2}}$$

The variation with $J^{\frac{1}{2}}$ is clearly confirmed in (a) and (c) but with the high doping levels in (b), the conditions for high injection were not established. With $\eta_{int} \simeq 1$ under high injection, r is seen to be about a factor of 4 larger for the quaternary material.

fered by this particular double heterostructure results from the low thermal conductivity of the GaAlAs confining layers. It is about one-third that of GaAs. There is thus a proportionate increase in the temperature rise of the junction for a given current density. It may be noted that the active layer itself can be made from GaAlAs having a smaller Al fraction than the confining layers. This affords some flexibility over the emission wavelength and reduces the dislocation density in the active layer.

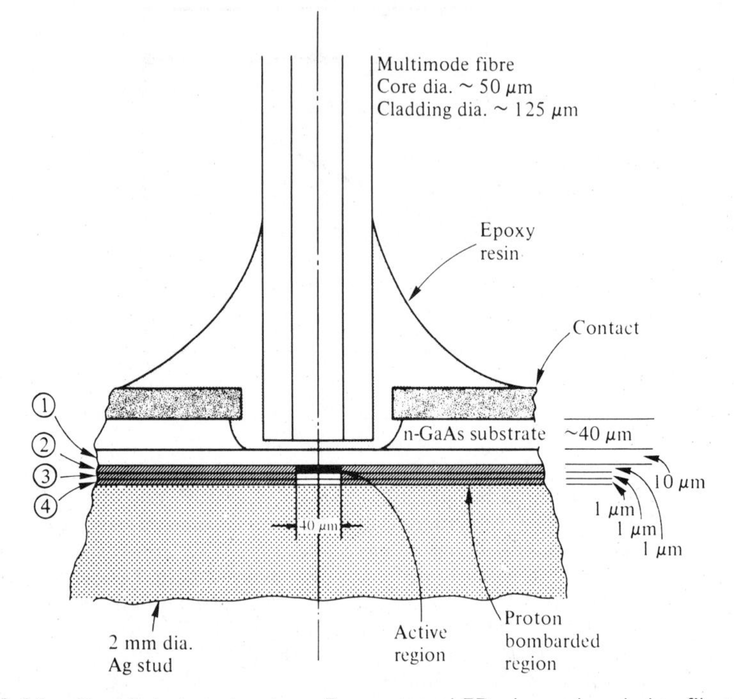

Fig. 9.14 Double heterostructure, Burrus-type LED, shown bonded to fiber with index-matching epoxy. The active region is defined by material made insulating by proton bombardment. The layers are: (1) n-GaAlAs; (2) n- or p-GaAs, active-layer; (3) p-GaAlAs; (4) p^+-GaAs contact layer.

In Fig. 9.14 the light-generating region is shown to be defined by proton-bombarded semiconductor rather than oxide isolation of the back contact. When a III–V semiconductor such as GaAs or GaAlAs is bombarded with high-energy protons, the damage caused to the crystal lattice produces a large increase in the electrical resistivity. A dose of 3×10^{19} protons/m^2 raises the resistivity by a factor of 100. The proton penetration depth is determined by the bombarding energy and is roughly 1 μm per 100 keV. The implantation of oxygen ions can also be used to produce a semi-insulating material. The use of proton bombardment rather than oxide isolation brings two benefits. In the first place the silica has a relatively low thermal conductivity, so its removal increases the lateral heat loss from the active region and reduces the temperature rise for a given current density. Secondly, with oxide isolation there remains a large area of inactive p–n junction which contributes to the depletion layer capacitance, C_J. Proton bombardment prevents this parasitic capacitance from being charged and discharged during high frequency operation.

The type of DH LED shown in Fig. 9.14 has proved to be reliable on life test and to show negligible degradation at room temperature with active-area current densities of 50 A/mm^2. Following the argument given in Section 8.6 we should expect this to produce an output radiance of 0.2–0.3 W/mm^2/sr into air. Thus a 50 μm diameter LED would emit 1–2 mW. When butt-jointed to a 0.17 NA fiber of greater core diameter it would be expected to launch about 50 μW. Such values are normally realized in practice.

It is clearly of the greatest importance to minimize the coupling losses. But with the active-area current density limited by the allowable temperature rise these are determined primarily by the numerical aperture and the core diameter of the fiber. Lens couplers are only of benefit when the fiber core diameter exceeds the LED diameter. However, as a full range of optical fiber systems develops, it is likely that GaAlAs/GaAs LED sources, emitting light at wavelengths around 0.85 μm, will be most suited to links of moderate bandwidth over moderate distances, say <100 (Mb/s)km. In that case they may be used in conjunction with fibers of higher numerical aperture and larger core diameter than those envisaged for long-range, high-data-rate telecommunications. It may then be beneficial to make small-area LEDs operating at high current density and use lenses to optimize the coupling efficiency. This is particularly so when lateral heat loss from the active region is significant, because the thermal impedance then decreases with diameter. As a result higher current densities may be used with active regions of smaller diameter.

It has been found in a particular practical case that the use of fibers with spherical ends, formed by controlled melting, increased the coupling efficiency up to four times. In this example an LED with a 35 μm diameter emissive region was coupled to a fiber with an 85 μm diameter core, a numerical aperture of 0.14 and a fiber end radius of 75 μm. This particular method is, however, sensitive to the precise alignment between LED and fiber. An example of the use of a self-aligned spherical lens on a GaAlAs/GaAs DH LED with a 35 μm diameter active area is shown in Fig. 9.15. With a 100 μm diameter spherical lens of refractive index 2.0, some 100 μW could be launched onto a fiber of core diameter 80 μm and numerical aperture 0.14 with 50 mA bias current. It will be seen that this is not a 'Burrus type' of LED and the thermal impedance of the active layer is high. Nevertheless, degradation is small over 1000 h at 25°C ambient temperature with this 50 A/mm^2 current density.

An example of a 1.3 μm wavelength DH LED using GaInAsP/InP is shown in Fig. 9.16. Here the InP substrate is transparent to the emitted radiation and a monolithic lensed surface could, in principle, have been used. As can be seen

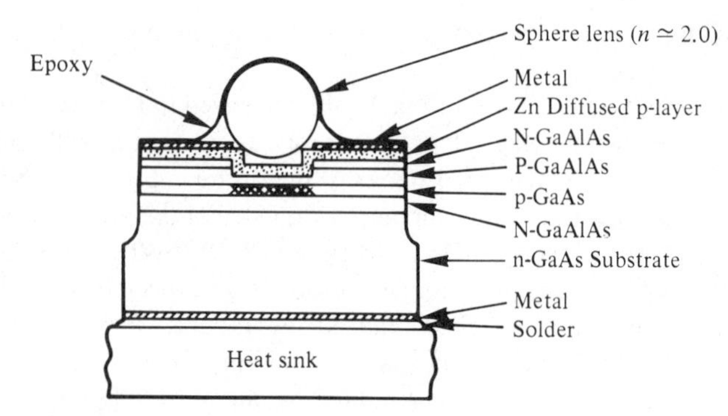

Fig. 9.15 A schematic cross-section of a DH LED with a self-aligned spherical lens. [Taken from S. Horiuchi *et al.*, A new LED structure with a self-aligned sphere lens for efficient coupling to optical fibers, *IEEE Trans. on Elect. Devices*, **ED-24**, 986–90; © 1977 IEEE.]

from the figure, this particular device has employed instead a truncated spherical lens made from glass of high refractive index. The coupling efficiency depends on the alignment and on the correct truncation of the lens. About 100 μW may be launched onto 85 μm core diameter 0.16 NA fiber with a device current density of 50 A/mm^2. The spectral wavelength spread of this LED is about 100 nm at 1.3 μm, corresponding to a spread of about $3kT$ in photon energy.

The importance of LED sources at this wavelength of minimum fiber dispersion will be clear from the discussions in Section 3.3 and Section 17.3.2. The role that LED sources may take in fiber optic systems using the wavelengths of minimum attenuation, 1.5–1.6 μm, is less clear. It has, however, been found possible to fabricate InGaAsP/InP DH LED sources with a range of different active-layer compositions, so that they emit at wavelengths ranging from 1.05 μm to 1.6 μm. Sources of the type shown schematically in Fig. 9.17(a) gave the spectral characteristics shown previously in Fig. 8.5. Because the transparency of the InP substrate to these wavelengths makes the etched well of the Burrus design unnecessary, the opportunity has been taken here to form a monolithic microlens by argon-ion beam etching. A 40 μm diameter contact was used in these devices and the active layer was 1.5 μm thick. Their basic performance characteristics with 100 mA bias current (80 A/mm^2) are set out in Table 9.1. At 1.5 μm the output power is markedly nonlinear as can be seen in Fig. 9.17(b). The reason for this has been hotly debated: it may be due to the onset of an Auger recombination process (see Fig. 8.1); it may be due to the onset of population inversion (see Section 10.1); it may be the result of the leakage of carriers over potential barriers; it may be an abnormal thermal

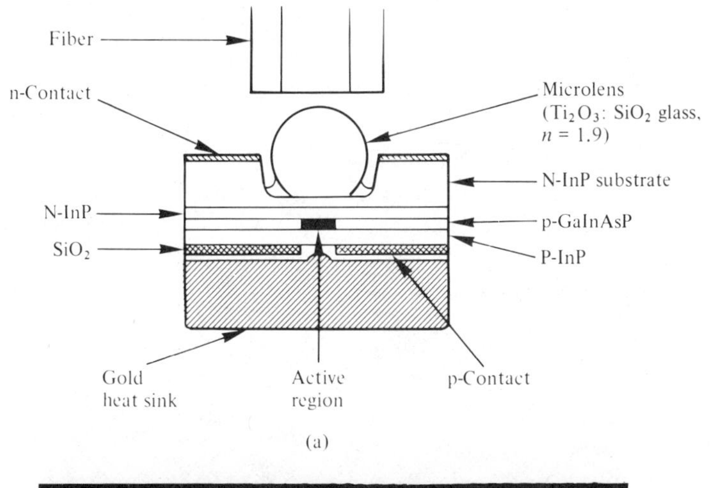

(a)

(b)

Fig. 9.16 A DH LED for longer wavelengths: (a) schematic cross-section; (b) scanning electron photomicrograph. [Reproduced by permission of Plessey Research (Caswell) Ltd.]

effect. This is the type of problem that will become resolved as the technology of longer wavelength devices is improved. It is thought to be related to the observed temperature dependence of the optical output power at a given bias current. This has been found to follow approximately a relationship of the form

$$\Phi(T)/\Phi(T_1) = \exp\{-(T - T_1)/T_0\} \qquad (9.3.1)$$

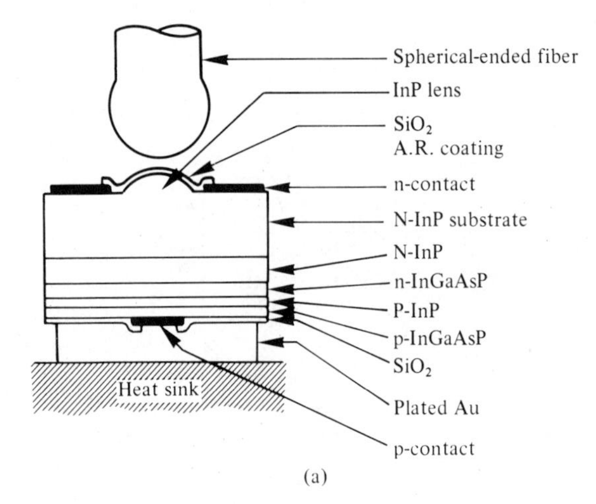

(a)

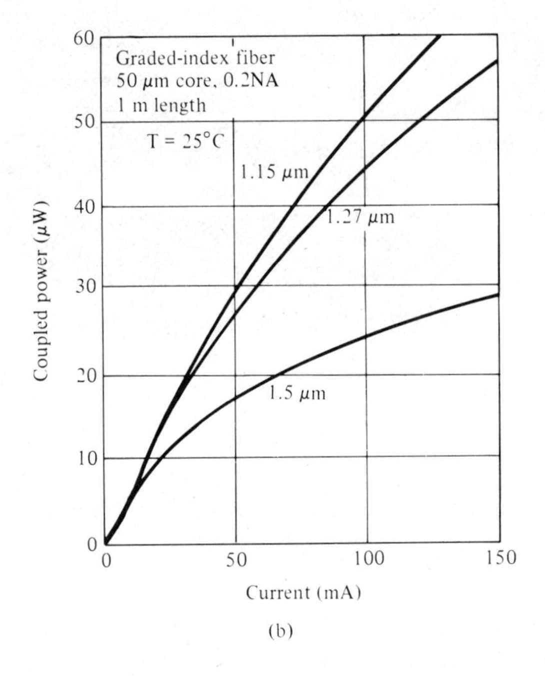

(b)

Fig. 9.17 An alternative design of DH LED for longer wavelengths: (a) schematic cross-section; (b) coupled power as a function of current. [Taken from O. Wada *et al.*, Performance and reliability of high radiance InGaAsP DH LEDs operating in the 1.15–1.5 μm wavelength region, *IEEE Jnl. of Qu. Ets.* **QE-18**, 368–74; © 1982 IEEE.]

Table 9.1 Basic performance characteristics of LEDs of the type shown in Fig. 9.17

λ/[μm]	$\Delta\lambda$/[nm]	Power coupled into fibers [μW] GI 50 μm, 0.2 NA	SI 85 μm, 0.16 NA	$f_m = 1/2\pi\tau$ [MHz]
1.15	93	50	188	30
1.27	110	44	170	40
1.50	145	24	79	50

The values refer to a bias current of 100 mA.

The value of T_0 has varied from 80 K to 145 K for 1.3 μm LEDs of different manufacture. It decreases for longer wavelength devices, indicating a greater temperature sensitivity.

For devices that survive an initial 'burn-in' period and are then subject to life test at elevated temperatures, the average degradation of the optical output power Φ_T has been observed to fit an exponential decay curve:

$$\Phi_T(t) = \Phi_T(0) \exp(-\beta t) \qquad (9.3.2)$$

The decay time constant in turn varies exponentially with temperature, T, as

$$\beta = \beta_0 \exp(-\mathcal{E}_a/kT) \qquad (9.3.3)$$

where k is Boltzmann's constant and β_0 and $\mathcal{E}_a$ are empirical constants. Such expressions imply that the degradation mechanism is thermally activated and has an activation energy $\mathcal{E}_a$. Tests on GaAlAs/GaAs diodes indicate values of $\mathcal{E}_a \simeq 0.6$ eV, $\beta_0 \simeq 10^2$ h^{-1}, whereas with InGaAsP/InP devices the curves fit $\mathcal{E}_a \simeq 1$ eV, $\beta_0 \simeq 2 \times 10^7$ h^{-1}. This high value of $\mathcal{E}_a$ augurs well for the quaternary material. Insofar as these results may be extrapolated to lower temperatures they predict operating lives to half-power of $\tau_{1/2} = 0.69/\beta \gtrsim 10^9$ h for InGaAsP devices compared with $\tau_{1/2} \simeq 10^6$–10^7 h for GaAlAs devices at 60°C.

The edge-emitting, double-heterostructure LED shown in Fig. 9.18 is an alternative design that gives an order-of-magnitude increase in radiance from a very small emissive area. It has a number of interesting features. The optical power is guided by internal reflection at the heterojunctions along a path parallel to the junction and is brought out of the end-face of the diode. The active region is defined by the stripe contact and by the slot at the rear of the active layer. This enables the length of the active region to be kept short (too short to support laser oscillation—see Section 10.3) without making the chip size too small to handle conveniently. The active layer itself tends to self-absorb the light generated, but it is made very thin with the result that much of the guided power is carried in the confining layers which do not absorb, being a

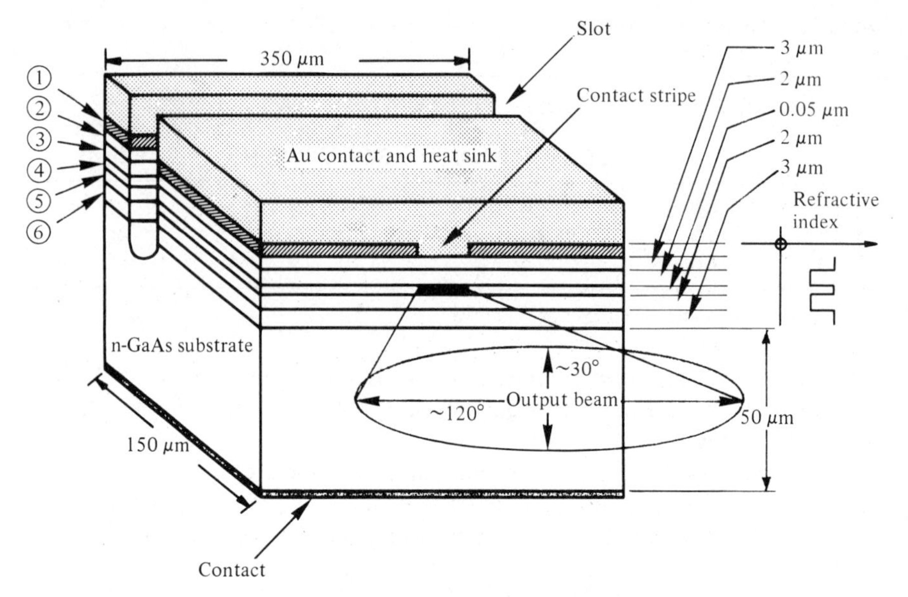

Fig. 9.18 Schematic illustration of a DH, edge-emitting LED; the layers are (1) silica insulator (defining the contact stripe); (2) p^+-GaAs; (3) P-$Ga_{0.6}Al_{0.4}As$; (4) active layer, n^--$Ga_{0.9}Al_{0.1}As$; (5) N-$Ga_{0.6}Al_{0.4}As$; (6) n-GaAs, epitaxial layer.

material of wider band gap. The absorption that does occur is greatest for the shorter wavelengths. This effectively narrows the spectral linewidth, from 35 nm to 25 nm at 0.9 μm and from 100 nm to 70 nm at 1.3 μm in particular examples. One effect of the optical guidance is that the beam width of the light emitted is quite narrow in the direction perpendicular to the junction, typically no more than 30°. This and the small emissive area make the edge-emitting LED well suited to lens couplers and good coupling efficiencies can be obtained even to fibers having a relatively small core diameter, say around 50 μm. This type of device is, however, more difficult to heat-sink than the surface-emitting LED, it is more difficult to handle mechanically and it has not yet come into general use.

9.4 FABRICATION OF HETEROSTRUCTURES

In this section we will give a brief description of the techniques used in the fabrication of heterostructures from III–V compound semiconductors. Readers familiar with silicon device technology will know that most silicon devices are made within epitaxial layers, of high purity and carefully controlled dimensions, grown on a prepared single-crystal substrate. Epitaxial techniques are also used in the fabrication of III–V devices, but the chemistry of these materials causes different methods to be used both for the preparation of the substrate and the growth of the epitaxial layers.

Single-crystal silicon for the device substrate is normally produced by crystallization onto a seed (Czrochralski method). The resulting cylindrical boule may be further purified by zone-refining. It is sliced into wafers which are polished. The epitaxial layers are formed by chemical vapor deposition processes very similar to those described in Chapter 4 for the production of optical fiber. There are two important distinctions. First the chemical reactions are such that vapors such as SiH_4 and $POCl_3$ are completely reduced, depositing silicon, doped as required, onto the substrate. Secondly, the deposited material continues the single-crystal structure of the substrate. The quality of the epitaxial layer depends both on the quality of the substrate and the conditions of deposition.

Substrate material for III–V semiconductors is obtained either by the Czrochralski method or by recrystallization from the liquid phase. However, compared to silicon a relatively high level of crystal defects and dislocations occurs. Luminescent devices have been produced using *vapor phase epitaxy* onto heated substrates. Both hydride and halide vapors have been used, but in neither case has the quality of the layers been able to match that required for efficient luminescent diodes. This is essentially a matter of keeping the rate of nonradiative decay sufficiently low, that is, of keeping τ_{nr} long enough. Better results have been obtained by deposition from organometallic vapors. With chemical vapor deposition (CVD) precise control can be maintained over the compositions and thicknesses of the various epitaxial layers as they are put down in sequence. Growth rates of the order of 20 μm/h can be obtained. A technique that gives similar control over the epitaxial layer structure and has been used to produce high-quality material is *molecular beam epitaxy*. Essentially the materials needed to form the layer are evaporated in the required proportions from heated filaments. The growth rates are slow (of the order of 1 μm/h) and the technique requires high vacuum conditions. It has not come into general use.

The technique that has been most widely used for III–V luminescent devices is *liquid phase epitaxy*. This gives fast but controllable growth rates, and is particularly suited to the production of thin, heavily doped layers with abrupt transitions. The process takes place in a temperature-controlled furnace and makes use of a 'boat' made from graphite and fitted with a slider. A schematic illustration of an arrangement which enables a number of different epitaxial layers to be produced in succession on a substrate is shown in Fig. 9.19. The

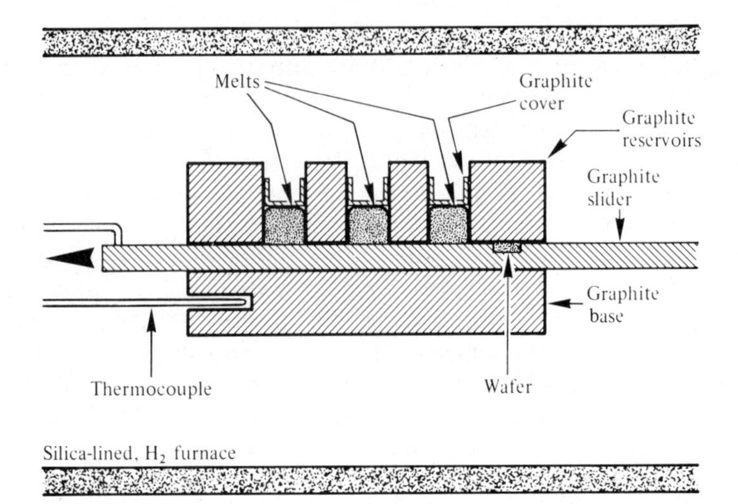

Fig. 9.19 A schematic cross-section of the graphite boat used for the liquid phase epitaxial growth of the three layers of a heterostructure.
The boat has a sliding base into which the substrate slice is recessed. This is moved in turn into position under the reservoirs containing the saturated solutions of the materials to be deposited. The furnace may then be cooled slightly to permit the layers to crystallize out onto the substrate.

substrate wafer is held in a recess in the slider and is drawn successively under the reservoirs in the boat. These contain saturated solutions of the materials to be deposited. By allowing the temperature to fall slightly while the solution is in contact with the substrate, crystallization is brought about. The reaction kinetics are complicated in that different materials and different dopants present in the liquid phase are deposited at different rates. However, control of the concentrations present in the melt enables a layer of precisely controlled, repeatable composition to be deposited. This applies even when the layer may be as complex as a quaternary semiconductor doped with a small concentration of group VI or group II material. Although most successful heterostructure devices have been produced by liquid phase epitaxy, this position could well change.

It is the use of CVD techniques together with photolithography and oxide masking that has been responsible for the development of mass-produced silicon integrated circuits. The lack of a similarly successful technique for III–V materials is a considerable barrier to the development and large-scale production of more complex integrated III–V devices.

PROBLEMS

9.1 Calculate the width of the depletion layer in an ideal, unbiassed n–N heterojunction, like that shown in Fig. 9.2, for the case $\chi_1 = 3.10$ eV, $\mathcal{E}_{g1} = 1.40$ eV, $\chi_2 = 2.50$ eV, $\mathcal{E}_{g2} = 2.10$ eV, $\mathcal{E}_{c1} - \mathcal{E}_F = 0.20$ eV and $\mathcal{E}_{c2} - \mathcal{E}_F = 0.40$ eV. The relative permitivities of the semiconductors are $\varepsilon_{r1} = 13$ and $\varepsilon_{r2} = 12$, and the doping concentration in each semiconductor is 10^{22} m^{-3}. Sketch the energy level diagram across the junction. Assume that the vacuum level is continuous at the junction and neglect any change of potential in the accumulation layer.

9.2 Derive expressions corresponding to eqns (7.6.7) and (7.7.2) for the width and self-capacitance of a p–N heterojunction.

9.3 Derive an expression similar to eqn. (9.1.7) for the ratio of the hole to electron current densities crossing a forward biassed n–P heterojunction as in Fig. 9.6(b). Hence calculate the injection efficiency at 100°C for holes entering the narrow band gap region of a forward biassed n$^+$P heterojunction in which $n_A = 10^{21}$ m^{-3} and $n_D = 10^{24}$ m^{-3}. Assume $\mathcal{E}_{g1} = 0.95$ eV, $\mathcal{E}_{g2} = 1.35$ eV, $\tau_p/[\mathrm{s}] = 10^{16}[\mathrm{m}^{-3}]/n_A$, $\tau_n/[\mathrm{s}] = 10^{16}[\mathrm{m}^{-3}]/n_D$, $D_e = D_h$ and $K_1 = K_2$.

9.4 Derive the expression corresponding to eqn. (9.2.14) for the ratio of recombination current to leakage current for the holes crossing the forward-biassed N–n–P heterostructure shown in Fig. 9.9(b).

9.5 The current density injected into an N–p–P double heterostructure is modulated at an angular frequency ω. That is, $J = J_0 + J_1 \exp j\omega t$. Assume,

(a) the leakage and hole currents are small enough to be negligible;
(b) the radiative and nonradiative recombination time constants in the active p-layer are τ_{rr} and τ_{nr} respectively;
(c) the recombination velocity at each heterojunction is s;
(d) the active layer thickness is d.

Show that the modulated optical power density generated is

$$P = P_0 + P_1 \exp j(\omega t + \phi)$$

where

$$(P_0/J_0) = \eta_{int}(hc/e\lambda) \quad \text{and} \quad (P_1/J_1) = \{\eta_{int}/(1 + j\omega\tau)\}(hc/e\lambda)$$

where

$$\eta_{int} = \tau/\tau_{rr}, \; 1/\tau = 1/\tau_{rr} + 1/\tau_{nr} + 2s/d$$

and (hc/λ) represents the average photon energy.

9.6 In a double heterostructure diode the low-injection radiative and nonradiative recombination times in the active layer, which is 0.5 μm thick, are 10 ns and 30 ns respectively, and the heterojunction recombination velocities are both 10 m/s. If the radiative recombination coefficient for the material of the active layer is $r = 2 \times 10^{-16}$ m^3/s, calculate:

(a) the doping concentration in the active layer;
(b) the internal quantum efficiency;
(c) the modulation cutoff frequency;
(d) the current density needed to establish conditions of high injection.

9.7 Two different types of LED are specified for an optical fiber system designed to operate over the temperature range 0 to 60°C. They are found to have values of the temperature parameter T_0 in eqn. (9.3.1) of 80 K and 120 K. In each case calculate the ratio of minimum to maximum output power at the extremes of temperature.

9.8 Samples of GaAlAs/GaAs and InGaAsP/InP LEDs were lifetested at 200°C. The time taken for the optical output to decay to 50% was 10 000 hours in the case of the former devices, and 1000 hours for the latter. The decay processes are assumed in each case to be governed by an activation-energy, as in eqns (9.3.2) and (9.3.3), the values being 0.6 eV and 1.0 eV respectively. Calculate the half-power lives to be expected at a temperature of 100°C.

SUMMARY

n–N and p–P heterojunctions are normally found to be ohmic. n–P and N–p heterojunctions have characteristics similar to p–n homojunctions but the injection efficiency of minority carriers entering the material of narrow band gap is enhanced.

The double heterostructure tends to confine both minority carriers and optical radiation within the active layer.

Light generated by recombination in the material of narrow band gap is not absorbed in that of wide band gap.

In a double heterostructure diode the internal quantum efficiency is $\eta_{int} = \tau/\tau_{rr1}$, where $1/\tau = 1/\tau_{rr1} + 1/\tau_{nr1} + 2s/d$. The modulation bandwidth is $f_m = 1/2\pi\tau$. 'High injection' reduces τ_{rr1} when $J > ern_{Al}^2 d/\eta_{int}$. Then $\tau_{rr1} \simeq (ed/rJ)^{1/2}$.

Various types of surface-emitting and edge-emitting double-heterostructure LEDs have been made. For 0.85 μm GaAlAs/GaAs may be used; for longer wavelengths (1–1.6 μm) the InGaAsP/InP system is preferred. Fabrication is normally by liquid phase epitaxy.

Coupling efficiency can be increased using lensed surfaces if fibers with larger core diameter and higher numerical aperture are used. Between 20 μW and 100 μW may be launched onto 50 μm core diameter, 0.17 NA fibers, and between 100 μW and 300 μW onto 85 μm core diameter, 0.17 NA fibers.

10

Laser Action in Semiconductors

10.1 THE BASIC PRINCIPLES OF LASER ACTION

10.1.1 Spontaneous Emission, Stimulated Emission and Absorption

The radiative recombination that is responsible for injection luminescence is the result of *spontaneous*, band-to-band electronic transitions. Transitions between electronic states may also be *stimulated* by the presence of electromagnetic radiation of the correct wavelength. Between states of energy $\mathcal{E}_1$ and $\mathcal{E}_2$ ($\mathcal{E}_2 > \mathcal{E}_1$), this requires radiation of frequency $f_{21} = (\mathcal{E}_2 - \mathcal{E}_1)/h$ and hence of free-space wavelength $\lambda_{21} = hc/(\mathcal{E}_2 - \mathcal{E}_1)$, where $h = 6.626 \times 10^{-34}$ J.s is Planck's constant. When the radiation interacts with an atom in the lower energy state, then a single quantum of radiation may be *absorbed* and the atom transferred to the upper level. When it interacts with an atom in the upper energy state, then a stimulated downward transition may result, as an alternative to the normal spontaneous emission. The mean lifetime of the excited state is therefore decreased in the presence of radiation. Any radiation quanta emitted under stimulated emission have the same frequency and are in phase with the stimulating radiation. They are *coherent* with it.

The double heterostructure semiconductor lasers normally used in optical communication links are similar in structure to the edge-emitting LED described in Section 9.3. Their emitted radiation, like that of other lasers, is almost entirely produced by stimulated emission. This endows it with three important properties. Compared with LED emission (a) it spreads over a narrower band of wavelengths; (b) it is more directional, with the result that the external quantum efficiency may be improved; (c) its modulation bandwidth is greater.

To start with we will develop some important relationships describing these emission and absorption processes by appealing first to the idealized and very simplified atomic energy-level system shown in Fig. 10.1, in which we have identified just two electronic states having energies $\mathcal{E}_1$ and $\mathcal{E}_2$. We will then

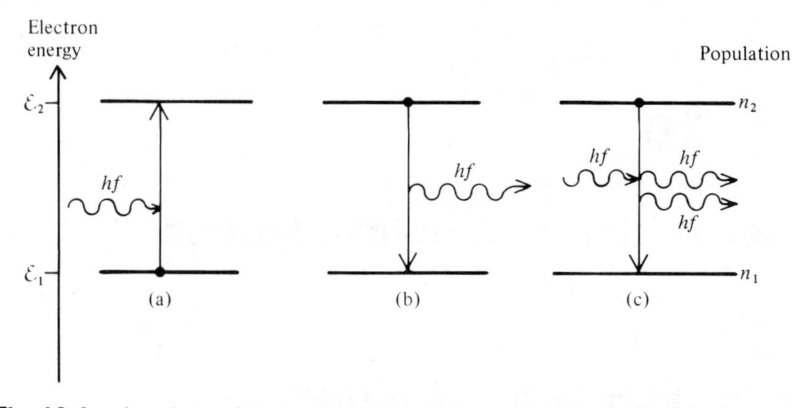

Fig. 10.1 A schematic representation of the processes of: (a) absorption; (b) spontaneous emission; (c) stimulated emission, for an idealized atomic system having two allowed electron energy states, $\mathcal{E}_1$ and $\mathcal{E}_2$.

modify these relationships with a view to taking some account of the much more complex electronic structure of a semiconductor. A semiclassical approach is used in both cases. On the basis of this we will finally consider the necessary conditions for the production of laser radiation. More detailed treatments of these matters will be found in specialist texts such as |7.1|, |10.1| and |10.2|.

First consider a population of the two-level atoms of Fig. 10.1 inside an enclosure in total thermodynamic equilibrium. The numbers of atoms per unit volume, n_1 and n_2, whose electrons are at any instant at the levels $\mathcal{E}_1$ and $\mathcal{E}_2$, respectively, are related by the Boltzmann equation

$$\frac{n_1}{n_2} = \exp\left[\frac{\mathcal{E}_2 - \mathcal{E}_1}{kT}\right] \tag{10.1.1}$$

The radiation density inside the enclosure is given by the Planck black-body radiation law. Expressed as the electromagnetic energy, $\rho(f)$, per unit volume, per unit range of spectral frequency about the frequency, f, this takes the form:

$$\rho(f) = \rho_{BB}(f) = \frac{8\pi h f^3}{c^3 |\exp(hf/kT) - 1|} \tag{10.1.2}$$

where the subscript BB refers to the case of black-body equilibrium. In eqns. (10.1.1) and (10.1.2), h is Planck's constant, c the velocity of light, k the Boltzmann constant and T the temperature of the enclosure.

The probability that an atom with its electron in the lower energy state, $\mathcal{E}_1$, will absorb a photon of energy $\mathcal{E}_{ph} = \mathcal{E}_2 - \mathcal{E}_1$ in a time interval, δt, is proportional to $\rho(f_{21})$ and to δt, and so may be expressed as $B_{12}\rho(f_{21})\,\delta t$. Similarly, the probability that an electron in the upper state, $\mathcal{E}_2$, will be stimulated to make the downward transition to $\mathcal{E}_1$ may be expressed as $B_{21}\,\rho(f_{21})\,\delta t$. Finally the electron in the upper state may decay spontaneously with a probability $A_{21}\;\delta t$. The coefficients, A_{21}, B_{21}, B_{12} are all properties of the atom and the energy states involved. In particular, the mean lifetime of the upper excited state against spontaneous emission, τ_{sp}, is simply

$$\tau_{sp} = 1/A_{21} \tag{10.1.3}$$

Under conditions of thermodynamic equilibrium, when $\rho(f) = \rho_{BB}(f)$, the rate of excitation from state 1 to state 2 must exactly balance the total rate of decay from 2 to 1. These rates, referred to unit volume, are obtained by multiplying the probabilities of decay per unit time by the numbers of atoms per unit volume at the respective levels, so that the balance between the rates of excitation and de-excitation is given by

$$A_{21}n_2 + B_{21}\rho_{BB}(f_{21})n_2 = B_{12}\rho_{BB}(f_{21})n_1 \tag{10.1.4}$$

$$\therefore \quad \rho_{BB}(f_{21}) = \frac{A_{21}/B_{21}}{B_{12}n_1/B_{21}n_2 - 1} = \frac{A_{21}/B_{21}}{\dfrac{B_{12}}{B_{21}}\exp\dfrac{(\mathcal{E}_2 - \mathcal{E}_1)}{kT} - 1} \tag{10.1.5}$$

where we have made use of eqn. (10.1.1). For both eqn. (10.1.2) and eqn. (10.1.5) to be valid independent of the temperature, it is necessary that

$$B_{12} = B_{21} = B \tag{10.1.6}$$

and

$$\frac{A_{21}}{B_{21}} = \frac{A}{B} = \frac{8\pi h f_{21}^3}{c^3} \tag{10.1.7}$$

These related coefficients are known as the Einstein A and B coefficients.

10.1.2 The Condition for Laser Action

Next we shall consider a beam of radiation passing in the z-direction through this assembly of atoms. Assume that it carries a power density of P [watts/m^2] and extends over a narrow range of frequencies about f_{21} as shown in Fig.

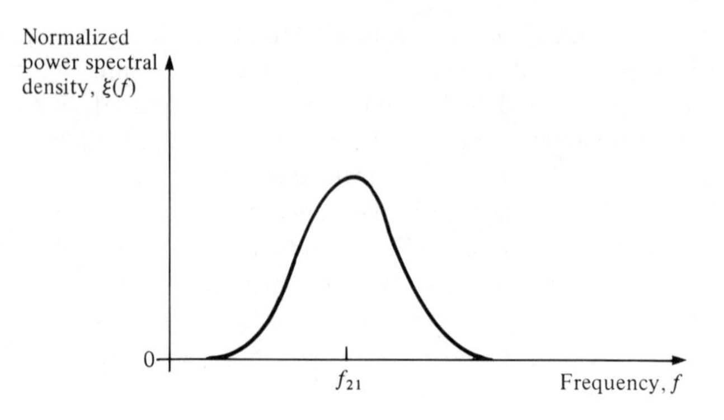

Fig. 10.2 Power spectral density function of a beam of radiation. The function is normalized such that $\int_0^\infty \xi(f)\mathrm{d}f = 1$.

10.2. Its normalized spectral distribution is $\xi(f)$ with

$$\int_0^\infty \xi(f)\,\mathrm{d}f = 1 \tag{10.1.8}$$

so the spectral electromagnetic energy density in the beam is

$$\rho(f) = \frac{P\xi(f)}{c} \tag{10.1.9}$$

We assume that at its maximum this is significantly greater than $\rho_{BB}(f)$. The beam suffers net absorption and hence a loss of power as it travels, and this loss equals the energy absorbed per second per unit volume by the atoms. Thus

$$-\frac{\mathrm{d}P}{\mathrm{d}z} = B\rho(f_{21})(n_1 - n_2)hf_{21} \tag{10.1.10}$$

We may define a beam-attenuation coefficient, α_{12}, as

$$\alpha_{12} = -\frac{1}{P}\frac{\mathrm{d}P}{\mathrm{d}z} = \frac{B(n_1 - n_2)hf_{21}\xi(f_{21})}{c} \tag{10.1.11}$$

using eqn. (10.1.9). With α_{12} independent of z, the power in the beam attenuates exponentially:

$$P(z) = P(0)\exp(-\alpha_{12}z) \tag{10.1.12}$$

In saying this we have neglected as small any spontaneous emission at the frequency f_{21}.

After a period of time many of the atoms exposed to the beam will have redistributed themselves from the lower to the upper state. Some people like to think of this as a change of temperature in the enclosure affecting just the frequencies in the beam. In that case we would think of temperature as determined by the local value of $\rho(f)$ through eqn. (10.1.2). It is greatly increased by the presence of the beam. This is by the way, but what is important is that a new equilibrium is established when the net absorption is balanced by spontaneous emission and still n_1 is greater than n_2. However, when there are external means available whereby the upper state may be populated preferentially, it is sometimes possible to create situations in which n_2 can be made greater than n_1. Then a ***population inversion*** is said to exist. A direct consequence is that then the rate of stimulated emission exceeds the rate of absorption. The beam is supplemented as it passes through the medium and an optical amplifier has been produced. Its gain coefficient, g_{12}, is given by

$$g_{12} = -\alpha_{12} = \frac{1}{P}\frac{\mathrm{d}P}{\mathrm{d}z} \tag{10.1.13}$$

Using eqns. (10.1.11), (10.1.7) and (10.1.3),

$$g_{12} = \frac{Ac^2(n_2 - n_1)\xi(f_{21})}{8\pi f_{21}^2} = \frac{c^2}{8\pi}\frac{(n_2 - n_1)\xi(f_{21})}{f_{21}^2\tau_{sp}} \tag{10.1.14}$$

Now the power density at frequency, f_{21}, will *grow* exponentially as it passes through the medium:

$$P(z) = P(0)\exp(g_{12}z) \tag{10.1.15}$$

Creating a population inversion and producing optical gain represents the first of the two essential steps needed to obtain laser action. The second step is to provide some positive feedback to turn the optical amplifier into a laser oscillator. This may simply take the form of a pair of mirrors which reflect the amplified light back and forth through the active medium. Then an optical cavity is formed. The cavity has a particular set of characteristic resonant frequencies and the radiation is characterized by these rather than by the normal spectral properties of emission from the two atomic energy levels. The optical power density at any of these cavity resonant frequencies comes into equilibrium such that the optical power loss during each transit of the radiation through the medium just balances the optical power gain. Included as a loss to the cavity must be the power which is transmitted through the mirrors and which forms the output beam of the laser. Self-oscillation cannot start before

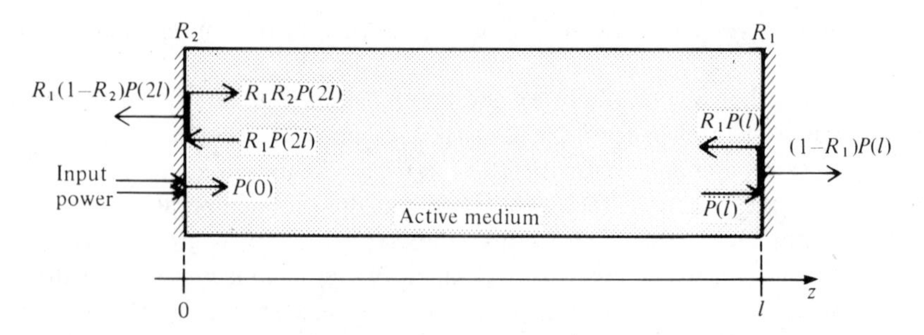

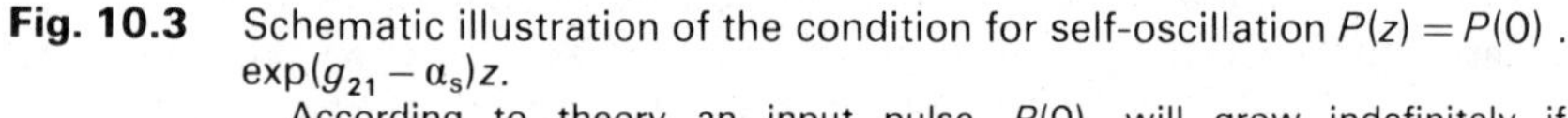

Fig. 10.3 Schematic illustration of the condition for self-oscillation $P(z) = P(0) \, . \, \exp(g_{21} - \alpha_s)z$.
According to theory an input pulse, $P(0)$, will grow indefinitely if $R_1R_2 \, . \, P(2l) > P(0)$. In practice the power level is limited by saturation effects.

the gain exceeds the losses. This requirement sets a threshold population inversion, $(n_2 - n_1)_{th}$, which has to be exceeded before laser action can occur. Some of the light generated by stimulated emission is scattered out of the active medium during transit. This can be represented by a loss coefficient, α_s, analogous to the absorption coefficient, α_{21}. The variation with distance of the optical power in the beam then becomes

$$P(z) = P(0) \exp [(g_{12} - \alpha_s)z] \tag{10.1.16}$$

so that the gain in a two-way transit of the active medium is

$$P(2l)/P(0) = \exp [2(g_{12} - \alpha_s)l] \tag{10.1.17}$$

where l is the effective length of the active medium as shown in Fig. 10.3. At the mirrors fractions R_1 and R_2 of the incident power are reflected. Then the condition for self-oscillation becomes

$$R_1R_2P(2l) > P(0)$$

$$\therefore \quad R_1R_2 \exp [2(g_{12} - \alpha_s)l] > 1$$

$$\therefore \quad g_{12} > \alpha_{tot} = \alpha_s + \frac{1}{2l} \ln (1/R_1R_2) \tag{10.1.18}$$

where α_{tot} is an effective total loss coefficient, which takes into account the transmission through the mirrors. If the mirrors are good, then almost $(1 - R_1)$ and $(1 - R_2)$ is transmitted as the optical power at each end. When both reflection coefficients are equal, $R_1 = R_2 = R$, the threshold condition simplifies to

$$g_{12} > \alpha_s + \frac{1}{l} \ln (1/R) \qquad (10.1.19)$$

with g_{12} determined by eqn. (10.1.14).

It will be noted that g_{12} is proportional to $\xi(f)$ and therefore inversely proportional to the spectral line width of the beam of radiation passing through the assembly of atoms. Normally the radiation that initiates the laser action is the spontaneous emission from the two lasing energy levels. The wavelength of this radiation spreads over a range as a result of a number of well established effects such as natural line width (which derives from the uncertainty principle), Doppler broadening, and pressure broadening. Thus, in order to establish the threshold condition for laser action, $\xi(f)$ is normally taken to be the normalized spectral line intensity for the spontaneous emission between the two lasing energy levels.

10.2 OPTICAL GAIN IN A SEMICONDUCTOR

10.2.1 The Condition for Gain

Population inversion between appropriate levels is established in different ways in different types of laser. In solid-state lasers like the ruby laser or the neodymium laser, light from a powerful source is absorbed in the active medium and increases the population of a number of higher energy levels. These decay spontaneously and rapidly down to a relatively long-life (metastable) excited state. This fills up more quickly than it can decay and a population inversion is set up between this level and lower lasing levels. In gas lasers a similar metastable level is preferentially populated with the help of electronic excitation. With semiconductors the injection of carriers at a forward-biassed p–n junction can sometimes lead to an inversion between the populations of the energy levels in the conduction band and those in the valence band. The energy-level structure inside a semiconductor is, of course, a great deal more complicated than the simple two-level system we have discussed so far. Nevertheless, the same ideas can be applied almost exactly. In particular it is possible to derive a very simple condition that must be satisfied before the rate of stimulated emission can exceed the rate of absorption.

We shall consider semiconductor laser action specifically in the context of the forward-biassed, N–p–P, double heterostructure that was illustrated by Fig. 9.12. In order to model the behavior of this structure under forward bias we suggested that the free electrons in the active layer, the p-region, would be maintained in approximate thermal equilibrium with the free electrons in the N-region. This implies that the collision and relaxation processes that maintain the energy distribution are so rapid that high current densities and large optical emission do not significantly disturb it. Then, in the active layer, the probability that a state in the conduction band is occupied is determined by the energy of the state, $\mathcal{E}_2$, relative to the Fermi energy of the N-region, $\mathcal{E}_{FN}$. The probability is given by F_2 (see eqn. (7.2.3)), where

$$F_2 = F_N(\mathcal{E}_2) = \frac{1}{\exp(\mathcal{E}_2 - \mathcal{E}_{FN})/kT + 1} \tag{10.2.1}$$

Conversely, the holes in the active layer are considered to be in approximate thermal equilibrium with the holes of the P-region. Then the probability that an energy level at an energy, $\mathcal{E}_1$, in the valence band is unoccupied (i.e. that there is a hole there) is determined by the difference between $\mathcal{E}_1$ and the Fermi energy of the P-region, $\mathcal{E}_{FP}$. This probability is given by $(1 - F_1)$, where

$$F_1 = F_P(\mathcal{E}_1) = \frac{1}{\exp(\mathcal{E}_1 - \mathcal{E}_{FP})/kT + 1} \tag{10.2.2}$$

The fact that this is in reality a highly nonequilibrium situation is brought out by this separation of the effective Fermi energies for electrons and holes. They are sometimes referred to as quasi-Fermi levels. We shall show that stimulated emission can exceed absorption only for those photons whose energy is less than the separation of the quasi-Fermi levels. Since the photon energy must also be greater than the band-gap energy, this means that at least one and preferably both of the quasi-Fermi levels in the active region have to lie outside the band gap and within the conduction or valence bands. This requirement can be brought about by high doping levels and by sufficiently heavy forward biassing of the N–p junction, as shown in Fig. 10.4.

The rate, r_{12}, at which electrons in an energy state, $\mathcal{E}_1$, in the valence band can be excited up to a state of energy $\mathcal{E}_2$ in the conduction band depends on four factors:

(a) The probability that the transition can occur, expressed as a coefficient, B_{12}.
(b) The electromagnetic energy density at the frequency, $f_{21} = (\mathcal{E}_2 - \mathcal{E}_1)/h$, represented as $\rho(f_{21})$.
(c) The probability that the state at $\mathcal{E}_1$ is occupied, F_1.
(d) The probability that the state at $\mathcal{E}_2$ is unoccupied, $1 - F_2$.

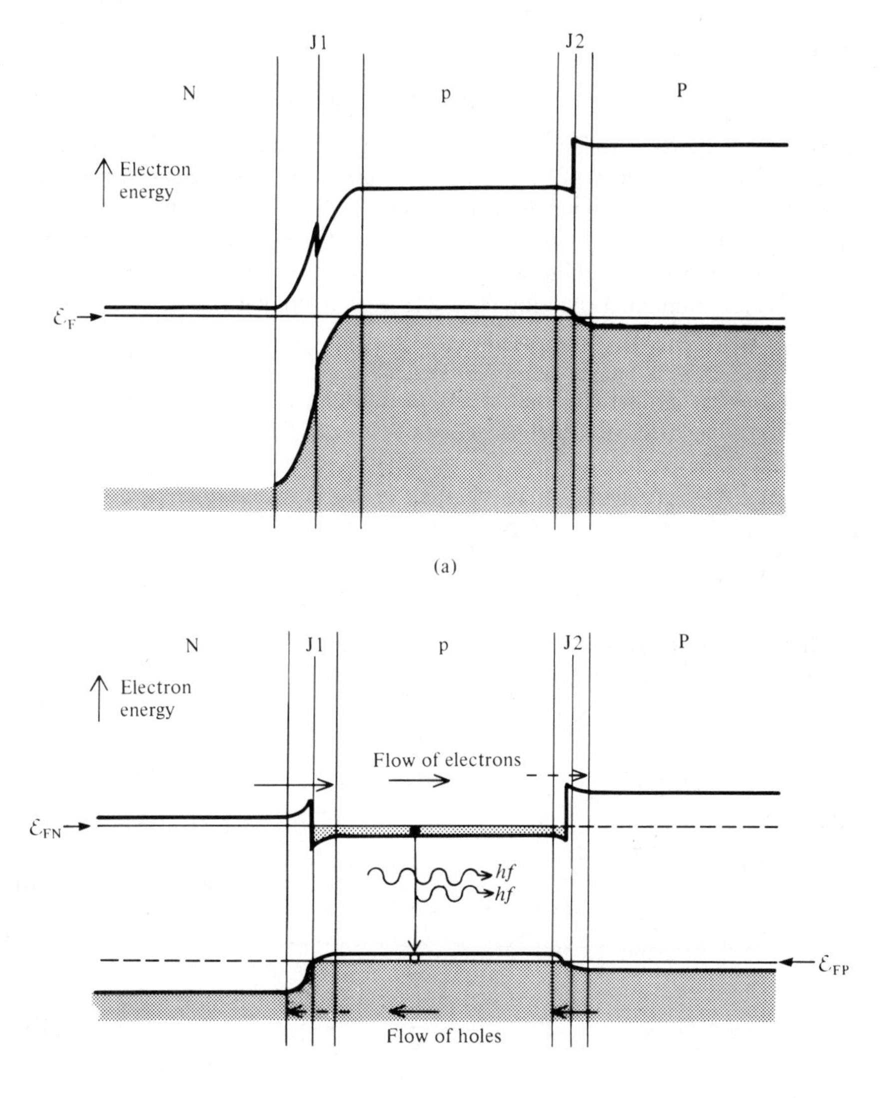

Fig. 10.4 Energy band structure across an N–p–P double heterostructure in which the p-region is heavily doped: (a) equilibrium energy levels, showing $\mathcal{E}_{FP}$ to lie within the valence band of the p-region; (b) strongly forward-biassed so that the electrons injected over the N–p barrier cause $\mathcal{E}_{FN}$ to lie within the conduction band of the p-region. As a result, in the p-region, population inversion has taken place between the energy levels at the bottom of the conduction band and those at the top of the valence band.

Thus

$$r_{12} = B_{12}\rho(f_{21})F_1(1 - F_2) \tag{10.2.3}$$

It is most important to be quite clear about the dimensions of the various quantities related by eqn. (10.2.3). We have defined $\rho(f)$ as the spectral energy density, which in SI units would be measured in [$\mathrm{J.m^{-3}.Hz^{-1}}$]. The Fermi functions F_1 and F_2 are numbers, so with r_{21} being the number of transitions per second, B_{12} must have units of [$\mathrm{m^3\ J^{-1}\ s^{-2}}$]. By a similar argument, the rate, r_{21}, at which stimulated emission from a level at $\mathcal{E}_2$ to one at $\mathcal{E}_1$ can take place is determined by:

(a) The probability that the transition can occur, $B_{21} = B_{12} = B$.
(b) The electromagnetic energy density, $\rho(f_{21})$.
(c) The probability that the state at $\mathcal{E}_1$ is empty, $1 - F_1$.
(d) The probability that the state at $\mathcal{E}_2$ is occupied, F_2.

Thus

$$r_{21} = B_{21}\rho(f_{21})(1 - F_1)F_2 \tag{10.2.4}$$

Stimulated emission will exceed absorption when $r_{21} > r_{12}$, that is,

$$(1 - F_1)F_2 > F_1(1 - F_2)$$

or (10.2.5)

$$F_2 > F_1$$

Put

$$x = \frac{\mathcal{E}_1 - \mathcal{E}_{FP}}{kT} \tag{10.2.6}$$

and

$$y = \frac{\mathcal{E}_2 - \mathcal{E}_{FN}}{kT} \tag{10.2.7}$$

so that

$$F_1 = \frac{1}{e^x + 1}, \qquad (1 - F_1) = \frac{e^x}{e^x + 1}, \qquad F_2 = \frac{1}{e^y + 1}, \qquad (1 - F_2) = \frac{e^y}{e^y + 1}$$

Then, condition (10.2.5) becomes

$$\frac{1}{(e^y + 1)} > \frac{1}{(e^x + 1)}$$

or $x > y$, which is

$$\mathcal{E}_1 - \mathcal{E}_{FP} > \mathcal{E}_2 - \mathcal{E}_{FN}$$

or

$$\mathcal{E}_{ph} = \mathcal{E}_2 - \mathcal{E}_1 < \mathcal{E}_{FN} - \mathcal{E}_{FP} = \Delta\mathcal{E}_F \qquad (10.2.8)$$

The semiconductor will act as an optical amplifier for radiation whose photon energy exceeds the band-gap energy but is less than the difference between the quasi-Fermi levels. The net rate of stimulated emission, r_{st}, between the two levels at $\mathcal{E}_1$ and $\mathcal{E}_2$ may be written as

$$r_{st} = r_{21} - r_{12} = B\rho(f_{21})[(1-F_1)F_2 - F_1(1-F_2)]$$

$$= B\rho(f_{21})(F_2 - F_1) \qquad (10.2.9)$$

Finally, the spontaneous emission rate, r_{sp}, between these levels is

$$r_{sp} = AF_2(1 - F_1) \qquad (10.2.10)$$

In eqn. (10.2.10), r_{sp} and A both have the units [s^{-1}], as does r_{st} in eqn. (10.2.9).

10.2.2 Rates of Spontaneous and Stimulated Emission

In a semiconductor the energy levels in the valence and conduction bands are so many in number and so closely spaced in energy that we are accustomed to treating them as a continuum. In Section 8.2 the number of energy levels per unit volume in the energy range $d\mathcal{E}_2$ about $\mathcal{E}_2$ in the conduction band was expressed as $S_c(\mathcal{E}_2)\,d\mathcal{E}_2$. The equivalent level concentration about $\mathcal{E}_1$ in the valence band was given as $S_v(\mathcal{E}_1)\,d\mathcal{E}_1$. Typical forms for these density of states functions, S_c and S_v, were shown in Fig. 8.2. The rates of emission between the levels around $\mathcal{E}_2$ and those around $\mathcal{E}_1$ are decided by the same factors that determine the rates between individual levels, as in eqns. (10.2.9) and (10.2.10), but in addition they are proportional to the densities of states at $\mathcal{E}_2$ and $\mathcal{E}_1$. Thus the net rate of stimulated emission per unit volume between bands of energies $d\mathcal{E}_1$ and $d\mathcal{E}_2$ in width is

$$r'_{st}\, d\mathcal{E}_1\, d\mathcal{E}_2 = B'\rho(f_{21})S_c(\mathcal{E}_2)\, d\mathcal{E}_2 S_v(\mathcal{E}_1)\, d\mathcal{E}_1(F_2 - F_1) \qquad (10.2.11)$$

by analogy with eqn. (10.2.9). The term B' represents the strength of the interaction between the states. It has the rather unlikely units of [$m^6\ s^{-2}\ J^{-1}$]. Similarly the net rate of spontaneous emission per unit volume is

$$r'_{sp}\, d\mathcal{E}_1\, d\mathcal{E}_2 = A'S_c(\mathcal{E}_2)\, d\mathcal{E}_2 S_v(\mathcal{E}_1)\, d\mathcal{E}_1 F_2(1 - F_1) \qquad (10.2.12)$$

The constant A' is again a property of the states involved and is thus expected to be related to B'. Its units are $[\mathrm{m}^3\ \mathrm{s}^{-1}]$.

In equilibrium there is just one Fermi energy, $\mathcal{E}_{FN} = \mathcal{E}_{FP} = \mathcal{E}_F$, the radiation density is that of a black body $\rho(f_{21}) = \rho_{BB}(f_{21})$, and the interchange between the energy levels must balance, $r'_{st} + r'_{sp} = 0$. Thus,

$$A'F_2(1 - F_1) + B'\rho_{BB}(f_{21})(F_2 - F_1) = 0$$

$$\therefore \quad \rho_{BB}(f_{21}) = \frac{8\pi h f_{21}^3}{c^3[\exp(hf_{21}/kT) - 1]} = \frac{A'/B'}{(F_1 - F_2)/F_2(1 - F_1)} \tag{10.2.13}$$

using eqn. (10.1.2) and rearranging. Now

$$\frac{F_1 - F_2}{F_2(1 - F_1)} = \frac{1/(e^x + 1) - 1/(e^y + 1)}{[1/(e^y + 1)][e^x/(e^x + 1)]} = \frac{e^y - e^x}{e^x} = e^{(y-x)} - 1$$

using the substitutions of eqns. (10.2.6) and (10.2.7). But,

$$y - x = \frac{(\mathcal{E}_2 - \mathcal{E}_{FN}) - (\mathcal{E}_1 - \mathcal{E}_{FP})}{kT} = \frac{\mathcal{E}_2 - \mathcal{E}_1}{kT} = \frac{hf_{21}}{kT}$$

since in equilibrium $\mathcal{E}_{FP} = \mathcal{E}_{FN}$. Substituting these results into eqn. (10.2.13) gives

$$\frac{A'}{B'} = \frac{8\pi h f_{21}^3}{c^3} \tag{10.2.14}$$

10.2.3 The Effect of Refractive Index

We have up to now neglected a factor that is quite significant in a semiconductor laser. This is the refractive index. In many scientific subjects the choice of symbols is a pain. In a book such as this, which draws on a number of subject areas that have each developed separately, it is particularly difficult. We have probably offended a number of readers by insisting on the use of f for frequency, as electrical engineers do, even for optical frequencies. And no doubt there are many other examples. We are now going to add insult to injury by changing the symbols for the phase and group refractive indices. In Chapter 2 and subsequently we used n and N respectively, as is the custom in fiber optics. Now, while we are concerned with semiconductors this would be most inconvenient, so we shall use μ and μ_g, respectively, and apologize!

In an enclosure containing a refractive medium the black-body radiation density, eqn. (10.1.2), becomes

$$\rho_{BB}(f) = \frac{8\pi\mu^2\mu_g hf^3}{c^3[\exp(hf/kT) - 1]} \quad (10.2.15)$$

The relation between the radiation spectral density and the power density, eqn. (10.1.9), becomes

$$\rho(f) = \frac{\mu P \xi(f)}{c} \quad (10.2.16)$$

Equations (10.2.13) and (10.2.14) require modification with the result that

$$\frac{A'}{B'} = \frac{8\pi\mu^2\mu_g hf_{21}^3}{c^3} = \frac{8\pi\mu^2\mu_g \mathcal{E}_{21}^3}{h^2c^3} \quad (10.2.17)$$

Some values for the refractive indices of semiconductors were included in Table 7.2. Not surprisingly in view of the proximity of the absorption band edge, the refractive index varies rapidly for photon energies around the band-gap energy, as the results for GaAs shown in Fig. 10.5 demonstrate.

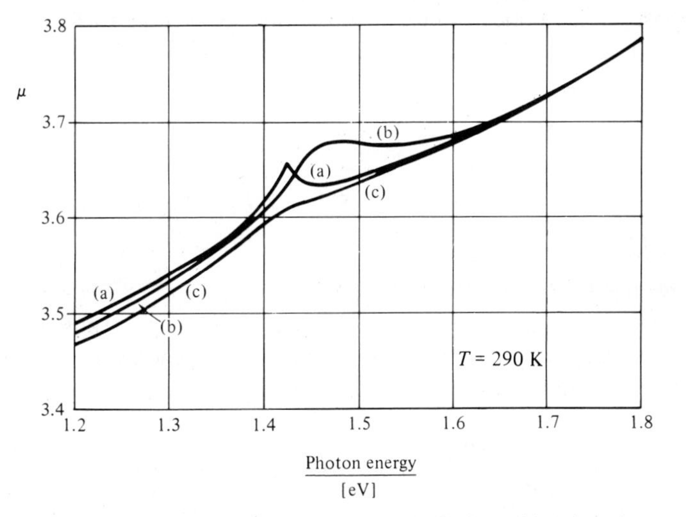

Fig. 10.5 Variation of the refractive index of GaAs with photon energy at different doping concentrations, $T = 290$ K: (a) high purity material, $n_D = 5 \times 10^{19}$m^{-3}; (b) n-type material, $n_D = 2 \times 10^{24}$m^{-3}; (c) p-type material, $n_A = 1 \times 10^{25}$m^{-3}. [Data derived from D. D. Sell *et al.*, Concentration dependence of the refractive index for n- and p-type GaAs between 1.2 and 1.8 eV, *J. Appl. Phys.* **45**, 2650–7 (1974).]

10.2.4 Calculation of the Gain Coefficient

Let us now consider a nonequilibrium situation such as that which may arise from carrier injection at a p–n junction and which we may recognize by the separation of the quasi-Fermi levels. The radiation power density resulting from spontaneous emission from the band of energy levels $d\mathcal{E}_2$ around $\mathcal{E}_2$ to the band $d\mathcal{E}_1$ around $\mathcal{E}_1$ we shall write as

$$P_{21} = \frac{c}{\mu} \rho(f_{21})\, df \tag{10.2.18}$$

Let

$$d\mathcal{E}_1 = d\mathcal{E}_2 = d\mathcal{E} = h\, df \tag{10.2.19}$$

then

$$\rho(f_{21}) = \frac{\mu h P_{21}}{c\, d\mathcal{E}} \tag{10.2.20}$$

The net rate of stimulated emission between these bands of levels causes a rate of increase of the optical power density with distance given by

$$\begin{aligned} \frac{dP_{21}}{dz} &= \mathcal{E}_{ph} r'_{st} (d\mathcal{E})^2 \\ &= \mathcal{E}_{ph} B' \frac{\mu h P_{21}}{c} S_c(\mathcal{E}_2) S_v(\mathcal{E}_1)(F_2 - F_1)\, d\mathcal{E} \end{aligned} \tag{10.2.21}$$

where we have used eqn. (10.2.11), substituted for $\rho(f_{21})$ using eqn. (10.2.20), and expressed the photon energy as $\mathcal{E}_{ph} = \mathcal{E}_{21} = \mathcal{E}_2 - \mathcal{E}_1$. Thus the contribution to the gain coefficient produced by these sets of energy levels in the range $d\mathcal{E}$ around $\mathcal{E}_1$ and $\mathcal{E}_2$ is given by

$$g_{21}\, d\mathcal{E} = \frac{1}{P_{21}} \frac{dP_{21}}{dz} = \frac{\mu h}{c} \mathcal{E}_{ph} B' S_c(\mathcal{E}_2) S_v(\mathcal{E}_1)(F_2 - F_1)\, d\mathcal{E} \tag{10.2.22}$$

To obtain the total gain coefficient for light of this wavelength, $g(\mathcal{E}_{ph})$, we have to integrate over all such pairs of energy bands in the valence and conduction bands subject to the condition $\mathcal{E}_1 = \mathcal{E}_2 - \mathcal{E}_{ph}$. Then,

$$g(\mathcal{E}_{ph}) = \int g_{21}\,d\mathcal{E} = \frac{\mu h \mathcal{E}_{ph}}{c} \int_{\mathcal{E}_c}^{\mathcal{E}_v + \mathcal{E}_{ph}} B' S_c(\mathcal{E}_2) S_v(\mathcal{E}_2 - \mathcal{E}_{ph})(F_2 - F_1)\,d\mathcal{E}_2 \tag{10.2.23}$$

In eqn. (10.2.23) we have chosen to express the integration in terms of the conduction band energies, $\mathcal{E}_2$, although it would have been equally valid to write the integral in terms of the valence band energies, $\mathcal{E}_1$. We shall find it more convenient to use the coefficient, A', rather than B', so we may substitute eqn. (10.2.17) into eqn. (10.2.23) to give

$$g(\mathcal{E}_{ph}) = \frac{h^3 c^2}{8\pi \mu \mu_g \mathcal{E}_{ph}^2} \int_{\mathcal{E}_c}^{\mathcal{E}_v + \mathcal{E}_{ph}} A' S_c(\mathcal{E}_2) S_v(\mathcal{E}_2 - \mathcal{E}_{ph})(F_2 - F_1)\,d\mathcal{E}_2 \tag{10.2.24}$$

As before, when we were considering just two levels, the term $F_2 - F_1$ is greater than zero only when

$$\mathcal{E}_{ph} = \mathcal{E}_2 - \mathcal{E}_1 > \mathcal{E}_{FN} - \mathcal{E}_{FP} = \Delta\mathcal{E}_F \tag{10.2.8}$$

When this condition is satisfied, $F_2 - F_1 > 0$ over the whole range of conduction-band energies and $g(\mathcal{E}_{ph}) > 0$, so that there is net gain. When the condition (10.2.8) is not satisfied, $g(\mathcal{E}_{ph}) < 0$ and there is net absorption.

10.2.5 Relation of the Gain Coefficient to the Current Density

Equation (10.2.24) expresses the gain/absorption coefficient in terms of the interaction parameter, A', and the density-of-states functions, S_c and S_v which are properties of the semiconductor, and the Fermi functions, $F_2 - F_1$, which depend on the injection level. With a suitably simple model, it would be possible to calculate theoretical values. What we really want, though, is to be able to relate the gain coefficient to the electrical current density in the diode. This can be done in terms of the spontaneous emission rate, as we shall show, but it does not lead to closed solutions.

We recall from Section 9.2.1 that with a double heterostructure having a high degree of carrier confinement almost all the diode current is expended in recombination in the active area. Of this, a fraction η_{int} is radiative recombination, and this represents the total of all spontaneous emission at all frequencies, which is given by

$$R_{sp} = \frac{\eta_{int} J}{ed} \tag{10.2.25}$$

where J is the diode current density and d is the depth of the active region. The units of R_{sp} are $|\mathrm{m^{-3}\ s^{-1}}|$. Now this total rate of spontaneous emission may also be obtained by integrating $r'_{sp}(\mathrm{d}\mathcal{E})^2$ of eqn. (10.2.12) first over all the pairs of energy bands, $\mathrm{d}\mathcal{E}$, separated by $\mathcal{E}_{ph}$ and then over all values of $\mathcal{E}_{ph}$. Thus

$$R_{sp} = \int\int A' S_c(\mathcal{E}_2) S_v(\mathcal{E}_1) F_2(1 - F_1)(\mathrm{d}\mathcal{E})^2$$

If A' is independent of the energy levels involved, the integrations can be performed separately over the two bands with the result that they reduce to

$$R_{sp} = A' \int_c F_2 S_c(\mathcal{E}_2)\,\mathrm{d}\mathcal{E}_2 \int_v (1 - F_1) S_v(\mathcal{E}_1)\,\mathrm{d}\mathcal{E}_1$$

$$= A'np \tag{10.2.26}$$

Comparison with eqn. (8.4.7) shows that

$$A' \equiv r \simeq 10^{-16}\ \mathrm{m^3/s} \tag{10.2.27}$$

for direct band-gap semiconductors.

10.2.6 A Specific, Simplified, Worked Example

In order to give some idea of how these calculations may be carried out, let us take a specific example of a double heterostructure having a 0.5 μm wide, p-type, GaAs active region. We shall take the acceptor concentration to be $10^{24}\ \mathrm{m^{-3}}$, and the internal quantum efficiency to be 0.8. With a current density of $10^7\ \mathrm{A/m^2} = 10\ \mathrm{A/mm^2}$ flowing across the junction we shall seek an estimate of the gain coefficient at a particular photon energy, which we shall specify later. In order to simplify the calculations we shall assume the temperature to be low so that the Fermi functions may be taken to be equal to 1 at all energies below $\mathcal{E}_F$ and equal to 0 at all energies above $\mathcal{E}_F$.

The spontaneous emission rate is given by eqn. (10.2.24) as

$$R_{sp} = \frac{\eta_{int} J}{ed} = \frac{0.8 \times 10^7}{1.6 \times 10^{-19} \times 0.5 \times 10^{-6}} = 10^{32}\ \mathrm{m^{-3}\ s^{-1}} \tag{10.2.28}$$

The product np is given by eqn. (10.2.26) as

$$np = \frac{R_{sp}}{r} = \frac{10^{32}}{10^{-16}} = 10^{48}\ \text{m}^{-6} \tag{10.2.29}$$

With $p - n = n_A = 10^{24}\ \text{m}^{-3}$ this gives $n = 0.6 \times 10^{24}\ \text{m}^{-3}$ and $p = 1.6 \times 10^{24}\ \text{m}^{-3}$. Now we may estimate the positions of the quasi-Fermi levels using

$$\int_{\mathcal{E}_c}^{\mathcal{E}_{FN}} S_c\, d\mathcal{E} = n \qquad \text{and} \qquad \int_{\mathcal{E}_{FP}}^{\mathcal{E}_v} S_v\, d\mathcal{E} = p \tag{10.2.30}$$

We shall ignore any band-tailing effects and put the values appropriate to GaAs into eqns. (8.2.5) and (8.2.6):

In SI units,

$$S_c = 2 \times 10^{54}(\mathcal{E}_2 - \mathcal{E}_c)^{1/2} \qquad \text{and} \qquad S_v = 4 \times 10^{55}(\mathcal{E}_v - \mathcal{E}_1)^{1/2}$$

so that after integration eqns. (10.2.30) give

$$(\mathcal{E}_{FN} - \mathcal{E}_c)^{3/2} = \frac{3}{2} \times \frac{0.6 \times 10^{24}}{2 \times 10^{54}} = 4.6 \times 10^{-31}\ |\text{J}^{3/2}|$$

$$\therefore\quad (\mathcal{E}_{FN} - \mathcal{E}_c) = 6.0 \times 10^{-21}\ |\text{J}| = 0.037\ \text{eV}$$

and

$$(\mathcal{E}_v - \mathcal{E}_{FP})^{3/2} = \frac{3}{2} \times \frac{1.6 \times 10^{24}}{4 \times 10^{55}} = 6 \times 10^{-32}\ |\text{J}^{3/2}|$$

$$\therefore\quad (\mathcal{E}_v - \mathcal{E}_{FP}) = 1.5 \times 10^{-21}\ |\text{J}| = 0.010\ \text{eV}.$$

The range of photon energies over which a positive gain coefficient is to be expected is from $\mathcal{E}_g$ to ($\mathcal{E}_g$ + 0.047 eV). This is illustrated in Fig. 10.6. Let us seek the value of $g(\mathcal{E}_{ph})$ at $\mathcal{E}_{ph} = \mathcal{E}_g + 0.02$ eV, somewhere near to the center of this range of wavelengths. In order to obtain $g(\mathcal{E}_{ph})$ we have to evaluate eqn. (10.2.24)

$$g(\mathcal{E}_{ph}) = \frac{h^3 c^2 A'}{8\pi\mu\mu_g\, \mathcal{E}_{ph}^2} \int_{\mathcal{E}_c + 0.01\,\text{eV}}^{\mathcal{E}_c + 0.02\,\text{eV}} 8 \times 10^{109} \times (\mathcal{E}_2 - \mathcal{E}_c)^{1/2}(\mathcal{E}_v - \mathcal{E}_1)^{1/2}\, d\mathcal{E}_2$$

(10.2.31)

The upper limit represents the point at which $\mathcal{E}_2 - \mathcal{E}_{ph}$ lies above $\mathcal{E}_v$, so that there are no energy states for the downward transition to fill. The lower

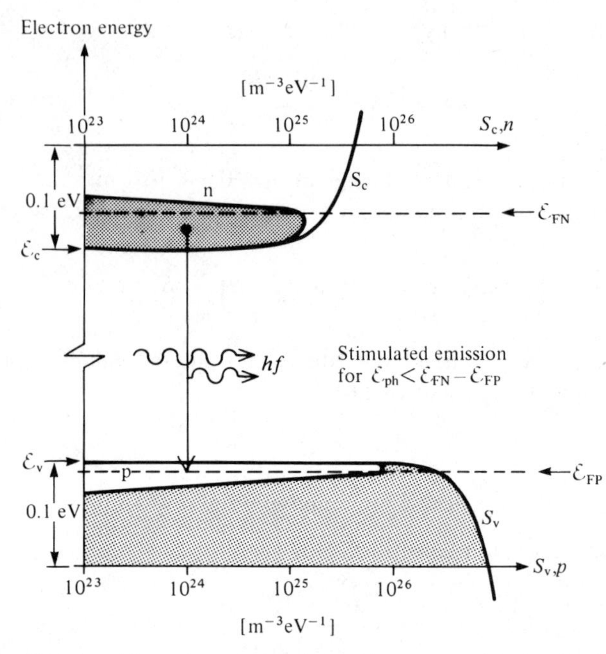

Fig. 10.6 The densities of states and the electron and hole distribution functions at an excitation level such that $n = 0.6 \times 10^{24}\text{m}^{-3}$ and $p = 1.6 \times 10^{24}\text{m}^{-3}$.

In the diagram the temperature is taken to be low but finite, say about 30 K. Otherwise the diagram corresponds to the example given with

$$\frac{S_v}{[\text{m}^{-3}\text{eV}^{-1}]} = 2.6 \times 10^{27} \frac{(\mathcal{E}_v - \mathcal{E}_1)^{\frac{1}{2}}}{[\text{eV}^{\frac{1}{2}}]}$$

$$\frac{S_c}{[\text{m}^{-3}\text{eV}^{-1}]} = 1.3 \times 10^{26} \frac{(\mathcal{E}_2 - \mathcal{E}_c)^{\frac{1}{2}}}{[\text{eV}^{\frac{1}{2}}]}$$

integrating limit on $\mathcal{E}_2$ lies $\mathcal{E}_{ph}$ above $\mathcal{E}_{FP}$. For values below this $F_1 = F_2 = 1$, so that all states in both bands are filled and $(F_2 - F_1) = 0$. This is a property of the low-temperature assumption we have made in order to simplify the calculations. It is quite artificial and normally the integral would have to be taken down to $\mathcal{E}_c$ and would have to take account of the full Fermi functions. Put

$$w = \mathcal{E}_2 - \mathcal{E}_c \tag{10.2.32}$$

so that

$$(\mathcal{E}_v - \mathcal{E}_1) = \mathcal{E}_v - (\mathcal{E}_2 - \mathcal{E}_{ph}) = \mathcal{E}_v - \mathcal{E}_2 + \mathcal{E}_g + 0.02 \text{ eV}$$
$$= \mathcal{E}_c - \mathcal{E}_2 + 0.02 \text{ eV}$$
$$= 0.02 \text{ eV} - w \qquad (10.2.33)$$

We shall express all energies in electron volts and use the values $\mathcal{E}_g = 1.42$ eV, $\mathcal{E}_{ph} = 1.44$ eV, $A' = r = 10^{-16}$ m^3/s, $\mu = 3.6$, $\mu_g = 4$. These also represent GaAs. Then

$$\frac{g(\mathcal{E}_{ph})}{|\mathrm{m}^{-1}|} = \left[\frac{8 \times 10^{109} \times 10^{-16} \times (6.626 \times 10^{-34})^3 \times (3 \times 10^8)^2}{8\pi \times 3.6 \times 4 \times (1.44)^2}\right]$$
$$\times \int_{w=0.01}^{0.02} w^{1/2}(0.02 - w)^{1/2}\, \mathrm{d}w \qquad (10.2.34)$$

With these particular limits the value of the integral is simply $(\pi/4)(0.01)^2$ as can be seen from Fig. 10.7.

$$\therefore \quad g(\mathcal{E}_{ph}) = 2.8 \times 10^8 \times \frac{\pi}{4} \times 10^{-4} = 2.2 \times 10^4 \text{ m}^{-1} = 22 \text{ mm}^{-1} \qquad (10.2.35)$$

Typical values for the scattering loss coefficient, α_s, are found to lie in the range 10^3–10^4 m^{-1}, increasing with the doping concentration. In our example the lower figure is the more likely so the net gain coefficient is 21 mm^{-1} and the optical gain in a single transit of an active region 0.4 mm long is $e^{8.4} = 4447$. This brings out one of the characteristics of semiconductor lasers: because of the high carrier concentrations, very large gain coefficients can be obtained.

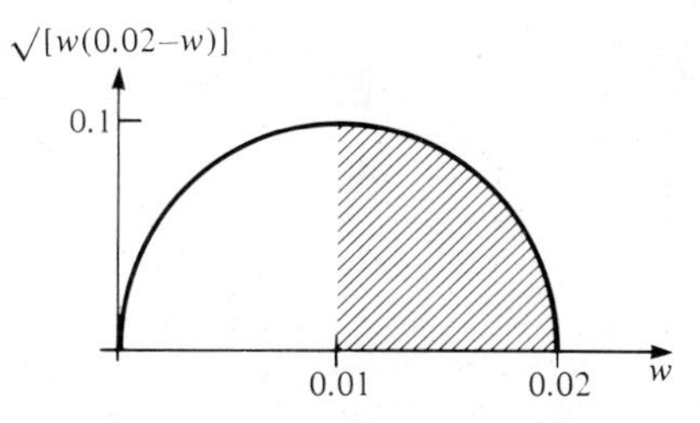

Fig. 10.7 Plot of $w^{\frac{1}{2}}\,(0.02 - w)^{\frac{1}{2}}$ against w; the integral of eqn. (10.2.34) represents the area of the shaded quadrant.

10.2.7 Semi-empirical Analysis

Readers having access to a small computer and with time to spare should have no great difficulty in extending the results just obtained to cover the variation of the gain coefficient with photon energy and with current density. However, for these calculations to be worthwhile three things which have been neglected need to be included in the analysis. These are

(a) More sophisticated density-of-states functions, S_c and S_v, which take into account band-tailing at high doping concentrations.

(b) The full Fermi distribution functions for nonzero temperatures.

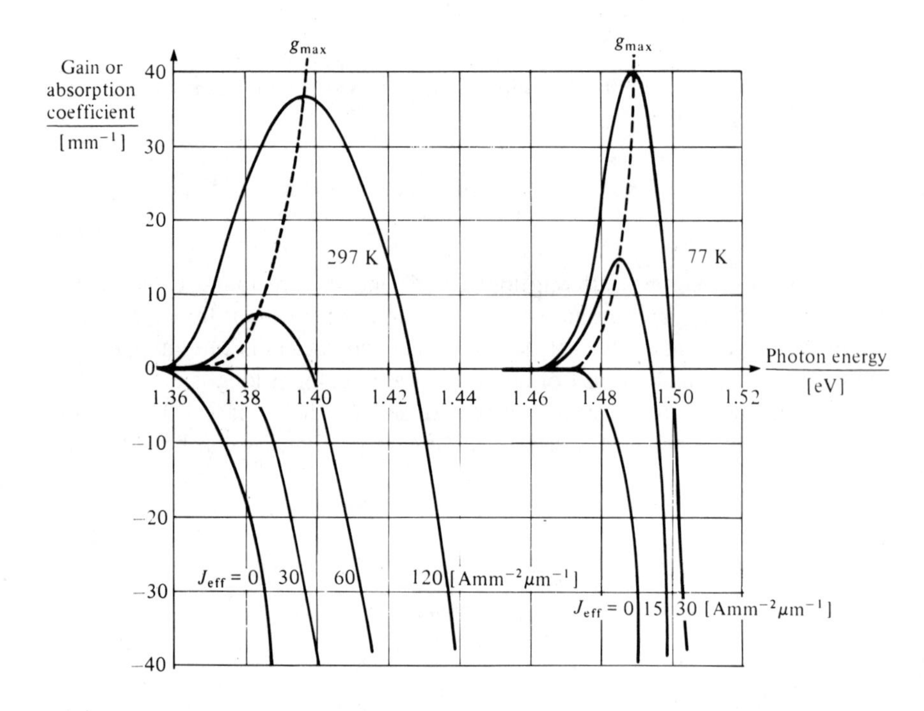

Fig. 10.8 Theoretical calculations of the gain coefficient in p-type GaAs as a function of photon energy with J_{eff} as a parameter. [Data taken from F. Stern, Calculated spectral dependence of gain in excited GaAs, *J. Appl. Phys.* **47**, 5382–6 (1976).]

The values of $J_{eff} = \eta_{int}J/d$ are given in units of [A mm^{-2} μm^{-1}]. The acceptor concentration was taken to be 4×10^{23}m^{-3}.

(c) The selection rules which govern the probability of electron transfers between states in the valence and conduction bands.

Some results from such analyses are shown in Figs 10.8 and 10.9. It would not be appropriate to pursue these fundamental matters in such detail here and readers who wish to see how the curves were obtained are referred to one of the specialist texts: Part A of [7.1], [10.1] or [10.2], and thence to the literature.

It is clear from eqn. (10.2.25) that the key parameter determining the gain coefficient is the effective current density, J_{eff}, which we may define as

$$J_{eff} = \frac{\eta_{int} J}{d} \tag{10.2.36}$$

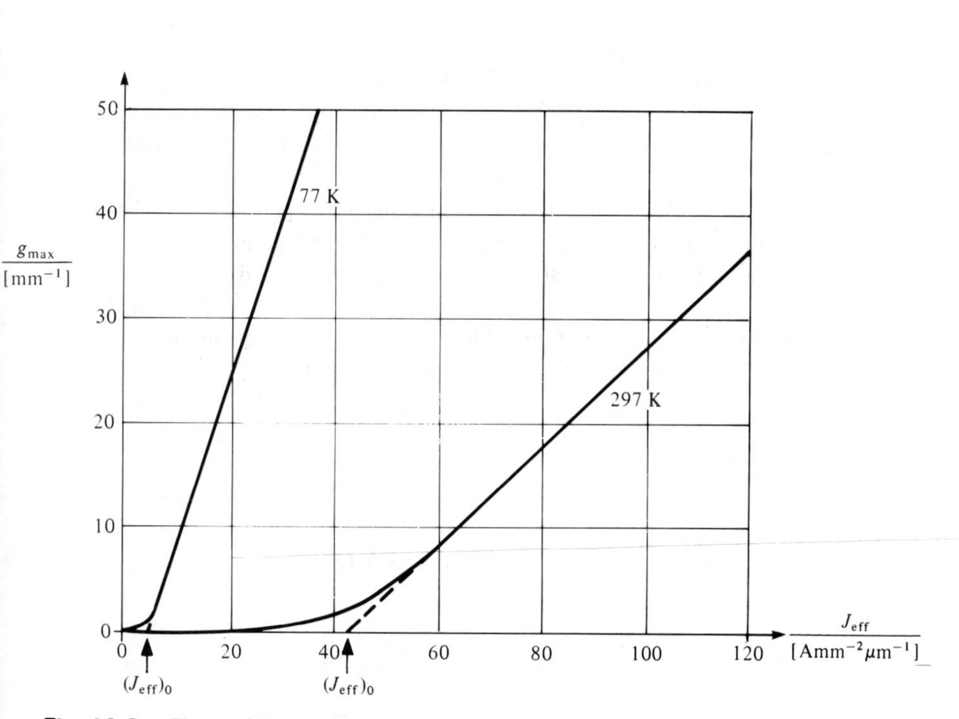

Fig. 10.9 Theoretical calculations of the variation of the maximum gain coefficient with J_{eff} at different temperatures. [Data taken from F. Stern, *J. Appl. Phys.* **47**, 5382–6 (1976).]

The calculations, like those of Fig. 10.8, have taken into account the effect of band-tailing and the reduction in the band-gap at the acceptor concentration of $4 \times 10^{23}\text{m}^{-3}$. The arrows indicate the values of the parameter $(J_{eff})_0$ at 77 and 297 K, see eqn. (10.2.37).

and this is the parameter that has been used in Figs. 10.8 and 10.9. From the curves in these figures it can be seen that in practice a certain minimum effective current density, $(J_{eff})_0$, has to be reached before a positive gain coefficient can be obtained. It is, however, the maximum value of the gain coefficient, g_{max}, that is of greatest importance, because it is this that decides whether or not the lasing threshold is reached and hence determines the principal lasing frequency. Above $(J_{eff})_0$, g_{max} is seen to increase almost linearly with J_{eff} and this relationship may be expressed in the form

$$g_{max} = \beta[J_{eff} - (J_{eff})_0] \tag{10.2.37}$$

An increase in temperature causes the positive gain coefficient to extend over a wider range of optical frequencies, but its maximum value is reduced. The slope, β, of the g_{max} vs. J_{eff} curve above $(J_{eff})_0$ is inversely proportional to temperature. This is mainly the result of the wider distributions of the carrier energies at higher temperatures. The variation of $(J_{eff})_0$ with temperature can be fitted approximately to a $T^{3/2}$ law. There are theoretical grounds based on the simplified band structure of Section 10.2.3 for expecting the value of $(J_{eff})_0$ to increase as T^3 in compensated material and as $T^{3/2}$ in heavily doped material.

We shall obtain these last results by making the assumption that in p-type material a minimum requirement for population inversion, and hence a positive gain coefficient, is that the electron quasi-Fermi-level should lie in the conduction band. That is, $\mathcal{E}_{FN} > \mathcal{E}_c$. Let n_0 be the value of n required to satisfy this condition. Its value may be obtained by integrating the electron concentration distribution $n(\mathcal{E}_2)$ over the conduction band, subject to $\mathcal{E}_{FN} = \mathcal{E}_c$. Now,

$$n(\mathcal{E}_2)\,d\mathcal{E}_2 = S_c(\mathcal{E}_2)F_2(\mathcal{E}_2)\,d\mathcal{E}_2 \tag{8.2.3}$$

where

$$S_c(\mathcal{E}_2) = K_c(\mathcal{E}_2 - \mathcal{E}_c)^{1/2}$$

with $K_c = 2 \times 10^{54}\,[\mathrm{m^{-3}\,J^{-3/2}}]$ for GaAs, and $F_2(\mathcal{E})$ is given by eqn. (10.2.1). In order to simplify the algebra, put

$$\nu = \frac{w}{kT} = \frac{\mathcal{E}_2 - \mathcal{E}_c}{kT} = \frac{\mathcal{E}_2 - \mathcal{E}_{FN}}{kT} \tag{10.2.38}$$

Then, integrating eqn. (8.2.3) gives

$$n_0 = K_c(kT)^{3/2}\int_0^\infty \frac{\nu^{1/2}\,d\nu}{\exp(\nu) + 1}$$

$$= K_c(kT)^{3/2}F_{1/2}(0) \tag{10.2.39}$$

The Fermi integral $F_{1/2}(0)$ has the value 0.68.
Referring to eqns. (10.2.28) and (10.2.29) we see that

$$np = \frac{R_{sp}}{r} = \frac{J_{eff}}{er} \tag{10.2.40}$$

In highly doped p-type material we may assume that $p \simeq n_A$, the acceptor concentration. Then

$$n = \frac{J_{eff}}{ern_A} \tag{10.2.41}$$

and

$$(J_{eff})_0 = n_A ern_0 = 0.68 K_c n_A ern(kT)^{3/2} \tag{10.2.42}$$

In compensated material we may assume that $p \simeq n$, so that

$$n^2 = J_{eff}/er \tag{10.2.43}$$

and

$$(J_{eff})_0 = ern_0^2 = 0.46 K_c^2 er(kT)^3 \tag{10.2.44}$$

A cross-check with the results of the more sophisticated calculations given in Fig. 10.9 will show that the values of $(J_{eff})_0$ obtained at 77 K and 297 K do indicate a $T^{3/2}$ law. However, the absolute values of $(J_{eff})_0$ are nearly an order of magnitude larger than those given by eqn. (10.2.42).

10.3 THE LASING THRESHOLD

The gain coefficient needed in order to initiate laser action between two energy levels was given in eqn. (10.1.18) as

$$g_{21} > \alpha_{tot} = \alpha_s + \frac{\ln(1/R_1R_2)}{2l} \tag{10.1.18}$$

This can, in principle, be applied directly to the semiconductor laser simply by writing g_{max} instead of g_{21}. However, we ought to consider more carefully the formation of the laser cavity in a double heterostructure semiconductor laser. Because of the high gain coefficients that can be obtained, the cavity may be made very short in comparison with other laser types, and 0.2–1.0 mm is typical. Furthermore, the reflectivities of the mirrors are not critical and it is usual simply to make use of the Fresnel reflection that arises at the semiconductor–air interface. As we saw in Section 8.5, this reflection coefficient is

$$R = \frac{(\mu - 1)^2}{(\mu + 1)^2} = \left(\frac{2.7}{4.7}\right)^2 = 0.33 \tag{10.3.1}$$

in the case of GaAs. The cavity is formed by aligning the p–n junction along one particular crystal direction, usually the [100] plane, and cleaving the wafer along a natural cleavage plane at right angles, usually the [110] plane. The cleaved faces will form the laser end faces. They may be coated in order to increase reflectivity but this is not necessary. The cleaved slices are then sawn into laser chips. With broad devices it is normal to leave the sawn edges rough so that laser action in the lateral direction is discouraged.

We mentioned in Section 9.1.2 that the double heterostructure offered the possibility of optically confining the radiation within the active layer because of the refractive index difference between this and the surrounding layers. The situation is analogous to that of the clad fiber. We have seen in the last section that in order to minimize the current density required for population inversion the active layer should be as thin as possible. Indeed, the values we quoted in the examples were comparable with the optical wavelengths. The situation is similar to that of an optical fiber with a small core diameter that is able to support only a few low-order modes of propagation. We will return to consider some of the properties of these modes, but for the moment we note that a fraction of the electromagnetic power associated with any given mode is carried outside the active layer. Thus only the fraction, Γ, that lies within the active layer can participate in the stimulated emission processes and so benefit from the optical gain. The parameter, Γ, is known as the confinement factor, and it causes the lasing condition to be modified from that of eqn. (10.1.18) to

$$\Gamma g_{\text{max}} > \alpha_{\text{tot}} \tag{10.3.2}$$

Substituting eqn. (10.2.37), this becomes

$$\Gamma\beta[J_{\text{eff}} - (J_{\text{eff}})_0] > \alpha_s + \frac{\ln(1/R_1R_2)}{2l} \tag{10.3.3}$$

$$\therefore \quad J_{\text{eff}} = \frac{\eta_{\text{int}}J}{d} > (J_{\text{eff}})_0 + \frac{1}{\Gamma\beta}\left[\alpha_s + \frac{\ln(1/R_1R_2)}{2l}\right] \tag{10.3.4}$$

and the lasing threshold current density is given by

$$J_{\text{th}} = (J_{\text{eff}})_0\frac{d}{\eta_{\text{int}}} + \frac{d}{\Gamma\beta\eta_{\text{int}}}\left[\alpha_s + \frac{\ln(1/R_1R_2)}{2l}\right] \tag{10.3.5}$$

Because of the steep rise in g_{max} above $(J_{eff})_0$, there is often not a great difference between the current density needed to give a positive gain coefficient and the lasing threshold. A worked example based on a GaAs laser will serve to illustrate this. Take the following values:

$$d = 0.5\ \mu\text{m}, \qquad R_1 = 1, \qquad \Gamma = 0.8$$

$$l = 0.4\ \text{mm}, \qquad R_2 = 0.33, \qquad \eta_{int} = 0.9$$

$$\alpha_s = 1\ \text{mm}^{-1} = 10^3\ \text{m}^{-1}$$

$$(J_{eff})_0 = 43\ \text{A mm}^{-2}\ \mu\text{m}^{-1} = 4.3 \times 10^{13}\ \text{A.m}^{-3}$$

$$\beta = 0.48 \frac{\text{mm}^{-1}}{\text{A mm}^{-2}\ \mu\text{m}^{-1}} = 4.8 \times 10^{-10}\ \text{A}^{-1}\ \text{m}^2$$

These last two values are derived from the room temperature curve in Fig. 10.9. Then,

$$\alpha_{tot} = \alpha_s + \frac{\ln(1/R_1R_2)}{2l}$$

$$= 10^3 + 1.1/0.8 \times 10^{-3} = 10^3(1 + 1.38) = 2380\ [\text{m}^{-1}]$$

and,

$$J_{th} = (J_{eff})_0 \frac{d}{\eta_{int}} + \frac{\alpha_{tot}}{\Gamma\beta}\frac{d}{\eta_{int}}$$

$$= \frac{4.3 \times 10^{13} \times 0.5 \times 10^{-6}}{0.9} + \frac{2.38 \times 10^3 \times 0.5 \times 10^{-6}}{0.8 \times 4.8 \times 10^{-10} \times 0.9}$$

$$= (23.9 + 3.4) \times 10^6 = 27.3 \times 10^6\ [\text{A m}^{-2}]$$

Thus a lasing threshold of 27 A/mm^2 may be compared to a positive gain threshold of 24 A/mm^2. For a broad area laser with a junction area of, say, $w \times l = 0.1 \times 0.4$ mm^2 a threshold current density of 27 A/mm^2 corresponds to a threshold current of 1.1 A.

It is desirable for the laser cavity to be long enough for the scattering loss rather than the facet reflectivities to be the dominant factor in determining α_{tot}. In this case eqn. (10.3.5) reduces to

$$J_{th} \simeq \frac{d}{\eta_{int}}[(J_{eff})_0 + \alpha_s/\Gamma\beta] \qquad (10.3.6)$$

With the given values of the material parameters this takes the form

$$\frac{J_{\text{th}}}{[\text{A/mm}^2]} \simeq \frac{1}{\eta_{\text{int}}} \frac{d}{[\mu\text{m}]} (43 + 2.1/\Gamma)$$

At first sight we might expect the variation of J_{th} with temperature to follow that of $(J_{\text{eff}})_0$, and on the basis of the analysis of Section 10.2.4 we might expect this to take the form of a power law. However, that analysis was based on a very simplified model, and there are several effects which may influence $J_{\text{th}}(T)$. These include the variation with temperature of the internal quantum efficiency and the electronic and optical confinement factors at the heterojunctions. It has, in fact, become customary to fit experimentally determined values

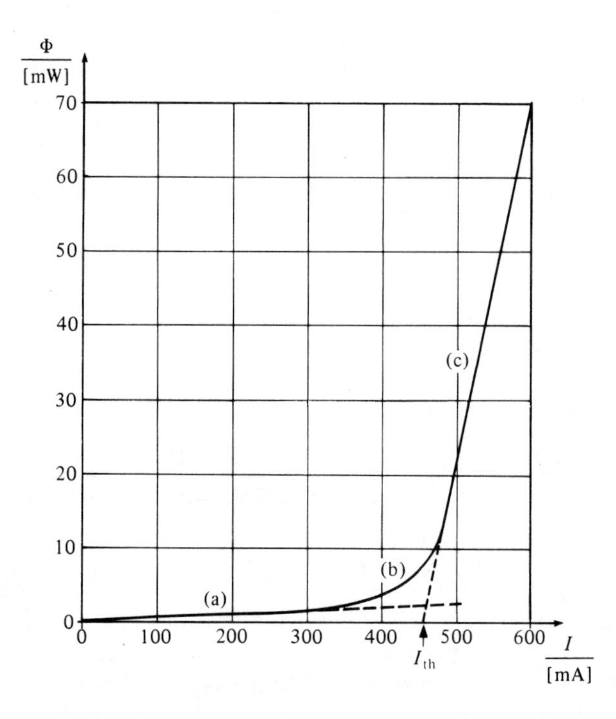

Fig. 10.10 A typical, experimental curve showing the total, pulsed, output power from a semiconductor laser as a function of the diode current.
At low current, (a), the light is generated by spontaneous emission and the diode behaves like an edge-emitting LED. Around (b) there is a significant increase in the proportion of stimulated emission; diodes operated in this region are called superradiant. At currents well above threshold, (c), laser action occurs with the output radiation dominated by stimulated emission. The spectral characteristics of the three regions are shown in Fig. 10.11.

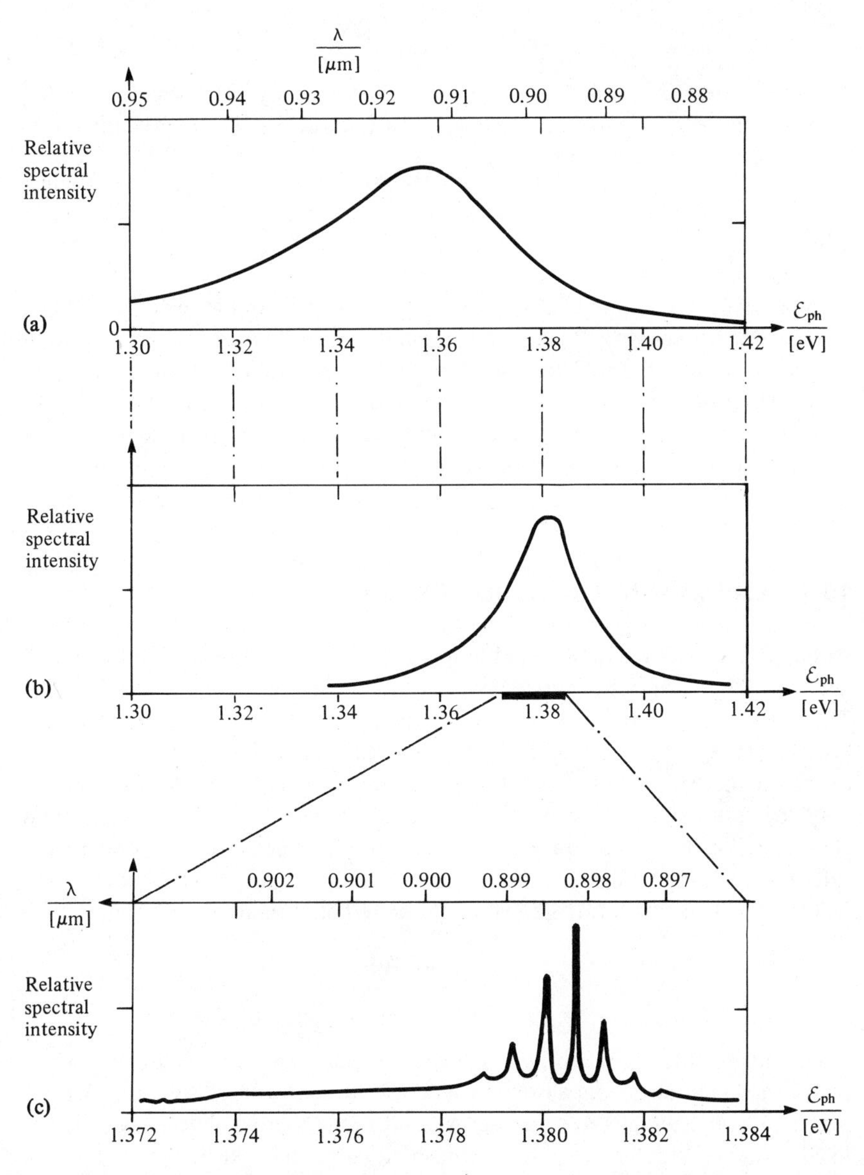

Fig. 10.11 Spectral characteristics of the output radiation from a semiconductor laser at different current levels (The characteristics correspond to the regions (a), (b) and (c) shown in Fig. 10.10): (a) spontaneous emission; (b) superradiance; (c) lasing (note the change of horizontal scale).

of $J_{th}(T)$ to the form

$$J_{th}(T) = J_0 \exp(T/T_0) \tag{10.3.7}$$

The parameters J_0 and T_0 are determined empirically. The smaller T_0, the steeper is the temperature variation of $J_{th}(T)$, the fractional change of threshold current density with temperature being

$$\frac{1}{J_{th}} \frac{dJ_{th}}{dT} = \frac{1}{T_0} \tag{10.3.8}$$

Typical values of T_0 for GaAs/GaAlAs DH lasers lie in the region of 150 K, whereas for InGaAsP/InP lasers T_0 is usually about 70 K over the wavelength range 1.1 μm to 1.6 μm. This low value of T_0 represents a serious problem in the development of long-wavelength laser sources. Its origin may be related to the effects that cause the internal quantum efficiency of long-wavelength LEDs to decrease with increasing temperature.

10.4 LASER CHARACTERISTICS

Although the laser threshold condition is important, we shall be mainly concerned with the properties of semiconductor lasers biassed well beyond this point. The spectral characteristics of the laser light then become a property of the laser cavity. Figures 10.10 and 10.11 illustrate the behavior of a semiconductor laser on either side of threshold. Figure 10.10 shows how the threshold current may be estimated experimentally. The slope of the power vs. current characteristic in the spontaneous emission region should correspond to the external quantum efficiency defined in Section 8.5. The slope in the lasing region is related to what is known as the differential quantum efficiency, η_D:

$$\eta_D = \frac{e}{\mathcal{E}_{ph}} \frac{\Delta\Phi}{\Delta I} \tag{10.4.1}$$

One matter that needs to be made clear is that increasing the diode current above threshold does not cause the gain coefficient, $g(\mathcal{E}_{ph})$, to continue increasing. This stabilizes at the value necessary to maintain equilibrium and is not likely to be very different from the threshold value. It is set by the same equation:

$$g(\mathcal{E}_{ph}) = \alpha_s + \frac{\ln(1/R_1R_2)}{2l} = \alpha_s + \frac{1}{2l}\ln(1/R) \tag{10.4.2}$$

in the case that $R_2 = 1$ and $R_1 = R$. The implication is that the gain coefficient and thus the carrier concentrations saturate above threshold and the additional optical power is generated almost entirely by stimulated emission processes as the radiation density builds up.

In Section 8.5 we saw that only a fraction of the light generated by spontaneous emission was able to get out of the semiconductor surface. This fraction was written as $(1-R)f'$. The transmitted fraction of the light generated by stimulated emission in a laser is much higher, because most of the power is directed at right angles to the output face and because it simply suffers multiple reflections until it does emerge. We shall write the transmission factor for stimulated emission as f''. The total emitted power above the lasing threshold may then be expressed as

$$\Phi = (1-R)f'\eta_{sp}\frac{I\mathcal{E}_{ph}}{e} + f''\eta_{st}\frac{\mathcal{E}_{ph}}{e}(I - I_{th}) \tag{10.4.3}$$

Here η_{sp} and η_{st} are the internal quantum efficiencies for the generation of spontaneous and stimulated radiation, respectively. In Section 8.5 η_{sp} was written as η_{int}; but it is necessary now to distinguish the two emission processes. Differentiation of eqn. (10.4.3) makes it clear that

$$\eta_D = \frac{e}{\mathcal{E}_{ph}}\frac{\Delta\Phi}{\Delta I} = (1-R)f'\eta_{sp} + f''\eta_{st} \simeq f''\eta_{st} \tag{10.4.4}$$

We may also relate f'' to the gain and loss coefficients, $g(\mathcal{E}_{ph})$ and α_s. The stimulated emission power generated per unit length of active medium is proportional to $g(\mathcal{E}_{ph})$ as was expressed in eqn. (10.1.13). Of this a fraction proportional to α_s is scattered out of the beam and lost, while the rest, which is proportional to $g(\mathcal{E}_{ph}) - \alpha_s$, eventually forms the laser output. Thus

$$f'' = \frac{g(\mathcal{E}_{ph}) - \alpha_s}{g(\mathcal{E}_{ph})} \tag{10.4.5}$$

$$= \frac{1}{[2\alpha_s l/\ln(1/R) + 1]} \tag{10.4.6}$$

using eqn. (10.4.2)

$$\therefore \quad \eta_D \simeq \eta_{st}\left[1 + \frac{2\alpha_s l}{\ln(1/R)}\right]^{-1} \tag{10.4.7}$$

Equation (10.4.7) shows that the differential quantum efficiency may be increased by increasing η_{st} or decreasing α_s. These are both desirable. Decreasing l or R also increases η_D but at the same time increases J_{th}. As a result the overall power conversion efficiency of the laser, Φ/I, may not improve.

In practice the optical power versus diode current characteristic is less well-behaved than Fig. 10.10 implies. The lasing part of the characteristic is often 'kinked', and the optical power may tend to pulsate. A great deal of development work has gone into eliminating these undesirable features, which are discussed further in Chapter 11.

PROBLEMS

10.1 Prove eqn. (10.1.3).

10.2 A double heterostructure GaAs/GaAlAs laser has a cavity length of 0.3 mm, a scattering loss coefficient $\alpha_s = 1\ \text{mm}^{-1}$, and uncoated facet reflectivities of 0.33.
(a) Calculate the reduction in the threshold gain coefficient that occurs when the reflectivity of one of the facets is increased to 100%.
(b) Using the curves shown on Fig. 10.9, estimate the reduction in J_{eff} that will be brought about at room temperature by this change.
(c) The active layer of the laser is 10 μm wide and 0.4 μm deep and its internal quantum efficiency is 0.7. Calculate the effect on the threshold current when the facet reflectivity is changed in the manner described.

10.3 If the emission wavelength of the laser of Problem 10.2 is 0.86 μm and its quantum efficiency for stimulated emission is 0.9, calculate the change in the slope of the output characteristic, $\Delta\Phi/\Delta I$, brought about by the reflectivity change.

10.4 A particular GaAs laser having a cavity length of 0.3 mm ($\mu = 4$) requires a gain coefficient of $6\ \text{mm}^{-1}$ in order to reach the lasing threshold. Using the theoretical gain curves shown in Fig. 10.8, estimate the number of longitudinal lasing modes excited by an effective current density, $J_{eff} = 60\ \text{A/(mm}^2\ \mu\text{m)}$ at 297 K. Explain qualitatively why it is that the increasing the current density does not cause the number of modes to increase in the way that Fig. 10.8 would indicate.

REFERENCES

10.1 H. Kressel and J. K. Butler, *Semiconductor Lasers and Heterojunction LEDs*, Academic Press (1978).

10.2 G. H. B. Thompson, *Physics of Semiconductor Laser Devices*, Wiley (1980).

SUMMARY

Laser action between two discrete, atomic energy levels requires

(a) a population inversion to give optical gain through stimulated emission.
(b) an optical cavity to provide positive optical feedback.

When the gain exceeds the transmission and scattering losses, self-oscillation, or lasing, occurs:

$$g_{12} = \frac{c^2}{8\pi}\frac{(n_2 - n_1)}{f_{21}^2 \tau_{sp}} \xi(f_{21}) > \alpha_s + \frac{1}{2l}\ln(1/R_1R_2)$$

The nature of the radiation then changes from being characteristic of the atomic energy levels to being characteristic of the cavity. It is dominated by stimulated rather than spontaneous emission and becomes highly monochromatic and coherent.

Optical gain may occur in any region of a semiconductor where the separation of the quasi-Fermi energies for electrons and holes exceeds the band-gap energy. This requires the injection of a high current density of minority carriers into a heavily doped region. Empirically,

$$g_{max} = \beta[J_{eff} - (J_{eff})_0] \qquad \text{where } J_{eff} = \eta_{int}J/d$$

There is a threshold current density, J_{th}, slightly greater than $(J_{eff})_0 d/\eta_{int}$, below which optical emission is predominantly spontaneous emission and above which it rises rapidly as a result of stimulated emission (see Fig. 10.10). The spectral characteristics change at the same time, as shown in Fig. 10.11

$$J_{th} = (J_{eff})_0 d/\eta_{int} + (\alpha_{tot}/\Gamma\beta)(d/\eta_{int}) = J_0 \exp(T/T_0)$$

T_0 is about 150 K for GaAlAs/GaAs lasers and about 70 K for InGaAsP/InP devices. In the lasing region $\Delta\Phi/\Delta I = \eta_D \mathcal{E}_{ph}/e$, where $\eta_D \simeq \eta_{st}[1 + 2\alpha_s l/\ln(1/R)]^{-1}$.

11

Semiconductor Lasers for Optical Communication

11.1 THE DEVELOPMENT OF STRIPE-GEOMETRY LASERS

In Chapter 10 we have given a basic, theoretical discussion of the principles of laser action in semiconductors. The intention of the present chapter is to describe in more detail some of the designs of semiconductor injection laser that have been proposed for use as optical communication sources. We shall wish to examine their electrical and optical characteristics. The stringency of the requirements put on a laser source for communications should not be underestimated. They may be summarized as follows:

(a) continuous or almost continuous operation at room temperature and above;
(b) an operating life of the order of 100 000 hours (12 years) during which the characteristics should not degrade excessively, and the probability of failure should be low;
(c) a low threshold current (I_{th});
(d) a high modulation bandwidth;
(e) linearity of the optical output power with current above threshold;
(f) a small emissive area so that efficient coupling into the fiber may be obtained;
(g) short-term and long-term temporal stability of the optical output power;
(h) high radiance;
(i) narrow spectral width.

We shall be concerned to examine to what extent it is possible for these requirements to be met.

It has to be said that the reliability of the optical source has been the cause of the greatest uncertainty during the initial development period of optical fiber communication systems and has been the factor most likely to delay the introduction of fibers into public telephone networks. The earliest semiconductor injection lasers failed after only a few minutes, and even when the first double heterostructure designs were produced, lives of some tens of hours were normal. Anyone who has been involved with the design or development of electronic devices will know that there are always many ways in which any given device may fail or be somehow unsatisfactory. The goal is always to find a design and a manufacturing method that will produce a high yield of 'good' devices. What has been unusual about the development of semiconductor lasers has been the extent to which the many failure mechanisms have been analyzed in detail in the published literature. There has been no shortage either of detailed designs of double heterostructure lasers aimed at eliminating one or other of the failure modes or otherwise improving the operating characteristics. This is perhaps a measure of the seriousness of the problem. At the time of writing, this story is far from complete, so it would be inappropriate for us either to pursue all of the many possible designs and their properties, or to emphasize one in such a way as to imply that it was the definitive form. We shall try to give a general idea of the problems and some of the proposed solutions. Readers who wish to pursue these matters further are referred again to the specialist texts [7.1], [10.1] or [10.2], or to one of the many review articles. Reference [11.1] is one of the many that may be found helpful.

Semiconductor lasers were first fabricated in 1962 using GaAs. They took the form of diffused p–n homojunctions. There was, thus, no external means of providing optical or carrier confinement and the carrier diffusion lengths determined the active-layer thickness. The room-temperature threshold current density of these early devices was of the order of 300–500 A/mm^2, and only low duty-cycle pulsed operation was possible. Lowering the junction temperature reduced the threshold current density, which became 5–10 A/mm^2 at liquid nitrogen temperature (77 K). The highest ambient temperature at which a well-cooled device could be operated continuously was found to be in the region of 200 K.

During the period 1968–70 a number of workers developed the techniques described in Chapter 9, whereby GaAlAs/GaAs heterojunctions could be formed on GaAs substrates. The quality of these junctions was sufficiently good for the internal quantum efficiency to remain high and for laser action to be possible. A typical, single-heterojunction (SH) laser consisted of a p–n junction formed in GaAs with a p–P heterojunction between GaAs and GaAlAs spaced at a distance of a few microns from it. As a result of the carrier and optical confinement that occurred at the heterojunction, the room-temperature threshold current density was reduced to about 100 A/mm^2.

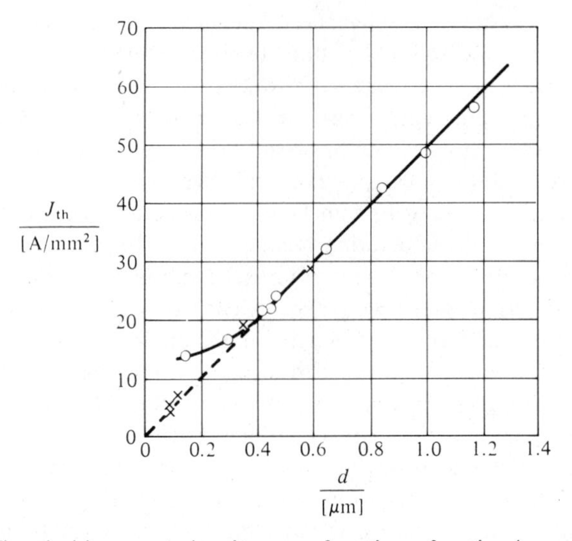

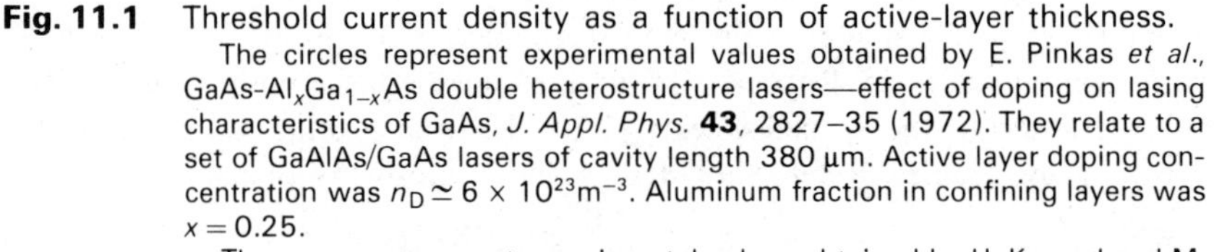

Fig. 11.1 Threshold current density as a function of active-layer thickness.
The circles represent experimental values obtained by E. Pinkas *et al.*, GaAs-Al_xGa_{1-x}As double heterostructure lasers—effect of doping on lasing characteristics of GaAs, *J. Appl. Phys.* **43**, 2827–35 (1972). They relate to a set of GaAlAs/GaAs lasers of cavity length 380 μm. Active layer doping concentration was $n_D \simeq 6 \times 10^{23} m^{-3}$. Aluminum fraction in confining layers was $x = 0.25$.
The crosses represent experimental values obtained by H. Kressel and M. Ettenberg, Low-threshold double heterojunction AlGaAs/GaAs laser diodes: theory and experiment, *J. Appl. Phys.* **47**, 3533–7 (1976). They relate to diodes of active area approximately 500 × 100 μm, having a doping concentration of about $5 \times 10^{22} m^{-3}$ (undoped). In the confining layers $x = 0.65$.

The ability to make double heterostructures soon followed, and their improved confinement caused a further reduction of the threshold current density. Equation (10.3.5) indicated that J_{th} should be proportional to the thickness of the active layer, d. The circles in Fig. 11.1 show how this was borne out in practice in a range of lasers that were otherwise similar. As the ratio d/λ is reduced below about 0.5, the optical confinement factor, Γ, starts to decrease. This gives rise to a minimum value of J_{th} which in this example would be about 12 A/mm² at a thickness of about 0.1 μm. Such reduced current densities make continuous operation possible at room temperature and above, and thus satisfy the first of the requirements listed earlier. Even smaller values of J_{th} can be obtained if the refractive index step at the heterojunctions is made larger so that still thinner active layers can be used without decreasing Γ. Values as low as 5 A/mm², with d about 0.1 μm, have been reported to occur in diodes using $Ga_{0.35}Al_{0.65}As$ for the confining layers. Then the refractive-index change at the active layer/confining layer heterojunction is 0.4. Some results for such diodes are shown by the crosses in Fig. 11.1.

The discussion given in Chapter 10 was essentially a one-dimensional theoretical analysis of semiconductor laser action in the sense that we neglected totally any lateral variation in the active layer in the plane parallel to the junctions. This would seem to be reasonable for a typical, broad-area, sawn-cavity device in which the junction area might be 0.1 mm wide by 0.5 mm long ($w \times l$). However, there are several factors which make it desirable to reduce the active layer width. First the operating current is reduced in direct proportion. Thus, if J_{th} remained constant at 20 A/mm^2 and $l = 0.5$ mm, a reduction of w from 100 μm to 10 μm would reduce the threshold current from 1 A to 100 mA. At this level the driving circuitry is greatly simplified, particularly at high frequency. Secondly, the reduced emissive area ($w \times d$) better matches the fiber core diameter with the result that the coupling efficiency is improved. Thirdly, it is found that the linearity of the output power with driving current can be improved. These advantages have led to the widespread use of the 'stripe geometry' for communications lasers. A similar technique is used for the edge-emitting LED illustrated in Fig. 9.18.

The stripe contact may be brought about in a number of ways and some are illustrated in Fig. 11.2. When oxide isolation, proton bombardment or zinc diffusion is used to define the stripe, as in Fig. 11.2(a)–(c), there is only limited confinement in the lateral direction of either the optical power or the carriers. This gives rise to a significant increase in J_{th} for stripe widths less than about 10 μm. Such lasers are said to be *gain-guided*, because the light is channelled in the region of greatest population inversion, which in turn is set up by the distribution of the current density. Much stronger lateral confinement is provided by the configuration shown in Fig. 11.2(d) and (e) which are types of *buried heterostructure* (BH) laser. These are said to be *index-guided*, because the lower refractive index of the regrown layer of GaAlAs acts as a lateral waveguide and confines the optical power just as the lateral heterojunction confines the carriers. The structure shown in Fig. 11.2(d) may be made by producing a normal DH laser, etching back to the n-GaAlAs layer on either side of the active stripe and then re-growing n- and p-GaAlAs layers onto the etched-away region. With buried heterostructures, stripe widths as small as 2 μm have been produced and threshold currents of less than 10 mA have been obtained, but then the total output power into air is limited to 1–2 mW. Buried heterostructure lasers, index-guided into a stable lasing mode, offer better temporal stability and greater linearity of the optical output power. It is easier to operate them in a single longitudinal mode. For these reasons they have become important for fiber-optic communication even though they are technologically more complex and difficult to make.

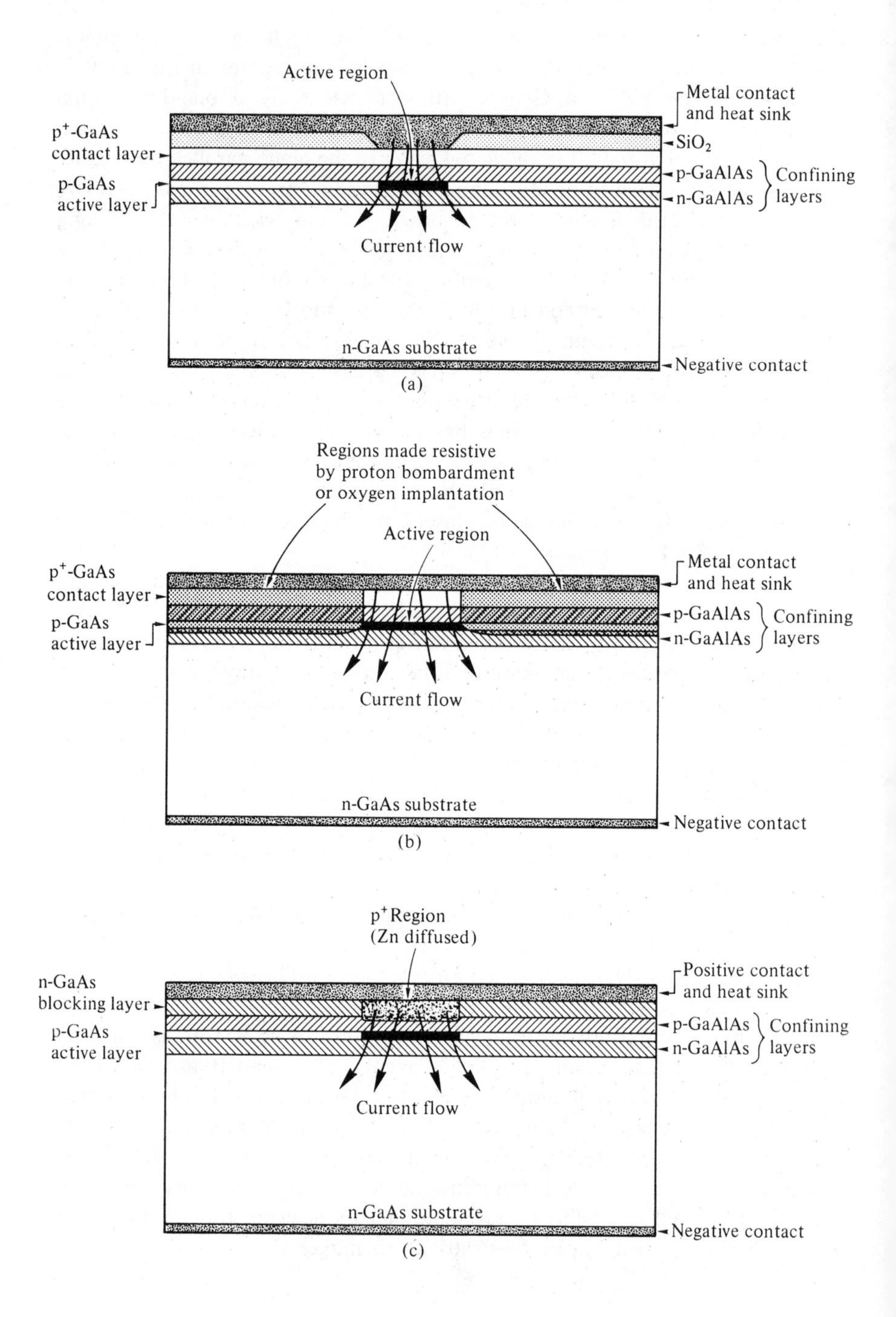
Active region
Metal contact and heat sink
p⁺-GaAs contact layer
SiO_2
p-GaAs active layer
p-GaAlAs
n-GaAlAs
Confining layers
Current flow
n-GaAs substrate
Negative contact
(a)
Regions made resistive by proton bombardment or oxygen implantation
Active region
Metal contact and heat sink
p⁺-GaAs contact layer
p-GaAs active layer
p-GaAlAs
n-GaAlAs
Confining layers
Current flow
n-GaAs substrate
Negative contact
(b)
p⁺ Region (Zn diffused)
n-GaAs blocking layer
p-GaAs active layer
Positive contact and heat sink
p-GaAlAs
n-GaAlAs
Confining layers
Current flow
n-GaAs substrate
Negative contact
(c)

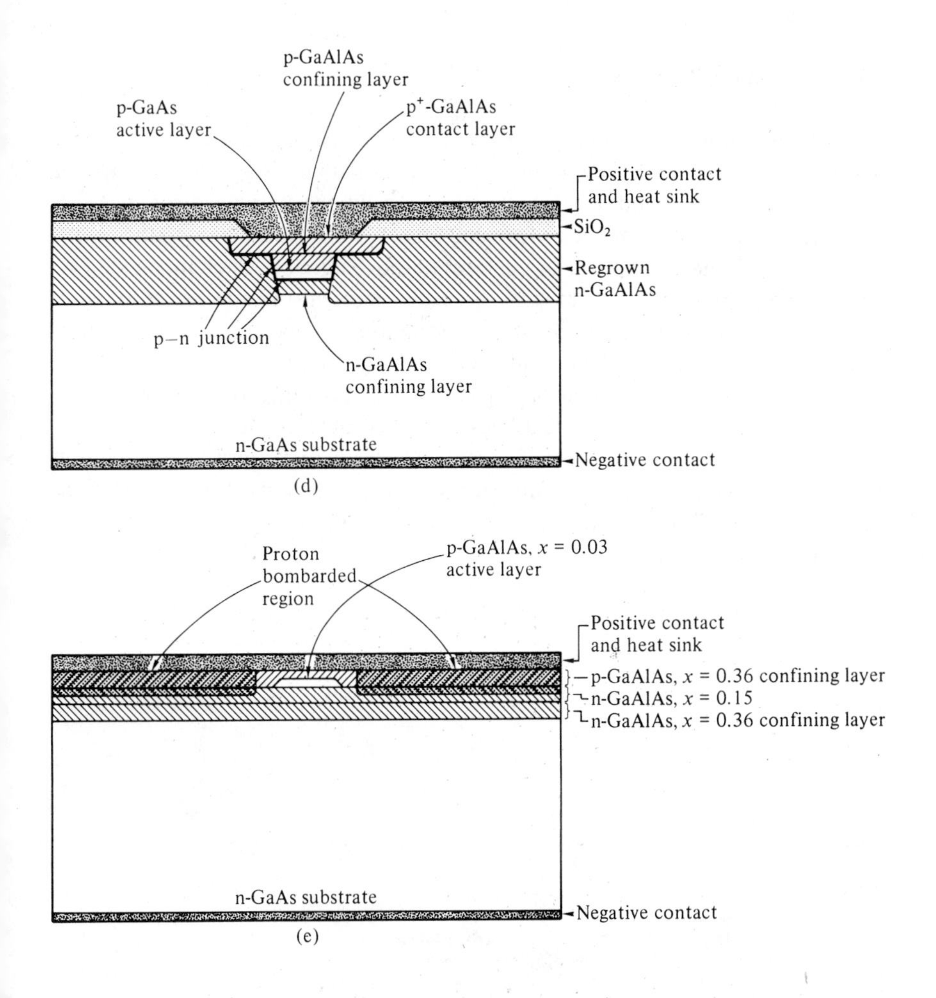

Fig. 11.2 Schematic cross-sections illustrating some of the techniques used to produce narrow stripe-geometry lasers: (a) oxide isolation; (b) isolation by proton bombardment or oxygen implantation; (c) use of a deep zinc diffusion; (d) buried heterostructure; (e) modified-strip buried heterostructure.

Although all the illustrations are based on GaAlAs/GaAs lasers, the methods can be adapted to produce long wavelength lasers using the InGaAsP/InP system.

11.2 OPTICAL AND ELECTRICAL CHARACTERISTICS OF STRIPE-GEOMETRY AND BURIED HETEROSTRUCTURE LASERS

11.2.1 Laser Modes

The active region of a stripe-geometry laser is essentially a rectangular box, and this forms a resonant cavity for the laser radiation. As such it will support a number of cavity modes of oscillation, each having its own resonance frequency. Those mode frequencies for which there is net overall gain in the cavity will be excited and will appear in the laser output radiation. Each mode may be characterized by three integers (i, j, k) which represent the number of maxima in the electromagnetic energy-density distribution in the three perpendicular directions of the cavity. This is illustrated in Fig. 11.3.

The break-up of the broad spectrum of the spontaneous emission into a number of narrow, evenly spaced lines was illustrated in Fig. 10.11. The lines represent a set of longitudinal mode frequencies. In simple terms the condition for resonance is that an integral number of half-wavelengths should fit between the cavity boundaries. In fact matters are complicated by the nature of the

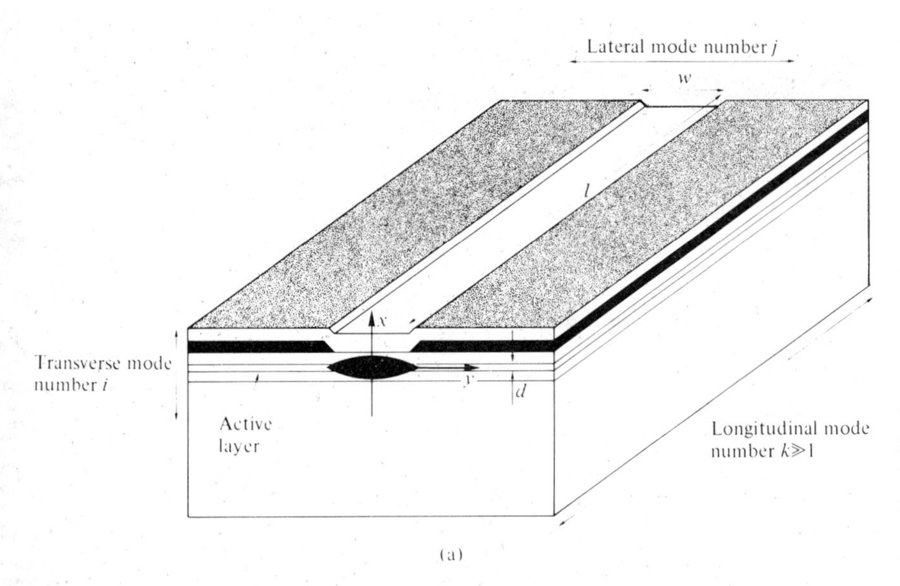

Fig. 11.3 Cavity modes: (a) a schematic diagram of a stripe-geometry laser showing the designation of and power density distribution of the lowest order of the cavity modes.

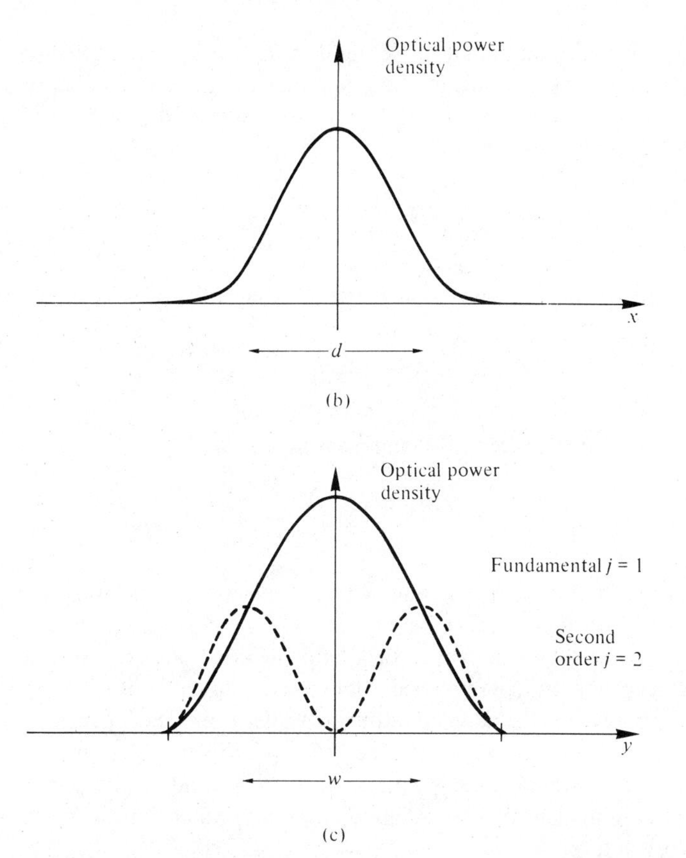

Fig. 11.3 Cavity modes: (b) schematic near-field distributions for the fundamental transverse mode, $i = 1$; (c) schematic near-field distribution for the fundamental and second-order lateral modes, $j = 1, 2$.

dielectric change at the boundary, by the gain profile within the cavity, by the presence of free carriers which cause a local reduction in the refractive index, and by local increases in temperature which increase the refractive index. Neglecting these effects it is easy to show that the mode frequency separation will be inversely proportional to the cavity length. The condition for longitudinal resonance is

$$k\frac{\lambda_k}{2} = \mu_g l \tag{11.2.1}$$

where k is the longitudinal mode number, λ_k the free-space wavelength of the radiation of the kth mode, μ_g the group index of the semiconductor material at this wavelength and l the length of the laser cavity. Putting $f_k = c/\lambda_k$ gives the frequency of the kth mode as

$$f_k = \frac{kc}{2\mu_g l} \tag{11.2.2}$$

Thus the frequency separation between adjacent modes is

$$\Delta f = f_k - f_{k-1} = \frac{c}{2\mu_g l} \tag{11.2.3}$$

In terms of free-space wavelength assuming $\Delta f \ll f_k$,

$$\Delta\lambda = \lambda_k \frac{\Delta f}{f_k} = \frac{\lambda_k^2}{2\mu_g l} \tag{11.2.4}$$

With $\mu_g = 4$ for GaAs, a laser with a cavity length of 500 μm would have $\Delta f = 75$ GHz and $\Delta\lambda \simeq 0.2$ nm.

The active-layer thickness of double-heterostructure lasers is invariably much less than 1 μm. As a result, only the fundamental transverse cavity mode can normally be excited. The number of lateral modes depends principally on the width of the cavity, but the method used to define the stripe width is also important: as we have seen, with oxide isolation and proton bombardment the lateral confinement of the radiation is much weaker than it is with the buried heterostructure.

When stripe widths exceed about 20–30 μm a fairly sharp lasing threshold is usually obtained and above threshold higher-order modes are excited in steadily increasing numbers. The result is a fairly linear rise of optical power with driving current until increasing temperature leads to some saturation. Matters are not quite as simple as implied here because the optical gain profile does cause some self-focussing of the radiation which tends to concentrate into a filament within the laser cavity. Reduction of the stripe width below about 20 μm increases the higher-order mode losses. Then, just above the lasing threshold only the fundamental lateral mode appears. After some further increase in the driving current higher-order lateral modes appear as their higher threshold current densities are reached. Further reduction of the stripe width below about 10 μm increases the threshold currents for the higher-order modes to levels beyond the capacity of the laser. The presence of higher-order lateral modes affects both the distribution and the spectrum of the laser radiation.

The distribution of the output radiation may be characterized in two ways.

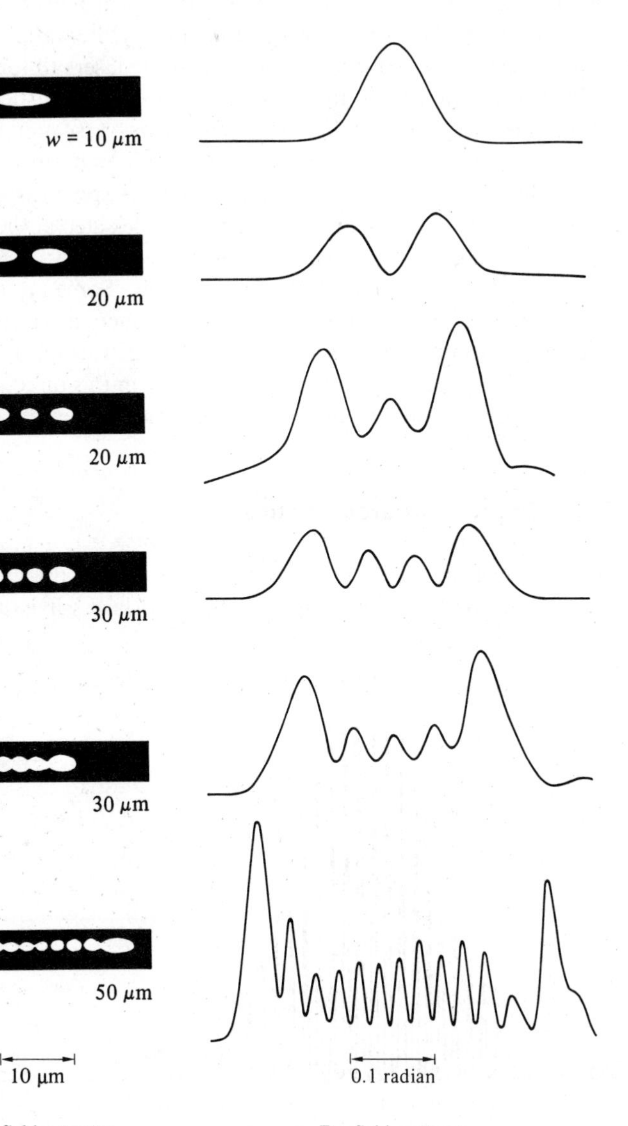

Fig. 11.4 Near-field and far-field intensity patterns as a function of stripe width. [Taken from H. Yonezu *et al.*, A GaAs-$Al_xGa_{1-x}As$ double heterostructure planar stripe laser, *Jpn. J. Appl. Phys.* **12**, 1585–92 (1973). Reproduced with the permission of the publisher.]

The *near-field* distribution refers to the variation of power density across the output facet of the laser as shown in Fig. 11.3. Observation of the near-field distribution shows the tendency of broader stripe lasers to form filaments. The *far-field* distribution refers to the directional characteristics of the emitted radiation. The fact that only the fundamental transverse mode is excited is confirmed by single-peaked near- and far-field patterns in the direction perpendicular to the junction plane. The angular spread is a function of the thickness of the active layer and the refractive index step at the heterojunctions. Typically, the range of directions over which the intensity exceeds half the peak intensity encloses an angle of 40° approximately. Some typical near- and far-field distributions for directions parallel to the junction when the fundamental and higher-order lateral modes are excited are illustrated in Fig. 11.4. It will be observed that the beam width is much narrower in this direction, usually 5° to 10° to half the maximum intensity in the lowest-order modes.

11.2.2 Spectral Characteristics

The major effect on the spectral characteristics is the number of longitudinal modes excited. With narrow-stripe, gain-guided lasers it is usual to find that

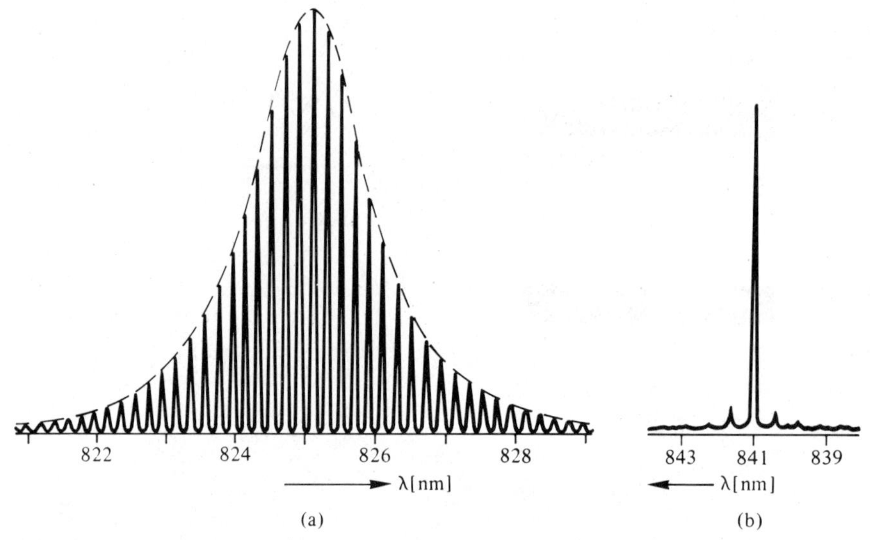

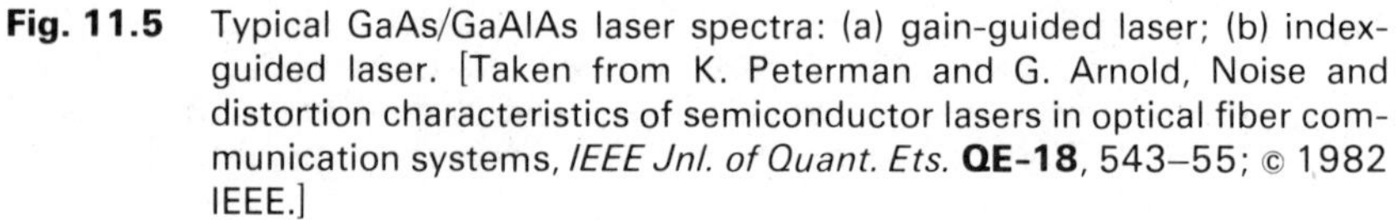

Fig. 11.5 Typical GaAs/GaAlAs laser spectra: (a) gain-guided laser; (b) index-guided laser. [Taken from K. Peterman and G. Arnold, Noise and distortion characteristics of semiconductor lasers in optical fiber communication systems, *IEEE Jnl. of Quant. Ets.* **QE-18**, 543–55; © 1982 IEEE.]

many modes are excited and a rather broad spectral line-width is obtained. This is shown in Fig. 11.5(a). With index-guided lasers several modes are usually present just above the lasing threshold, but as the driving current is increased, one or two modes come to dominate the others, as shown in Fig. 11.5(b). Reducing the cavity length increases the spacing between the mode frequencies so that fewer can be supported within the envelope of optical gain. For this reason shorter cavities of length 100 μm or less are favored, and there is then the possibility of operating in a single longitudinal mode at the higher output levels.

The spectral line width of an individual single mode has frequently been quoted as being 0.01 nm or less. However, in many cases the value has been governed by the limiting resolution of the measuring instrument! More sophisticated measurements, based on coherence length,† have indicated linewidths of this order (i.e. 0.01 nm or a few GHz) with gain-guided lasers, but as low as 10^{-4} nm (30 MHz) with index-guiding. There are theoretical grounds for predicting these orders of magnitude. Multiple lateral modes would be expected to give rise to a general broadening of each of the longitudinal mode lines, or perhaps to split each into a subset of lines.

11.2.3 Power and Voltage Characteristics

Typical characteristics of output power and diode voltage versus current for a gain-guided, GaAlAs/GaAs laser of stripe width about 10 μm are shown in Fig. 11.6(a). The lasing threshold for such devices tends to become 'softer' as the stripe width decreases, and the laser output above threshold still contains a high proportion of spontaneous emission. This is a factor which tends to broaden the spectral width of the individual laser modes. Below threshold the V–I curves can normally be fitted to eqn. (9.1.9) with $a \simeq 2$, provided that an allowance is made for ohmic resistance within the diode. This is usually in the range 1–10 Ω. The voltage drop at normal bias currents is usually between 1.5 V and 2.0 V for GaAlAs/GaAs lasers and around 1.2 V for InGaAsP/InP devices. Detailed analysis of the V–I curves provides a good guide to the laser behavior, as can be seen from the sets of curves for an index-guided laser shown in Fig. 11.6(b). Above threshold the junction voltage tends to saturate, an effect that can be seen clearly in the curve of $I\,dV/dI$ versus I. Kinks in the output characteristic like those shown in Fig. 11.6(c) are also reflected in the characteristics of the voltage and its derivatives.

† The coherence length l_c is the distance over which the electromagnetic waves from a source remain in phase with one another. It is related to the spectral spread of the source, $\Delta f = (c/\lambda^2)\,\Delta\lambda$, by $l_c \simeq c/2\pi\,\Delta f = \lambda^2/2\pi\,\Delta\lambda$ (cf. Section 15.4.4).

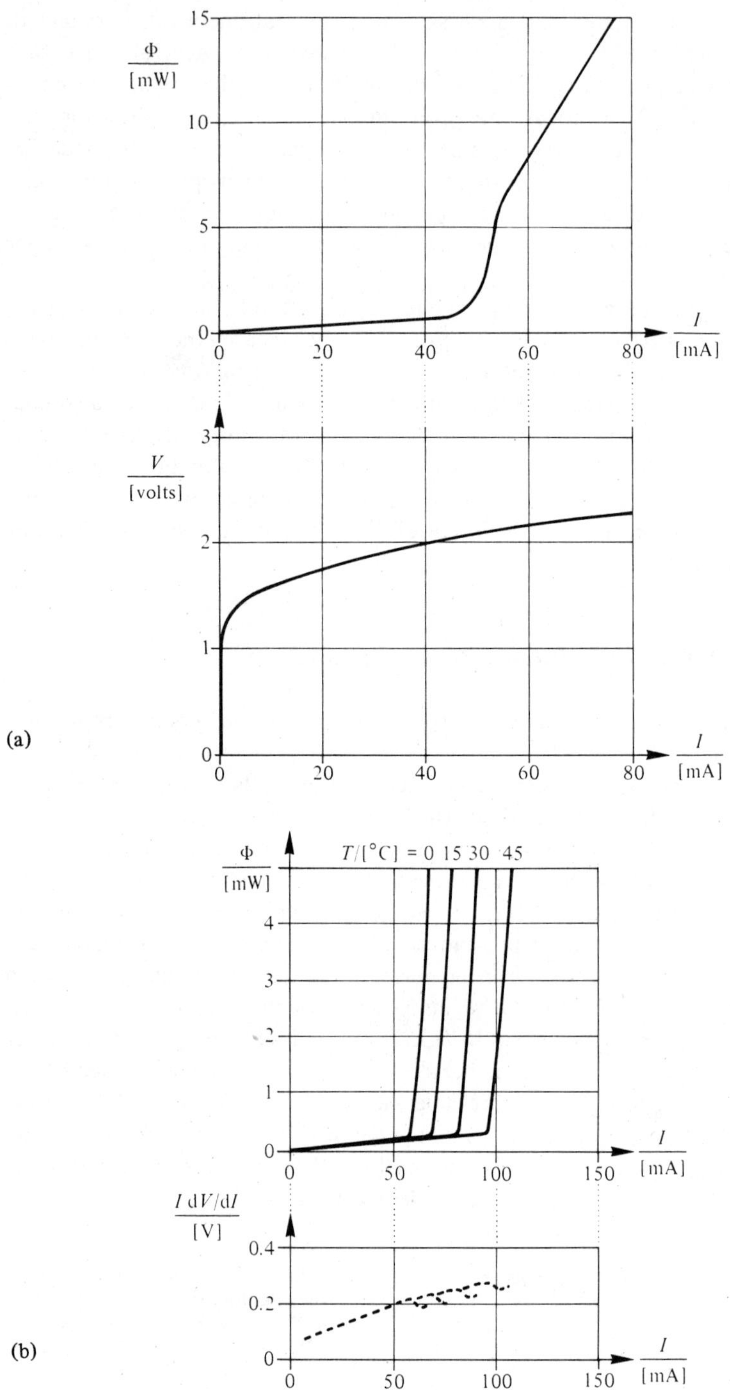

Φ
[mW]
15
10
5
0
0
20
40
60
80
I
[mA]
V
[volts]
3
2
1
0
0
20
40
60
80
I
[mA]
(a)
Φ
[mW]
T/[°C] = 0 15 30 45
4
3
2
1
0
0
50
100
150
I
[mA]
I dV/dI
[V]
0.4
0.2
0
0
50
100
150
I
[mA]
(b)

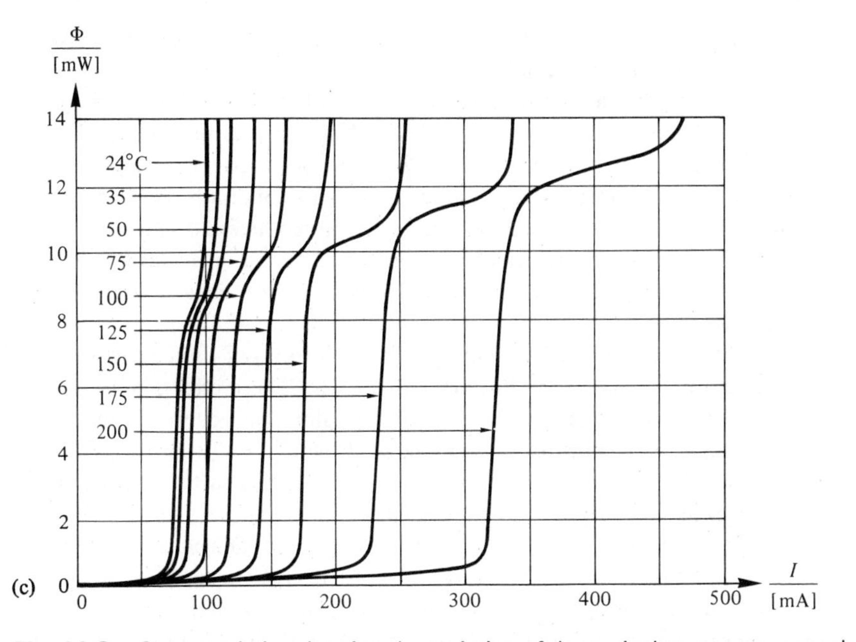

Fig. 11.6 Characteristics showing the variation of the optical output power and the diode voltage and its derivative with current for typical stripe-geometry AlGaAs/GaAs lasers: (a) a gain-guided laser with $l =$ 130 μm and $w = 12$ μm; (b) an index-guided laser; (c) a proton-bombarded laser of stripe-width 12 μm showing the effect of temperature on the 'kink'. [Part (c) taken from R. L. Hartman and R. W. Dixon, Reliability of DH GaAs lasers at elevated temperatures, *Appl. Phys. Letts*, **26**, 239–40 (1 Mar. 1975).]

The kinks of Fig. 11.6(c) are typical of gain-guided lasers having stripe widths approximately in the range 10–20 μm. They are a serious embarrassment when stable, linear operation is required. They make pulsed operation difficult and analogue working impossible. Their origin is associated with a gross change in the lateral mode distribution. At the kink a new lasing mode, one having more loss but a gain versus current characteristic of greater slope than the lower power mode, takes over the laser oscillation. During the transition the output power rises more slowly with increasing current, or it may even fall, until it again increases normally once the new mode is established. The kink occurs because with gain-guided lasers the channelling is very weak and can easily be overcome by the reduction of refractive-index caused by high carrier concentrations. The use of very narrow stripes (< 10 μm) overcomes the problem because the losses at points away from the center of the channel are increased. The basic cause of instability is not removed, but the level at which it occurs is shifted to higher power levels, ones lying above the normal operating range of the laser. Index-guiding, on the other hand, enables the laser oscillation to be positively channelled into the required fundamental mode. Because the lateral refractive-index steps in the buried heterostructure of Fig. 11.2(d) are quite large, very narrow stripes of 1–2 μm are required if operation is to be confined to the fundamental lateral mode only. This condition is relaxed somewhat to 4–5 μm in the so-called modified-strip buried-heterostructure (MSBH) of Fig. 11.2(e). In either case kinks are eliminated as in Fig. 11.6(b) and linear output characteristics are obtained—but this is at the cost of significantly reduced output power.

Lasers have also shown a tendency to develop self-oscillation or self-pulsation at frequencies from 200 MHz to 2 GHz. These effects must be distinguished from the relaxation oscillations or ringing that occurs at the start of a laser pulse, which will be discussed in the next section. Pulsations would often arise during life test and were associated with defects which caused local regions of excess absorption in the optical channel. These defects delay the onset of laser oscillation until high radiation and carrier densities have built up in the channel. When lasing does start the excess energy is rapidly dumped into the laser output. This leads to a depopulation of the excited states, which in turn quenches the output until the overall channel gain builds up once more to the lasing threshold. Then the cycle is repeated. In laser jargon this may be thought of as a regular, self-imposed Q-switching process. Improved control over material quality has minimized the occurrence of such defects, and hence of the pulsations. However, the fact that pulsations do occur highlights the inherent instability of the laser action against non-uniformities within the laser cavity.

A problem that remains is the effect of temperature changes in the active region. The effect on the lasing threshold was discussed in Section 10.3. This and other changes in the characteristics can be seen in Figs. 11.6(b) and (c). At sufficiently high temperature the diode fails to lase at all. Change of temperature also causes the individual longitudinal mode wavelengths and the power distribution among the modes to change. Thus the spectral characteristics of the light can be seen to vary both as the average power level varies and with time during pulsed operation. Often the bulk of the optical power transfers from one longitudinal mode to another during a pulse, an effect known as mode-hopping.

11.2.4 Frequency Response

The final matter that needs to be discussed is the response of the laser to high-frequency modulation of the driving current. We shall first consider the case of a laser biassed beyond threshold, but having a sinusoidal modulation superimposed on the steady current. In most types of laser, apart from those having very narrow stripes, the modulated optical power is found to go through a resonance like the ones shown in Fig. 11.7. As can be seen the resonant frequency depends on the bias level, but is usually in the frequency range 0.2–2.0 GHz. The cause of this resonance is the interaction between the optical power density and the excess carrier density in the cavity. Both of these represent ways in which energy may be stored, and this energy may be exchanged between them. The similarity to other resonant systems will be obvious. In particular we may make a comparison with an electrical resonance circuit in which stored energy is exchanged between the inductive components and the capacitative components. Indeed the characteristic responses shown in Fig. 11.7 are similar to those of the parallel resonance circuit of Fig. 11.8.

The origin of the laser resonance may be understood as follows. An increasing current leads to an increase in the carrier concentrations after a short time delay. This in turn leads to an increase in the recombination radiation density, which, after a further delay, stimulates recombination and causes the carrier concentrations to fall. Because of the time delays, the fall overshoots the equilibrium value and an oscillation occurs. The natural frequency of the

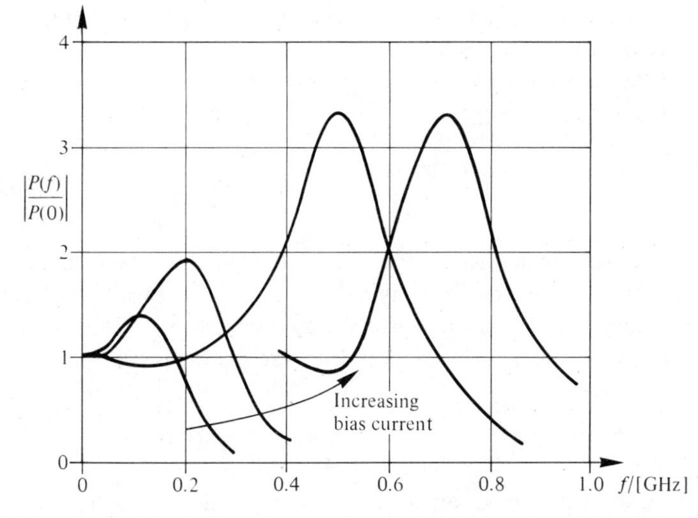

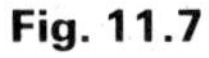
Fig. 11.7 Typical variation of modulated optical power with modulating frequency.

The resonance increases and occurs at higher frequencies as the bias level is increased above the lasing threshold.

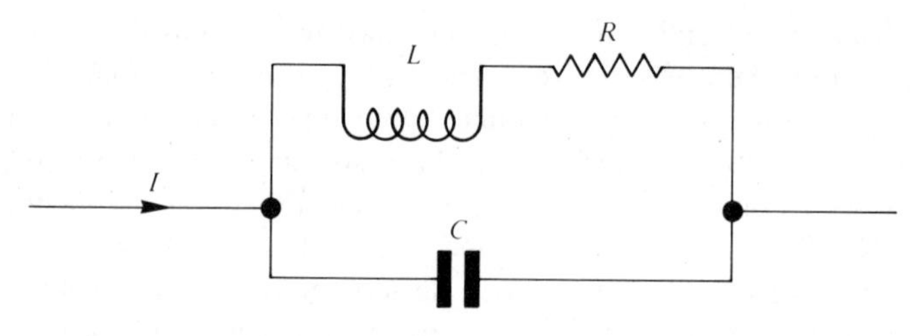

Fig. 11.8 Parallel resonance circuit.

system, f_0, is a function of the optical decay time constant, τ_{ph}, and the carrier recombination time constant, τ_{sp}. However, the interactions are nonlinear, so that the analysis is complicated and the resonant frequency depends, as can be seen in Fig. 11.7, on the amount by which the laser bias level, I_0, exceeds the threshold current, I_{th}. The curves can be fitted approximately to a theoretical relation of the form

$$\frac{P(\omega)}{P(0)} = \frac{\omega_0^2}{(\omega_0^2 - \omega^2) + j\beta\omega} \tag{11.2.5}$$

where $\omega = 2\pi f$, $\omega_0^2 = (I_0 - I_{th})/\tau_{sp}\tau_{ph} I_{th}$, and $\beta = I_0/\tau_{sp} I_{th}$.

A brief derivation of this equation is given in Appendix 6 and fuller discussions will be found in Chapter 17 of Ref. [10.1] and in Ref. [11.2].

The optical time constant, τ_{ph}, is essentially the average photon lifetime within the cavity and may be derived as follows: In eqn. (10.1.18) the photon losses were expressed in terms of the effective fractional loss of optical power per unit length of the cavity:

$$\alpha_{tot} = -\frac{1}{P}\frac{dP}{dz} = \alpha_s + \frac{1}{2l}\ln(1/R_1 R_2) \tag{10.1.18}$$

where α_s is the scattering loss coefficient, l is the length of the cavity, R_1 and R_2 are the facet reflectivities. We may express this just as easily as a fractional loss per unit time, the reciprocal of which would represent the exponential time constant, τ_{ph}, of the decay of radiation if it were not being replenished. Thus,

$$\frac{1}{\tau_{ph}} = -\frac{1}{P}\frac{dP}{dt} = \frac{c}{\mu}\alpha_{tot} = \frac{c}{\mu}\left(\alpha_s + \frac{1}{2l}\ln(1/R_1 R_2)\right) \tag{11.2.6}$$

From the discussion in Chapter 10 we may take as a typical value $\alpha_{tot} = 2000\ \text{m}^{-1}$. Then, with $\mu = 3.6$, $\tau_{ph} = 6$ ps. From the discussion in

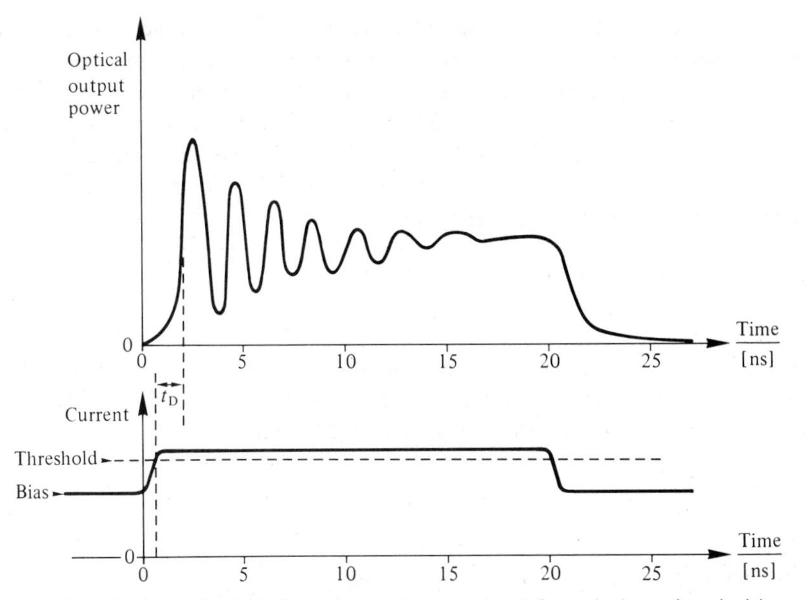

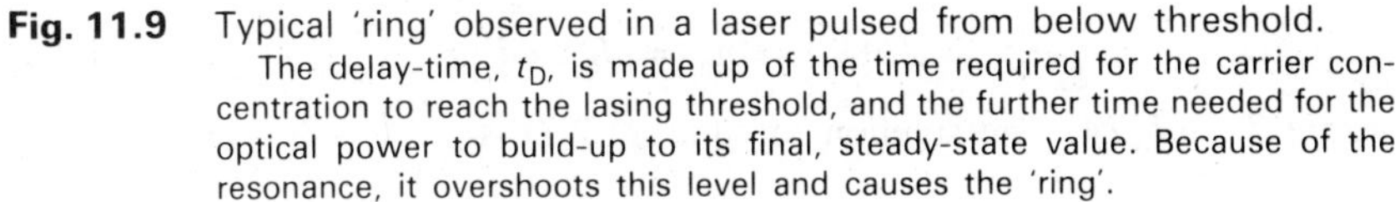

Fig. 11.9 Typical 'ring' observed in a laser pulsed from below threshold. The delay-time, t_D, is made up of the time required for the carrier concentration to reach the lasing threshold, and the further time needed for the optical power to build-up to its final, steady-state value. Because of the resonance, it overshoots this level and causes the 'ring'.

Chapter 8 we may take 10 ns to be a typical value for τ_{sp} so that with $I_0 = 1.2I_{th}$ we would expect the resonant frequency, $f_0 = \omega_0/2\pi$ to be about 0.3 GHz.

The second situation in which the radiation/carrier resonance is important is during switch-on. We shall examine the case where an instantaneous current step is applied to the laser diode. This may be from zero current or it may be superimposed on a steady bias current. In turn this bias may be either below or just above the lasing threshold. Because of the laser resonance just discussed it is no surprise that following such a current step the laser output normally exhibits a 'ring', as shown in Fig. 11.9. Although the upper current level has some effect in increasing the ring frequency and narrowing the optical spectrum, the level of the bias is much more significant. As the bias current is raised from zero towards and through threshold, the delay-time, t_D, is decreased, the level of ringing is steadily damped out, and the average spectral width observed during a pulse is much reduced. For all these reasons it has become normal practice to provide a bias level that is maintained on or just above the lasing threshold.

A simple theory can be used to relate the delay time t_D to the bias current I_0, the threshold current I_{th} and the effective recombination time constant, τ, defined in eqn. (9.2.6). In fact measurement of t_D has been used as a means of estimating τ. The current is assumed to increase sharply from I_0 to some new constant value I above the lasing threshold. The electron current density $J = I/A$ entering the active layer of the double heterostructure causes the electron concentration to increase according to

$$\frac{dn}{dt} = \frac{J}{ed} - \frac{n}{\tau} \tag{11.2.7}$$

A is the area of the active layer and d its thickness, and we assume no leakage over the second heterojunction and a negligible hole current. The initial equilibrium value of the electron concentration will be

$$n(0) = \frac{J_0 \tau}{ed} \tag{11.2.8}$$

and the concentration at threshold will be

$$n(t_D) = n_{th} = J_{th} \tau / ed \tag{11.2.9}$$

where $J_{th} = I_{th}/A$. The solution to eqn. (11.2.7) is

$$n(t) \frac{ed}{\tau} = J - (J - J_0) \exp(-t/\tau) \tag{11.2.10}$$

Substituting eqn. (11.2.9) into eqn. (11.2.10) gives

$$J_{th} = J - (J - J_0) \exp(-t_D/\tau) \tag{11.2.11}$$

That is,

$$t_D = \tau \ln\left[\frac{(J - J_0)}{(J - J_{th})}\right] = \tau \ln\left[\frac{(I - I_0)}{(I - I_{th})}\right] \tag{11.2.12}$$

Clearly as $I_0 \to I_{th}$ we shall expect that $t_D \to 0$. This analysis has assumed that τ is constant during the increase in electron concentration. If the rate of recombination is proportional to n^2 rather than n, as under the conditions of high injection discussed in Section 9.2, then it may be shown that

$$t_D = \tau \left(\frac{I_{th}}{I}\right)^{1/2} \left[\tanh^{-1}\left(\frac{I_{th}}{I}\right)^{1/2} - \tanh^{-1}\left(\frac{I_0}{I}\right)^{1/2}\right] \tag{11.2.13}$$

It has been found that gain-guided lasers with a very narrow stripe width, say less than 5 μm, do not show the relaxation oscillations of Fig. 11.9. It has already been stated that these have a much softer lasing threshold and that the output radiation contains a much higher proportion of spontaneous emission. Such devices behave rather more like superluminescent diodes. As a result the resonance is largely damped out by the spontaneous emission losses and the step characteristic is more like that of the DH LED, with the rise-time governed by the carrier lifetime, τ_{sp}. This is illustrated in Fig. 11.10. Even with narrow stripe lasers which do not 'ring', it is desirable to provide a bias near to the lasing threshold in order to minimize the voltage excursions and hence the drive currents needed to charge the diode capacitance.

In all types of laser the magnitude of the temperature variation of the output characteristic (as shown, for example, in Fig. 11.6) and similar variations that occur during the life of the laser, mean that some care has to be taken in order to maintain the bias level near to the lasing threshold and the upper level at the required output power. In optical fiber systems it has become normal practice to allow the unwanted radiation from the rear facet of the laser to fall onto a photodiode, and to use the current generated there to provide control of both the bias and the output levels. A typical feedback system is shown in Fig. 11.11. The pulse currents required to drive narrow-stripe, DH laser diodes biassed near to threshold are of the order of 10 mA or so and may be provided by conventional bipolar integrated circuits and transistors for all except the highest frequencies. Then, GaAs MESFETs may be used at the output stage, allowing bit rates in excess of 1 Gb/s.

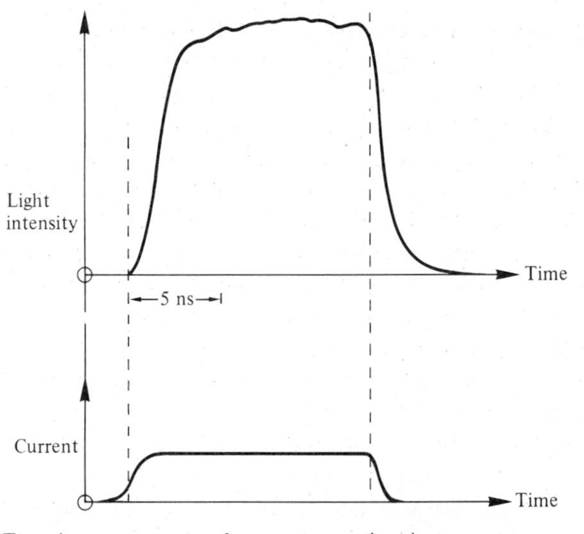

Fig. 11.10 Transient response of a narrow stripe laser.

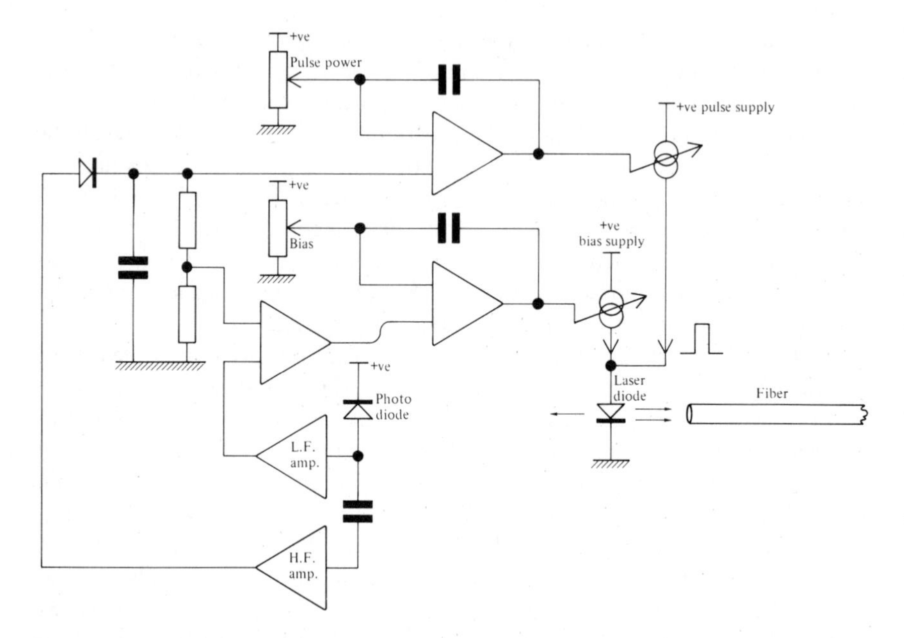

Fig. 11.11 A typical drive circuit for a laser diode with direct feedback control of the threshold level and the pulse power.

11.3 SOURCES FOR LONGER WAVELENGTHS

The discussion in this chapter, as in the immediately previous ones, has concentrated on GaAlAs/GaAs sources. This reflects their predominance in the first generation of optical communication systems. They emit over the wavelength range 0.8–0.9 μm as the proportion of Al, if any, in the active layer is varied. In order to make use of the beneficial dispersion and attenuation minima that occur in silica fibers, longer wavelengths are required and this means that sources made from the narrower band-gap semiconductors discussed in Chapter 7 have to be developed. The system that has been given the greatest attention so far, for lasers as for LEDs, uses lattice-matched layers of InGaAsP grown by liquid-phase epitaxy onto substrates of InP. Successful double heterostructure lasers with low threshold currents have also been obtained using layers of GaAlAsSb grown on GaSb, but they are limited to wavelengths less than about 1.2 μm.

Substrate InP material is available commercially and the InGaAsP system has the advantage that no aluminum is present. The high chemical reactivity of this material causes complications in the fabrication of devices which require it: it is necessary to keep the system quite free of oxygen and water vapor, for example. Perfect lattice matching may be obtained by keeping the component concentrations, $In_{1-x}Ga_xAs_{1-y}P_y$, such that $y \simeq 2.2x$. Double heterostructure and buried heterostructure lasers have been made successfully for wavelengths in the ranges 1.2–1.3 μm and 1.5–1.7 μm. The confining layers are made from InP and the active layer from the quaternary material with the component proportions adjusted to give the required wavelength. Threshold current densities as low as 10 A/mm^2 with active-layer thicknesses of about 0.2 μm have been obtained.

One problem is the sensitivity of the threshold current to temperature. Typical values for the parameter, T_0, of eqn. (10.3.7) lie in the range 50–70 K. Thus, an increase from 20°C to 60°C causes I_{th} to double. It is hoped that this may improve as the quality of grown material improves. In all other respects the laboratory-produced lasers have given a performance comparable to or better than GaAs devices.

11.4 THE RELIABILITY OF DH SEMICONDUCTOR LEDs AND LASERS

Many failure mechanisms and causes of degradation have been discovered as a result of extensive studies of the behavior of semiconductor optical sources during extended life tests. With components that are required to operate for ten or twenty years it is necessary to establish some means of accelerating the degradation processes and to look for ways of detecting at an early stage those devices that may fail during their operational life. As a result there has been much emphasis on life tests carried out at high ambient temperature and under conditions of high humidity. However, before these results can be meaningful, laws governing the failure processes have to be established with some degree of confidence. Classification of the mechanisms of failure of semiconductor lasers and LEDs is not easy, but here we shall try to distinguish between effects that occur in the bulk of the active layer and those that are specific to the output facets. We should also separate those processes that are going on all the time from those that arise at a critical value of the current density or the output power.

When the power density emitted from the output facet exceeds about 10 kW/mm^2, catastrophic damage occurs. This is associated with absorption of the optical power in the vicinity of the surface causing local melting of the semiconductor. For a stripe geometry laser having a stripe width of about 10 μm and thus an emission cross-section of a few square microns, this corresponds to an output power of some tens of milliwatts. The exact value depends on the near-field distribution. Facet coatings, for example a film of alumina, increase the threshold for this effect. Longer-term erosion of the facets has been observed, possibly the result of a photo-oxidation process, but the extent to which this leads to a significant degradation of the laser performance remains a matter of dispute.

Bulk effects may be divided into those that occur locally and those that seem to be uniformly distributed throughout the active region. The former have become known as ***dark-line defects***. Observation of both has been aided by the production of special experimental laser diodes in which a window left in the substrate metallization allows the radiation generated below threshold by spontaneous recombination to be monitored throughout the life of the device.

Dark-line defects have been shown to be regions of strong nonradiative recombination caused by the accumulation of a network of dislocations. These may originate from a single defect in the original diode, and their effect is to increase the threshold current and to reduce the output power. Device failure follows rapidly. The elimination of dark-line defects depends on the use of high-quality, dislocation-free substrate material (<10 dislocations per mm^2) and on the minimization of strain in the finished device. Here the care with which the diode is bonded onto the header and the contact wire is bonded onto the metallization is crucial and special, soft, low melting point, indium solders are normally used, and mechanical pressures are kept to a minimum. If these precautions are taken, if devices are screened initially for visible damage and are subjected to a 'burn-in' period long enough for any incipient dark-line defects to appear, then this particular failure mechanism ceases to be a serious problem.

Independent of the formation of dark-line defects the threshold current is still found to increase, and the laser power at a given bias level to decrease, steadily during life. Associated with this is a steady general decrease in the level of spontaneous radiation seen in the 'windowed' devices biassed just below threshold which is similar to the degradation observed in LEDs. It does not depend on whether the device is operating as a laser or an LED, but is rather a function of current density and temperature. The decay follows eqns. (9.3.2) and (9.3.3). It is thought that the energy released in nonradiative carrier recombination aids either the creation or the displacement of point defects in the crystal, and that these act as trapping levels. As a result, there is an increase in the rate of nonradiative recombination (a reduction in τ_{nr}) and hence a reduction in internal efficiency. The rate of creation of such traps would be expected

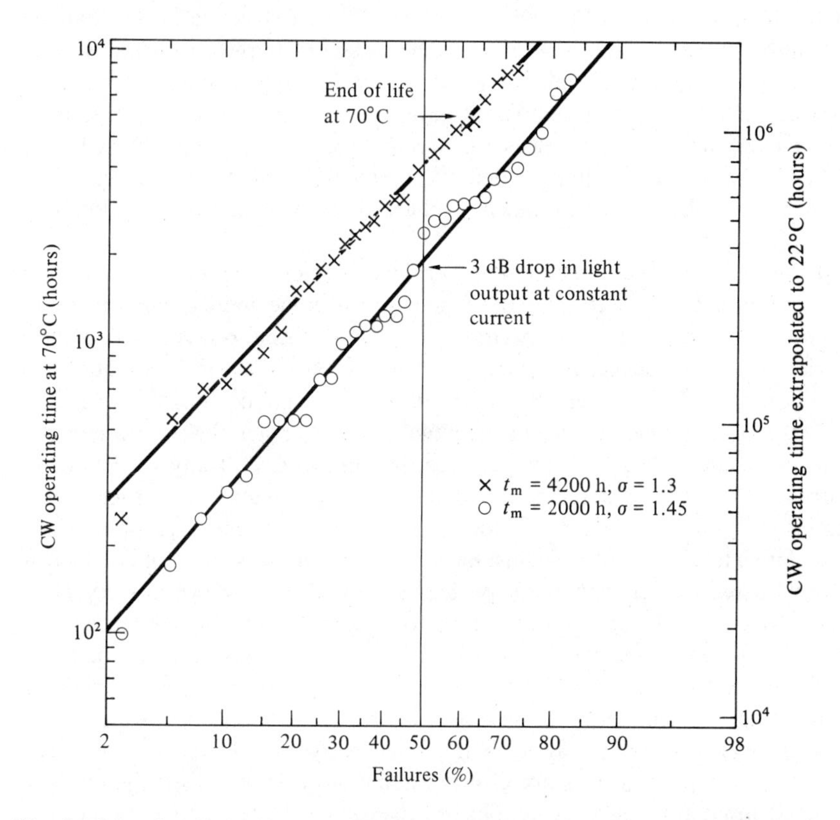

Fig. 11.12 A plot of device life on a log scale against the cumulative percentage of devices which have failed on a normal probability scale. [These results are taken from M. Ettenberg, A statistical study of the reliability of oxide-defined stripe CW lasers of (AlGa)As, *J. Appl. Phys.* **50**, 1195–1202 (1979).]

The data refer to GaAlAs lasers with a 12 μm oxide-defined stripe, operated continuously at 70°C. The upper curve refers to the time after which the lasers were unable to produce 1.25 mW emission. The lower curve shows the time at which the output at constant current had fallen to half its initial value. The median times to 'failure', t_m, were 4200 hours and 2000 hours, respectively, with log-normal standard deviations of 1.3 and 1.45. The right-hand scale shows the room temperature lifetimes to be expected if $t_m \propto \exp\{0.95\ e/kT\}$. This assumes a failure mechanism with an activation energy of 0.95 eV.

to depend on an activation energy, $\mathcal{E}_a$, and thus to be a function of temperature of the form $\exp(-\mathcal{E}_a/kT)$. The indications from LED life tests are that the activation energy for GaAlAs/GaAs devices is in the region of 0.6 eV. Life tests on lasers have given values from 0.5–1 eV, but they tend to be carried out over a much more limited temperature range. With longer wavelength sources the fact that less energy is released on recombination may mean that a higher activation energy is required for defect generation. Certainly life-test results on InGaAsP/InP LEDs, which indicate values for $\mathcal{E}_a$ around 1.0 eV, would tend to support this.

These results do not depend in any critical sense on the precise definition of the end of laser life. It may be a given percentage increase in the room temperature value of I_{th} or a specified reduction in output power at a fixed bias level. There is evidence, for example, that a 50% reduction in spontaneous emission at a fixed current below threshold corresponds roughly to a 20% increase in the threshold current. Eventually devices reach the state where at a given ambient or heat-sink temperature they fail to lase at any level of drive current. When a large population of similar and 'well made' lasers is life-tested, end-of-life, suitably defined, is found to occur statistically. A plot of the cumulative total of failures against life fits a log-normal distribution curve. This is quite usual for semiconductor devices. An example is shown in Fig. 11.12. The device manufacturer has to aim not only to get a high median lifetime, t_m, but also to minimize the standard deviation, σ. Steady improvements in material quality and technological control of the growth processes have led to steady improvements of both these parameters. In particular the inclusion of a small proportion (about 8%) of Al in the active layer of GaAs lasers has been found to be beneficial. It is not clearly established whether this is the result of reduced strain at the heterojunctions or because the highly reactive Al acts as a getter and helps to remove impurities from the bulk semiconductor. On the basis of results like those shown in Fig. 11.12, extrapolated median lifetimes in excess of 10^5 hours at room temperature are predicted. The reader will appreciate the caution with which such claims should be treated, until they can be positively confirmed.

PROBLEMS

11.1 Calculate the wavelength separation between the longitudinal modes of a semiconductor laser in which the active layer is 0.2 mm long and has a refractive index of 4.0.

11.2 Consider the optical cavity of a semiconductor laser operating at wavelengths close to 1.30 μm to be a rectangular box having the dimensions $l = 150$ μm, $w = 20$ μm, $d = 1.0$ μm in a material of refractive index $\mu = 4$. Assume for the purposes of this question that the cavity walls are perfectly reflecting so that the resonant modes satisfy the condition

$$\left(\frac{2\mu}{\lambda_{i,j,k}}\right)^2 = \left(\frac{i}{d}\right)^2 + \left(\frac{j}{w}\right)^2 + \left(\frac{k}{l}\right)^2$$

where i, j and k are integers.

(a) Calculate the approximate magnitude of the longitudinal mode number.
(b) Calculate the longitudinal mode separation, assuming $i = 1, j = 1$.
(c) Hence calculate the separation between the first and second order lateral modes.
(d) Calculate the separation between first and second order lateral modes when w is reduced to 10 μm. (Note that higher order lateral modes are not normally excited in lasers with stripe widths less than about 15 μm.) A more rigorous analysis of the mode structure of semiconductor lasers is given, for example, in Chapter 5 of Ref. [10.1].

11.3 Estimate the coherence length of the light emitted by each of the two lasers whose spectral emission characteristics are shown in Fig. 11.5. Compare these values with the coherence length of the light emitted by each of the three light emitting diodes whose spectra are shown in Fig. 8.5, p. 235. Base the estimations on the spectral width at half power, and take $\Delta\lambda \simeq 0.1$ nm in Fig. 11.5(b).

11.4 Derive eqn. (11.2.13).

11.5 For the laser of Problem 10.2, calculate the effect of the change in facet reflectivity on the photon lifetime. Hence

(a) Calculate the modulation resonance frequency f_0 and the ratio $P(f_0)/P(0)$ in each case when the lasers are biassed (i) 1% and (ii) 10% above threshold. Assume $\mu = 4.0$ and the effective carrier lifetime to be 20 ns.
(b) Calculate in each case the delay time t_D following a step change of current from zero to (i) 1% and (ii) 10% above threshold.
(c) Calculate in each case the delay time t_D following a step change of current from $0.9I_{th}$ to (i) 1% and (ii) 10% above threshold.

REFERENCES

11.1 R. W. Dixon, Current Developments in GaAs Laser Device Development, *The Bell Syst. Tech. Jnl.* **59** (5), 669–722 (1980).

11.2 G. Arnold, P. Russer and K. Petermann, Modulation of laser diodes, H. Kressel (ed), *Semiconductor Devices for Optical Communication* (Ch. 7) Topics in Applied Physics, Vol. 39, Springer-Verlag (1980).

SUMMARY

Laser threshold current can be minimized by the use of a narrow contact stripe. This provides gain guiding for the laser oscillation mode and improves the lateral mode stability. Structures like the buried heterostructure achieve the same results by index guiding. Such lasers have linear power output characteristics above threshold, but limited power capability—a few milliwatts. Examples of each type are shown in Fig. 11.2.

Longitudinal mode separation is $\Delta f = c/2\mu_g l$, $\Delta\lambda = \lambda^2/2\mu_g l$. Buried heterostructures with short cavity lengths offer the possibility of single longitudinal mode operation with a spectral spread of a few megahertz.

Most types of laser 'ring' when pulsed on from below threshold. This is associated with a resonance in the frequency characteristic at 0.2–2 GHz. The exception is the narrow-stripe, gain-guided laser which has a 'soft' threshold and a high proportion of spontaneous emission in its output.

It is normal in pulsed operation to maintain the laser at a bias close to threshold. This minimizes the switch-on 'ring', reduces the switch-on delay time, t_D, and minimizes the capacitative current that has to be switched.

$$t_D = \tau \ln [(I - I_0)/(I - I_{th})]$$

Laser sources operating over the wavelength range 1.0–1.6 μm can be made using InGaAsP/InP materials. These suffer from a rapid variation of threshold current with temperature ($T_0 = 50$–70 K) but otherwise behave as well as or better than GaAlAs/GaAs sources.

Several catastrophic and gradual failure mechanisms have been identified. Output power densities in excess of 10 kW/mm^2 cause facet damage. Dark-line defects can be eliminated by not allowing defects or mechanical strain on the laser chip. The remaining gradual degradation mechanism is similar to that occurring in LEDs and results from a decrease in the nonradiative recombination lifetime, τ_{nr}. Point defects may be created in conjunction with nonradiative recombination events. This would account for lifetimes controlled by an activation energy, $\mathcal{E}_a$, that is higher for longer wavelength sources: $t_m \propto \exp(-\mathcal{E}_a/kT)$.

12

Semiconductor p–i–n Photodiode Detectors

12.1 GENERAL PRINCIPLES

In Chapters 12 and 13 we shall examine the basic processes underlying the operation of the two types of detector used in optical fiber communication systems, the p–i–n photodiode and the avalanche photodiode (APD). We shall be concerned to identify those features which may limit their performance, especially their sensitivity and frequency response, and we shall cite some representative examples of particular designs. Detector design has been the subject of a number of comprehensive review articles of which Refs. [12.1] and [12.2] are among the best and the most accessible.

Both types of detector are based on the reverse-biassed p–n junction. Unilluminated, the voltage–current characteristic of a p–n junction diode would be expected to follow eqn. (9.1.8). But when light of wavelength less than some threshold wavelength, λ_{th}, is shone onto the junction, the effect is to shift the characteristics by a constant current, I_{ph}, as shown in Fig. 12.1. This is the result of a quantum interaction between the radiation and the valence band electrons in which an electron is excited across the band gap and an electron–hole pair is created. This process is illustrated in Fig. 12.2(a). Ideally the threshold wavelength is related to the band-gap energy, $\mathcal{E}_g$, by

$$\lambda_{th} = \frac{hc}{\mathcal{E}_g} = \frac{1.24}{\mathcal{E}_g} \,[\mu\text{m.eV}] \tag{12.1.1}$$

We shall examine this optical absorption process in greater detail in the next section.

The photogenerated current, I_{ph}, increases linearly with the optical power over a range of many orders of magnitude, and the ratio of I_{ph} to the incident optical power, Φ, is known as the ***responsivity***, $\mathcal{R}$ of the photodiode. Because of the quantum nature of the interaction it is customary to define a quantum

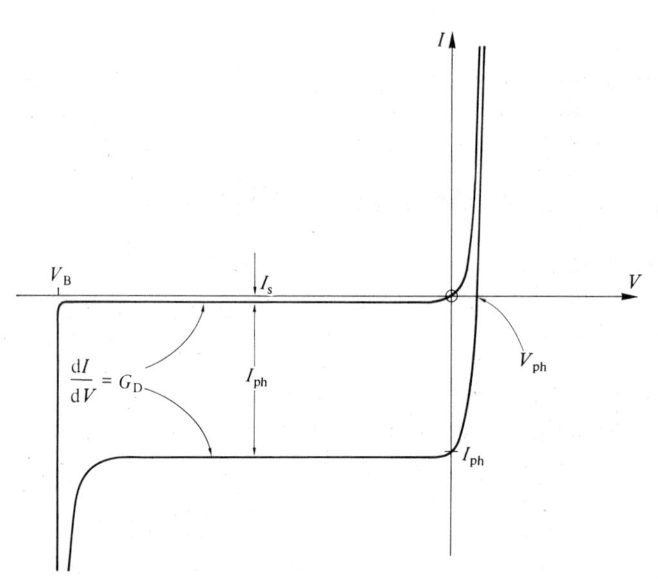

Fig. 12.1 Voltage–current characteristics of a p–n junction photodiode showing the dark current, I_s, and the additional current, I_{ph}, generated when the diode is illuminated.

efficiency, η, for a given device. This is the ratio between I_{ph}, measured in electrons per second, and the incident optical flux, measured in photons per second. Thus:

$$\eta = \frac{I_{ph}}{e} \div \frac{\Phi}{\mathcal{E}_{ph}} = \frac{I_{ph}}{\Phi}\frac{hc}{e\lambda} = \mathcal{R}\frac{hc}{e\lambda} \tag{12.1.2}$$

Note that this definition of η is the converse of that given in Section 8.3 for the internal quantum efficiency of an optical source. There we were concerned to maximize the number of photons created per electron; here we wish to maximize the number of electrons generated per incident photon. In either case it is assumed that η cannot exceed unity.

Reference to Fig. 12.1 makes it clear that a junction photodiode may be used to detect optical radiation in several different ways. The one that is perhaps the simplest is sometimes called the photovoltaic or open-circuit mode. Here the diode is connected directly across the input of a voltage amplifier with a high input impedance, and the variation of V_{ph} (Fig. 12.1) is measured. An alternative is the short-circuit mode, in which the diode current is amplified by a current amplifier having a very low input resistance, with the result that the diode voltage is maintained near to zero. This, in theory, causes least random

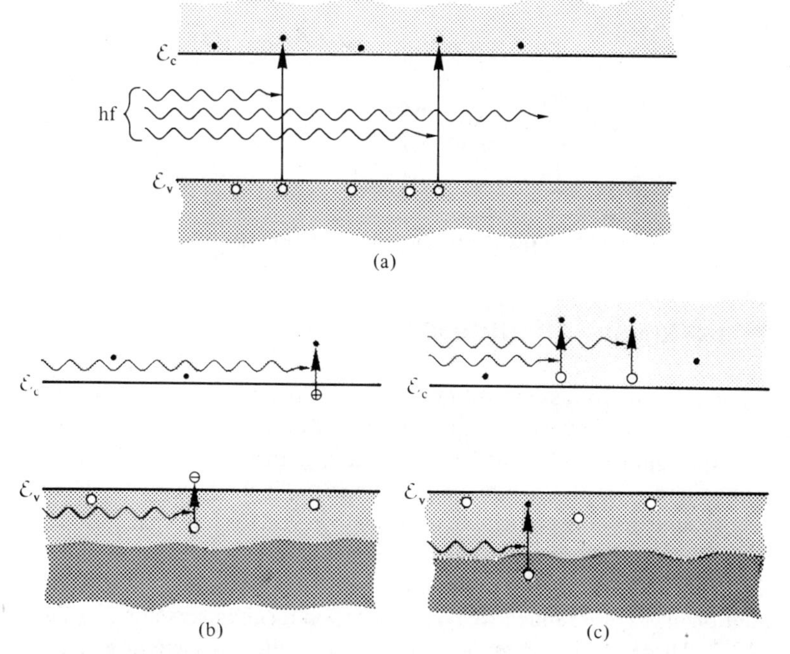

Fig. 12.2 Electron energy band diagrams illustrating optical absorption processes: (a) as a result of band-to-band excitation; (b) following excitation of donor and acceptor impurities; (c) as a result of free particle transitions.

noise to be superimposed on the diode current. In practice, however, photodiodes for optical communication use neither of these modes, but are always reverse-biassed. This is because the quantum efficiency and the bandwidth are then significantly improved for reasons we shall discuss later. When the reverse bias is increased to values which approach the breakdown voltage, V_B, the photogenerated current is multiplied as a result of the same carrier avalanche ionization processes that give rise to breakdown. This region is also shown on Fig. 12.1. The process forms the basis of the avalanche photodiode detectors which we shall discuss in Chapter 13.

When reverse-biassed, the small-signal characteristics of the illuminated photodiode are those of the ideal current generator shown in Fig. 12.3. The small shunt conductance, G_D, represents the slope of the reverse-bias characteristics. The series resistance, R_S, is that of the bulk semiconductor and contacts, and is normally less than 10 Ω. The capacitance, C_D, is that of the p–n junction, the leads and the packaging, and can usually be kept below 1 pF. This is very important for high-frequency operation.

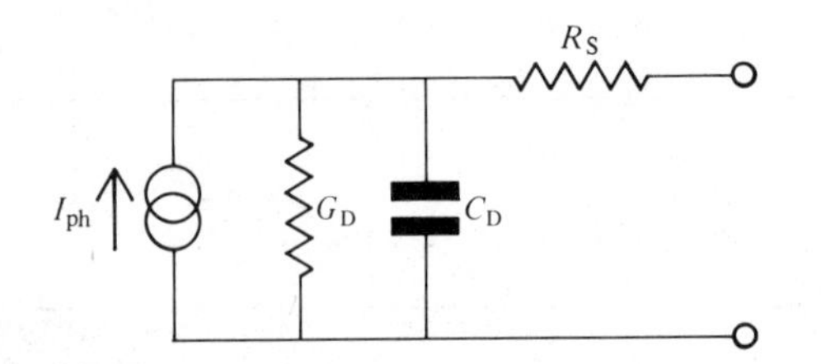

Fig. 12.3 Small-signal equivalent circuit for a reverse-biassed photodiode.

12.2 INTRINSIC ABSORPTION

In Chapter 10 we discussed a situation that is created in semiconductor lasers where a population inversion is produced and optical gain occurs. This is, of course, very unusual. Normally, light passing through a semiconductor is attenuated and, as with the attenuation in optical fibers discussed in Chapter 3, many mechanisms may contribute to this. However, for free-space wavelengths satisfying $\lambda < \lambda_{th}$, where λ_{th} is given by eqn. (12.1.1), the quantum interaction shown in Fig. 12.2(a) totally dominates the others and gives rise to very rapid attenuation. Two other much weaker absorption mechanisms are also shown in Fig. 12.2. These may become significant when the wavelength exceeds λ_{th}.

The attenuation leads to an exponential decay of the propagating radiation power density, $P(x)$:

$$P(x) = P(0) \exp(-\alpha x) \qquad (12.2.1)$$

The attenuation coefficient, α, is characteristic of the material and is a strong function of wavelength near to threshold. Some representative values for direct and indirect band-gap semiconductors are shown in Fig. 12.4. With a direct band gap there is no constraint on the carrier–pair creation process and the absorption coefficient increases sharply for wavelengths just shorter than λ_{th}. With indirect band-gap materials, such as Si and Ge, a simultaneous interaction with the crystal lattice is required so that momentum may be conserved in the carrier-generation process. This may involve a lattice vibration, an impurity, or some other form of lattice disorder, but the result is that α increases only gradually as the wavelength reduces below threshold. At shorter wavelengths where excitation across the direct band gap becomes possible (3.4 eV for Si and 0.8 eV for Ge), α increases again. The threshold wavelength is found to vary with temperature and with doping concentration because of the effects these parameters have on the band-gap energy and the band edges, as discussed in Chapters 7 and 10.

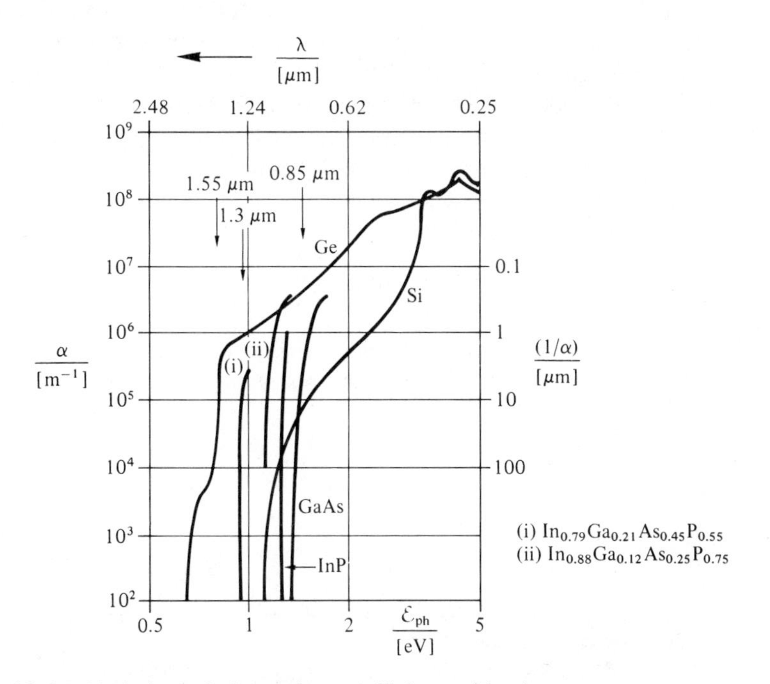

Fig. 12.4 Variation of absorption coefficient with photon energy.

12.3 QUANTUM EFFICIENCY

12.3.1 The Ideal Photodiode

The excess, photogenerated carriers recombine on average after the mean minority carrier lifetime, τ_n or τ_p. This time depends on the material type, quality and doping concentration. For the incoming optical power to be converted in the detector into an electrical signal, it is necessary that the electrons and holes should suffer some net separation before recombination takes place. This means establishing a nonuniformity of some kind in this part of the semiconductor. One of the simplest ways is to set up an electric field to separate the carriers. As we have already said, in all the detectors we shall be concerned with this is done by ensuring that some of the photogeneration occurs within the depletion layer of a reverse-biassed p–n junction. The excess electrons and holes are then swept away and collected at the contacts before they have time to recombine and in consequence a current is induced in the external biassing circuit.

In the ideal photodiode all the incident light would be absorbed in this way in the depletion layer and all the carriers generated would be collected. The quantum efficiency would then be unity, and the induced photodiode current with an incident optical power Φ would be

$$I_{ph} = (I_{ph})_{ideal} = \frac{e\Phi}{\mathcal{E}_{ph}} \tag{12.3.1}$$

Of course in practice some of the incident light is reflected at the semiconductor surface and not all of the remainder is absorbed in the depletion layer. This is illustrated in Fig. 12.5.

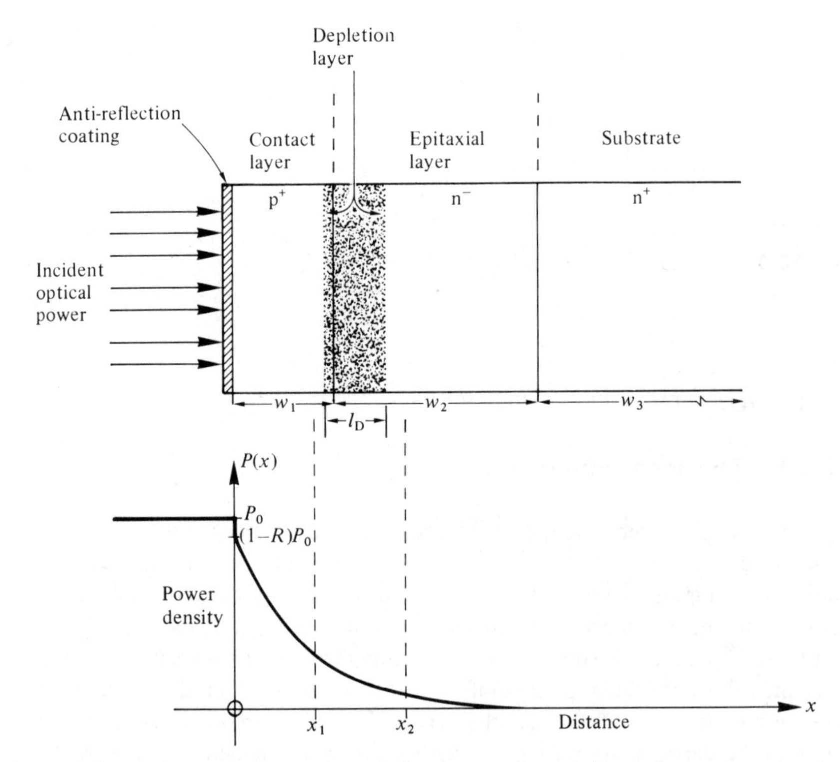

Fig. 12.5 Schematic illustration of the attenuation of the optical power density with depth below the incident surface. The diagram illustrates the importance of minimizing surface reflection, minimising the distance x_1, and maximizing the distance $(x_2 - x_1)$ by increasing the depletion-layer thickness.

Obtaining a high value of quantum efficiency depends on:

(a) minimizing reflection at the incident surface;
(b) maximizing the absorption within the depletion layer;
(c) avoiding recombination before the carriers are collected.

We will deal first in general terms with the design criteria that these requirements give rise to, before going on in subsequent sections to consider more specific photodetector designs.

12.3.2 Surface Reflection

The question of surface reflection is the same problem that we met in Chapter 8. Just as about 33% of the light generated within a semiconductor source suffers Fresnel reflection at the semiconductor–air interface, so too is about 33% of the light incident onto the surface from the outside reflected. To minimize this it has become normal practice to coat the surface with a film of transparent dielectric about a quarter of a wavelength thick. Ideally the film refractive index should be the square root of the semiconductor refractive index, as mentioned in the discussion in Sections 8.5 and 8.6. In practice thin films of silica ($\mu = 1.46$) are more convenient and give a very worthwhile improvement in the transmitted optical power. Sometimes Si_3N_4 ($\mu \simeq 2.0$) is used. It will be assumed throughout that such films are employed, and in general after Section 12.3.3 we shall neglect any reflection that may occur.

12.3.3 Maximizing Absorption within the Depletion Layer

Reference to Fig. 12.5 enables us to estimate the fraction of the incident power that will be absorbed usefully, that is, within the depletion layer or within about a diffusion length of the depletion-layer edge at either end. Let these bounds to the 'useful region' be at depths x_1 and x_2 below the surface. Let the incident power density be P_0 and the surface reflection coefficient be R. Then the power density entering the semiconductor is $(1 - R)P_0$ and the power density, P, usefully absorbed is given by the difference between that entering the useful region at x_1 and that leaving at x_2, namely,

$$P = (1 - R)P_0[\exp(-\alpha x_1) - \exp(-\alpha x_2)] \qquad (12.3.2)$$

Thus, the quantum efficiency is

$$\eta = \frac{P}{P_0} = (1 - R)\exp(-\alpha x_1)[1 - \exp\{-\alpha(x_2 - x_1)\}] \tag{12.3.3}$$

At first sight it would seem obvious that x_1 has to be made as small, and x_2 as large, as possible. In practice this leads to difficulties, so that compromises have to be reached.

12.3.4 Minimizing Recombination

Carrier lifetime is normally much greater than the carrier transit time across the depletion layer so that a negligible fraction of carriers generated there are lost by recombination. Carriers generated within a diffusion length of the depletion-layer edges may be collected over a period of the order of the recombination lifetime as a result of diffusion. Should these become an appreciable fraction of the total, the quantum efficiency would be reduced by the recombination that does occur, and the frequency response of the diode would be impaired. It is important, then, that the minimization of x_1 and the maximization of x_2 should not be dependent on the effective extension of the depletion layer by the diffusion regions at either end. The p–n junction should be formed close to the surface, and the depletion-layer width should be much greater than the attenuation distance. Thus, with reference to Fig. 12.5, we require

$$w_1 \ll 1/\alpha \ll l_D \tag{12.3.4}$$

If the first of these conditions is not met, surface recombination causes the quantum efficiency to be reduced. The surface recombination velocity should always be minimized, but when α is large it may, in addition, be necessary to resort to one of the types of device discussed in Section 12.4.3.

12.4 MATERIALS AND DESIGNS FOR p–i–n PHOTODIODES

12.4.1 General

The choice of detector materials and designs for optical communication is fairly clear cut. For use with GaAs/GaAlAs sources emitting in the wavelength range 0.8–0.9 μm it will be hard to improve on the best silicon detectors. They require a depletion-layer thickness of a few tens of microns and as discussed later are normally made in the form of a 'p–i–n' diode. For longer wavelengths, up to 1.8 μm, germanium detectors have been available for some time. At wavelengths longer than about 1.55 μm they, too, require depletion-layer widths of a few tens of microns. For use around 1.3 μm the active region should be an order of magnitude or more thinner. Detectors using direct band gap, ternary and quaternary semiconductors are under intensive development, particularly for use at these longer wavelengths. Because their attenuation depths are so small they are likely to use either the Schottky barrier or the heterojunction structures described in Section 12.4.3.

One problem with the narrower band-gap devices is the high level of the reverse current in the absence of incident radiation. This is known as the *dark current* and is shown as I_s in Fig. 12.1. The factors contributing to I_s will be discussed in Section 12.6, but we may note here that I_s tends to increase rapidly with temperature. For this reason narrow band-gap detectors for longer wavelengths may require cooling if they are to retain a reasonable sensitivity.

12.4.2 The Silicon p–i–n Photodiode

Up to now we have tacitly implied that the depletion layer would be wholly contained within the n^--layer of Fig. 12.5. A device made in this way is known as a p^+n diode. (With the dopant type and polarities reversed, it would become an n^+p diode.) This structure might seem to offer the flexibility of matching the depletion-layer width to the absorption coefficient simply by adjusting the bias voltage according to eqn. (7.6.7). In practice it is found that device performance is more readily optimized by the use of the p–i–n structure, in which the lightly doped n^--layer is made sufficiently thin and is sufficiently lightly doped that the normal bias voltage depletes it entirely. The depletion layer then 'reaches through' to the heavily doped substrate material.

With the p–i–n structure the optimum wavelength, the operating voltage, the device capacitance and its frequency response all may be predetermined and specified during manufacture. An example of the structure and the field and potential distributions is shown in Fig. 12.6. In the limit an almost uniform electric field,

$$E \simeq \frac{V_0}{w_2} \tag{12.4.1}$$

stretches between the p^+ and n^+ regions. V_0 is the applied bias voltage and w_2 is the width of the lightly doped n^- layer. Following convention this has been

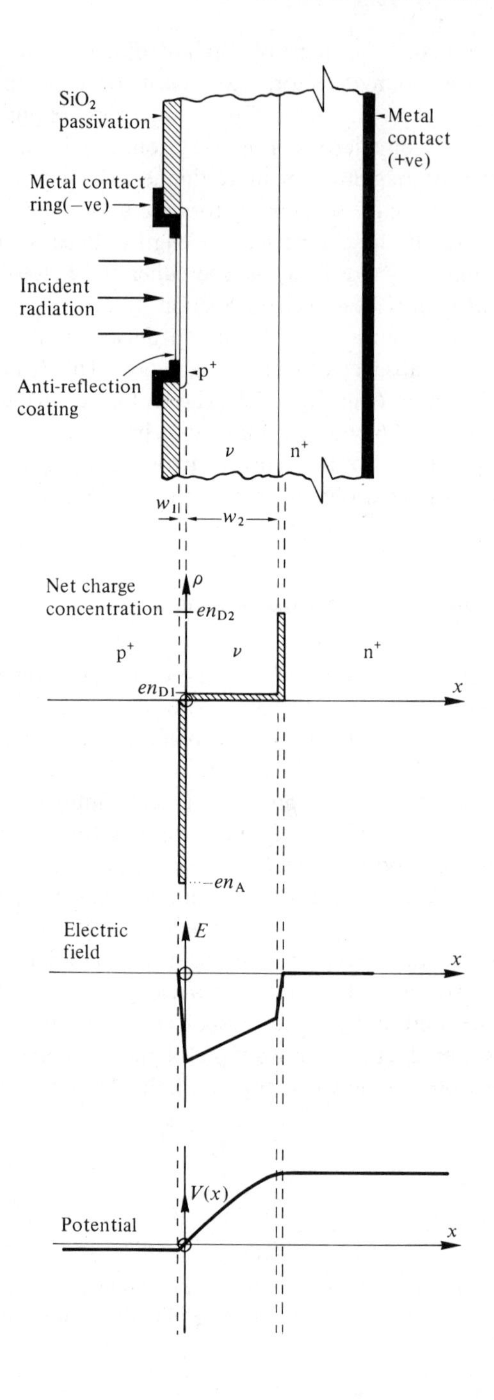
SiO2
passivation
Metal contact
ring(−ve)
Incident
radiation
Anti-reflection
coating
p+
Metal
contact
(+ve)
ν
n+
w1
w2
Net charge
concentration
ρ
enD2
p+
ν
n+
enD1
x
−enA
Electric
field
E
x
V(x)
Potential
x

labelled the ν-layer. The self-capacitance of the diode then approaches that of a plane-parallel capacitor:

$$C_D = \frac{\varepsilon_0 \varepsilon_r A}{w_2} \tag{12.4.2}$$

where A is the junction area, ε_0 the free-space permittivity and ε_r the relative permittivity. With $\varepsilon_r = 12$, $w_2 = 50\ \mu m$ and $A = 10^{-7}\ m^2$, $C_D = 0.2$ pF. Quantum efficiencies of 0.8 or more can readily be obtained at wavelengths of 0.8–0.9 μm, and dark currents are $\ll 1$ nA at room temperature.

12.4.3 Heterojunction and Schottky-barrier Diodes

When direct band-gap materials or indirect band-gap materials far from threshold are used, values of the attenuation coefficient may be very large, $\alpha > 10^6\ m^{-1}$. Good photodiode design then demands a very thin and heavily doped (therefore highly conducting) surface layer. Problems then arise from the proximity of the free surface with its relatively high surface recombination velocity. A large proportion of the carriers generated in the surface layer recombine at the surface rather than diffuse to the contact at the periphery or into the depletion layer. In consequence the quantum efficiency is impaired. Two types of photodiode structure seek to minimize this problem: the Schottky-barrier diode shown in Fig. 12.7(a) and the heterojunction diode shown in Fig. 12.7(b).

The Schottky-barrier diode does away altogether with the surface semiconductor layer by making use of a reverse-biassed, rectifying, metal–semiconductor contact. This is not always possible: in germanium, for example, the reverse current rises too rapidly with bias voltage. The metal layer must, of course, be thin enough to be transparent to the radiation. In practice this means a metal film thickness of less than about 10 nm.

Heterojunction diodes are more likely to be used for long-wavelength optical communication. The semiconductor forming the surface layer must have the wider band gap so that there is negligible absorption. As the light enters the

---

Fig. 12.6 A cross-section of the active region of a p–i–n photodiode together with diagrams showing the net charge distribution, the electric field strength and the potential variation when the diode is reverse biassed sufficiently to deplete the whole of the lightly doped ν-layer.

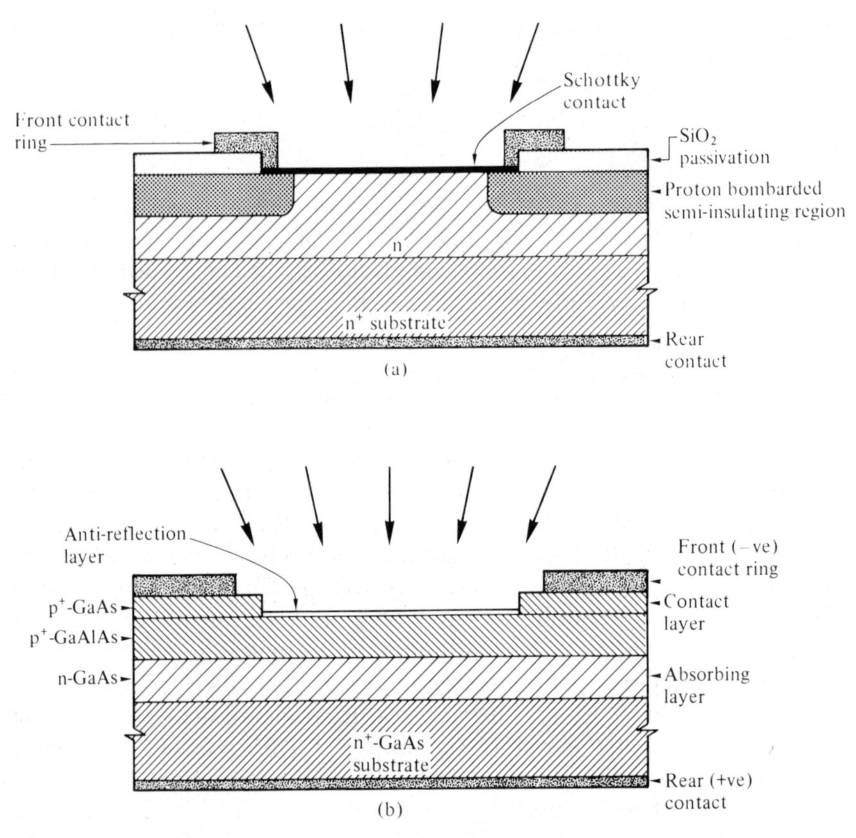

Fig. 12.7 Schematic cross-section of possible GaAs photodiodes: (a) using a rectifying, metal–semiconductor (Schottky barrier) contact. Surface leakage is minimized by the use of SiO_2 passivation and by surrounding the high-field, active region with material made semi-insulating by proton bombardment; (b) using a GaAs/GaAlAs heterojunction in which the p^+–GaAs contact layer is selectively etched in the active area so that the incident radiation enters directly into the transparent p^+–GaAlAs layer.

narrower band-gap material at the heterojunction, absorption becomes strong. And this corresponds with the region where the electric field is maximum. Provided the recombination velocity at the heterojunction is not excessive, high quantum efficiency can be obtained. Work has concentrated on two systems, namely,

$$(\mathrm{InGaAsP})_1/(\mathrm{InGaAsP})_2/\mathrm{InP}$$

and

$$(\mathrm{GaAlAsSb})_1/(\mathrm{GaAlAsSb})_2/\mathrm{GaSb}$$

where the three regions are, respectively,

(surface layer)/(drift region)/(substrate).

In the InGaAsP system the possible surface-layer compositions include InP. A number of other variants to these structures have been tried, some of which are discussed in Ref. [12.2].

Two forms of a small-area photodiode, which give a good response over the range from 1.0–1.5 μm, are shown in Fig. 12.8. Ternary $In_{0.53}Ga_{0.47}As$ is grown on an InP substrate to which it is lattice matched. Its band-gap energy is 0.75 eV, giving a long wavelength cutoff at 1.65 μm. The p–n homojunction is formed by diffusion of zinc, and the diode may be designed to be either front- or back-illuminated.

12.4.4 Photodiodes Designed for Operation at Wavelengths near to Threshold

In photodiodes made from indirect band gap semiconductors which are to be used to detect radiation close to the threshold wavelength the absorption is weak, say $\alpha < 10^4$ m^{-1}, and a very wide drift region is required. This applies particularly to silicon diodes to be used in the wavelength range 1.0–1.1 μm. High bias voltages and very lightly doped material are required. Two possible ways of avoiding the wide depletion layer are as follows:

(a) To use a back contact which reflects the radiation passing right through the diode back into the drift region.
(b) To use sideways illumination so that the optical path is parallel to the junction.

Each of these solutions brings technological problems.

12.5 IMPULSE AND FREQUENCY RESPONSE OF A p–i–n PHOTODIODE

12.5.1 Photodiode Response

Figure 12.9 shows a normal biassing arrangement for a photodiode with the output signal fed into an amplifier. The components, C_A and R_A, represent the amplifier input impedance. In general, a time-varying incident optical power, $\Phi(t)$, generates a time-varying photodiode current, $i_{ph}(t)$, and produces a

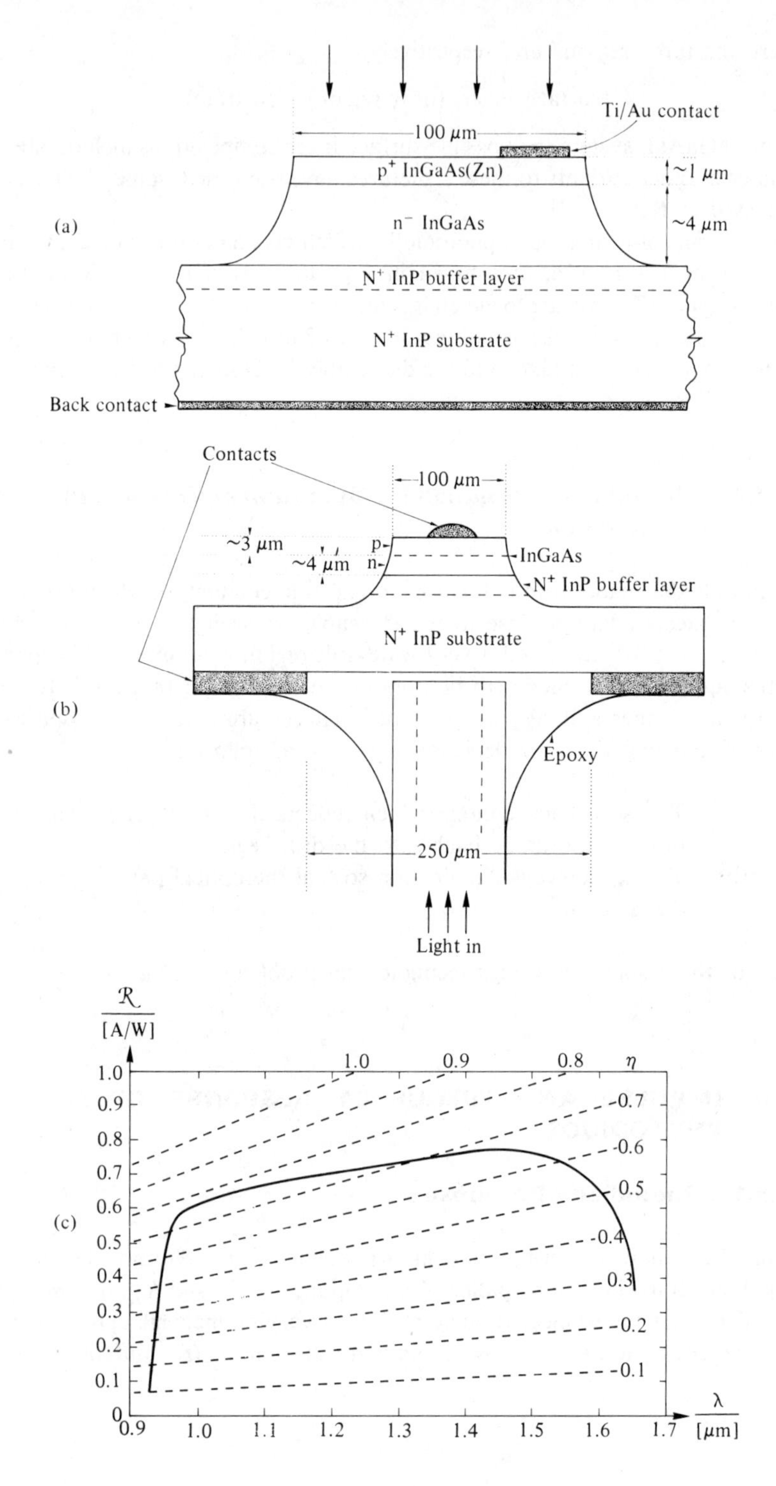
Ti/Au contact
100 μm
p⁺ InGaAs(Zn)
~1 μm
(a)
n⁻ InGaAs
~4 μm
N⁺ InP buffer layer
N⁺ InP substrate
Back contact
Contacts
100 μm
~3 μm
~4 μm
p
n
InGaAs
N⁺ InP buffer layer
N⁺ InP substrate
(b)
Epoxy
250 μm
Light in
R [A/W]
1.0
0.9
0.8
η
0.7
0.6
0.5
0.4
0.3
0.2
0.1
(c)
λ [μm]
0.9 1.0 1.1 1.2 1.3 1.4 1.5 1.6 1.7

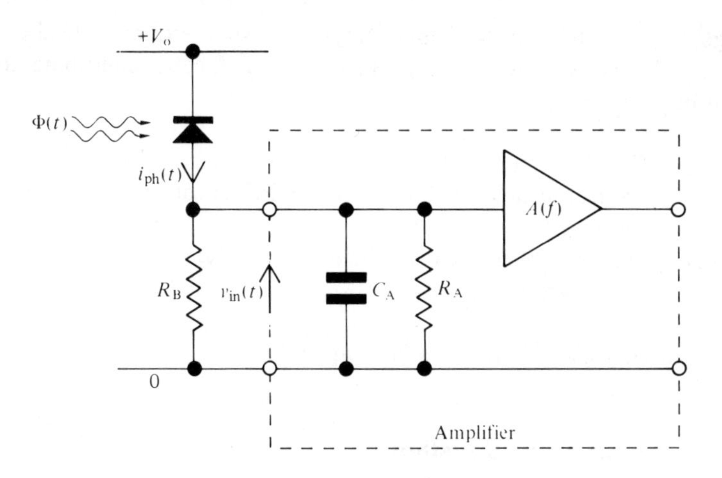

Fig. 12.9 Normal photodiode biassing arrangement.

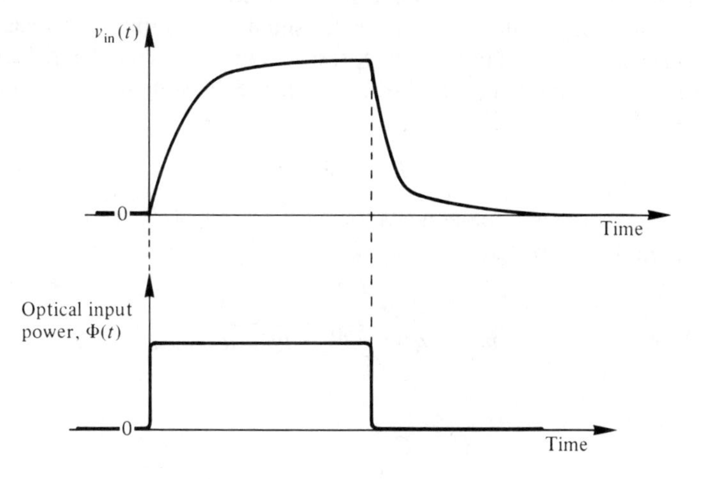

Fig. 12.10 Typical photodiode response.

Fig. 12.8 Examples of small area, InGaAs homojunction photodiodes: (a) front-illuminated; (b) back-illuminated. [Taken from T.-P. Lee *et al.*, InGaAs/InP p–i–n photodiodes for lightwave communications at the 0.95–1.65 µm wavelength, *IEEE J. Quantum Ets.*, **QE-17**, 232–8; © 1981, IEEE]; (c) responsivity of the diode shown in (b).

voltage, $v_{in}(t)$, at the amplifier input. A typical response to an optical pulse of some 10–20 ns is shown in Fig. 12.10. Three factors contribute to this response:

(a) The RC time constant of the diode and its load.
(b) The transit time resulting from the drift of carriers across the depletion layer.
(c) The delay resulting from the diffusion of carriers generated outside the depletion layer.

We will discuss each of these briefly, in turn.

12.5.2 Circuit Time Constant

An equivalent circuit for the photodiode was given in Fig. 12.3. We now assume that the diode current is fed into a resistive load, R_A, which is shunted by a capacitance, C_A, these representing the input impedance of an amplifier. There is likely to be some additional distributed stray capacitance, C_S, and the bias resistance, R_B, will also appear to shunt the output terminals. The photodiode and its load may then be represented by the circuit of Fig. 12.11 (a), which reduces to that of Fig. 12.11(b) if we make the simplifying assumption that

$$R_S \ll R_A \qquad (12.5.1)$$

and then lump the shunt components together. Usually $1/R_B$ and G_D are much smaller than $1/R_A$ so that

$$\frac{1}{R_A} + \frac{1}{R_B} + G_D = \frac{1}{R} \simeq \frac{1}{R_A} \qquad (12.5.2)$$

and

$$C = C_D + C_S + C_A \qquad (12.5.3)$$

Following a step change in the photogenerated current, the load voltage, V_{in}, will rise or fall exponentially with a time constant RC. When the photogenerated current is sinusoidally varying at angular frequency $\omega = 2\pi f$, and is represented by $I_{ph}(f)$, then the response of the load voltage, $V_{in}(f)$, will be given by

$$\frac{V_{in}(f)}{I_{ph}(f)} = \frac{R}{(1 + j2\pi fCR)} \qquad (12.5.4)$$

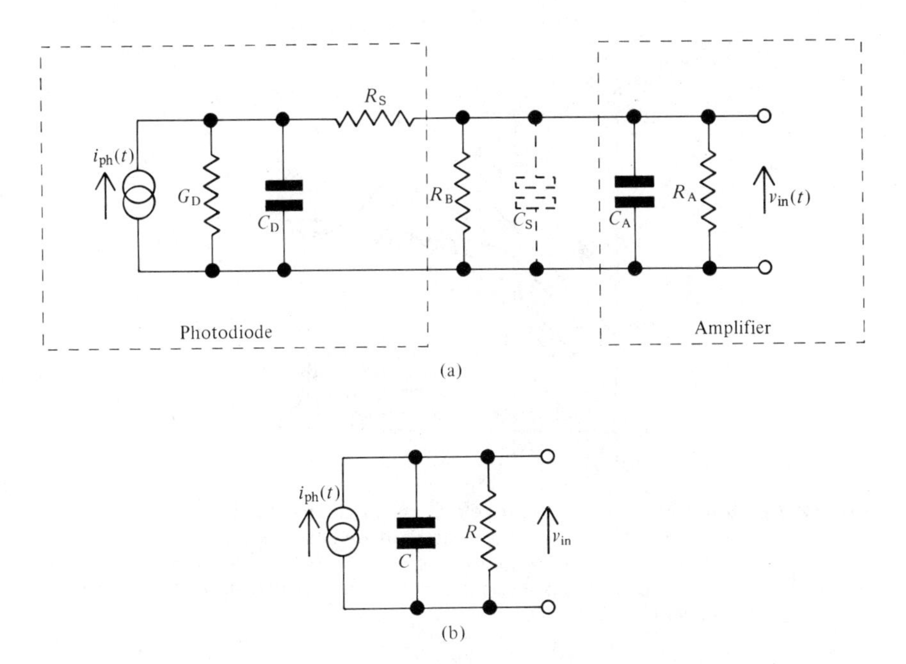

Fig. 12.11 Small-signal equivalent circuit for a normally biassed photodiode and amplifier: (a) complete circuit; (b) reduced circuit obtained by neglecting R_s and lumping together the parallel components.

For a good high-frequency response it is essential that C be kept as low as possible and the contribution from the photodiode can normally be kept below 1 pF. After that it is necessary either to reduce R or to provide high-frequency equalization. The merits of these two techniques have been the subject of much debate and will be discussed further in Chapter 14.

12.5.3 Transit Time

The effect of the carrier transit time is to prevent the photodiode current, i_{ph}, from following modulations of the optical power at high frequency. We will discuss this briefly for some special cases, but first we need to consider carrier motion in the drift region.

Carriers generated in an electric field acquire an average drift velocity, v_d, in a very short time, usually within a fraction of a picosecond. At low fields the drift velocity is proportional to the field strength E, but at high fields ($E > E_S$) a saturation velocity v_s is reached. The saturation field, E_S, is usually in the region of 10^6 V/m. Figure 12.12 shows the variation of v_d with E for a number

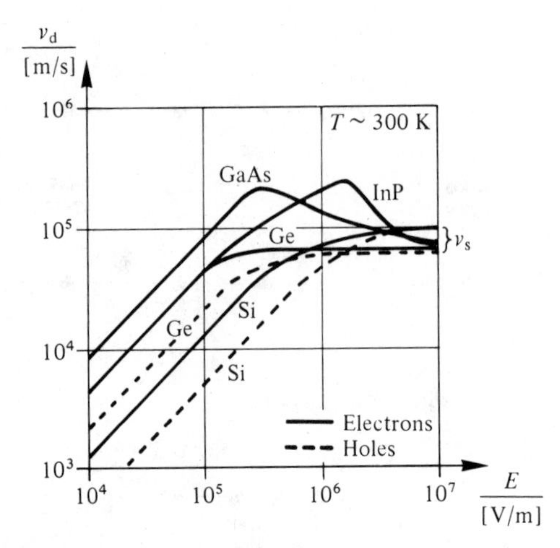

Fig. 12.12 Variation of carrier drift velocity with electric field strength, showing the saturation that occurs at high field strengths.

It should be understood that v_d and v_s are reduced in material that is heavily doped or contains a high concentration of defects, and also at higher temperatures.

of semiconductors. We shall assume that in a correctly biassed p–i–n photodiode the depletion-layer field is large enough to maintain the carrier saturation velocity throughout the drift region, w_2, in Fig. 12.6. The implications of this assumption for the bias voltage, V_0, and the doping concentration in the drift region, n_D, are as follows. Clearly, we require

$$V_0 > E_S w_2 \tag{12.5.5}$$

so for a silicon diode with a 50 μm drift region the bias voltage must exceed 10^6[V/m] × 50[μm] = 50[V]. According to eqn. (7.6.2), adapted to take account of n-type doping in region w_2, the electric field varies linearly:

$$\frac{dE}{dx} = \frac{en_D}{\varepsilon_r \varepsilon_0} = \text{constant} \tag{12.5.6}$$

and the total change in E is given by

$$\Delta E = \frac{en_D w_2}{\varepsilon_r \varepsilon_0} \tag{12.5.7}$$

Again with our silicon diode in mind let us say that we wish to limit the variation in E so that it falls from a maximum of 1.5×10^6 V/m at the $p^+\nu$ junction to not less than 1.0×10^6 V/m at the νn^+ junction, and so maintains the carrier saturation velocities everywhere within w_2. Then $\Delta E < 0.5 \times 10^6$ V/m and with $\varepsilon_r = 12$ and $w_2 = 50\ \mu m$ eqn. (12.5.7) gives $n_D < 6.5 \times 10^{18}\ m^{-3}$. A bias voltage of 62.5 V will be needed.

In order to discuss transit-time effects consider first an impulse of optical power which causes N electron-hole pairs to be generated, all at the p^+ edge of the depletion layer. The holes are collected immediately by the p^+ material while the electrons drift through the depletion layer for a time, t_{tr}, which depends on the electron saturation velocity, v_{se}:

$$t_{tr} = \frac{l}{v_{se}} \tag{12.5.8}$$

where $l \simeq w_2$ is the total depletion-layer width. They are then collected at the n^+ layer. The constant charge in transit at a constant speed causes a constant current to flow in the circuit throughout this period. The total charge which flows is Ne, so that

$$i(t) = \frac{Ne}{t_{tr}} = \frac{Nev_{se}}{l} \tag{12.5.9}$$

for $0 < t < t_{tr}$, as shown in Fig. 12.13(a).

If the carriers were generated at the n^+ edge of the depletion layer, the transit time and the current would be determined by the hole saturation velocity, as in Fig. 12.13(b). Next consider what happens when the electron–hole pairs are generated in the middle of the depletion layer. The electrons give rise to a current, (Nev_{se}/l) which flows for $(w_2/2v_{se})$, and the holes give rise to a current (Nev_{sh}/l) which flows for $(l/2v_{sh})$, where v_{sh} is the hole saturation velocity. Similar arguments can be made for generation anywhere else in the depletion layer. When the generation is uniform throughout the region a triangular response is expected as in Fig. 12.13(c). When account is taken of the exponential decay of optical power, and hence carrier generation rate, then impulse responses of the form shown in Fig. 12.13(d) are to be expected. We should expect the current pulse width at half-height to be rather less than the electron transit time across the whole of the depletion layer. For a silicon diode with $v_{se} = 10^5$ m/s and $w_2 = 50\ \mu m$, $(t_{tr})_e = 0.5$ ns. Carriers diffusing into the depletion layer from the n^+ and p^+ regions extend this impulse response by a time of the order of a recombination lifetime. This is unlikely to be less than about 10 ns. It is thus essential to keep the current in this 'diffusion tail' to negligible proportions by minimizing absorption outside the depletion layer.

Let us now consider the way that the transit-time affects the response of the photodiode to a sinusoidal modulation of the optical flux

$$\Phi(t) = \Phi_0(1 + m \sin \omega t) \tag{12.5.10}$$

where m is a modulation index. We will do a formal analysis for the simple case

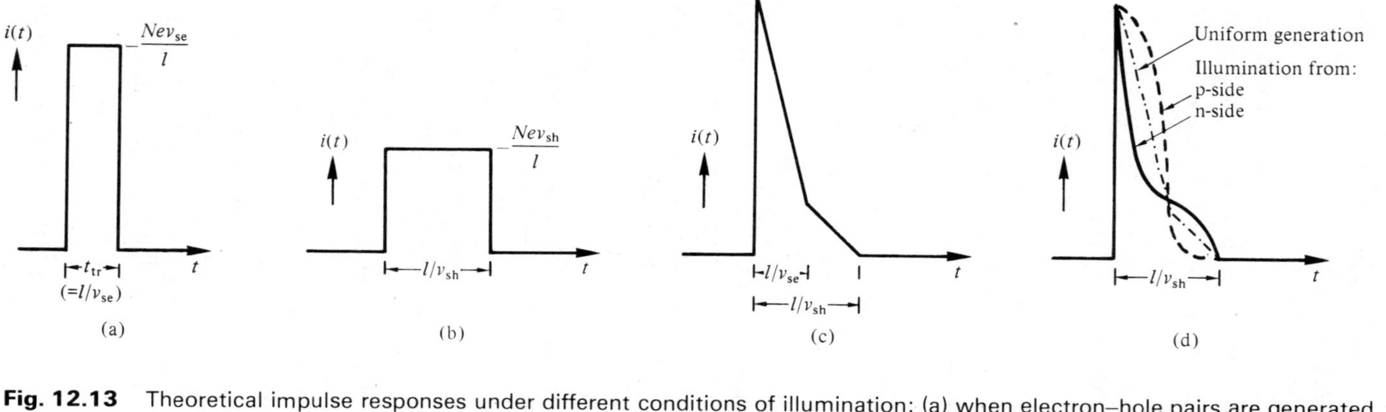

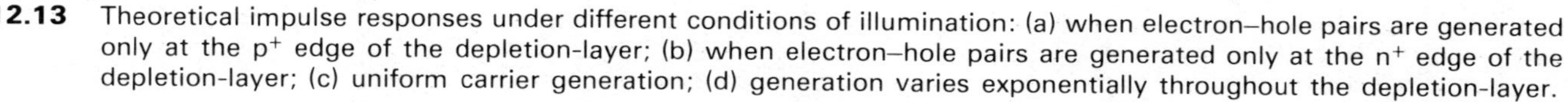

Fig. 12.13 Theoretical impulse responses under different conditions of illumination: (a) when electron–hole pairs are generated only at the p^+ edge of the depletion-layer; (b) when electron–hole pairs are generated only at the n^+ edge of the depletion-layer; (c) uniform carrier generation; (d) generation varies exponentially throughout the depletion-layer.

in which carrier generation takes place only at the p^+ edge of the depletion layer. We will now make this point the origin of the x-coordinate in Fig. 12.5. Let the rate of carrier generation be $\dot{N}_0(1 + m \sin \omega t)$. These carriers (electrons) just drift across the depletion layer at the electron saturation velocity v_{se}. This means that in the element dx at x we would expect to find those electrons that were generated at the time (x/v_{se}) earlier during the time interval $dt = dx/v_{se}$. These will number

$$\dot{N}_0[1 + m \sin \omega(t - x/v_{se})](dx/v_{se}),$$

so that the total number, N, in transit at any time, t, will be given by

$$N = \int_0^l \left(\frac{\dot{N}_0}{v_{se}}\right) [1 + m \sin \omega(t - x/v_{se})] \, dx$$

$$= \frac{\dot{N}_0 l}{v_{se}} + \frac{m\dot{N}_0}{\omega} \{\cos \omega(t - l/v_{se}) - \cos \omega t\}$$

$$= \frac{\dot{N}_0 l}{v_{se}} \left[1 + \frac{2mv_{se}}{\omega l} \sin \frac{\omega l}{2v_{se}} \sin \omega(t - l/2v_{se})\right] \qquad (12.5.11)$$

where l is the total width of the depletion region. The total current is

$$i = Nev_{se}/l \qquad (12.5.9)$$

and may be written as

$$i = I_0[1 + m' \sin(\omega t + \phi)] = \dot{N}_0 e\left[1 + m \frac{\sin(\pi f t_{tr})}{\pi f t_{tr}} \sin 2\pi f(t - t_{tr}/2)\right] \qquad (12.5.12)$$

where we have substituted $f = \omega/2\pi$ and $t_{tr} = l/v_{se}$. Clearly the response at high frequency falls off as

$$\frac{I(f)}{I(0)} = \frac{m'}{m} = \frac{\sin(\pi f t_{tr})}{\pi f t_{tr}} \qquad (12.5.13)$$

which is plotted in Fig. 12.14. This is, of course, just the Fourier transform of the impulse response. In the example given earlier where $t_{tr} = 0.5$ ns, the modulated current would fall by a factor of $\sqrt{2}$ (-3 dB$_{el}$) at a frequency f_m given by

$$f_m \simeq \frac{0.44}{t_{tr}} = 880 \text{ MHz} \qquad (12.5.14)$$

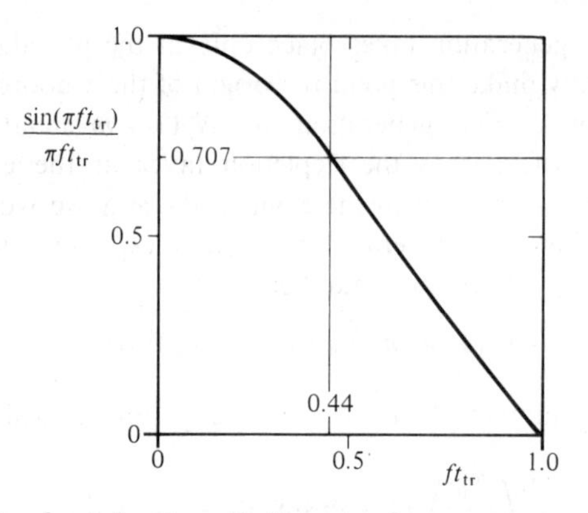

Fig. 12.14 Effect of carrier transit time on frequency response of a p–i–n photodiode. The graph shows the theoretical response when all the carriers are generated at one edge of the depletion-layer and suffer a constant transit time delay of t_{tr}.

When carriers are generated throughout the depletion layer two conflicting effects arise. The drift of holes with their lower saturation velocity tends to reduce the bandwidth, but the carriers have less distance to travel, which reduces t_{tr} and tends to increase the bandwidth. Detailed calculations can only be made if the carrier generation distribution and their drift velocities are specified.

12.6 NOISE IN p–i–n PHOTODIODES

The ultimate sensitivity of a photodiode is determined by the random voltage and current fluctuations that occur at its output terminals, both in the presence and in the absence of an optical signal. The problem lies in the need to recognize the signal as something more than one of the random fluctuations. This is a matter we shall wish to discuss at length in Chapters 14 and 15, but it is appropriate here to mention the sources of noise that arise in the photodiode itself. Essentially these are inherent in the statistical nature of the quantum detection process. The result is that an average photodetector current, $\bar{I}$, always shows a random fluctuation about this mean value and this is known as *shot noise*. The mean square value of the variation, I_{Sh}^2 is proportional to $\bar{I}$ and to the photodiode bandwidth, Δf. Thus,

$$I_{Sh} = (2e\bar{I}\,\Delta f)^{1/2} \tag{12.6.1}$$

A certain level of photodiode current is present even when there is no intended optical signal. This, too, has shot noise associated with it. Although we will refer to this current as the ***dark current***, I_d, it may comprise some photogenerated current, I_b, resulting from background radiation entering the diode in addition to the junction saturation current, I_s. Three components make up I_s. These are the carriers generated in the depletion layer (I_{s2} in eqn. (7.3.5)), those which diffuse into the depletion layer from the p^+ and n^+ regions (I_{s1} in eqn. (7.3.4)), and any surface currents that are set up under the action of the bias field. The surface currents can be minimized by careful attention to processing and surface passivation so that the surface state and impurity ion concentrations are reduced. An expression for the diffusion current, I_{s1}, was derived in Chapter 7: eqn. (7.5.10(a)). In that equation n_A and τ_p now refer to the p^+ region and n_D and τ_n to the n^+ region. The fact that both these regions are heavily doped minimizes this current. Carrier generation within the depletion layer, which we take to be the ν-region of Fig. 12.6, gives rise to a current, $I_{s2} = n_i eAw_2/\tau_v$, where A is the area, w_2 is the thickness and τ_v the minority carrier lifetime of the ν-region and n_i is the intrinsic carrier concentration as given by eqn. (7.2.1). The need to make τ_v as large as possible requires high-quality, defect-free material.

Traditionally a number of figures of merit have been used in order to assess the quality of photodetectors. These were mainly developed as a guide to the weakest, steady or slowly varying, infra-red radiation source that could be detected. They are, therefore, of little significance or use for evaluating the high bandwidth detectors required in optical communication systems. The three parameters most often used are the *noise equivalent power* (*NEP*), the *detectivity* (*D*), and the *specific detectivity* (*D**). For the record we will show how these are related to the wavelength to be detected and the quantum efficiency and dark current of the detector. The detector will be assumed to be working into a very high impedance so that the principal source of noise is the shot noise associated with the dark current and the signal current. The *NEP* is defined as the optical power (of specified wavelength or spectral content) required to produce a detector current equal to the r.m.s. noise current in unit bandwidth, that is, with $\Delta f = 1$ Hz. In order to evaluate *NEP* at a fixed wavelength we may rearrange eqn. (12.1.2) to give

$$\Phi = \frac{I_{ph}}{\eta}\frac{hc}{e\lambda} \tag{12.6.2}$$

and then put

$$I_{ph} = I_{Sh}$$

in eqn. (12.6.1). Thus

$$I_{ph} = I_{Sh} = \{2e(I_{ph} + I_d)\,\Delta f\}^{1/2} \tag{12.6.3}$$

When $I_d \ll I_{ph}$,

$$I_{ph} \simeq 2e\,\Delta f \tag{12.6.4}$$

and substituting this into eqn. (12.6.2) with $\Delta f = 1$ Hz,

$$\Phi = NEP \simeq \frac{2hc}{\eta\lambda} \tag{12.6.5}$$

With $\eta = 1$ this represents the *NEP* of an ideal quantum detector. When $I_d \gg I_{ph}$,

$$I_{ph} \simeq (2eI_d\,\Delta f)^{1/2} \tag{12.6.6}$$

and

$$\Phi = NEP \simeq \frac{hc(2eI_d)^{1/2}}{\eta e\lambda} \tag{12.6.7}$$

Detectivity, D, is defined as

$$D = \frac{1}{NEP} \tag{12.6.8}$$

For a photodiode dominated by dark current and detecting monochromatic radiation,

$$D = D_\lambda = \frac{\eta e\lambda}{hc(2eI_d)^{1/2}} \tag{12.6.9}$$

Finally the specific detectivity, D^*, takes account of the fact that in such photodiodes I_d is often proportional to the area of the detector. This is the case when background radiation and thermal generation are the dominant causes of I_d rather than surface conduction. Thus, D^* is defined as

$$D^* = DA^{1/2} = \frac{\eta e\lambda}{hc(2eI_d/A)^{1/2}} \tag{12.6.10}$$

where A is the detector area.

The wide bandwidth detectors required for optical communication often work into a low resistance and require a minimum signal current which greatly exceeds I_d. Then the load resistance, the amplifier and the signal current itself all introduce additional sources of noise which complicate these parameters and render them inappropriate for our use. We will return to a detailed discussion of noise-limited detection in Chapters 14 and 15.

PROBLEMS

12.1 (a) A Si p–i–n photodiode has a quantum efficiency of 0.7 at a wavelength of 0.85 μm. Calculate its responsivity.
(b) Calculate the responsivity of a Ge p^+n diode at 1.6 μm where its quantum efficiency is 0.4.
(c) A particular photodetector has a responsivity of 0.6 A/W for light of wavelength 1.3 μm. Calculate its quantum efficiency.

12.2 The p^+ contact layer of a Si p–i–n photodiode is 1 μm thick. Using the absorption data given in Fig. 12.4, and assuming that only radiation absorbed in the lightly doped ν-region is effective in contributing to the current generated,
(a) estimate the maximum quantum efficiency that can be expected at a wavelength of 0.9 μm, and
(b) estimate the minimum thickness of the ν-layer needed to ensure a quantum efficiency of 0.8 at this wavelength.
Neglect any reflection losses.

12.3 A silicon p^+–ν–n^+ photodiode is 0.1 mm square. Its ν-layer is 30 μm thick and has a doping concentration of 10^{19} m^{-3}.
(a) Neglecting surface reflection and absorption in the contact layer, and taking $\alpha = 7 \times 10^4$ m^{-1} at $\lambda = 0.82$ μm, calculate the maximum quantum efficiency and responsivity at this wavelength.
(b) Calculate the reverse bias voltage required to "reach through" to the n^+ substrate, and calculate the incremental self-capacitance of the photodiode after reach-through. Assume $\varepsilon_r = 12$.
(c) Calculate the reverse bias voltage required to establish an electric field greater than 10^6 V/m everywhere in the ν-region.
(d) Calculate the transit times for electrons and holes across the ν-region assuming that their average drift velocities are 7×10^4 and 4×10^4 m/s respectively.
(e) Hence obtain the impulse response when carriers are uniformly generated throughout the ν-region.

12.4 Use Table 7.2 to obtain values for n_i^2 and for the electron and hole diffusion coefficients in silicon by means of the Einstein relations, eqns (7.4.7) and (7.4.8). Considering the photodiode of Problem 12.3, and taking $n_A = n_D = 10^{23}$ m^{-3} as the doping concentrations in the p^+ and n^+ regions, respectively, and the carrier lifetime as $\tau = 10$ μs, compare the magnitude of the diffusion current I_{s1} generated in the heavily doped regions with the magnitude of the generation–recombination current I_{s2} generated in the ν-region. Make use of the expressions

$$I_{s1} = en_i^2\{(D_e/\tau)^{1/2}/n_A + (D_h/\tau)^{1/2}/n_D\}$$

and

$$I_{s2} = en_i w/\tau$$

and assume room temperature. Explain how this comparison would be affected at higher temperatures and in diodes made from semiconductors having smaller band gap energies.

12.5 (a) Prove eqn. (12.5.4).

(b) Calculate the maximum values of load resistance that may be used when a photodiode having 1 pF self-capacitance is required to have an upper cutoff frequency of (i) 1 MHz; (ii) 1 GHz.

(c) Derive an expression for $v_{in}(t)$ in Fig. 12.11 when the photogenerated current $i_{ph}(t)$ approximates to a step function $(0 \to I_0)$ at $t = 0$.

REFERENCES

12.1 H. Melchior, M. B. Fisher and F. R. Arams, Photodetectors for optical communication systems, *Proc. I.E.E.E.* **58**, 1466–86 (1970).

12.2 J. Muller, Photodiodes for optical communication, *Advances in Electronics and Electron Physics*, **55**, 189–308 (1981).

SUMMARY

Both p–i–n and avalanche photodiodes require

$$\mathcal{E}_g < \frac{1.24[\mu\text{m eV}]}{\lambda}$$

Responsivity $\mathcal{R} = I_{ph}/\Phi = \eta e\lambda/hc$, where η is the quantum efficiency of detection.

High quantum efficiency requires low surface reflectivity, a depletion layer matched in thickness and position to the absorption region, and negligible carrier recombination. Values between 0.5 and 0.95 can be obtained in practical detectors.

Silicon detectors are used in conjunction with GaAlAs/GaAs sources. Either InGaAs/InP heterostructure, or Ge n^+p detectors are used at longer wavelengths.

Response time is determined by the *CR* time constant of the diode capacitance and its load, by the carrier transit time within the diode, and by the local recombination time if any significant diffusion current is collected from regions outside the depletion layer.

The circuit-limited frequency response is $V_{in}(f)/I_{ph}(f) = R(1 + j2\pi fCR)^{-1}$. In a particular case the transit-time-limited response is

$$I(f)/I(0) = (\sin \pi f t_{tr})/(\pi f t_{tr}).$$

Bandwidths exceeding 1 GHz are practical.

Sensitivity is ultimately limited by the shot noise inherent in the quantum detection process: $I_{Sh} = (2e\bar{I}\,\Delta f)^{1/2}$. For an ideal quantum detector the noise equivalent power, $NEP = 2hf$.

13

Avalanche Photodiode Detectors

13.1 THE MULTIPLICATION PROCESS

13.1.1 Introduction

In Chapter 12 we discussed the noise introduced into the receiver system by the photodiode. In Chapter 14 we shall find that with p–i–n diode detectors this shot noise, which originates in the incoming signal and the dark current, is totally dominated by the additional electronic noise introduced into the system by the load resistor and the rest of the amplifier circuits. It would therefore benefit the signal-to-noise ratio if the signal could be multiplied within the detector itself. The shot noise would, of course, be multiplied as well, but the proportional effect of the additional noise sources would be reduced. As we have mentioned already in Chapter 12, the opportunity to do this is afforded by the carrier avalanche excitation that occurs in semiconductors at high values of the electric field. The multiplication process is not itself totally free of noise. Let us say that on average, each photogenerated carrier leads to the generation of M carriers at the end of the multiplication process. Any one carrier initiating an avalanche may produce more than M carriers or fewer than M carriers and this statistical nature of the process introduces further noise. The result is that whereas the signal current is multiplied by the factor M, the r.m.s. noise level increases by $MF^{1/2}$. The *noise factor*, F, is always greater than unity and is an increasing function of M. A consequence of this is that for any given avalanche photodiode in a particular receiver there is an optimum value of M which gives the best overall signal-to-noise ratio. These are matters we will examine quantitatively and in detail in Chapter 14. Here we will concentrate on the physical mechanisms involved in the multiplication process and on the device structures that have been evolved to exploit it to best effect.

Two different ways of illustrating the process of avalanche carrier generation are shown in Fig. 13.1. In Fig. 13.1(a) the electron energy band diagram is used to show how electron–hole pairs may be produced in an electric field. Although the average carrier drift velocities remain at the saturation values, v_{se}

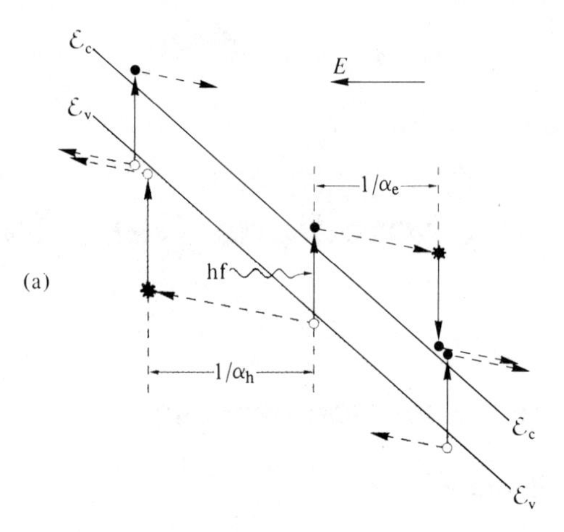

$\mathcal{E}_c$
$\mathcal{E}_v$
E
$1/\alpha_e$
hf
(a)
$1/\alpha_h$
$\mathcal{E}_c$
$\mathcal{E}_v$

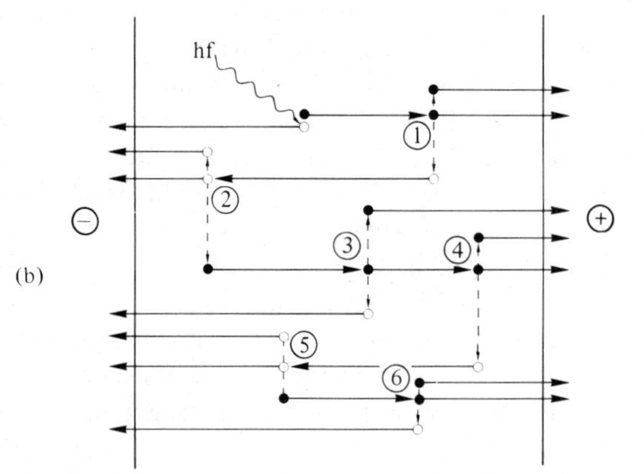

hf
1
2
−
+
3
4
(b)
5
6

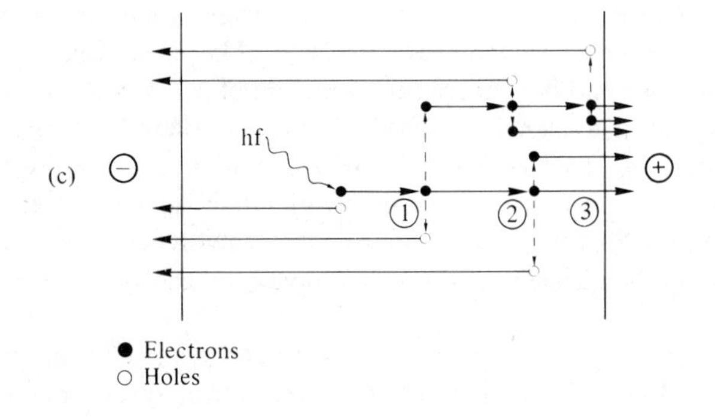

hf
(c)
−
+
1
2
3
● Electrons
○ Holes

and v_{sh}, their energy distributions acquire high-energy tails as a result of the increased acceleration between scattering collisions. In a sufficiently strong field an appreciable number of carriers acquire enough energy (about $\frac{3}{2}\mathcal{E}_g$) to enable them to interact with a suitable valence band electron in such a way as to excite it across the band gap. Both the new electron and the new hole may then go on to produce further excitations. In Fig. 13.1(b) the way in which a single photogenerated electron–hole pair may lead to the production of six further pairs is represented geometrically.

We may define *ionization coefficients* for electrons and holes, α_e and α_h respectively, as the probability that a given carrier will excite an electron–hole pair in unit distance. The coefficients increase so rapidly with increasing electric field strength that it is often convenient to think in terms of a breakdown field, E_B, at which avalanche excitation becomes critical, say α becomes of the order of 10^5–10^6 m^{-1}. Graphs of α_e and α_h versus electric field are plotted in Fig. 13.2 for a number of those semiconductors which we know to be of interest as detector materials. The curves refer to room temperature. As the temperature increases, the ionization coefficients decrease, because the greater number of scattering collisions reduces the high-energy tail of the carrier energy distribution and hence reduces the probability of excitation. In some materials $\alpha_e > \alpha_h$, in others $\alpha_h > \alpha_e$, while in gallium arsenide and gallium phosphide the two coefficients are approximately the same. The ratio

$$k = \alpha_h/\alpha_e \tag{13.1.1}$$

is found to lie in the range 0.01 to 100. In making the assumption that α_e and α_h are functions only of the electric field strength E, we are implying that

---

Fig. 13.1 The principle of avalanche multiplication in semiconductors: (a) ionizations initiated by a pair of photogenerated carriers in a region of high electric field are illustrated on the electron energy band diagram. The carriers lose some energy in scattering collisions as they move through the semiconductor, but eventually, in an average distance $1/\alpha_e$ or $1/\alpha_h$, acquire enough kinetic energy to make an ionizing collision (*); (b) a schematic illustration of the way in which a single photogenerated carrier pair may lead to six further ionizations ($M = 7$). This brings out the element of positive feedback in the avalanche process and shows how it may occupy several electron and hole transit times. In this illustration the value of $k = \alpha_h/\alpha_e$ is taken to be about 0.5; (c) when $k = 0$, only electrons cause ionizing collisions and the avalanche proceeds without feedback ($M = 5$).

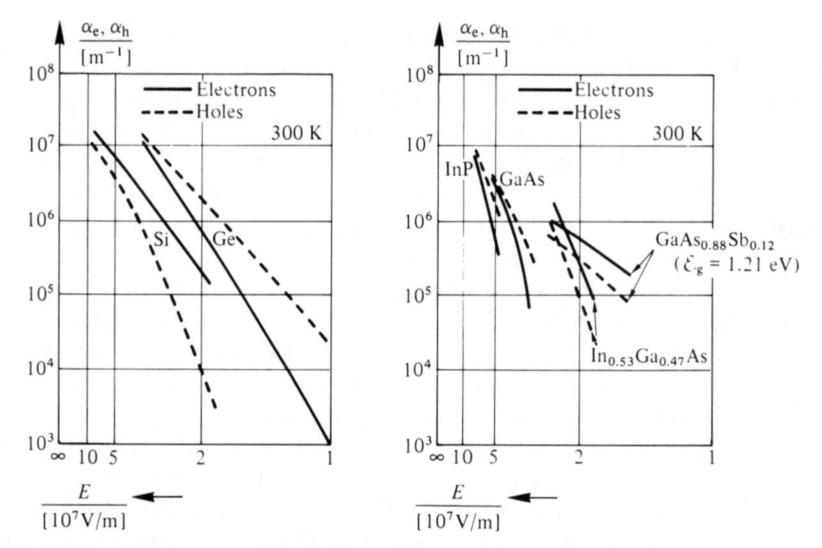

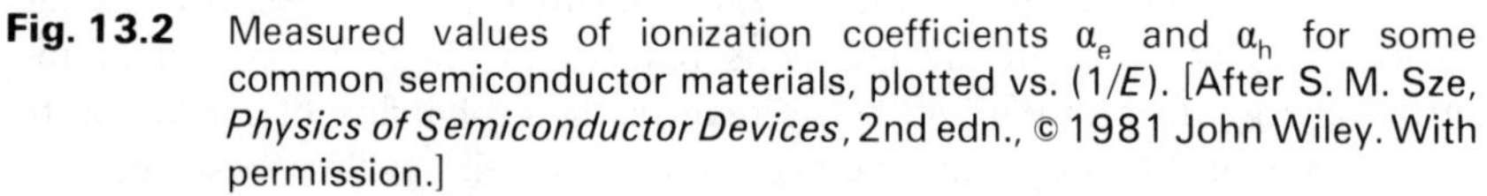

Fig. 13.2 Measured values of ionization coefficients α_e and α_h for some common semiconductor materials, plotted vs. $(1/E)$. [After S. M. Sze, *Physics of Semiconductor Devices*, 2nd edn., © 1981 John Wiley. With permission.]

within the average distance between ionizing collisions ($1/\alpha_e$ or $1/\alpha_h$)

(a) the fractional change in E is small;

(b) the number of elastic collisions is large so that an equilibrium velocity distribution is established;

(c) the potential energy lost by the carrier is large compared to the energy given up in the ionizing collision; thus $eE \gg \frac{3}{2}\alpha\mathcal{E}_g$.

Were only one type of carrier able to ionize, the electrons say, implying $k = 0$, then the avalanche would develop in the rather simpler manner shown in Fig. 13.1(c). Over a distance long compared with $(1/\alpha_e)$ the number of electrons would grow exponentially. In real materials the fact that $k \neq 0$ introduces an element of positive feedback into the multiplication process with the result that it is possible in theory for the number of electron–hole pairs to grow indefinitely within a finite distance. (This process is analogous to Townsend breakdown in a gas, but is in fact slightly more complicated. In the gas an initial electron generates ions and electrons. The ions fall back to the negative electrode where they have a finite probability of causing another electron to be emitted. When one initial electron produces enough ions to lead on average to the emission of one further electron, then the current may grow without limit and breakdown occurs. In practice a new form of discharge stabilizes at a higher current and a lower voltage.)

13.1.2 Avalanche Multiplication Theory

We will examine avalanche multiplication in a semiconductor by considering a steady-state situation in which an electron current $i_e(0)$ is injected into a depletion layer of width w at the plane $x = 0$. We will assume that the field strength is high enough to produce an avalanche, but that there is negligible thermal- or photo-generation of carriers in this region. As we shall see later, these assumptions are quite realistic for well designed APDs made from materials like silicon in which $\alpha_e \gg \alpha_h$. At any point x such that $0 < x < w$, the rate of generation of carriers is given by

$$\frac{\mathrm{d}i_e(x)}{\mathrm{d}x} = \alpha_e i_e(x) + \alpha_h i_h(x) \tag{13.1.2}$$

where $i_e(x)$ and $i_h(x)$ are the electron and hole currents, respectively, at the point x. At every point

$$i_e(x) + i_h(x) = I = \text{constant} \tag{13.1.3}$$

If there is no injection of holes at $x = w$, $i_h(w) = 0$ and $i_e(w) = I$. Substitution of eqn. (13.1.3) into eqn. (13.1.2) gives

$$\frac{\mathrm{d}i_e(x)}{\mathrm{d}x} - (\alpha_e - \alpha_h) i_e(x) = \alpha_h I \tag{13.1.4}$$

Equation (13.1.4) is a standard form† which integrates to give

$$i_e(x) = \frac{i_e(0) + \int_0^x \alpha_h I \exp\left\{-\int_0^x (\alpha_e - \alpha_h)\,\mathrm{d}x'\right\} \mathrm{d}x}{\exp\left\{-\int_0^x (\alpha_e - \alpha_h)\,\mathrm{d}x'\right\}} \tag{13.1.5}$$

---

† Equation (13.1.4) has the form

$$\frac{\mathrm{d}y}{\mathrm{d}x} + Py = Q$$

Multiplying through by the integrating factor $R = \exp\{\int P\,\mathrm{d}x\}$ gives

$$\frac{\mathrm{d}(Ry)}{\mathrm{d}x} = RQ$$

so that

$$Ry = \int RQ\,\mathrm{d}x + \text{constant.}$$

We may define a multiplication factor for the injected electrons as

$$M_e = \frac{I}{i_e(0)} = \frac{i_e(w)}{i_e(0)} \tag{13.1.6}$$

$$= \frac{i_e(0) + i_e(w)\int_0^w \alpha_h \exp\left\{-\int_0^x (\alpha_e - \alpha_h)\,dx'\right\} dx}{i_e(0)\exp\left\{-\int_0^w (\alpha_e - \alpha_h)\,dx\right\}} \tag{13.1.7}$$

by direct substitution into eqn. (13.1.5). Rearranging gives

$$M_e = \frac{1}{\exp\left\{-\int_0^w (\alpha_e - \alpha_h)\,dx\right\} - \int_0^w \alpha_h \exp\left\{-\int_0^x (\alpha_e - \alpha_h)\,dx'\right\} dx} \tag{13.1.8}$$

Using the fact that‡

$$\exp\left\{-\int_0^w (\alpha_e - \alpha_h)\,dx\right\} = 1 - \int_0^w (\alpha_e - \alpha_h)\exp\left\{-\int_0^x (\alpha_e - \alpha_h)\,dx'\right\} dx \tag{13.1.9}$$

‡

$$\frac{d(e^y)}{dx} = e^y \frac{dy}{dx}$$

$$\therefore \quad e^y = \int d(e^y) = \int e^y \frac{dy}{dx}\,dx + \text{constant}$$

Put

$$y = -\int_0^x (\alpha_e - \alpha_h)\,dx'$$

Then

$$\frac{dy}{dx} = -(\alpha_e - \alpha_h)$$

and

$$\exp\left\{-\int_0^x (\alpha_e - \alpha_h)\,dx'\right\} = -\int_0^x (\alpha_e - \alpha_h)\exp\left\{-\int_0^x (\alpha_e - \alpha_h)\,dx'\right\} dx + \text{constant}$$

The constant may be shown to be 1 by evaluating these expressions at $x = 0$.

$$\therefore \quad \exp\left\{-\int_0^w (\alpha_e - \alpha_h)\,dx\right\} = 1 - \int_0^w (\alpha_e - \alpha_h)\exp\left\{-\int_0^x (\alpha_e - \alpha_h)\,dx'\right\} dx \tag{13.1.9}$$

we obtain

$$M_e = \frac{1}{1 - \int_0^w \alpha_e \exp\left\{ -\int_0^x (\alpha_e - \alpha_h)\, dx' \right\} dx} \tag{13.1.10}$$

The condition for breakdown, $M \to \infty$, is

$$\int_0^w \alpha_e \exp\left\{ -\int_0^x (\alpha_e - \alpha_h)\, dx' \right\} dx = 1 \tag{13.1.11}$$

If we include carrier generation, $G(x)$, and hole injection, $i_h(w)$, in the analysis, expressions similar to but rather more complicated than eqn. (13.1.10) are obtained. However, the breakdown condition (13.1.11) remains the same. When $\alpha_e = \alpha_h$, that is $k = 1$, the equations become very straightforward: then, the total current

$$I = \frac{i_e(0) + i_h(w) + \int_0^w eG(x)\, dx}{1 - \int_0^w \alpha_e\, dx} \tag{13.1.12}$$

However, under these conditions ($k = 1$) values of $M > 10$ become critically dependent on the electric field strength and thus vary across the area of the diode if there is the slightest variation of the doping profile. This effect becomes even more marked when the 'wrong' type of carrier initiates the avalanche, that is, electrons in a material where $k > 1$ or holes when $k < 1$.

The analysis is also greatly simplified when it can be assumed that the avalanche occurs in a region of uniform electric field. Then α_e and α_h are independent of x and eqn. (13.1.10) takes on the form

$$\begin{aligned} M_e &= \frac{1}{1 - \alpha_e \int_0^w \exp\{-(1-k)\alpha_e x\}\, dx} \\ &= \frac{1}{1 - \frac{1}{(1-k)}[1 - \exp\{-(1-k)\alpha_e w\}]} \\ &= \frac{(1-k)}{\exp\{-(1-k)\alpha_e w\} - k} \end{aligned} \tag{13.1.13}$$

Then, as $k \to 1$, $M_e \to 1/(1 - \alpha_e w)$.

13.1.3 Experimental Behavior

Two factors limit the increase of M_e and hence I as the applied voltage approaches the breakdown voltage, V_B, at which the values of α_e and α_h satisfy condition (13.1.11). The first is the series resistance of the bulk semi-conductor, R_S, between the junction and the diode terminals. The second is the effect of the rise in temperature resulting from the increased dissipation as the current rises. This reduces the values of α_e and α_h and raises the breakdown voltage. It also increases the rate of thermal generation of carriers and hence the dark current. Multiplication factors measured as a function of the applied terminal voltage, V, can usually be fitted to the form

$$M = \frac{1}{[1 - (V - IR')/V_B]^n} \tag{13.1.14}$$

where $R' = R_S + R_{Th}$ is the sum of the series resistance, R_S, and an effective resistance, R_{Th}, which derives from the rise in temperature. The index, n, is a function of the detailed design and the material of the diode. Some typical curves of $M(V)$ for a silicon APD are shown in Fig. 13.4(b) in the next section.

13.2 APD DESIGNS

It is just as necessary to obtain the highest possible quantum efficiency in an avalanche photodiode as it is in a nonavalanching device, and all the requirements discussed in Section 12.3 remain important. In addition it is essential that carrier multiplication should take place uniformly across the whole area illuminated by the incident radiation. High quality material with a minimum of defects and dislocations must be used because these imperfections tend to cause a local enhancement of the electric field and produce premature avalanches or *microplasmas* in their immediate vicinity.

Formation of localized microplasmas is to be expected even in perfect material when it is subjected to a uniform field approaching the breakdown field. This makes the simple p–i–n structure unsatisfactory for APDs. The reason is that a negative resistance develops and an instability grows in any region where an avalanche is initiated. It is therefore best for the peak field, where the avalanche multiplication will occur, to be confined to a very thin layer. This is kept separate from the main optical absorption region, which is still required to have a thickness of the order of the absorption depth ($\sim 1/\alpha$)

and to have established across it an electric field strong enough to maintain carriers at their saturation drift velocities. The *reach-through* structure shown in Fig. 13.3 satisfies all these requirements. The avalanche should be initiated by the carrier with the higher ionization coefficient, because otherwise the APD bandwidth is reduced and its noise factor is increased. The $n^+p{-}\pi{-}p^+$ structure illustrated is therefore most suited to a material like silicon where $k \ll 1$. The photogenerated electrons initiate the avalanche and the holes then produced give rise to little further carrier generation.

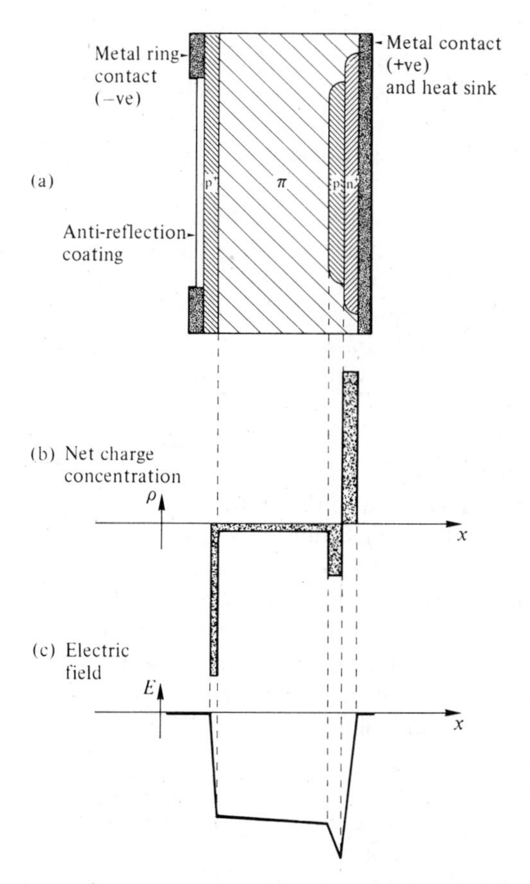

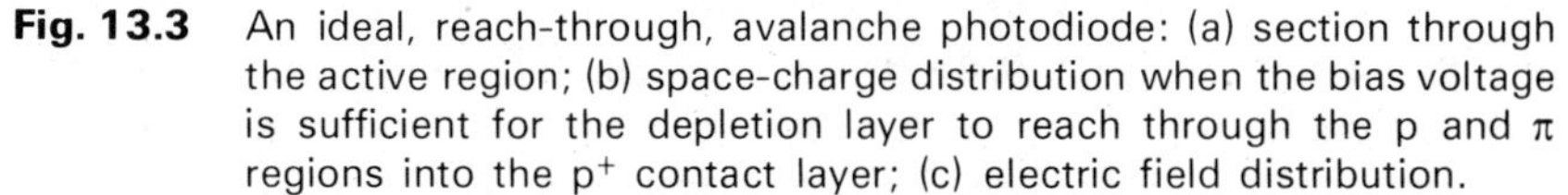
Fig. 13.3 An ideal, reach-through, avalanche photodiode: (a) section through the active region; (b) space-charge distribution when the bias voltage is sufficient for the depletion layer to reach through the p and π regions into the p^+ contact layer; (c) electric field distribution.

The APD shown in Fig. 13.3 might be fabricated by diffusing n-type and p-type dopants into a p-type slice of suitably high resistivity from either side. It has the particular merit that the n^+p junction where dissipation is greatest can be kept very close to the positive contact and heat sink. This minimizes the thermal impedance and helps maintain the thermal stability of the detector.

An alternative design, which sacrifices this last advantage but which can make use of the very high quality material available with silicon planar epitaxial technology, is shown in Fig. 13.4. The π-layer is grown epitaxially on the p^+ substrate. The p-layer, which must be thin and very uniformly doped, may be introduced by diffusion or by ion implantation. The n-type guard ring, the n^+ contact layer and the p^+ channel stop may be formed by further diffusions. This rather complicated structure is designed to ensure that the avalanche multiplication takes place uniformly across the active part of the device and not elsewhere. If the p- and n^+-layers were diffused in the simple form shown in Fig. 13.5, there would be an enhancement of the electric field at the corners, so that most of the avalanche multiplication would be confined to the periphery of the absorption region and would be relatively ineffective. Two features shown in Fig. 13.4, the guard-ring diffusion (B) and the extension of the ring-contact (C), have been introduced in order to minimize the electric fields around the p–n junction edge. The guard ring extends the n-type layer outside the p-diffusion. It has a lower n-type doping concentration and ideally forms a graded n–π junction with a breakdown voltage much higher than that of the n^+p junction of the central area. The radius of curvature of the n–π junction is increased over what would have been possible had the n^+-layer penetrated straight into the π-layer. This again helps to keep the edge breakdown voltage high. The contact ring is extended over the silica passivation so that it covers the n–π junction. This has the effect of widening the depletion layer at the surface and so minimizing the surface field. The effect is analogous to that of a depletion-mode MOSFET. Silicon APDs of the types shown in Figs. 13.3 and 13.4 can give multiplication factors of several hundred before microplasmas develop and the excess noise increases sharply in consequence. The quantum efficiency for wavelengths in the region of 0.85 μm can exceed 0.9, and the unmultiplied dark current can be reduced to the picoampere level at room temperature.

Germanium APDs suffer from a number of inherent problems which cause the useful gain to be limited to about 10–20 times. Dark current is high (μA) because of the high thermal generation rate and as a result of surface leakage. Defect-free substrate material is not readily available, and surface passivation is a problem. Most germanium photodiodes take the form of an abrupt n^+p junction made by diffusing donors into a p-type substrate. An example is shown in Fig. 13.6. This arrangement is used because the diffusion of p-type

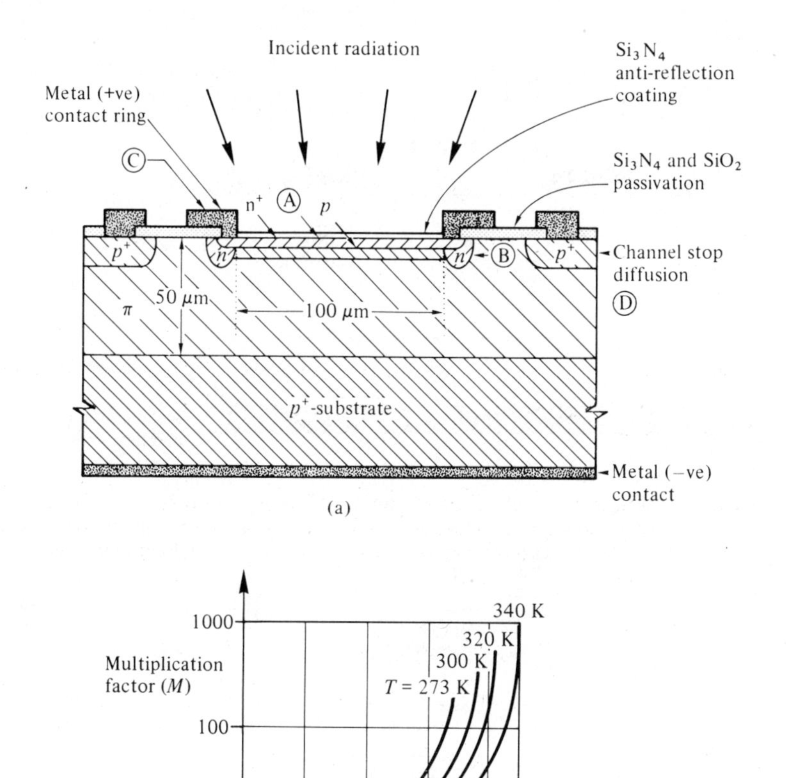

Fig. 13.4 A planar epitaxial silicon reach-through APD: (a) schematic cross-section; (b) variation of the avalanche multiplication factor, M, with applied voltage and temperature [taken from H. Melchoir *et al.*, Planar epitaxial silicon avalanche photodiode, *Bell Syst. Tech. Jnl.* **57**, 1791–7 (1978)].

Attention is drawn to the following features of (a): Ⓐ thickness of n^+-layer may be further reduced by etching; Ⓑ guard-ring diffusion to increase periphery breakdown voltage; Ⓒ extension of ring contact outside n-layer; Ⓓ channel-stop diffusion to limit lateral spread of depletion layer.

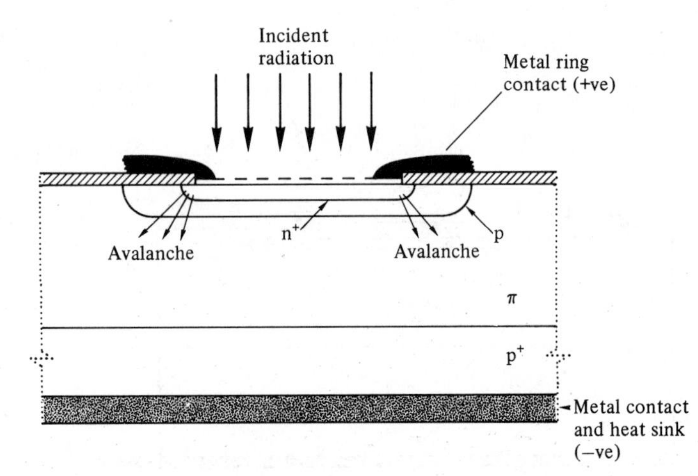

Fig. 13.5 A simple type of diffused reach-through structure.
This shows how field-enhancement at the periphery of the diffused n^+-layer may lead to premature avalanche breakdown taking place away from the active part of the device.

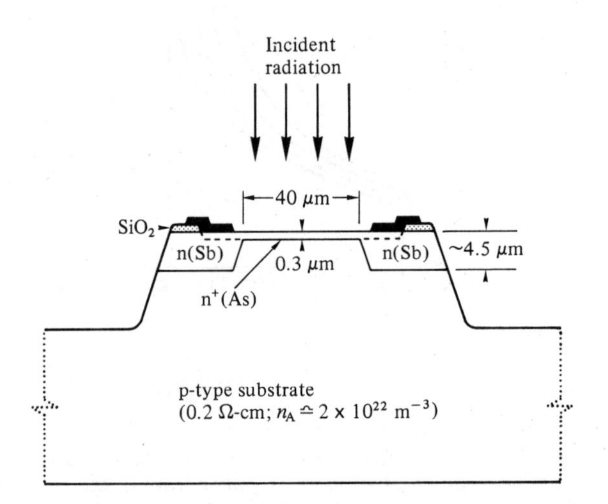

Fig. 13.6 An example of a mesa-etched, n^+p germanium APD, having 16 V breakdown voltage and a dark current less than 0.1 μA. This device was described in H. Melchoir and W. T. Lynch, Signal and noise response of high speed germanium avalanche photodiodes, *IEEE Trans. on Electron Devices*, **ED-13**, 829–38 (1966).

impurities in germanium is difficult to manage, but it means that the 'wrong' carriers, electrons, initiate the avalanche. As $k \gtrsim 1$ for germanium, bandwidth and noise are sacrificed with $F \simeq M$. More recently p^+n germanium APDs have been produced using ion implantation. With the diameter of the active region reduced to 30 μm, the dark current is about 0.1 μA and the capacitance about 0.5 pF, and quantum efficiencies of 0.9 and noise factors $F \lesssim M$ can be obtained.

Heterojunction APDs using III–V compound semiconductors are under intense development for wavelengths longer than 1 μm. They, too, suffer from difficulties with surface passivation and the quality of bulk material. The result is that only modest multiplication factors can be obtained. Furthermore, the ion and electron ionization coefficients tend to be comparable in these materials so for the reasons discussed in Sections 13.3 and 13.4 it is more difficult to achieve high bandwidth and low noise. Figure 13.7 shows a

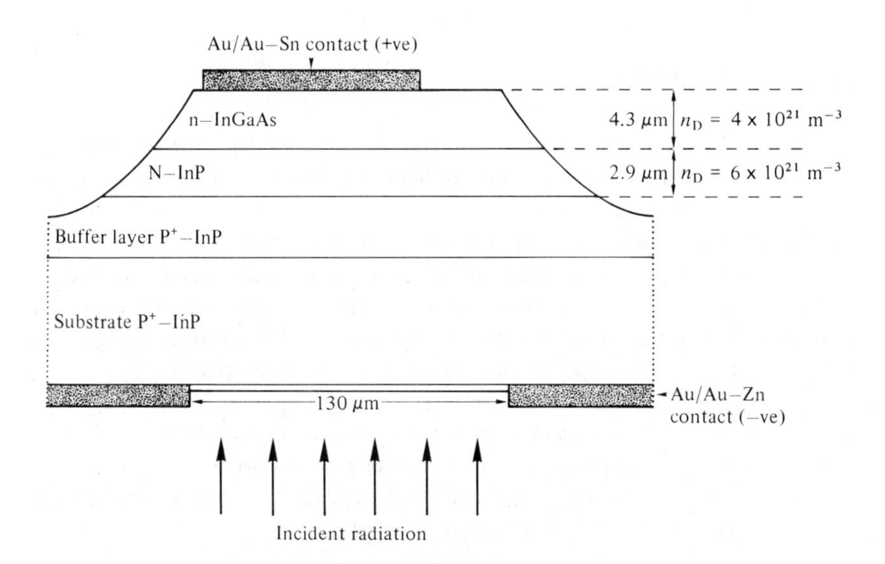

Fig. 13.7 A heterojunction APD.
In this particular example the p–n junction is formed in InP and an N—n heterojunction is formed between epitaxial layers of InP and lattice-matched InGaAs. Radiation is absorbed in the narrower band-gap ternary material. The thickness and doping concentration of the N–InP layer are controlled so as to ensure that at breakdown when the electric field at the P–N junction is about 4.5×10^7V/m, the field at the heterojunction does not exceed 1.5×10^7V/m. This field is sufficient to establish saturation carrier drift velocities, but not so great as to induce excess leakage by tunnelling through the heterojunction.

particular example of an $In_{0.53}Ga_{0.47}As$ heterojunction APD formed by liquid-phase epitaxy on an InP substrate. Attention may be drawn to a number of its features. The P^+–InP buffer layer is used to isolate the active region from the dislocations and defects present in the substrate. The P^+N junction is formed in epitaxial InP, followed by the N–n heterojunction between InP and the lattice-matched ternary material. The diode is then mesa-etched in order to minimize self-capacitance and to avoid passivation problems. Illumination is brought in through the InP substrate, which is transparent to wavelengths longer than 0.92 μm. Absorption takes place in the ternary material and it is the photogenerated holes that initiate the avalanche. The thickness and doping concentration of the N–InP layer are carefully controlled to ensure that the depletion layer reaches through to the ternary material but that at the onset of avalanche the field at the heterojunction does not exceed about 1.5×10^7 V/m. In this particular structure it is found that higher fields than this induce tunnelling currents in the heterojunction and give rise to excessive dark current.

13.3 APD BANDWIDTH

In this section we will avoid a detailed analysis of the consequences of sinusoidal modulation of the incident light but concentrate instead on the response of an APD to an optical impulse, as in the first part of Section 12.5. The full theory, which has much in common with the theory of IMPATT and TRAPATT oscillators, is complex and difficult, so we shall limit the discussion to the general physical principles and to an order-of-magnitude estimate of the bandwidth limitation. In the n^+–p–π–p^+-type of APD illustrated in Figs. 13.3 and 13.4, the overall response time is made up of three parts:

(a) the electron transit time across the drift region, $(t_{tr})_e = w_2/v_{se}$,
(b) the time required for the avalanche to develop, t_A,
(c) the transit time of the last holes produced in the avalanche back across the drift space, $(t_{tr})_h = w_2/v_{sh}$.

Parts (b) and (c) represent delays additional to those experienced in a non-avalanching diode.

The avalanche delay time, t_A, is a function of the ratio of the ionization coefficients, k. The distance–time diagrams of Fig. 13.8 give a graphic illustration of this. When $k = 0$, the avalanche develops within the normal electron transit time across the avalanche region (w_A/v_{se}). We assume $w_A \ll w_2$. When $k > 0$, the avalanche develops in multiple passes across the avalanche region

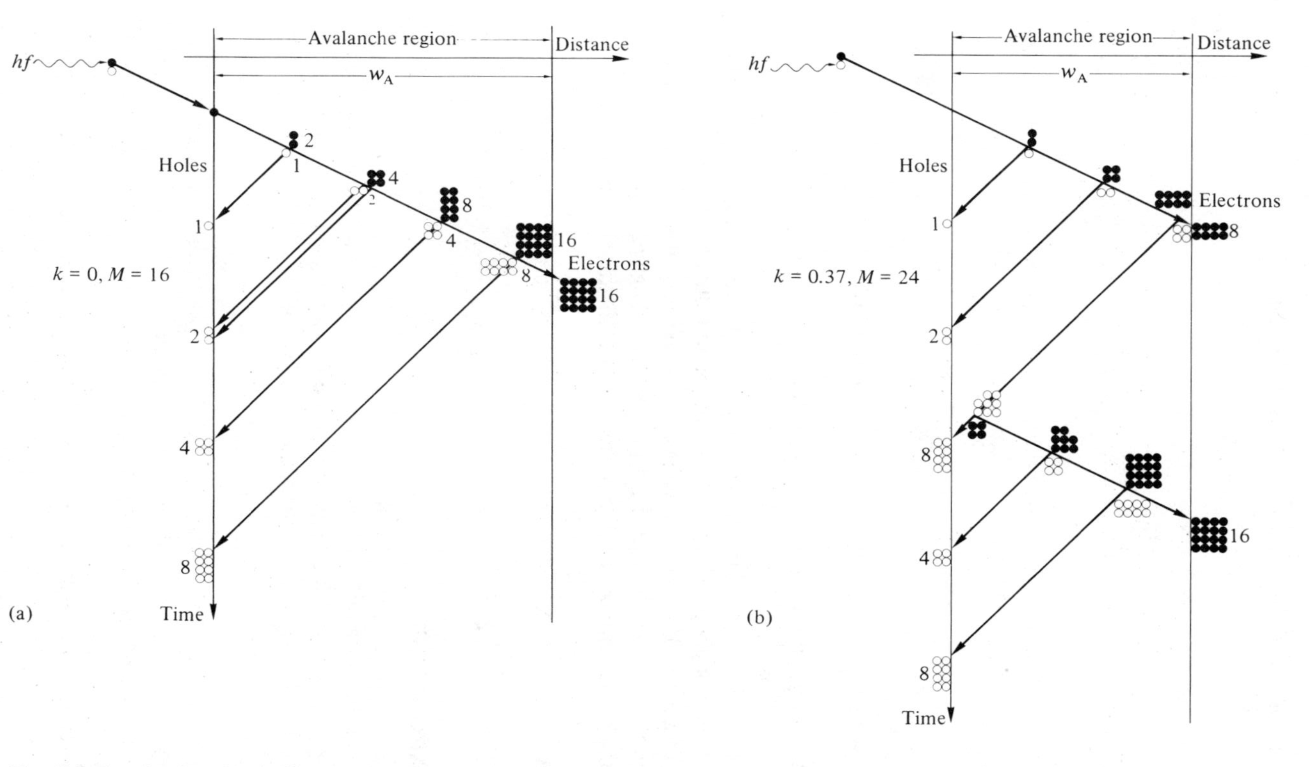

Fig. 13.8 Avalanche build-up shown on distance–time diagrams: (a) $k = 0$, $M = 16$; (b) $k = 0.37$, $M = 24$.

It will be appreciated that these diagrams have taken no account of the statistical nature of the ionization process. The drift velocities are assumed to be constant with the electron velocity twice the hole velocity, as indicated by the slopes of the lines on the diagrams.

and at high levels of multiplication, with $0 < k < 1$,

$$t_A \simeq Mkw_A/v_{se} \tag{13.3.1}$$

The overall response time, τ, then becomes

$$\tau \simeq \frac{(w_2 + Mkw_A)}{v_{se}} + \frac{(w_2 + w_A)}{v_{sh}} \tag{13.3.2}$$

and following the argument given in the second part of Section 12.5 we should expect the (−3 dB) bandwidth to be given approximately by

$$f_{(-3\,dB)} \simeq \frac{0.44}{\tau} \tag{13.3.3}$$

Let us extend the numerical example in Section 12.5 to a silicon diode of similar dimensions, designed to avalanche. Then, $v_{se} \simeq 10^5$ m/s, $v_{sh} \simeq 5 \times 10^4$ m/s, $k \simeq 0.1$. Assume $w_2 = 50$ μm, $w_A = 0.5$ μm and $M = 100$. Then, $t_A \simeq 50$ ps and $\tau \simeq 500 + 50 + 1010 = 1560$ ps $\simeq$ 1.6 ns, and $f_{(-3\,dB)} \simeq 280$ MHz. A similar nonavalanching diode would be expected to show a response time of about 0.5 ns and a −3 dB bandwidth of 880 MHz. The multiplication factor of 100 has been gained at the expense of a loss of a factor of about 3 in the bandwidth. The effect of eqn. (13.3.1) gives another reason for seeking materials with $k \ll 1$ for devices in which the electrons initiate the avalanche. Converse arguments apply throughout if holes initiate the avalanche; then it is desirable for k to be much greater than one.

13.4 APD NOISE

The value of the noise factor, F, and its variation with the multiplication factor, M, are clearly matters which bear on the optimization of the optical receiver. For purposes of system evaluation the approximation

$$F \simeq M^x \tag{13.4.1}$$

has often been used. The index, x, typically takes on values between 0.2 and 1.0, depending on the material and the type of carrier initiating the avalanche. As we shall see, eqn. (13.4.1) may be reasonably valid over a limited range of values of M.

A theoretical treatment by McIntyre [13.1], yields the following more complex expressions. When the multiplication is initiated by electrons

$$F_e = M_e[1 - (1 - k)(M_e - 1)^2/M_e^2] \tag{13.4.2}$$

When holes initiate the avalanche

$$F_h = M_h \left[1 + \frac{(1-k)}{k} \frac{(M_h^2 - 1)^2}{M_h^2} \right] \tag{13.4.3}$$

Equations (13.4.1) and (13.4.2) are plotted in Fig. 13.9 for comparison. It can be seen that when $k \ll 1$, then it is the electron current that should be multiplied in the avalanche, and when $k \gg 1$, the holes should be used.

We now have three very good reasons for avoiding materials in which $k \simeq 1$, and for choosing the carrier with the higher ionizing coefficient to initiate the avalanche. We saw in Section 13.1 that this would make the gain more stable and more uniform in a practical diode. We found in Section 13.3 that it gave a higher bandwidth. Now it is clear that this will also minimize the avalanche noise introduced.

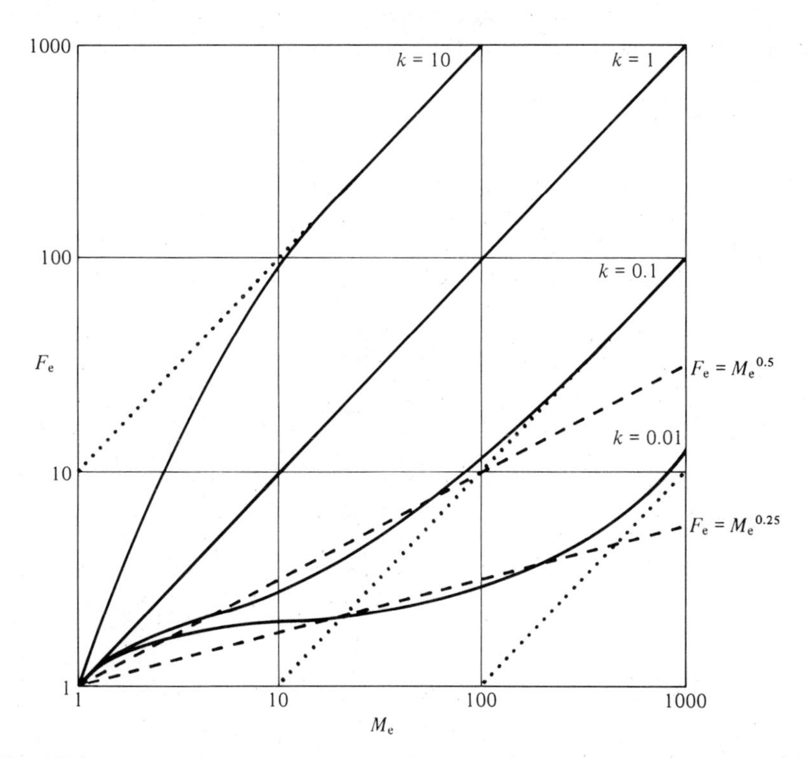

Fig. 13.9 Theoretical curves relating the noise factor, F_e, to the multiplication factor, M_e, when avalanche multiplication is initiated by electrons. The solid curves represent eqn. (13.4.2) for $k = 0.01$, 0.1, 1, 10. At large values of M_e they become asymptotic to $F_e = kM_e$ as shown by the dotted lines. The dashed lines represent eqn. (13.4.1) with $x = 0.25$ and 0.5.

We will next give a derivation of eqn. (13.4.2) as applied to the diode shown in Fig. 13.10, in which electrons generated at $x < 0$ initiate an avalanche in the region $0 < x < w$. We assume negligible hole injection at $x = w$ and negligible carrier generation between $x = 0$ and $x = w$ except as part of the avalanche process. That is, optical and thermal generation are neglected there. In Section 13.1 we derived eqn. (13.1.10) for the electron multiplication factor, M_e, that applies to electrons entering at $x = 0$. To evaluate the noise generation we now need to consider the multiplication factor, $M(x)$, that will apply to a carrier pair generated at some point x in the avalanche region. This is simply

$$M(x) = 1 + \int_0^x \alpha_h M(x')\,dx' + \int_x^w \alpha_e M(x')\,dx' \tag{13.4.4}$$

The second term on the right-hand side expresses the multiplication of the holes generated at x as they travel back to $x = 0$. The final term gives the electron multiplication between x and w. Differentiation of eqn. (13.4.4) gives

$$\frac{dM(x)}{dx} = -(\alpha_e - \alpha_h)M(x) \tag{13.4.5}$$

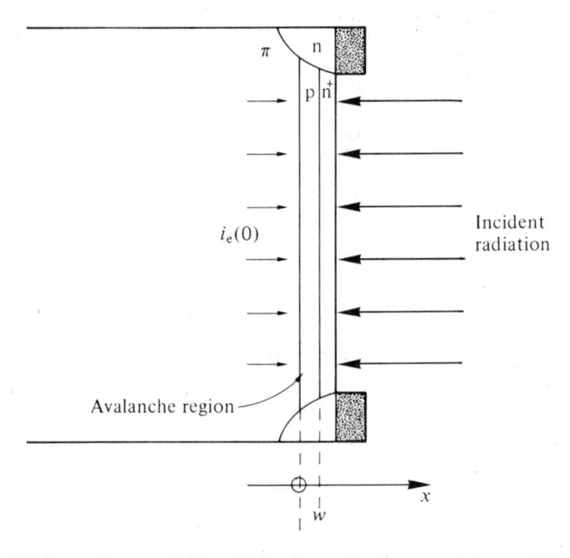

Fig. 13.10 Schematic cross-section through the active region of a reach-through APD showing the conditions under which eqn. (13.4.2) applies.

which may be solved to give

$$M(x) = M(0)\exp\left\{-\int_0^x (\alpha_e - \alpha_h)\,dx'\right\} \tag{13.4.6}$$

Note that $M(0) = M_e$.

From eqns. (13.4.4) and (13.4.5) we may derive two results which will be useful in the subsequent analysis. Substituting $x = w$ into eqn. (13.4.4) gives

$$M(w) = 1 + \int_0^w \alpha_h M(x)\,dx$$

$$= 1 + \frac{k}{(1-k)}\int_0^w (\alpha_e - \alpha_h)M(x)\,dx \tag{13.4.7}$$

with $k = \alpha_h/\alpha_e$ = constant. Substituting eqn. (13.4.5) gives

$$M(w) = 1 - \frac{k}{(1-k)}\int_0^w d\{M(x)\}$$

$$= 1 + \frac{k}{(1-k)}\{M(0) - M(w)\} \tag{13.4.8}$$

$$\left.\begin{aligned} \therefore\quad M(w) &= 1 - k + kM(0) \\ \text{or}\qquad M(w) - 1 &= k\{M(0) - 1\} \end{aligned}\right\} \tag{13.4.9}$$

An integral which we shall wish to evaluate later is $2\int_0^w \alpha_h M^2(x)\,dx$, and making similar substitutions this becomes:

$$2\int_0^w \alpha_h M^2(x)\,dx = \frac{2k}{(1-k)}\int_0^w (\alpha_e - \alpha_h)M^2(x)\,dx$$

$$= -\frac{k}{(1-k)}\int_0^w d\{M^2(x)\}$$

$$= \frac{k}{(1-k)}\{M^2(0) - M^2(w)\} \tag{13.4.10}$$

The theory of avalanche noise rests on two premises concerning the carrier pairs generated at x:

(a) that they will be multiplied by $M(x)$;

(b) that they are created in independent, random processes and thus carry full shot noise.

In this analysis we shall neglect all transit time effects and bandwidth limitations. We saw in Chapter 12 that the mean square shot noise per unit bandwidth associated with a randomly generated mean current, $\bar{I}$, was $2e\bar{I}$. We shall refer to this as the *mean square noise spectral density* and identify it symbolically with an asterisk. Thus

$$(I_{\mathrm{Sh}}^*)^2 = 2e\bar{I} \tag{13.4.11}$$

The current of electrons generated in the element $\mathrm{d}x$ at x is

$$\mathrm{d}i_{\mathrm{e}}(x) = \{\alpha_{\mathrm{e}} i_{\mathrm{e}}(x) + \alpha_{\mathrm{h}} i_{\mathrm{h}}(x)\}\ \mathrm{d}x \tag{13.4.12}$$

There is of course an equal current of holes produced, but we have chosen, quite arbitrarily, to work in terms of the electron currents. Associated with this current is a mean square shot noise spectral density of $2e\,\mathrm{d}i_{\mathrm{e}}(x)$. On the basis of the first premise we may assume that both the current $\mathrm{d}i_{\mathrm{e}}(x)$ and the r.m.s. noise associated with it $\{2e\,\mathrm{d}i_{\mathrm{e}}(x)\}^{1/2}$ are multiplied in the diode by the factor $M(x)$. Now, the noise generated in the element $\mathrm{d}x$ is independent of, and uncorrelated with, the noise generated elsewhere, so the total mean square noise spectral density (I_{M}^*) arising in the avalanche region is given by

$$(I_{\mathrm{M}}^*)^2 = \int_{x=0}^{w} M^2(x)\,.\,2e\,\mathrm{d}i_{\mathrm{e}}(x) \tag{13.4.13}$$

This may be integrated by parts:

$$\int_{x=0}^{w} M^2(x)\,\mathrm{d}i_{\mathrm{e}}(x) = [M^2(x) i_{\mathrm{e}}(x)]_{x=0}^{w} - \int_{x=0}^{w} i_{\mathrm{e}}(x)\,.\,2M(x)\frac{\mathrm{d}M(x)}{\mathrm{d}x}\,\mathrm{d}x \tag{13.4.14}$$

We may put in the limits on the first term on the right-hand side and use eqn. (13.4.5) to simplify the last term. Thus,

$$\int_{x=0}^{w} M^2(x)\,\mathrm{d}i_{\mathrm{e}}(x) = M^2(w) i_{\mathrm{e}}(w) - M^2(0) i_{\mathrm{e}}(0) + 2\int_{x=0}^{w} (\alpha_{\mathrm{e}} - \alpha_{\mathrm{h}}) i_{\mathrm{e}}(x) M^2(x)\,\mathrm{d}x \tag{13.4.15}$$

At this point we recall that the total current,

$$I = i_{\mathrm{e}}(x) + i_{\mathrm{h}}(x) = M(0) i_{\mathrm{e}}(0) = i_{\mathrm{e}}(w) \tag{13.4.16}$$

since $i_h(w) = 0$, and that

$$\frac{di_e(x)}{dx} = \alpha_e i_e(x) + \alpha_h i_h(x) \tag{13.4.12}$$

Thus,

$$(\alpha_e - \alpha_h) i_e(x) = \frac{di_e(x)}{dx} - \alpha_h I \tag{13.4.17}$$

Substitution of eqn. (13.4.17) into the final term of eqn. (13.4.15) enables the integral to be evaluated:

$$\begin{aligned}\int_0^w M^2(x)\, di_e(x) &= M^2(w) i_e(w) - M^2(0) i_e(0) \\ &\quad + 2\int_0^w M^2(x)\, di_e(x) - 2I\int_0^w \alpha_h M^2(x)\, dx \\ &= \{M(0) - M^2(w)\} I + \frac{Ik}{(1-k)}\{M^2(0) - M^2(w)\}\end{aligned} \tag{13.4.18}$$

where we have made use of eqn. (13.4.16) for the first term and eqn. (13.4.10) for the second term. Thus, finally,

$$\begin{aligned}(I_M^*)^2 &= 2eI\left[\{M(0) - M^2(w)\} + \frac{k}{(1-k)}\{M^2(0) - M^2(w)\}\right] \\ &= 2eI\left\{M(0) + \frac{k}{(1-k)} M^2(0) - \frac{M^2(w)}{(1-k)}\right\}\end{aligned} \tag{13.4.19}$$

To the mean square noise generated in the multiplication process, $(I_M^*)^2$, must be added the multiplied mean square noise fed in on the initial electron current. This is

$$M^2(0)(I_{Sh}^*)^2 = M^2(0) \,.\, 2ei_e(0) = 2eIM(0) \tag{13.4.20}$$

The noise factor, $F = F_e$, was defined as the ratio of the total noise to this multiplied shot noise, so that:

$$F_e = 1 + \frac{(I_M^*)^2}{2eIM(0)} \tag{13.4.21}$$

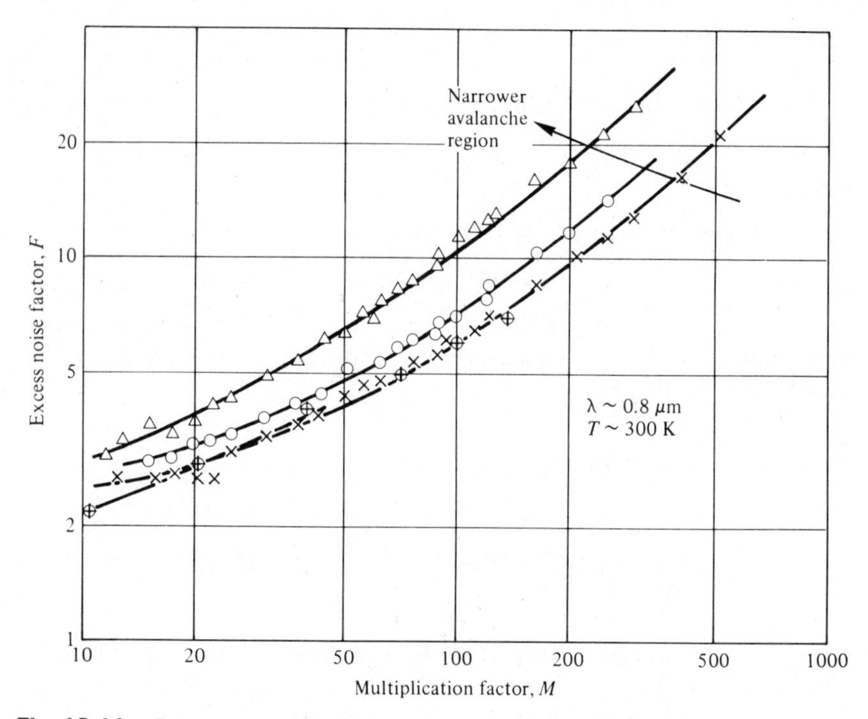

Fig. 13.11 Excess noise factor in Si-APDs. The data shown by ×, ○, △ refer to diodes in which the peak field and the depth of the avalanching region were varied. [Taken from T. Kaneda *et al.*, Excess noise in silicon avalanche photodiodes, *J. Appl. Phys.* **47**, 1605–7 (1976).] The data shown by ⊕ refer to the diodes of Fig. 13.4.

Thus

$$F_e = 1 + 1 + \frac{\{kM^2(0) - M^2(w)\}}{(1-k)M(0)} \tag{13.4.22}$$

Using eqn. (13.4.9)

$$\begin{aligned} kM^2(0) - M^2(w) &= kM^2(0) - (1-k)^2 - 2k(1-k)M(0) - k^2M^2(0) \\ &= k(1-k)M^2(0) - 2k(1-k)M(0) - (1-k)^2 \end{aligned}$$

Thus, writing M_e for $M(0)$

$$F_e = 2 + \{kM_e^2 - 2kM_e - (1-k)\}/M_e \tag{13.4.23}$$

$$= M_e\{1 - (1-k)(M_e - 1)^2/M_e^2\} \tag{13.4.2}$$

When M is large $F_e \rightarrow kM_e$, again stressing the desirability of a small value of k when electron injection is used. In general the ratio between the electron and hole ionization coefficients is greatest at low fields and decreases at high fields. This can be seen in Fig. 13.2 to be the case for silicon and germanium and it leads to some increase in the noise factor when very thin avalanche regions are used, because these necessarily require a higher average field strength in order to produce a given multiplication factor. The effect is illustrated in Fig. 13.11.

PROBLEMS

13.1 Obtain eqn (13.1.12).

13.2 When α_h and α_e are independent of position, and $\alpha_h/\alpha_e \rightarrow 1$, show that $M_e \rightarrow 1/(1-\alpha_e w)$ and $M_h \rightarrow 1/(1-\alpha_h w)$.

13.3 Using Fig. 13.2 sketch graphs of the variation of $k = \alpha_h/\alpha_e$ with electric field strength for the semiconductor materials shown. Hence indicate for each material whether electrons or holes should be used predominantly to initiate the avalanche.

13.4 An avalanche photodiode like the one shown in Fig. 13.4(a) is made with the following doping concentrations: in the n^+ and p^+ regions, 10^{24} m^{-3}; in the p-layer, 4×10^{21} m^{-3}; in the epitaxial π-layer, 10^{20} m^{-3}. The thickness of the p-layer is 2 μm and that of the π-layer is 20 μm. The diameter of the fully depleted π-region is limited by the channel-stop diffusion to 200 μm. The recombination lifetime is 10 μs throughout the active region. Assume uniform doping and abrupt junctions, and take $\varepsilon_r = 12$.

(a) Sketch the variation of the electric field across the device, indicating the values of dE/dx in each part.
(b) Calculate the voltage required to deplete the whole of the π-layer and the peak value of the electric field when this is just achieved.
(c) Calculate the bias voltage and the electric field at the $p^+\pi$ and $p\pi$ junctions when the peak field is 30 kV/mm.
(d) Discuss the expected variation of dark current with reverse bias voltage and explain the principal features of Fig. 13.4(b).
(e) Using eqns (13.3.2) and (13.3.3) calculate the response time and bandwidth when $M = 100$. Assume that the saturation drift velocities given in Section 13.3 apply and take $w_A = 1$ μm, $k = 0.1$.

13.5 By analogy with the derivation given in the text for F_e, the noise factor for electron multiplication (eqn. (13.4.2)), obtain eqn. (13.4.3) for F_h, the noise factor for hole multiplication.

13.6 Show that when $M_e \gg 1$, eqn. (13.4.2) approximates to $F_e = 2(1-k) + kM$, and obtain the corresponding expression for F_h when $M_h \gg 1$ using eqn. (13.4.3).

13.7 From the data shown in Fig. 13.11, it can be seen that at $M = 100$ $F = 6$ for one diode and $F = 10$ for another. Assuming that in each case electrons initiate the avalanche, calculate the implied values of k. Compare the results obtained from eqn. (13.4.2) with those obtained using the approximation derived in problem 13.6. In each case estimate the effective peak field in the avalanche region from Fig. 13.2.

REFERENCE

13.1 R. J. McIntyre, Multiplication noise in uniform avalanche diodes, *I.E.E.E. Transactions on Electron Devices*, **ED-13**, 164–8 (1966).

SUMMARY

At electric fields of a few tens of megavolts per meter avalanche ionization by electrons and holes multiplies the initial photogenerated current. The multiplication factor, M, is limited by the formation of microplasmas, but below this a noise factor, F, may be defined. A theory for F as a function of M is given in Section 13.4. For practical purposes the approximation $F = M^x$ may be used.

It is advantageous if the carrier with the higher ionization coefficient initiates the avalanche. This gives the most stable gain, the best frequency response, and the lowest noise factor.

In Si APDs $\alpha_e \gg \alpha_h$, especially at low fields. Thus the electron current should be multiplied as in the n^+–p–π–p^+ reach-through diodes shown in Figs. 13.3 and 13.4. Noise factors as low as $M^{0.3}$ may be obtained ($F = 4$ at $M = 100$) but the use of higher values of the peak electric field gives a better compromise with other parameters such as quantum efficiency (which can approach 100%), bandwidth (which can approach 1 GHz), and ease of fabrication. The noise factor may then rise to 6–10 at $M = 100$ ($x = 0.4$–0.5).

In Ge the electron and hole ionizing coefficients are of similar orders of magnitude ($k \simeq 1$) and theory then suggests $F \simeq M$, which is obtained in practice. Ge APDs can be made with $\eta > 80\%$ and bandwidths of several gigahertz, but they suffer from high values of dark current.

The multiplication factors that can be obtained with III–V semiconductor heterostructures are at present limited by leakage currents and microplasma formation.

14

The Receiver Amplifier

14.1 INTRODUCTION

Chapters 2–13 have dealt almost exclusively with the optical parts of optical fiber communication systems. At this point it becomes necessary to consider the electronic aspects of the signal-recovery process at the receiver.

The receiver may be represented by the generic block diagram of Fig. 14.1. As we have seen, the photodiode converts the received optical power into an electrical current, and the next requirement is the transformation of this current into an amplified voltage signal. As in all communication systems it is at this point, the receiver input where the signal level is weakest, that the system signal-to-noise ratio is determined and the system performance level established. In practice the need for a certain minimum signal-to-noise ratio sets a lower limit on the acceptable level of received optical power. This particular stage is therefore critical in the overall design of the system. In Section 14.2 the various sources of noise in the receiver will be discussed. Then in later sections we shall attempt to calculate the signal-to-noise ratio as a function of the received power level for different amplifier configurations. In subsequent chapters we will discuss the minimum signal-to-noise ratio that will give an acceptable system performance when different types of modulation are used.

The type of modulation does, of course, determine the form taken by the demodulator. In optical fiber systems we shall deal with two analogue and one digital modulation formats, the last being by far the most important. In each case it is the power of the optical carrier that is modulated. When we discuss

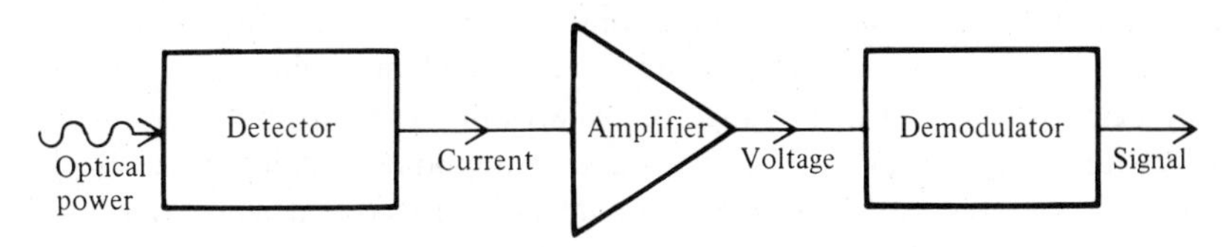

Fig. 14.1 Block diagram of receiver.

unguided systems in Chapter 16, we shall find that these offer a rather wider choice of modulation strategies. For fibers we shall consider:

(a) direct modulation of the optical carrier power level by the baseband signal;
(b) modulation of the optical carrier power by a frequency modulated subcarrier;
(c) pulse-code modulation of the optical carrier power.

The first of these is the simplest and may often be quite effective. The main problem is the difficulty of maintaining a high degree of linearity with either LED or laser sources. The second technique does permit such linearity to be obtained and is used, for example, in cable television applications where very low harmonic distortion and cross-modulation are essential. It is possible to use a frequency modulation format similar to the one normally used in electrical cable TV systems. We will discuss this application in Chapter 17. The third, and digital, technique will predominate in optical fibers used as part of the telephone system, especially those deployed in the high capacity parts of the system, and in all data transmission links.

14.2 SOURCES OF RECEIVER NOISE

We ought to be quite clear at the outset about what is to be understood by the term *noise*. It refers to those random fluctuations that are inherently generated in any electronic circuit or component, and over which the circuit designer has no direct control. The fluctuations are superimposed on, and tend to mask, any signals being processed in the circuit. We may make a distinction between noise and *interference*, which we take to refer to those unwanted signals that may be picked up from external sources or from other parts of the system and whose effects can always be minimized by good circuit layout and by proper screening of vulnerable parts.

We have already seen in Chapters 12 and 13 that the signal we are concerned with, because it originates in a random, quantum detection process in the photodiode, carries its own shot noise. The mean square fluctuation about an average, unmultiplied photodiode current, $\bar{I}$, is $2e\bar{I}\,\Delta f$, where Δf is the frequency range over which the fluctuations are observed. The fact that here, as with other noise sources, the mean square fluctuation increases in direct proportion to the frequency range or bandwidth (Δf) makes it convenient to designate noise sources by their mean square amplitude in unit bandwidth. This is usually referred to as the mean square noise spectral density and will be indicated by the use of an asterisk. Thus, in the case of shot noise, as we saw in Section 13.4, the mean square spectral density is

$$(I^*_{\mathrm{Sh}})^2 = 2e\bar{I} \tag{14.2.1}$$

for a p–i–n photodiode, and becomes

$$(I^*_{\mathrm{Sh}})^2 = 2eM^2F\bar{I} \tag{14.2.2}$$

after avalanche multiplication. When, as here, the noise spectral density is independent of frequency, the noise source is said to be *white*.

Any dissipative element in a system introduces noise. Thus in an electronic circuit any resistance gives rise to *thermal* or *Johnson* noise as a consequence of the random thermal motions of the charge carriers. These may be observed as current fluctuations in the resistive component or as the corresponding voltage fluctuations across its terminals. For a component of resistance, R, the mean square spectral densities are

$$(V^*_{\mathrm{Th}})^2 = 4kTR \tag{14.2.3}$$

and

$$(I^*_{\mathrm{Th}})^2 = 4kT/R \tag{14.2.4}$$

where k is Boltzmann's constant and T is the component temperature. Derivations of these expressions may be found in many texts. See, for example, Refs. |14.1|–|14.3|. Clearly thermal noise is also white.

The noise sources present in any active electronic device such as a transistor are many and complex. Their magnitudes depend on the material and the design of the device, and on the manner in which it is biassed. However, for any given device under specified bias conditions it is possible to represent its noise characteristics fully in the way illustrated in Fig. 14.2. The device is

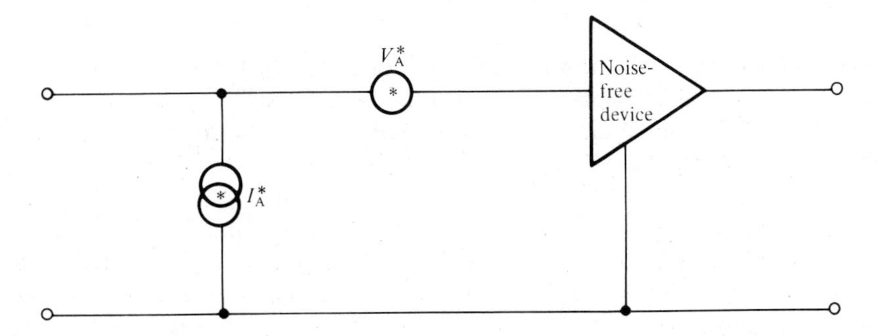

Fig. 14.2 General equivalent noise circuit for an active electronic device.

represented by a noise-free element (shown in the figure as a noise-free amplifier) having the same transfer characteristic. All the noise sources are represented by a noise voltage source, $V_A^*(f)$, and a noise current source, $I_A^*(f)$, at the input terminals.

In a field-effect transistor a major source of noise is the thermal noise associated with the channel resistance. This gives rise to a term $(\zeta 4kT/g_m)$ as a major component of $(V_A^*)^2$, where g_m is the forward transconductance of the device and the constant, ζ, takes on the approximate values of 0.7 for silicon FETs and 1.1 for gallium arsenide devices. A major contributor to the noise of bipolar junction transistors is the shot noise associated with the base and collector bias currents, I_B and I_C, respectively. The first of these contributes a term $(2eI_B)$ to $(I_A^*)^2$ while the second contributes a term $\{2(kT)^2/eI_C\} = \{2(kT)^2/eI_B\beta\}$ to $(V_A^*)^2$, β being the current gain of the transistor. It is always desirable to make β as large as possible, but there is then an optimum bias current which depends, among other things, on the required bandwidth.

Provided that each of these noise sources is independent of and uncorrelated with the others, the total noise is represented by the sum of the mean square values of each of the individual sources. This we shall assume.

14.3 CIRCUITS, DEVICES AND DEFINITIONS

In Sections 14.4 and 14.5 we shall attempt to calculate the signal-to-noise ratio to be expected at the output of the linear part of the receiver amplifier. We shall do this for two types of amplifier configuration. In both cases the amplifier normally consists of many stages, but provided that the first stage produces sufficient gain, the noise introduced by the remaining stages causes only a marginal increase in the total noise. For each amplifier we will consider three possible types of input transistor. They are:

(a) a silicon, junction field-effect transistor (Si-JFET);
(b) a silicon, bipolar junction transistor (Si-BJT);
(c) a gallium arsenide, metal–semiconductor, field-effect transistor (GaAs-MESFET).

The detector will be taken to be an avalanche photodiode giving a gain, M, and having a noise factor, F. The results will then apply to p–i–n photodiode detectors simply by setting $M = 1$ and $F = 1$. The input signal will be taken to be the Fourier transform $\Phi_R(f)$ of the modulated power received at the photodiode, $\Phi_R(t)$. This will give rise to a multiplied detector current which, following eqn. (12.1.2), may be written as

$$MI(f) = M\mathcal{R}\Phi_R(f) = M\eta \frac{e\lambda}{hc} \Phi_R(f) \tag{14.3.1}$$

Of course, there is no reason to assume that the received power modulation faithfully reflects the modulated power transmitted. Indeed if fiber dispersion is significant it will not. Similarly, as we have seen, both the quantum efficiency and the multiplication factor of the detector decrease at high frequencies: thus we should allow for $\eta(f)$ and $M(f)$ to be functions of frequency. Equalization to offset all of these effects as well as the reduced source modulation efficiency at high frequency can be provided by boosting the amplifier gain appropriately at high frequency. However, we shall assume that the range of modulation frequencies used lies within the frequency-independent capabilities of the source, the fiber and the detector. We shall, therefore, be concerned simply with the transfer function relating the amplifier output voltage, $V_{out}(f)$, to the detector current $MI(f)$, and will assume that this is required to be constant over a frequency range Δf.

With direct modulation Δf must cover the range of base-band frequencies. When a frequency-modulated subcarrier is used, Δf must extend over the range of the frequency modulation. In the case of pulse-code modulation, we shall show in Chapter 15 that the frequency range must extend at least from zero to one-half of the bit rate, that is

$$\Delta f = B/2 \tag{14.3.2}$$

We shall assume Δf to be defined by a separate filter unit whose normalized transfer function $H(f)$ is constant over the pass band and has infinitely sharp cutoff. This represents the ideal but is not physically realizable. In practice the response will roll-off gradually at the ends of the frequency range and a small correction may have to be made to the calculated signal-to-noise ratio in order to take account of this. Such correction factors have been calculated in the rather more sophisticated analyses given in Refs. [14.4] and [14.5].

14.4 THE VOLTAGE AMPLIFIER CIRCUIT

14.4.1 Signal-to-noise Ratio

First consider the signal transfer function. The small-signal equivalent circuit representing a photodiode detector feeding a voltage amplifier was discussed

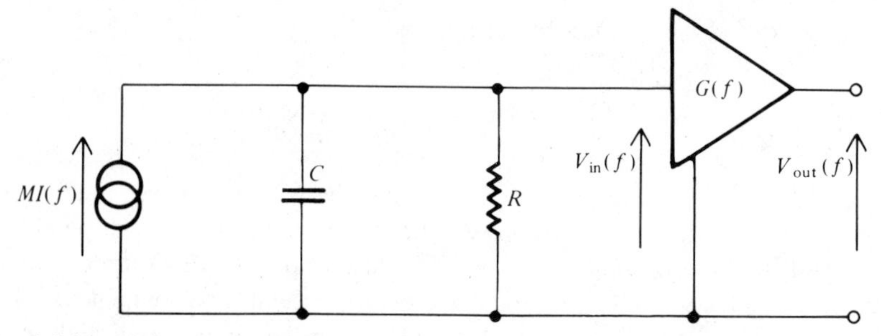

Fig. 14.3 Voltage amplifier circuit.

briefly in Section 12.5. It is shown again in Fig. 14.3. As before,

$$V_{\text{in}}(f) = \frac{RMI(f)}{(1 + \text{j}2\pi fCR)} \tag{14.4.1}$$

so that

$$V_{\text{out}}(f) = G(f)V_{\text{in}}(f) = G(f)\frac{RMI(f)}{(1 + \text{j}2\pi fCR)} \tag{14.4.2}$$

The total input capacitance, C, causes band limiting at frequencies higher than $(1/2\pi CR)$. Signals varying more rapidly than this are effectively integrated at the input. For the overall transfer function to remain frequency independent we require that

$$G(f) = G_0(1 + \text{j}2\pi fCR) \tag{14.4.3}$$

Then

$$V_{\text{out}} = G_0 MRI \tag{14.4.4}$$

We will assume that after some initial amplification such equalization is provided whenever it is required. Effectively at some point in the amplifier a differentiating circuit is needed. The frequency profiles are illustrated in Fig. 14.4, including the effect of the band-limiting filter $H(f)$.

An equivalent noise circuit for the photodiode and voltage amplifier is shown in Fig. 14.5. The various noise sources shown are as follows:

$I^*_{\text{Sh}} = (2eIM^2F)^{1/2}$ represents the multiplied shot noise originating in the photodiode.

$I^*_{\text{Th}} = \left(\frac{4kT}{R}\right)^{1/2} = \left\{4kT\left(\frac{1}{R_{\text{A}}} + \frac{1}{R_{\text{B}}} + G_{\text{D}}\right)\right\}^{1/2}$ represents the

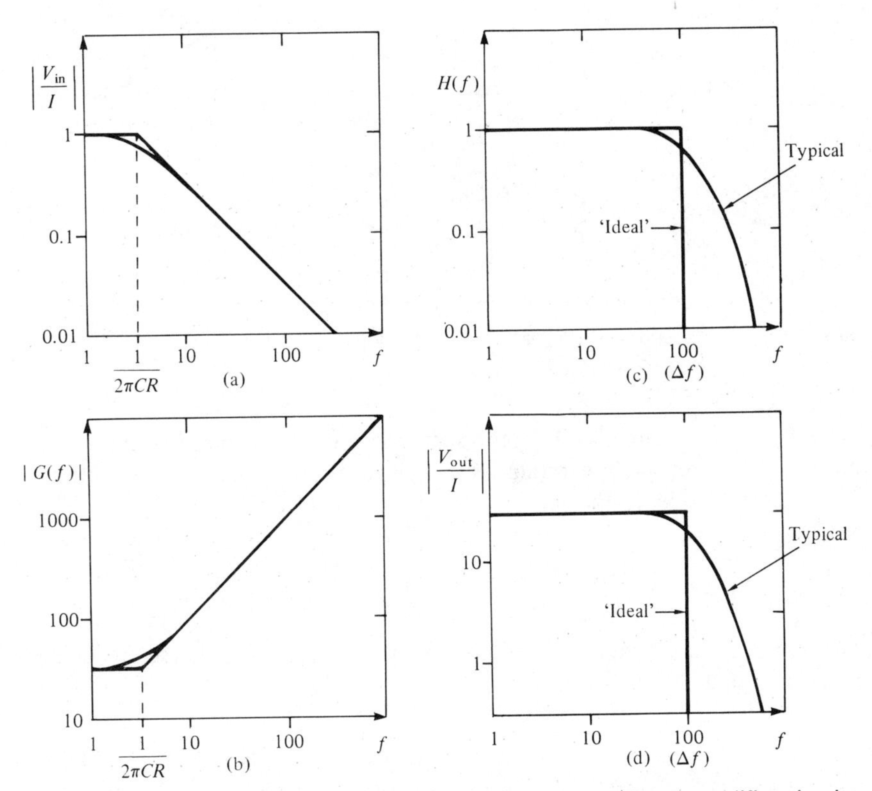

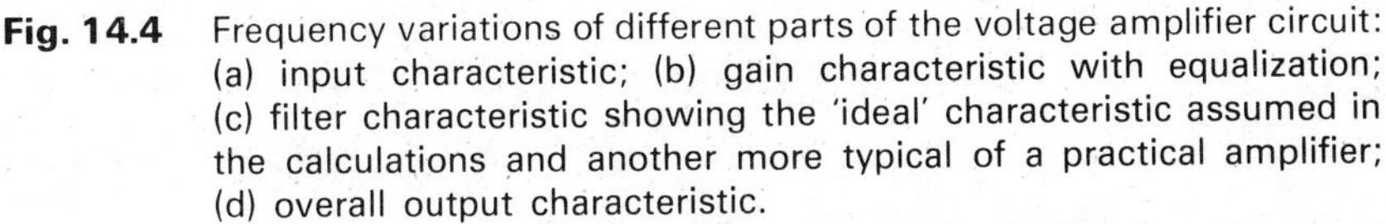

Fig. 14.4 Frequency variations of different parts of the voltage amplifier circuit: (a) input characteristic; (b) gain characteristic with equalization; (c) filter characteristic showing the 'ideal' characteristic assumed in the calculations and another more typical of a practical amplifier; (d) overall output characteristic.

The numerical values shown have no particular significance and are used only to indicate that in each case log–log scales have been used (Bode plots).

thermal noise originating in the resistive elements detailed in Fig. 12.11.

I_A^* represents the equivalent current noise source of the amplifier.
V_A^* represents the equivalent voltage noise source of the amplifier.

The circuit of Fig. 14.5 may be reduced to that of Fig. 14.6 by grouping together the three noise current sources such that

$$\begin{aligned}(I_T^*)^2 &= (I_{Sh}^*)^2 + (I_{Th}^*)^2 + (I_A^*)^2 \\ &= (2eIM^2F) + (4kT/R) + (I_A^*)^2 \qquad (14.4.5)\end{aligned}$$

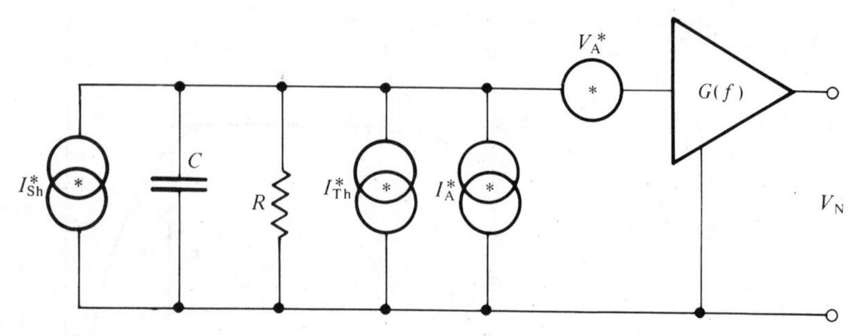

Fig. 14.5 Noise equivalent circuit for voltage amplifier.

At the output terminals V_N represents the total r.m.s. noise voltage. This may be calculated by integrating the amplified mean square noise amplitudes from I_T^* and V_A^* over the required frequency range Δf. Thus

$$V_N^2 = \int_{\Delta f} |G(f)|^2 (V_A^*)^2 \, df + \int_{\Delta f} \frac{|G(f)|^2 R^2 (I_T^*)^2 \, df}{|1 + j2\pi fCR|^2} \tag{14.4.6}$$

When we substitute eqn. (14.4.3) for $G(f)$ and take the required frequency range to extend from 0 to Δf, this becomes

$$V_N^2 = G_0^2 \int_0^{\Delta f} \{(1 + 4\pi^2 f^2 C^2 R^2)(V_A^*)^2 + R^2 (I_T^*)^2\} \, df \tag{14.4.7}$$

And when V_A^* and I_T^* are sensibly independent of frequency throughout Δf,

$$V_N = G_0 \{(1 + \tfrac{4}{3}\pi^2 (\Delta f)^2 C^2 R^2)(V_A^*)^2 + R^2 (I_T^*)^2\}^{1/2} (\Delta f)^{1/2} \tag{14.4.8}$$

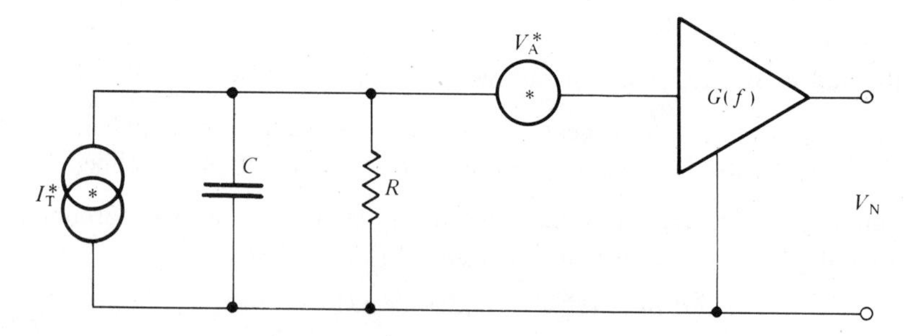

Fig. 14.6 Reduced noise equivalent circuit.

The r.m.s. signal-to-noise ratio, K, is simply

$$K = \frac{V_{\text{out}}}{|V_{\text{N}}|} = \frac{MIR}{\{(1 + \frac{4}{3}\pi^2(\Delta f)^2 C^2 R^2)(V_{\text{A}}^*)^2 + R^2(I_{\text{T}}^*)^2\}^{1/2}(\Delta f)^{1/2}} \tag{14.4.9}$$

This becomes

$$K = \frac{I}{\left\{\underset{(a)}{\frac{(V_{\text{A}}^*)^2}{M^2}\left(\frac{1}{R^2}\right.} + \underset{(b)}{\left.\frac{4\pi^2}{3}(\Delta f)^2 C^2\right)} + \underset{(c)}{2eIF} + \underset{(d)}{\frac{4kT}{M^2 R}} + \underset{(e)}{\frac{(I_{\text{A}}^*)^2}{M^2}}\right\}^{1/2}(\Delta f)^{1/2}} \tag{14.4.10}$$

when we expand I_{T}^* in order to make the dependence of K on the individual noise sources more explicit. For future convenience the five noise terms in the denominator have been separately identified, (a)–(e).

The value of K determines the quality of the communication channel and a minimum value is always specified in the design of the system. For example, in a pulse-code modulated (PCM) channel we might specify $K > 12$ ($S/N > 21.6$ dB), whereas in an analogue channel we might require $K > 200$ ($S/N > 46$ dB). Equation (14.4.10) is thus of the greatest importance in guiding the general design of the system and evaluating its expected performance.

Attention may be drawn to a number of features in eqn. (14.4.10):

1. The signal-to-noise ratio will be improved by increasing the multiplication factor, M, until the shot-noise term, (c), increased by the noise factor, F, which itself increases with M, comes to dominate the other terms. There is always an optimum value for M at this point.
2. Increasing the front-end resistance, R, improves the signal-to-noise ratio as long as terms (a) and (d) are significant. (It is interesting to note that this particular result has twice been 'rediscovered'. It was noted in passing in the 1930s when early work on amplifier noise was first published. It was rediscovered in the 1950s when low-noise amplifiers were required for low-light-level TV and photomultiplier signals. It was rediscovered again in the 1970s when low-noise amplifiers for optical communication receivers were being studied seriously. *Plus ça change*...!) Using a high input resistance brings with it a number of disadvantages, not the least being the need for considerable equalization at high frequencies. These problems will be discussed further in Section 14.4.3.

3. At high frequencies when equalization is required, term (b) comes to dominate and the noise increases as the square of the input capacitance. It is thus important that C be minimized. This also reduces the amount of equalization needed.
4. The presence of the shot-noise term (c) causes the total noise to be dependent on the level of the received signal. This is a feature which distinguishes optical from other types of communication system. It also means that eqn. (14.4.10) is quadratic in I. We shall first discuss situations in which term (c) is either significantly larger or significantly smaller than the other terms, and then return to the full solution in Section 14.4.5.
5. The five terms in the denominator of eqn. (14.4.10) have been added together as though they were uncorrelated Gaussian noise sources. In fact there are theoretical grounds for expecting the statistical properties of some of the terms to be non-Gaussian. Shot noise, for example, is a Poisson process following Poisson statistics and so too are some of the processes contributing to (I_A^*) and (V_A^*). The avalanche noise factor F may have a statistical distribution that is different again. As a working 'figure-of-merit' the signal-to-noise ratio, K, defined here is normally quite satisfactory, but in certain extreme cases, as we shall see in Section 15.2, the correct statistical noise distribution has to be used and the Gaussian and non-Gaussian terms have to be combined with care.

14.4.2 The Ideal Case

In an ideal receiver M would be sufficiently large for the shot-noise term, (c), to dominate the others, and the noise factor, F, would be unity. Then

$$K = \left(\frac{I}{2e\,\Delta f}\right)^{1/2} \qquad (14.4.11)$$

and we would require that

$$I > 2eK^2\,\Delta f \qquad (14.4.12)$$

This represents the ideal, quantum limit to detector sensitivity.

At wavelengths where a high-quality Si–APD can be used, and at not too high modulation frequencies, it is quite possible for term (c) to dominate the noise and for the quantum limit to be approached to within the APD noise factor. Then, we shall require that

$$I > 2eFK^2\,\Delta f \qquad (14.4.13)$$

and, using the photodiode responsivity, $\mathcal{R}$, defined in eqn. (12.1.2) ($\mathcal{R} = I/\Phi_R$), we may calculate the minimum received power to be given by

$$\Phi_R > 2eFK^2\,\Delta f/\mathcal{R} = 2(hc/\eta\lambda)FK^2\,\Delta f = 2(\mathcal{E}_{ph}/\eta)FK^2\,\Delta f \qquad (14.4.14)$$

For a Si–APD working at $\lambda = 0.85$ μm with a quantum efficiency of $\eta = 0.75$, $\mathcal{R} = 0.5$ A/W. So with $\Delta f = 1$ MHz, $F = 6$, $K = 12$, we shall require $\Phi_R > 0.5$ nW.

14.4.3 High Input Resistance or Integrating Amplifier

If we accept the need for equalization and are able to make R sufficiently large, the noise becomes dominated by terms (b), (c) and (e). Which of these is most significant then depends on the frequency range covered and on the type of input device used. The signal-to-noise ratio becomes:

$$K = \frac{I}{\left\{\underbrace{\frac{(V_A^*)^2}{M^2}\frac{4\pi^2}{3}(\Delta f)^2 C^2}_{(b)} + \underbrace{2eIF}_{(c)} + \underbrace{\frac{(I_A^*)^2}{M^2}}_{(e)}\right\}^{1/2}(\Delta f)^{1/2}} \qquad (14.4.15)$$

If we define R_A as

$$R_A = V_A^*/I_A^* \qquad (14.4.16)$$

then the voltage noise term (b) dominates the current-noise term (e) at high frequencies when

$$\Delta f > \sqrt{3}/2\pi C R_A = (\Delta f)_0 \qquad (14.4.17)$$

The relative magnitudes of the two amplifier noise terms are shown in Fig. 14.7. We have used the value $C = 5$ pF and assumed:

For the Si-FET,

$$V_A^* = 4\ \text{nV}/\sqrt{\text{Hz}}, \qquad I_A^* = 10\ \text{fA}/\sqrt{\text{Hz}}$$

giving

$$R_A = 400\ \text{k}\Omega \qquad \text{and} \qquad (\Delta f)_0 = 140\ \text{kHz}.$$

For the Si-BJT,

$$V_A^* = 2\ \text{nV}/\sqrt{\text{Hz}}, \qquad I_A^* = 2\ \text{pA}/\sqrt{\text{Hz}}$$

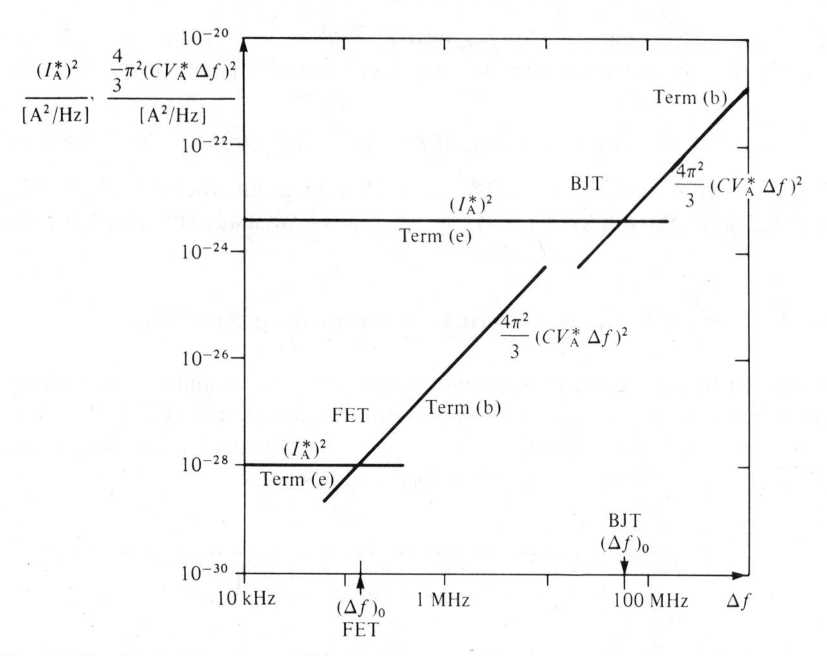

Fig. 14.7 Variation of amplifier voltage and current noise sources with frequency in the high input resistance amplifier.

giving

$$R_A = 1\ \text{k}\Omega \qquad \text{and} \qquad (\Delta f)_0 = 55\ \text{MHz}.$$

These noise spectral densities would be expected to rise at low frequencies because of $1/f$ or flicker noise, and also at high frequencies as the device bandwidth limits its gain performance. The third type of device mentioned in Section 14.3, the GaAs-MESFET, can be made with a high transconductance ($g_m \simeq 50$ mS), a very low gate capacitance (<0.5 pF) and an excellent bandwidth capability (>1 GHz). Its equivalent noise voltage source is similar to or lower than that of a Si-BJT ($\sim$1 nV/$\sqrt{\text{Hz}}$) and its equivalent noise current source is likely to lie between that of a Si-BJT and a Si-FET ($\lesssim$0.1 pA/$\sqrt{\text{Hz}}$).

The relative size of the shot-noise term (c) depends on the magnitudes of M and F. With a p–i–n photodiode, $M = 1$, $F = 1$, it is normally insignificant. When a good silicon APD is used with, for example, $M = 100$ and $F = 6$, then the reduction in terms (b) and (e) and the increase in term (c) cause it to dominate under a wide range of practical conditions. Then the receiver is at its quantum noise limit and eqns. (14.4.13) and (14.4.14) apply. The conditions for this may be obtained as follows. At low frequencies, $\Delta f < (\Delta f)_0$, when the amplifier current noise term (e) exceeds the voltage noise term (b), shot noise dominates when

$$2eIM^2F > (I_A^*)^2 \tag{14.4.18}$$

We will substitute into eqn. (14.4.18) the limiting value of I then satisfying the signal-to-noise ratio, namely

$$I = 2eFK^2 \, \Delta f \tag{14.4.19}$$

The condition for shot noise to dominate then becomes

$$\Delta f > \frac{(I_A^*)^2}{(2eMFK)^2} = (\Delta f)_1 \tag{14.4.20}$$

At high frequencies, $\Delta f > (\Delta f)_0$, comparison has to be made with the amplifier voltage noise term (b). Then the condition for the dominance of the shot noise becomes

$$2eIM^2F > \frac{4\pi^2}{3}(\Delta f)^2 C^2 (V_A^*)^2 \tag{14.4.21}$$

and, substituting eqn. (14.4.19), this gives

$$\Delta f < \frac{3(eMFK)^2}{(\pi C V_A^*)^2} = (\Delta f)_2 \tag{14.4.22}$$

Putting in the previously quoted values of I_A^* and V_A^*, assuming $K = 12$ and $C = 5$ pF, and using the values $M = 100$, $F = 6$ for a silicon APD, we obtain

with a Si-FET, $(\Delta f)_1 = 20$ Hz, $(\Delta f)_2 = 1$ GHz

with a Si-BJT, $(\Delta f)_1 = 750$ kHz, $(\Delta f)_2 = 4$ GHz.

Thus, both devices permit shot noise limited detection over most practical frequencies. It is important to stress the dependence of this result on the assumed values of M and F. Quite modest reductions would see the amplifier noise terms swamping the shot noise.

Although the use of a high input resistance helps to maximize the signal-to-noise ratio at the receiver, its use gives rise to two major disadvantages, both of which stem from the need for a considerable amount of equalization. The first is that the equalization has to be tailor-made for each individual circuit. It cannot simply be pre-set. The reason is that the gain has to take the form $G(f) = G_0(1 + j2\pi fCR)$, and both C and R vary from device to device, from circuit to circuit, and probably with temperature as well. Each circuit, then, has to be individually tuned. The second problem is that the wide variation of gain at different frequencies means that the dynamic range of the amplifier is reduced. (This is a measure of the maximum undistorted signal that can be accommodated, and is normally expressed by its ratio to the smallest acceptable signal.) The reason may be understood as follows. Say that the gain provided for high frequencies is required to be one hundred times that for low frequencies, as in Fig. 14.4. This will be achieved by attenuating the low frequencies after the initial amplification, which in this case would need to be at least 1000 times (60 dB). Having such a high gain presents no problem for the high frequencies which were already attenuated at the input, but it does mean that any low frequency signals are very large at this point and are liable to saturate the amplifier. It is this which restricts the overall dynamic range.

In Section 14.4.4 we shall examine a circuit with an input resistance low enough for equalization not to be needed. This will enable us to estimate the noise penalty incurred.

14.4.4 Low Input Resistance

No equalization is needed as long as the condition

$$R < \frac{1}{2\pi C \Delta f} \tag{14.4.23}$$

can be satisfied. This is plotted in Fig. 14.8 for some representative values of C in the range 1–10 pF. Which of the terms (a), (d) or (e) then dominates the noise depends on the value of R and the type of input device used. Their relative magnitudes are illustrated in Fig. 14.9 where the resistance values R_1 and R_2 are defined by

$$R_1 = \frac{(V_A^*)^2}{4kT} \tag{14.4.24}$$

and

$$R_2 = \frac{4kT}{(I_A^*)^2} \tag{14.4.25}$$

For

$$R_1 < R < R_2 \tag{14.4.26}$$

thermal noise exceeds amplifier noise.

When the limiting condition, $R = 1/(2\pi C\, \Delta f)$, is substituted into eqn.

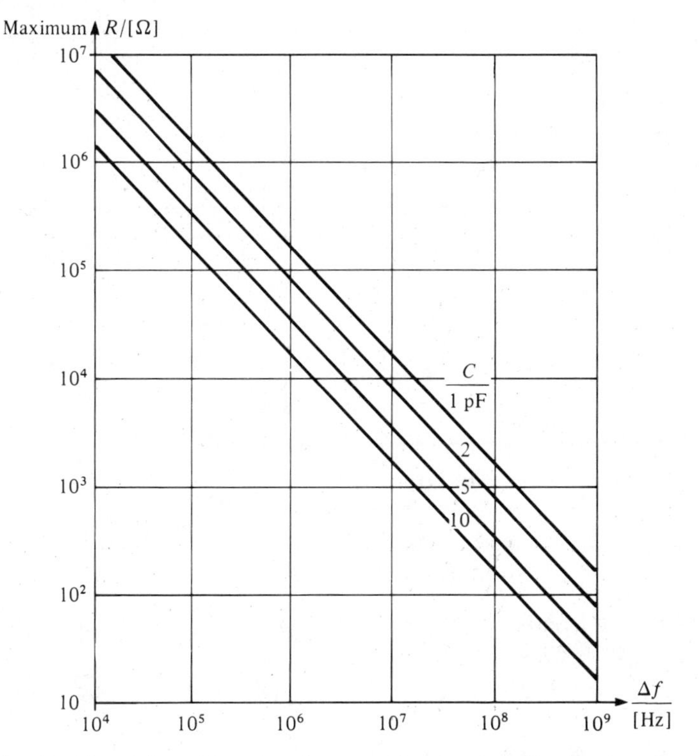

Fig. 14.8 Limiting value of input resistance when no equalization is to be used.

(14.4.10), the signal-to-noise ratio becomes

$$K = \frac{I}{\left\{ \underbrace{\frac{16\pi^2}{3} \frac{(V_A^*)^2}{M^2} C^2 (\Delta f)^2}_{(a)+(b)} + \underbrace{2eIF}_{(c)} + \underbrace{\frac{8\pi kTC\,\Delta f}{M^2}}_{(d)} + \underbrace{\frac{(I_A^*)^2}{M^2}}_{(e)} \right\}^{1/2} (\Delta f)^{1/2}} \tag{14.4.27}$$

The terms are shown as a function of frequency in Fig. 14.10 for Si-FET and Si-BJT input devices when $C = 5$ pF.

We have already examined for the case of the high input resistance amplifier the frequency range over which the shot noise term (c) may be expected to exceed the amplifier noise terms (a), (b) and (e). This was

$$(\Delta f)_1 < \Delta f < (\Delta f)_2 \tag{14.4.28}$$

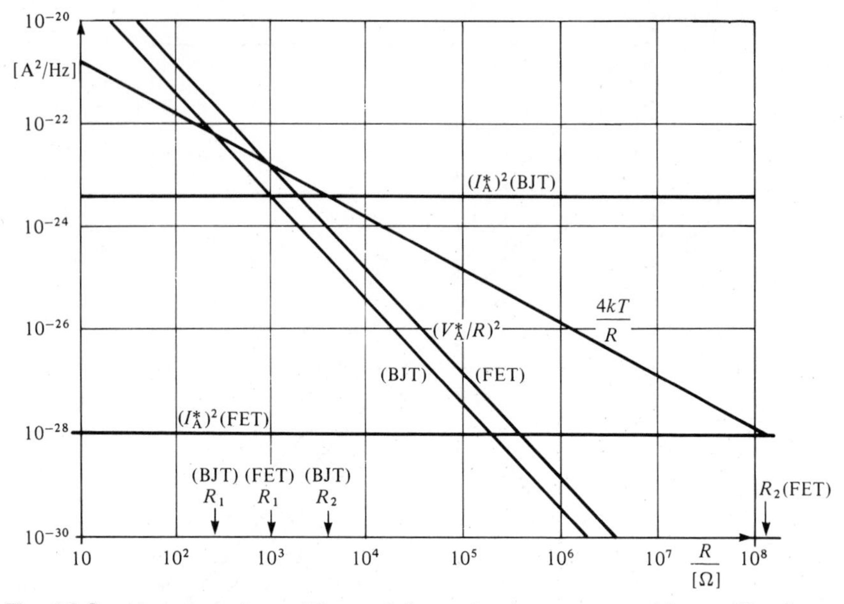

Fig. 14.9 Variation of amplifier and thermal noise sources with amplifier input resistance.

where $(\Delta f)_1$ and $(\Delta f)_2$ were given by eqns. (14.4.20) and (14.4.22) respectively. When the value of the input resistance is reduced to the limiting value of eqn. (14.4.23), the voltage noise term is increased by a factor of four and $(\Delta f)_2$ is correspondingly reduced by a factor of four compared to their values when R is large. However, if R lies within the limits of eqn. (14.4.26) we need also to take account of thermal noise. The condition for the shot noise term always to dominate the thermal noise when R is given by the limiting value of eqn. (14.4.23) is

$$2eIM^2F > \frac{4kT}{R} = 8\pi kTC\,\Delta f \tag{14.4.29}$$

If this applies, we may, as before, substitute the minimum quantum noise limited current for I, namely

$$I = 2e\,\Delta f\,FK^2 \tag{14.4.19}$$

so that

$$(2eMFK)^2\,\Delta f > 8\pi kTC\,\Delta f \tag{14.4.30}$$

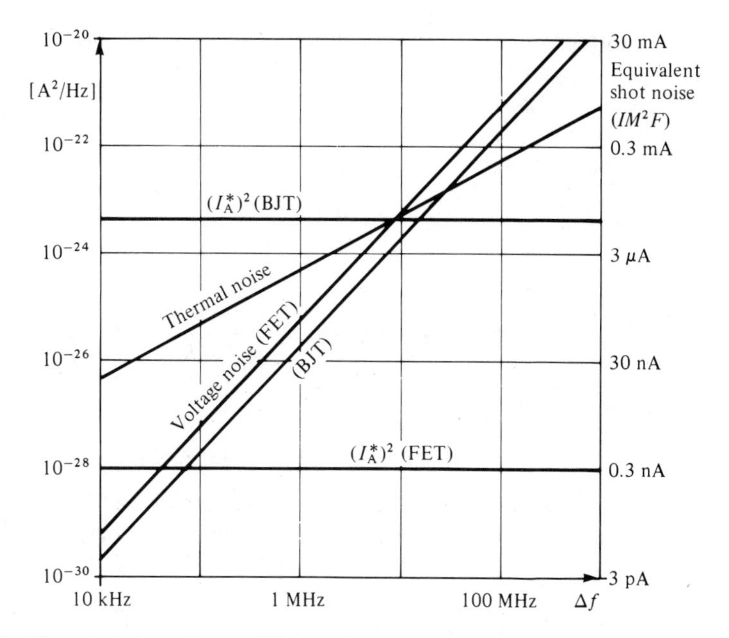

Fig. 14.10 Variation of amplifier and thermal noise sources with frequency when $R = 1/2\pi\Delta f C$ and $C = 5$ pF.

Voltage noise is given by $(16\pi^2/3)C^2(V_A^*)^2(\Delta f)^2$. Thermal noise is given by $8\pi kTC\Delta f$.

That is,

$$C < \frac{(eMFK)^2}{2\pi kT} = C_0 \qquad (14.4.31)$$

with $M = 100$, $F = 6$, $K = 12$, $C_0 = 50$ pF. Again the use of a good APD permitting a value of M of 20 or more ensures that the shot-noise detection limit can be obtained even when a low input resistance is used and no equalization is employed. However, this is not true with a p–i–n photodiode detector, and the sacrifice in increased noise may then be considerable.

14.4.5 Explicit Solution for Photodiode Current

Up to now we have concentrated on identifying the conditions under which the shot noise term (c) in eqn. (14.4.10) is either dominant or insignificant. We ought finally to recognize that this equation is quadratic in I and give the full solution. Squaring and rearranging eqn. (14.4.10) we obtain

$$I^2 - 2pI - q = 0 \tag{14.4.32}$$

where

$$p = e\,\Delta f\,FK^2 \tag{14.4.33}$$

and

$$q = \frac{K^2\,\Delta f}{M^2}\left\{\frac{(V_A^*)^2}{R^2}\left(1 + \frac{4\pi^2}{3}(\Delta f)^2 C^2 R^2\right) + \frac{4kT}{R} + (I_A^*)^2\right\} \tag{14.4.34}$$

The solution is

$$I = p\{1 + (1 + q/p^2)^{1/2}\} = p\{1 + (1 + 4\chi^2)^{1/2}\} \tag{14.4.35}$$

where $\chi^2 = \frac{1}{4}(q/p^2)$ is the ratio of the mean square amplifier and thermal noise terms to the mean square shot-noise term under quantum limited conditions. We have from equations (14.4.33) and (14.4.34),

$$\chi^2 = \tfrac{1}{4}(q/p^2) = \frac{\left\{\dfrac{(V_A^*)^2}{M^2}\left(\dfrac{1}{R^2} + \dfrac{4\pi^2}{3}(\Delta f)^2 C^2\right) + \dfrac{4kT}{M^2 R} + \dfrac{(I_A^*)^2}{M^2}\right\}}{4K^2 e^2 F^2\,\Delta f} \tag{14.4.36}$$

Referring to eqn. (14.4.13), we see that in the quantum limit

$$I = I_{QL} = 2e\,\Delta f\,FK^2 = 2p \tag{14.4.37}$$

so that

$$(I_{Sh}^*)^2_{QL} = 2eI_{QL}F = (2eFK)^2\,\Delta f \tag{14.4.38}$$

which is the denominator of eqn. (14.4.36). The numerator is the sum of the amplifier and thermal noise terms.

14.5 THE TRANSIMPEDANCE FEEDBACK AMPLIFIER

The voltage amplifier circuit discussed in Section 14.4, although important historically and theoretically, is not the configuration most commonly used in practice in optical communication receivers. Usually the transimpedance feedback amplifier shown in Fig. 14.11 is preferred. Its advantage lies principally in

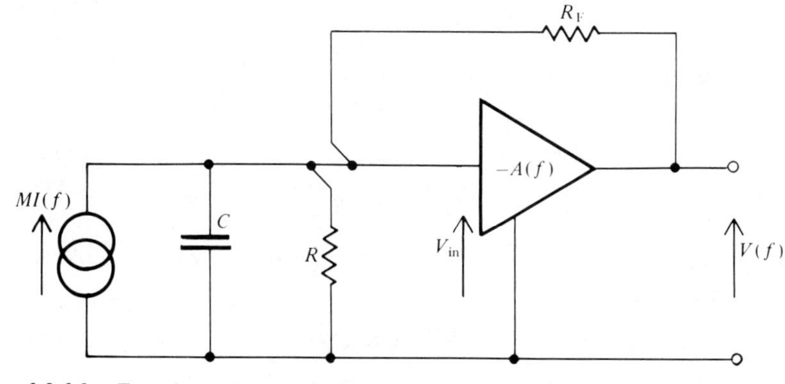

Fig. 14.11 Transimpedance feedback amplifier.

the fact that the need for equalization can be avoided, while at the same time the amplifier noise may be much less than would be the case in an unequalized voltage amplifier.

The transfer function may be derived by summing the currents at the amplifier input, remembering that the amplifier input resistance is included in R and that

$$V_{in} = -V/A \tag{14.5.1}$$

Then

$$MI + \frac{V - V_{in}}{R_F} = V_{in}\left(\frac{1}{R} + j2\pi fC\right)$$

$$\therefore \quad MI = -V\left(\frac{1}{R_F} + \frac{1}{AR_F} + \frac{1}{AR} + \frac{j2\pi fC}{A}\right)$$

$$\therefore \quad V = \frac{-R_F MI}{1 + \dfrac{1}{A} + \dfrac{R_F}{AR} + j\dfrac{2\pi fCR_F}{A}}$$

$$= \frac{-R_F MI/(1 + 1/A + R_F/AR)}{[1 + j2\pi fCR_F/(1 + R_F/R + A)]} \tag{14.5.2}$$

When

$$A \gg \left(1 + \frac{R_F}{R}\right) \tag{14.5.3}$$

$$V \simeq \frac{-R_F MI}{(1 + j2\pi f C R_F/A)} \tag{14.5.4}$$

Then, no further equalization for signals in the frequency range 0 to Δf will be needed, provided that

$$A \gg 2\pi\, C R_F\, \Delta f \tag{14.5.5}$$

Then

$$V \simeq -R_F MI \tag{14.5.6}$$

The equivalent noise circuit for the transimpedance feedback amplifier is shown in Fig. 14.12. The noise current source, I_T^*, includes the shot noise, the amplifier current noise, and the thermal noise of the biassing and amplifier input resistance, as set out in eqn. (14.4.5). The feedback noise, V_F^*, is simply the voltage noise generated in the feedback resistor:

$$(V_F^*)^2 = 4kTR_F \tag{14.5.7}$$

The spectral density, V_N^*, of the noise appearing at the output terminals may be calculated in the following way. We shall calculate separately the noise spectral density at the output resulting from each of the three sources, V_A^*, I_T^* and V_F^* (V_{N1}^*, V_{N2}^* and V_{N3}^*, respectively) and then sum the mean square values obtained using the principle of superposition. Thus for the voltage source

$$\left(V_A^* + \frac{V_{N1}^*}{(-A)}\right)\left(\frac{1}{R} + j2\pi f C\right) = \left(V_{N1}^* - V_A^* - \frac{V_{N1}^*}{(-A)}\right)\frac{1}{R_F} \tag{14.5.8}$$

$$\therefore \quad V_{N1}^* = V_A^* R_F \left(\frac{1}{R_F} + \frac{1}{R} + j2\pi f C\right) \Big/ \left(1 + \frac{1}{A} + \frac{R_F}{AR} + \frac{j2\pi f C R_F}{A}\right)$$

$$\simeq V_A^* \left(1 + \frac{R_F}{R} + j2\pi f C R_F\right) \tag{14.5.9}$$

For the current source

$$\left(V_{N2}^* - \frac{V_{N2}^*}{(-A)}\right)\frac{1}{R_F} + I_T^* = \frac{V_{N2}^*}{(-A)}\left(\frac{1}{R} + j2\pi f C\right)$$

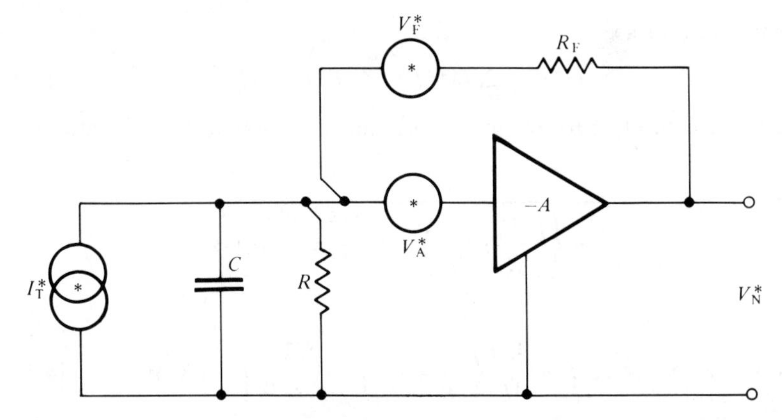

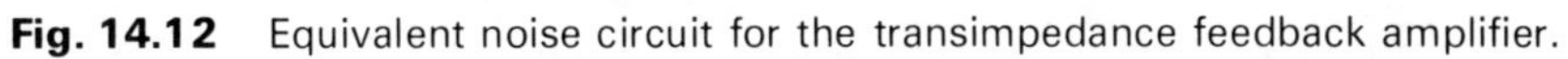
Fig. 14.12 Equivalent noise circuit for the transimpedance feedback amplifier.

$$\therefore \quad V^*_{N2}\left(\frac{1}{R_F}+\frac{1}{AR_F}+\frac{1}{AR}+\frac{j2\pi fC}{A}\right)=-I^*_T$$

$$\therefore \quad V^*_{N2}\simeq -R_F I^*_T \tag{14.5.10}$$

For the feedback source

$$V^*_{N3}=V^*_F+V^*_{N3}/(-A)$$

$$=V^*_F/(1+1/A)\simeq V^*_F \tag{14.5.11}$$

The total spectral density at the output is

$$(V^*_N)^2=(V^*_{N1})^2+(V^*_{N2})^2+(V^*_{N3})^2$$

$$=(V^*_A)^2\left\{\left(1+\frac{R_F}{R}\right)^2+4\pi^2 f^2C^2R_F^2\right\}+R_F^2(I^*_T)^2+(V^*_F)^2 \tag{14.5.12}$$

Integrating over the required frequency range we obtain for the total mean square noise at the output:

$$V_N^2=\int_0^{\Delta f}(V^*_N)^2\,\mathrm{d}f$$

$$\therefore \quad V_N=\left[(V^*_A)^2\left\{\left(1+\frac{R_F}{R}\right)^2+\frac{4\pi^2}{3}(\Delta f)^2C^2R_F^2\right\}\right.$$

$$\left.+R_F^2(I^*_T)^2+4kTR_F\right]^{1/2}(\Delta f)^{1/2} \tag{14.5.13}$$

Expanding (I_T^*) as

$$(I_T^*)^2 = 2eIM^2F + 4kT/R + (I_A^*)^2 \tag{14.4.5}$$

and using eqn. (14.5.6), we may write the r.m.s. signal-to-noise ratio as

$$K = \left|\frac{V}{V_N}\right| = \frac{I}{\left[\underset{(a)}{\frac{(V_A^*)^2}{M^2}\left\{\left(\frac{1}{R}+\frac{1}{R_F}\right)^2} + \underset{(b)}{\frac{4\pi^2}{3}C^2(\Delta f)^2\right\}} + \underset{(c)}{2eIF} + \underset{(d)}{\frac{4kT}{M^2}\left(\frac{1}{R}+\frac{1}{R_F}\right)} + \underset{(e)}{\frac{(I_A^*)^2}{M^2}}\right]^{1/2}(\Delta f)^{1/2}} \tag{14.5.14}$$

It will be noted that eqn. (14.5.14) is identical with eqn. (14.4.10) with $(1/R)$ replaced by $(1/R + 1/R_F)$ in terms (a) and (d). The important point is that we may now increase the values of R and R_F in order to reduce these terms without having to provide equalization. That is, as long as eqn. (14.5.5) remains satisfied.

The main problem with this circuit is one of amplifier stability. The use of a long feedback path around a high-gain, high-input impedance amplifier makes the circuit prone to high-frequency oscillations as a result of positive feedback via parasitic capacitance. Avoiding this requires careful layout and effective screening of the sensitive components.

14.6 A WORKED EXAMPLE

In order to put some numerical values into the noise analysis that has just been presented, we will put ourselves into the position of a system designer faced with evaluating the relative merits of a p–i–n diode and an avalanche photodiode detector. The problem with the APD is that it requires a high-voltage-bias supply and temperature compensation. Its advantage is that it will give greater sensitivity and thus allow a lower level of received power to be acceptable. This, in turn, may mean reduced transmitter power (perhaps an LED rather than a laser) or a longer repeater spacing and the reduced costs in equipment and maintenance that that brings. The question is how much less received power still gives an acceptable signal-to-noise ratio.

We will base the example on a system that is pulse modulated at 140 Mb/s and uses a source of wavelength 1.3 μm. As we shall see in Chapter 15, an r.m.s. signal-to-noise ratio of 20 at the receiver will give an adequate safety margin and a receiver bandwidth of 70 MHz is required. The receiver amplifier will be of the transimpedance feedback type, and we will assume that a silicon bipolar junction transistor is the preferred input device on the grounds of cheapness and availability. Its average noise over the frequency range is represented by $V_A^* = 2\ \text{nV}/\sqrt{\text{Hz}}$ and $I_A^* = 2\ \text{pA}/\sqrt{\text{Hz}}$ under suitable bias conditions. The circuit components are assumed to take on the following values for both p–i–n and APD detectors:

Total input capacitance:	3 pF
Input resistance :	40 kΩ
Feedback resistance :	10 kΩ

The APD will be assumed to have a maximum useful multiplication factor of $M = 20$ at which the noise factor is $F = 10$. At higher values of M we shall assume that increasing dark current and microplasma breakdown lead to a rapid increase in F. Then the magnitudes of the various noise sources will be as shown in Table 14.1.

Thus the use of the APD improves the receiver sensitivity by a factor of five. We are not suggesting that this is an optimized design from the point of view of the signal-to-noise ratio, although it is not far from being so. It may, for example be possible to revise the transistor bias such as to reduce the value of I_A^* even though this leads to an increase in V_A^*. Further increase in R and R_F

Table 14.1

	p–i–n $M = 1, F = 1$	APD $M = 20, F = 10$
Term (a)	6.25×10^{-26} [A²/Hz]	1.6×10^{-28} [A²/Hz]
Term (b)	2.3×10^{-24} [A²/Hz]	5.8×10^{-27} [A²/Hz]
Term (d)	2.1×10^{-24} [A²/Hz]	5.2×10^{-27} [A²/Hz]
Term (e)	4.0×10^{-24} [A²/Hz]	1.0×10^{-26} [A²/Hz]
Total, (a) + (b) + (d) + (e) = (A)	8.5×10^{-24} [A²/Hz]	2.1×10^{-26} [A²/Hz]
$\frac{1}{4}(I^*_{Sh})^2_{QL} = (KeF)^2\Delta f = $ (B)	7.2×10^{-28} [A²/Hz]	7.2×10^{-26} [A²/Hz]
$q/p^2 = $ (A)/(B)	11800	0.3
$I_{min} = p\{1 + (1 + q/p^2)^{\frac{1}{2}}\}$	490 nA	96 nA
$(\Phi_R)_{min}$ (with $\eta = 0.5$)	1 μW	200 nW

gives only a small improvement by the reduction in term (d). Term (a) is insignificant anyway. These matters are clearly of much greater importance when the p–i–n diode detector is used. Then the use of a GaAs–MESFET for the first stage of amplification is likely to be beneficial because of its low I_A^* and good high-frequency performance.

PROBLEMS •

14.1 For the following types of field-effect transistor calculate the magnitude of the component V_A^* associated with the channel resistance, assuming this to be given by $(\zeta 4kT/g_m)^{1/2}$. In each case take $T = 300$ K.
(a) An Si–FET in which $g_m = 5$ mS and $\zeta = 0.7$.
(b) A GaAs–MESFET in which $g_m = 50$ mS and $\zeta = 1.1$.

14.2 The gate leakage current I_G contributes a component $(2eI_G)^{1/2}$ to the equivalent input noise current of a field effect transistor. Calculate the magnitude of this component when $I_G = 1$ nA.

14.3 The noise in a particular optical communication receiver with an FET front-end is dominated by term (b) in eqn. (14.4.10). The voltage noise of the FET, V_A^*, may be assumed to be dominated by the noise of the channel resistance, given by $(\zeta 4kT/g_m)^{1/2}$. Show that under these circumstances the ratio of (input capacitance)^2 to transconductance (C^2/g_m) of the FET represents a 'figure-of-merit' which should be minimized.

14.4 At mid-band frequencies the main contribution to the equivalent input noise current source I_A^* for a bipolar junction transistor is a term $(2eI_B)^{1/2}$ which originates in the shot noise of the base current, I_B. At the same time the main contribution to the equivalent input noise voltage source, V_A^*, is a term $(2/eI_C)^{1/2}kT = (2/eI_B\beta)^{1/2}kT$, where I_B is the transistor base bias current, I_C the collector bias current and $\beta = I_C/I_B$ the current gain of the transistor.
(a) Show that the latter term may be expressed as $(2kT/g_m)^{1/2}$, where $g_m = dI_C/dV_{BE}$ is the forward transconductance of the transistor.
(b) Calculate values for the magnitudes of these components of I_A^* and V_A^* for a transistor with $\beta = 100$ when $I_B = 1$, 10 and 100 μA.

14.5 In a particular receiver having an integrating bipolar front-end amplifier, I_A^* may be taken to be $(2eI_B)^{1/2}$ and V_A^* to be $(2/eI_C)^{1/2}kT = (2/eI_B\beta)^{1/2}kT$ where I_B, I_C and $\beta = I_C/I_B$ are as defined in Problem 14.4. Assuming that the signal-to-noise ratio is given by eqn. (14.4.15):
(a) Show that the optimum base bias current is given by

$$I_{B\,opt} = 2\pi kTC\Delta f/e(3\beta)^{1/2}$$

(b) Show that with the bias current optimized for bandwidth, Δf, the sum of the amplifier noise terms, (b) and (e) in eqn. (14.4.15), is given by:

$$(I_C^*)^2 = \frac{8\pi kTC\Delta f}{(3\beta)^{1/2}M^2}$$

Note that $(C/\beta^{1/2})$ represents a 'figure-of-merit' that should be minimized for bipolar junction transistors.

(c) Hence show that the photogenerated current required to maintain a specified signal-to-noise ratio, K, increases in direct proportion to the bandwidth, Δf, when optimum bias conditions are maintained.

(d) If the transistor to be used in this receiver has a current gain $\beta = 120$, and $C = 5$ pF, calculate $I_{B\,opt}$ and I_C^* for $\Delta f = 10$ MHz and 100 MHz and $M = 1$ and 100. Assume that β is constant over the required frequency ranges and take $T = 300$ K. Hence calculate the signal current, I, required to give a signal-to-noise ratio $K = 20$ in each case assuming all other noise sources to be negligible in comparison with I_{Sh}^* and I_C^*.

14.6 (a) Compare the magnitudes of the five terms in the denominator of eqn. (14.4.10) when

$$V_A^* = 5\ \text{nV}/\sqrt{\text{Hz}},\quad I_A^* = 20\ \text{fA}/\sqrt{\text{Hz}},\quad M = F = 1,$$

$$R = 1\ \text{M}\Omega,\quad C = 5\ \text{pF},\quad \Delta f = 1\ \text{MHz},\quad I = 0.1\ \mu\text{A},\quad T = 300\ \text{K}.$$

(b) Indicate whether or not equalization is required.

(c) Calculate the corresponding signal-to-noise ratio.

(d) Calculate by what factor this is worse than the ideal quantum noise limit.

14.7 Investigate independently the effect of the following variations in the conditions specified in Problem 14.6 on the relative magnitudes of the five terms of eqn. (14.4.10):

(a) A decrease of R to 10 kΩ.

(b) A decrease of C to 2 pF.

(c) A decrease of I to 10 nA.

14.8 (a) Compare the magnitudes of the five terms in the denominator of eqn. (14.5.14) when

$$V_A^* = 2\ \text{nV}/\sqrt{\text{Hz}},\quad I_A^* = 2\ \text{pA}/\sqrt{\text{Hz}},\quad M = 100,\quad F = 5,$$

$$R = R_F = 1\ \text{k}\Omega,\quad C = 2\ \text{pF},\quad \Delta f = 100\ \text{MHz},\quad I = 1\ \mu\text{A},\quad T = 300\ \text{K}.$$

(b) Calculate the signal-to-noise ratio.

(c) Calculate by what factor the signal-to-noise ratio is worse than that under ideal quantum noise limited conditions.

14.9 Investigate the effect of the following changes to the conditions described in Problem 14.8 (considering each independently) on the relative magnitudes of the five terms of eqn. (14.5.14).

(a) An increase in R and R_F to 100 kΩ.

(b) An increase in C to 5 pF.

(c) A decrease in I to 0.1 μA.

(d) The use of a p–i–n photodiode, making $M = F = 1$.

14.10 Indicate the gain, A, required to satisfy conditions (14.5.3) and (14.5.5) for the four combinations of parameters of Problems 14.8 and 14.9, namely $C = 2$ pF and 5 pF, R $(= R_F) = 1$ kΩ and 100 kΩ.

REFERENCES

14.1 C. D. Motchenbacher and F. C. Fitchen, *Low-Noise Electronic Design*, Wiley/Interscience (1973).
14.2 A. van der Ziel, *Solid-State Physical Electronics*, Prentice-Hall, (1968).
14.3 F. N. H. Robinson, *Noise and Fluctuations in Electronic Devices and Circuits*, O.U.P. (1974).
14.4 R. G. Smith and S. D. Personick, Receiver design for optical fiber communication systems, in H. Kressel (ed.) *Semiconductor Devices for Optical Communication* (Ch. 4), Topics in Applied Physics, Vol. 39, Springer-Verlag (1980).
14.5 S. D. Personick, Receiver design for digital fiber optic communication systems, I and II, *Bell System Tech. Jnl.* **52**(6), 843–86 (July/August 1973).

SUMMARY

The need to preserve a certain minimum signal-to-noise ratio at the receiver input represents one of the chief design criteria of an optical, as of any, communication system. A value of 20–30 dB may be adequate in a digital channel, but values as high as 50–60 dB may be required in analogue links.

Noise is represented in terms of the r.m.s. spectral density. Shot noise is generated in the detection process: $(I^*_{\text{Sh}})^2 = 2eM^2F\bar{I}$. Thermal noise is generated in all the components which make up the receiver amplifier input resistance: $(V^*_{\text{Th}})^2 = 4kTR$ or $(I^*_{\text{Th}})^2 = 4kT/R$. Amplifier noise is represented by a noise current source I^*_{A} and a noise voltage source V^*_{A} at the input as shown in Fig. 14.2.

The input signal is $MI = M\mathcal{R}\Phi_{\text{R}} = M\eta(e\lambda/hc)\Phi_{\text{R}}$.

The signal-to-noise ratio for an equalized voltage amplifier is given by eqn. (14.4.10), that for a transimpedance feedback amplifier by eqn. (14.5.14).

For an ideal quantum limited detector the current required to give a signal-to-noise ratio of K is $I = 2eK^2\,\Delta f$.

The use of a large input resistance reduces the amplifier voltage noise and the thermal noise components. Unless the transimpedance feedback configuration is used, equalization may then be needed and dynamic range will be reduced.

When a high-gain APD is used, multiplied shot noise normally dominates, as it does also when large signal-to-noise ratios are required. With a p–i–n photodiode:

(a) Amplifier voltage noise tends to dominate at high frequencies, and a silicon bipolar transistor is the preferred input device. It is then most important to minimize capacitance.

(b) Amplifier current noise tends to dominate at low frequencies and a Si-FET is the preferred input device.

(c) Unless the input resistance is large, thermal noise tends to dominate the middle range of frequencies.

For the highest frequencies, around 1 GHz, the GaAs MESFET is the preferred input device. A package which integrates the detector and receiver amplifier so as to minimize the total input capacitance is then essential.

15

The Regeneration of Digital Signals

15.1 CAUSES OF REGENERATION ERROR

15.1.1 The Ideal Digital System

In the receiver of a digital system the linear amplification section is followed by a regenerator as shown in the block diagram of Fig. 15.1. We have included as part of the linear amplifier any equalization or filtering that may be provided, and the regenerator proper is preceded by a decision circuit. In an optical repeater the regenerated pulse train directly modulates the output from an optical source and thereby launches the onward transmission of the signal. In a terminal receiver the regenerated pulses require demultiplexing and demodulating in order that the original signal waveforms may be recovered. In the present chapter we shall be concerned to establish the least signal-to-noise ratio at the receiver required to ensure that these original waveforms are recovered with an adequate level of fidelity. In previous chapters we have already used two 'rule-of-thumb' estimates for this. In Chapters 1 and 3 we indicated that a typical, minimum received power requirement might be in the region of 0.1 to 1.0 nW/(Mb/s). In Chapter 14 we suggested that the signal-to-noise ratio should be in the region of 12 (21.6 dB). The worked example of Section 14.6

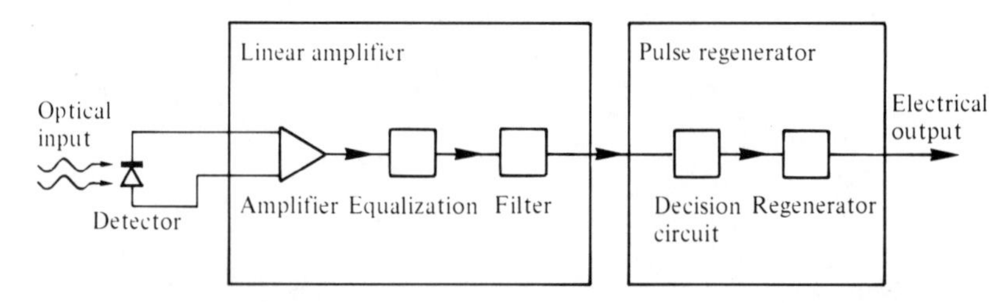

Fig. 15.1 Block diagram of a digital optical communication receiver.

used a value of 20 to give a power margin and this suggested the need for a rather higher level of received power. This was, however, based on a very conservative amplifier specification.

By a digital, optical communication system we mean one in which a stream of binary data is transmitted by modulating the optical source such that the energy emitted during each bit period is at one of two levels. That is, *high* or *low*, corresponding to 1 or 0. The emission may take the form of a short pulse of radiation during each bit period. This is known as *return-to-zero* (RZ) signalling. Alternatively the power level may be maintained more or less constant for the whole bit period, a technique known as *non-return-to-zero* (NRZ) signalling. In order to simplify the analysis we will define an ideal pulse-code modulated (PCM) system as one having the following characteristics.

(a) The bit period has the constant value T, so the bit rate is $B = 1/T$.
(b) The optical energy transmitted during a bit period when a 1 is sent is $\mathcal{E}_T$ and during a bit period when a 0 is sent is zero. The optical energies received during these respective bit periods are $\mathcal{E}_R$ and zero.
(c) The optical power takes the form of an impulse at some fixed time within the bit period, as shown in Fig. 15.2.
(d) In an extended bit stream there is an equal probability that a 1 or a 0 is transmitted. This is a normal feature of most signalling codes. We may then express the long-term average of the received power, $\bar{\Phi}_R$, in terms of the average power, Φ_R, received during any bit period when a 1 is sent. Thus

$$\bar{\Phi}_R = \tfrac{1}{2}\Phi_R = \tfrac{1}{2}\mathcal{E}_R/T = \tfrac{1}{2}\mathcal{E}_R B \qquad (15.1.1)$$

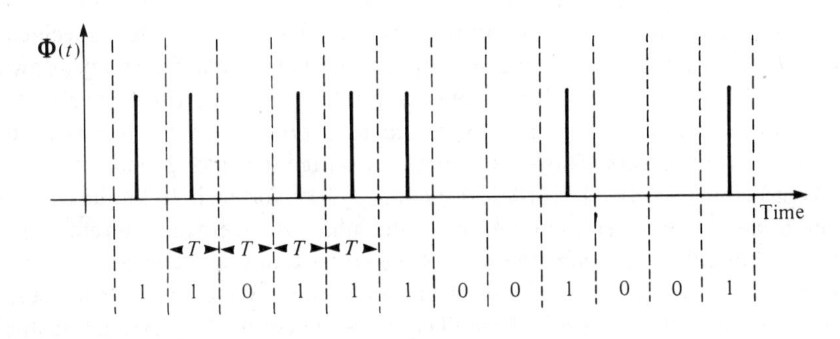

Fig. 15.2 Received power in an ideal system.

A real system departs from this ideal in a number of ways:

(a) The bit period is not constant—an effect known as *jitter*.
(b) The optical energy transmitted is not exactly the same for every 1 that is sent, nor for every 0: there will be transmitter noise giving random variations from pulse to pulse. Fluctuations in the energy received are further increased by variations in the channel attenuation.
(c) Aside from transmitter and channel noise, it is likely that when 0 is sent a small but finite level of optical energy is transmitted. The ratio of the average energy received which represents 0, $\mathcal{E}_R(0)$, to that received which represents 1, $\mathcal{E}_R(1)$, is known as the *extinction ratio*, r_e:

$$r_e = \mathcal{E}_R(0)/\mathcal{E}_R(1) \tag{15.1.2}$$

For the ideal system we assume $r_e = 0$, but this is not normally the case, especially with laser sources biassed near to threshold.
(d) The finite length of the transmitted pulses and the additional time dispersion during transit means that in practical systems some of the energy that belongs to one particular bit period actually arrives in one of the adjacent bit periods. This effect is known as *intersymbol interference*.

15.1.2 Causes of Regeneration Error

In the receiver, after the initial amplification, the decision circuit causes the waveform to be sampled at some point during each bit period, and compared to some pre-set threshold level. If its amplitude exceeds the threshold, a 1 is regenerated; if it does not, a 0 is assumed. When errors are made the regenerated waveform differs from the one originally transmitted. The definition of an acceptable error rate forms an essential part of any system specification. For digital trunk telephone systems an international standard lays down that in a loop of 2500 km there should be no more than 2 errors in every 10^7 bits transmitted. This is usually expressed as a *probability of error* (*PE*) of 2×10^{-7} over the loop. This means that on average the error probability must be kept below $(2 \times 10^{-7}) \times (10/2500) = 0.8 \times 10^{-9}$ for each 10 km link in the loop. It should be understood that this is the minimum average requirement for each 10 km link. In practice the bulk of the errors that occur can be attributed to only a very few of the many links in a long loop. What is more, the actual performance of the system is more likely to be determined by external disturbance, 'interference' in our terminology, than by the internal sources of noise that we discuss in Chapters 14 and 15. This often gives rise to bursts of errors rather than a steady random distribution. One of the great merits of fiber systems is that the transmission path itself is not normally susceptible to such interference, as it is with electrical systems. However, the terminal equipment is just as susceptible and so are any electrical power circuits that may form part of the optical fiber cable. With this in mind we shall take a value of 10^{-9} to be the normal PE requirement for a typical optical link. In other applications the specified PE might vary over the range 10^{-15}–10^{-6} but, as we shall see, at these levels the required signal power is relatively insensitive to the precise PE value to be achieved.

The probability of error may be broken down into two parts:

$$\begin{aligned} PE = & \text{ (probability of regenerating 0 when 1 is sent)} \\ & \times \text{(probability of sending 1)} \\ & + \text{(probability of regenerating 1 when 0 is sent)} \\ & \times \text{(probability of sending 0)} \end{aligned}$$

With normal notation this may be written as

$$PE = P(0 \mid 1)P(1) + P(1 \mid 0)P(0) \tag{15.1.3}$$

and when 0s and 1s are sent in equal numbers, $P(1) = P(0) = \frac{1}{2}$, so that

$$PE = \tfrac{1}{2}\{P(0 \mid 1) + P(1 \mid 0)\} \tag{15.1.4}$$

At first sight it might be thought that the linear amplifier section of the receiver should have as wide a bandwidth as possible so that when a 1 is received the signal entering the decision circuit remains as a short pulse within the bit period. The moment of decision would then be timed to coincide with this pulse. Errors would occur for two reasons.

(a) Noise gives rise to instantaneous variations in the amplitude of the voltage at the input to the decision circuit. Thus when a 0 has been received the voltage may be momentarily raised above the threshold level or when a 1 is received it may fall below, as shown in Fig. 15.3. The receiver noise level increases with the bandwidth.

(b) Any timing variations which affect the moment of arrival of the signal or the moment of sampling within the decision circuit may cause the waveform to be sampled at a point other than at its maximum amplitude. The narrower the pulse the more serious such variations are.

The use of an unnecessarily wide-band amplifier therefore increases the level of noise and makes the timing of the instant of decision more critical. However, a third source of error arises if the amplifier bandwidth is reduced. This is:

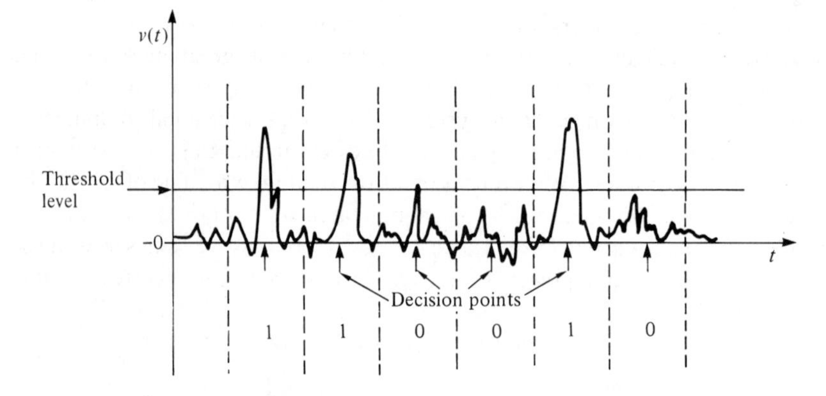

Fig. 15.3 Output waveform from a wide-band linear amplifier section showing possible causes of error through noise and jitter.

(c) Intersymbol interference, which, as we have said, occurs when the power received during one bit period affects the signal amplitude during any other bit period. The lower the amplifier bandwidth, the more likely this is, because the impulse response is then more spread out and extends into adjacent bit-periods. When the bandwidth limitation arises at the source or in the fiber, the result is that optical power from one bit period arrives at the receiver in adjacent bit periods. In addition to the normal intersymbol interference this causes, it also introduces extra noise, in the form of shot noise, into the adjacent periods.

Clearly the optimum bandwidth is a matter for compromise between these three factors. In the diagram of Fig. 15.1 the bandwidth is determined by the filter characteristic $H(f)$, and, as we shall see next, the exact shape of this filter characteristic can be important in minimizing regeneration errors.

15.1.3 Filter Characteristics Designed to Minimize Intersymbol Interference

It is possible to define a general class of filters with the property that their response to an impulse received at time t_0 is zero at all times $t_0 \pm nT$, where n is an integer. This property happens to be possessed by the ideal low-pass filter assumed in Chapter 14, which has the frequency response:

$$\begin{aligned} H(f) &= 1 \qquad \text{for } 0 < f < B/2 \\ &= 0 \qquad \text{for } B/2 < f \end{aligned} \tag{15.1.5}$$

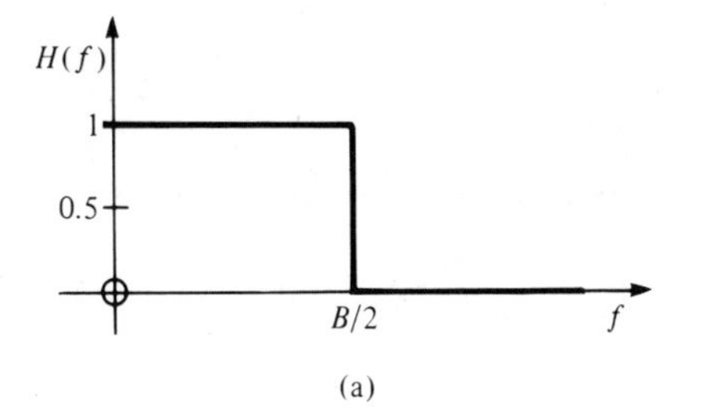

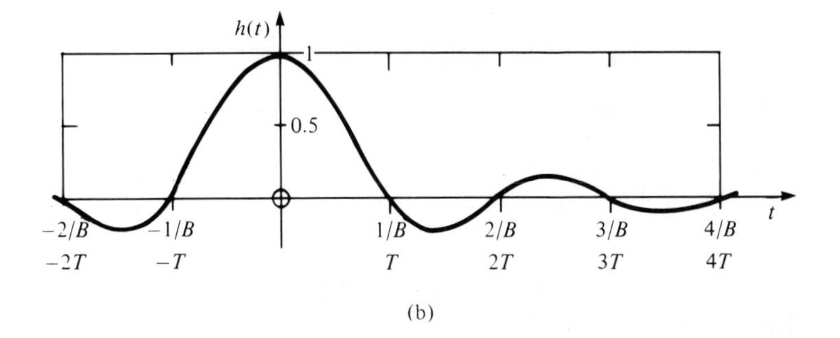

Fig. 15.4 Ideal low-pass filter:
(a) frequency response $H(f) \begin{array}{l} = 1 \text{ for } 0 < f < B/2 \\ = 0 \text{ for } B/2 < f \end{array}$

(b) impulse response $h(t) = \dfrac{\sin \pi t B}{\pi t \mathrm{B}}$

The corresponding impulse response is

$$h(t) = \frac{\sin(\pi t B)}{\pi t B} \tag{15.1.6}$$

which is zero at all times $\pm nT$, where $T = 1/B$, as shown in Fig. 15.4.

In fact it is shown in Appendix 7 that this property is shared by all filters whose frequency response is antisymmetric about the point $H(B/2) = \frac{1}{2}$. A particular example which has been much discussed is the *raised-cosine* filter:

$$\begin{aligned} H(f) &= \tfrac{1}{2}(1 + \cos \pi f/B) \qquad && \text{for } 0 < f < B \\ &= 0 && \text{for } B < f \end{aligned} \tag{15.1.7}$$

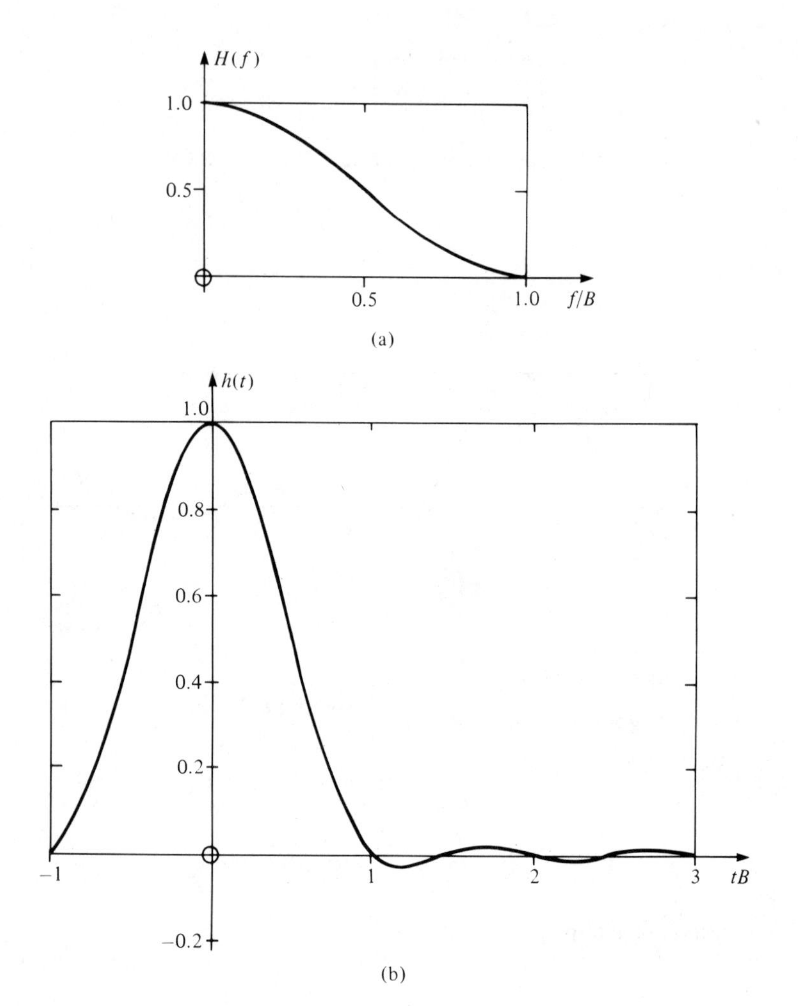

Fig. 15.5 Raised-cosine filter:

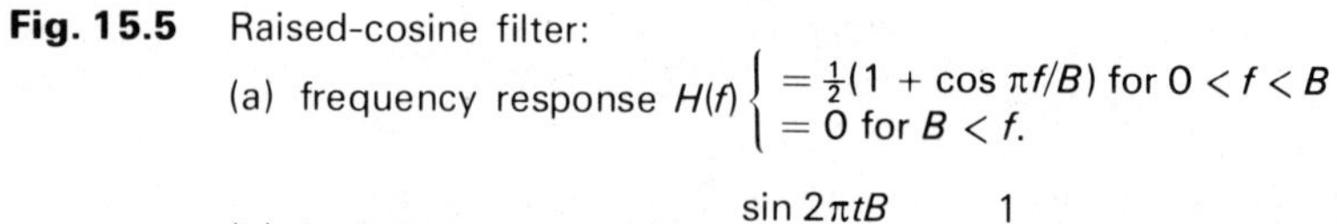

(a) frequency response $H(f) \begin{cases} = \frac{1}{2}(1 + \cos \pi f/B) \text{ for } 0 < f < B \\ = 0 \text{ for } B < f. \end{cases}$

(b) impulse response $h(t) = \dfrac{\sin 2\pi tB}{2\pi tB} \dfrac{1}{[1 - (2tB)^2]}$

This is shown in Fig. 15.5 together with its impulse response, which is

$$h(t) = \frac{\sin 2\pi tB}{2\pi tB} \frac{1}{[1-(2tB)^2]} \tag{15.1.8}$$

Although the ideal low-pass filter cannot be realized in practice, it does theoretically yield the best signal-to-noise ratio, and we shall continue to make it the basis of our 'ideal' system. However, it would give rise to relatively large amplitudes in adjacent bit periods, as can be seen in Fig. 15.4(b), and this means that quite small amounts of pulse-broadening or jitter produce a great deal of intersymbol interference. The raised-cosine filter introduces more noise but is much more tolerant of jitter and pulse broadening.

15.1.4 The Eye Diagram

The trade off between sensitivity to amplitude variations (noise) and time variations (jitter) is most readily demonstrated by means of the *eye diagram*. This is the technique used experimentally to assess system performance. The waveforms from a large number of random pulse sequences are superposed in one bit period, as shown in Fig. 15.6. For errors to be minimized the eye should be as open as possible, and with a sharply band-limited filter it is very narrow. A more gradual filter such as the raised cosine leaves an open eye but increases the noise. As a result, a higher received signal power level is needed in order to preserve a specified signal-to-noise ratio. We may think of this as a *noise penalty*. We shall return to consider this further in Section 15.4. Meanwhile Section 15.2 will discuss the error probability in an ideal system in which the only noise present is the shot noise on the optical signal. This represents the quantum limit to detection. In Section 15.3 the effect of amplifier noise on the error probability of an otherwise ideal system will be examined.

15.2 THE QUANTUM LIMIT TO DETECTION

A steady light flux incident onto a photodiode generates carrier pairs in independent, random events. This is known as a Poisson process. On average an optical energy, $\mathcal{E}_R$, arriving in a given period would be expected to produce N carrier pairs, where

$$N = \eta \frac{\mathcal{E}_R}{\mathcal{E}_{ph}} = \eta \frac{\mathcal{E}_R \lambda}{hc} \tag{15.2.1}$$

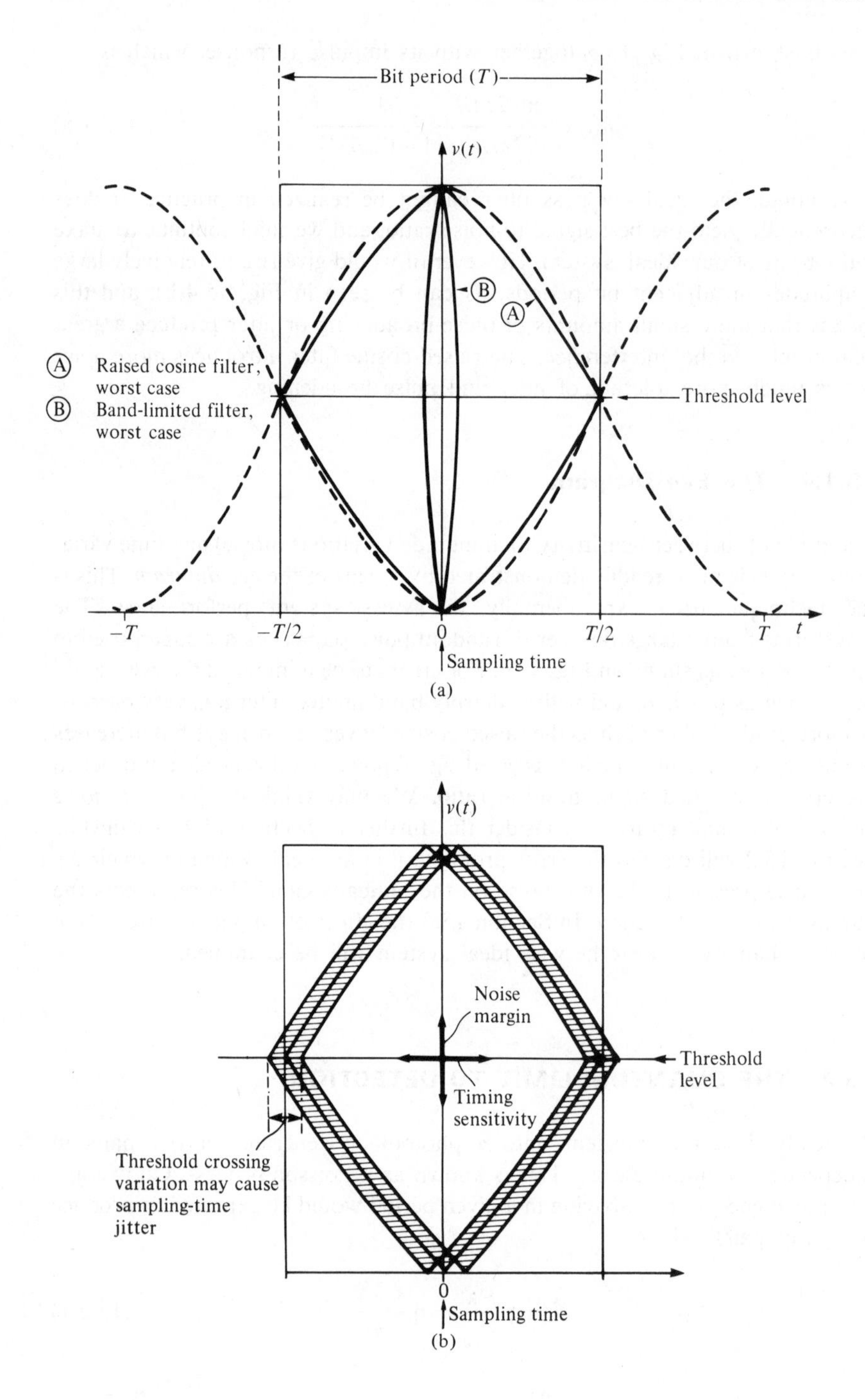
Bit period (T)
v(t)
B
A
A Raised cosine filter, worst case
B Band-limited filter, worst case
Threshold level
−T
−T/2
0
T/2
T
t
Sampling time
(a)
v(t)
Noise margin
Threshold level
Timing sensitivity
Threshold crossing variation may cause sampling-time jitter
0
Sampling time
(b)

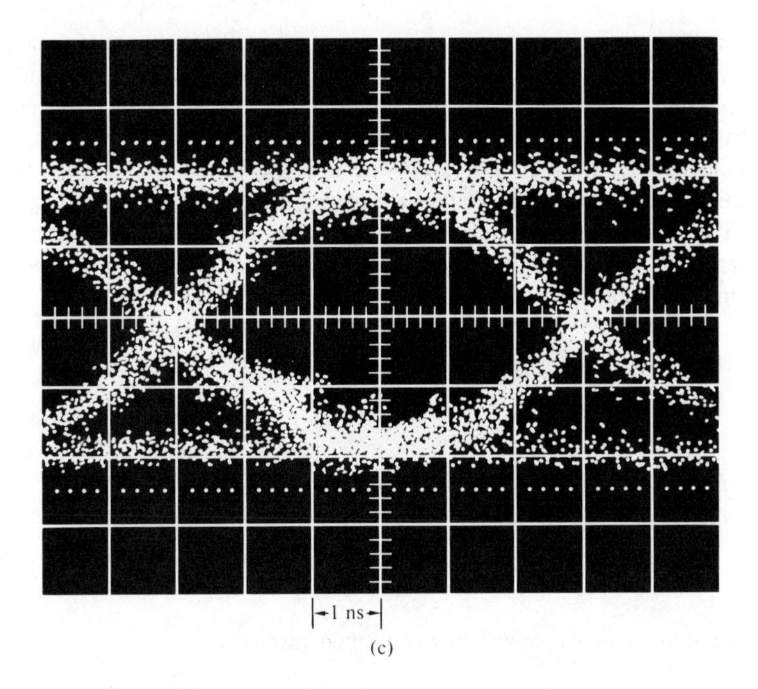

Fig. 15.6 Eye diagrams: (a) theoretical—the dashed curves represent the impulse-response of the raised-cosine filter. The full curves represent the cumulative effects of the worst combination of preceding and succeeding pulses on a received 0 and a received 1. With the raised-cosine filter the eye is narrowed only slightly, but with a sharply band-limited filter it becomes very narrow indeed; (b) schematic eye diagram for the raised-cosine filter showing the broadening caused by finite pulse width and jitter; (c) actual eye diagram obtained by superimposing many pulse-trains arriving in random sequence. [Redrawn from R. W. Berry and I. A. Ravenscroft, Optical-fiber transmission systems: the 140 M bit/s feasibility trial, *Post Office Elec. Engrs. J.* **70**, 261–8 (1978).]

As previously, η is the quantum efficiency of the interaction and $\mathcal{E}_{ph}$ the photon energy. Because of the statistical nature of the interaction, the actual number of carrier pairs generated by each optical pulse varies about the mean value. The probability that the number produced is k is given by the Poisson probability distribution:

$$P(k|N) = \frac{\exp(-N) \cdot N^k}{k!} \tag{15.2.2}$$

The mean square variation about the mean value, N, is then also N.

In an ideal system this variation in the number of carrier pairs generated would be the only source of noise. Furthermore, optical energy would be received and carriers generated only when a 1 was sent. If the receiver was sensitive enough to detect a single photogenerated electron–hole pair, the threshold could be set at this level. There would be no errors when 0 was transmitted since no power would be received and no signal generated. Only when the incident optical energy corresponding to 1 failed to generate any carriers at all, instead of the expected number N, would an error be recorded. Remembering that 0s and 1s are sent in equal numbers,

$$PE = \tfrac{1}{2}\{P(0 \mid 1) + P(1 \mid 0)\} \tag{15.1.3}$$

$$= \frac{1}{2}\left\{\frac{\exp(-N) \cdot N^0}{0!} + 0\right\} = \tfrac{1}{2}\exp(-N) \tag{15.2.3}$$

For $PE < 10^{-9}$ we require that $N > 20$ and thus

$$\mathcal{E}_R = \frac{N\mathcal{E}_{ph}}{\eta} > \frac{20\mathcal{E}_{ph}}{\eta} \tag{15.2.4}$$

The minimum mean received power is then given by

$$\bar{\Phi}_R = \tfrac{1}{2}\mathcal{E}_R B > \frac{10\mathcal{E}_{ph}B}{\eta} \tag{15.2.5}$$

This represents the absolute quantum limit to detectability. With $\eta = 1$ and $\lambda = 0.9$ μm so that $\mathcal{E}_{ph} = 1.38$ eV, $\bar{\Phi}_r > 2.2$ pW/(Mb/s). Comparison with the rules of thumb quoted earlier indicates that amplifier noise and other problems encountered in practical systems degrade their sensitivity such that the received power requirements are some two orders of magnitude above this quantum limit. A possibly more convenient way of representing this result is to do so in terms of the average received energy per bit transmitted. The absolute quantum limit with $\eta = 1$ and with 0s and 1s equally probable then stands at the average value of 10 received photons per bit.

15.3 THE EFFECT OF AMPLIFIER NOISE AND THERMAL NOISE ON THE ERROR PROBABILITY

15.3.1 Probability of Error when Shot Noise is Negligible

Section 15.2 dealt with an idealized situation in which the only noise present in the system was shot noise. Now, we will go to the other extreme and consider the case in which the shot noise is negligible in comparison with the thermal and amplifier noise present in the receiver. The random fluctuations of voltage or current associated with these latter noise sources are normally found to follow a Gaussian distribution. For our purposes it is convenient to relate the fluctuations seen at the output of the amplifier to the equivalent number of carriers which would have had to be generated in the photodiode in order to produce the same effect. In the case of thermal and amplifier noise these effective numbers of carriers take on a Gaussian distribution about the mean, or expected, value.

The instantaneous voltage which is applied to the regenerator at the sampling time within a given bit period comprises the amplified and filtered signal voltage, V_S, and a noise voltage, of r.m.s. amplitude, V_N. The signal voltage corresponds to the number N of carriers generated by the optical signal received during the bit period and the noise voltage corresponds to an effective fluctuation about N whose r.m.s. value we will take to be σ. When the noise distribution is Gaussian, the probability that the total output voltage in a given bit period, $V_S + V_N$, corresponds to k carrier pairs generated in the photodiode is given by

$$P(k|N) = \frac{1}{(2\pi)^{1/2}\sigma} \exp\left[-(N-k)^2/2\sigma^2\right] \tag{15.3.1}$$

We assume as before that on average N carrier pairs are generated when 1 is received and zero when 0 is received. For purposes of analysis we shall assume equal noise in both cases and set the threshold level half-way between, that is, at the voltage level $V_S/2$ corresponding to $N/2$ carrier pairs. The situation is illustrated in Fig. 15.7. As before, we shall define the signal-to-noise ratio, K, as the ratio of the peak signal level when a 1 is received to the r.m.s. noise level, that is

$$K = \frac{V_S}{V_N} = \frac{N}{\sigma} \tag{15.3.2}$$

The probability of error is given by

$$PE = \tfrac{1}{2}\{P(0|1) + P(1|0)\} = \frac{1}{2}\left\{ \sum_{k=0}^{N/2} P(k|N) + \sum_{k=N/2}^{\infty} P(k|0)\right\} \tag{15.3.3}$$

We have to substitute the Gaussian distributions (15.3.1) into (15.3.3). The calculation is made easier when $N \gg 1$ because the summations may then be replaced by integrals which are well known:

$$PE = \frac{1}{2}\frac{1}{(2\pi)^{1/2}\sigma}\left[\int_0^{N/2} \exp\left(-\frac{(N-k)^2}{2\sigma^2}\right) \mathrm{d}k + \int_{N/2}^{\infty} \exp\left(-\frac{k^2}{2\sigma^2}\right) \mathrm{d}k\right] \tag{15.3.4}$$

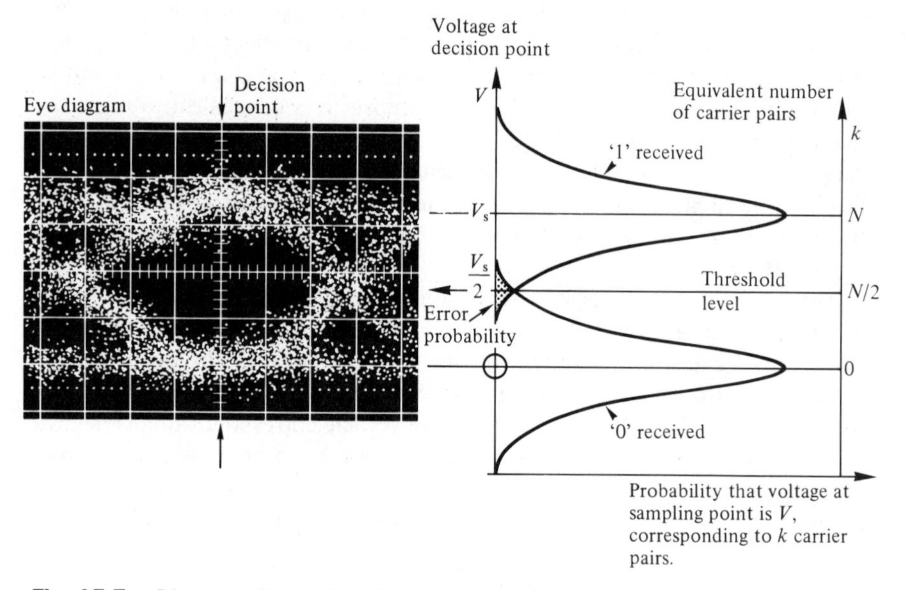

Fig. 15.7 Diagram illustrating the voltage probability distribution at the decision point.

The two integrals are equal, by symmetry, so that

$$PE = \frac{1}{(2\pi)^{1/2}\sigma} \int_{N/2}^{\infty} \exp\left(-\frac{k^2}{2\sigma^2}\right) \mathrm{d}k$$

$$= \operatorname{erfc}\,(N/2\sigma) \tag{15.3.5}$$

$$= \operatorname{erfc}\,(K/2) \tag{15.3.6}$$

The function erfc (x) is the *complementary error function*

$$\operatorname{erfc}\,(x) = \frac{1}{(2\pi)^{1/2}} \int_{x}^{\infty} \exp\,(-t^2/2)\,\mathrm{d}t \tag{15.3.7}$$

which may be obtained from tables.† It is plotted in Fig. 15.8, from which it can be seen that a signal-to-noise ratio of $K > 12$ is required to ensure that $PE < 10^{-9}$. The erfc curve is very steep in this region and it is of course absurd to think of designing a system to give a particular PE value. What is done is to estimate the signal-to-noise ratio required to meet the PE specification in the

† See for example Table 26.2 in M. Abramowitz and I. A. Stegun, *Handbook of Mathematical Functions*, Dover (1965).

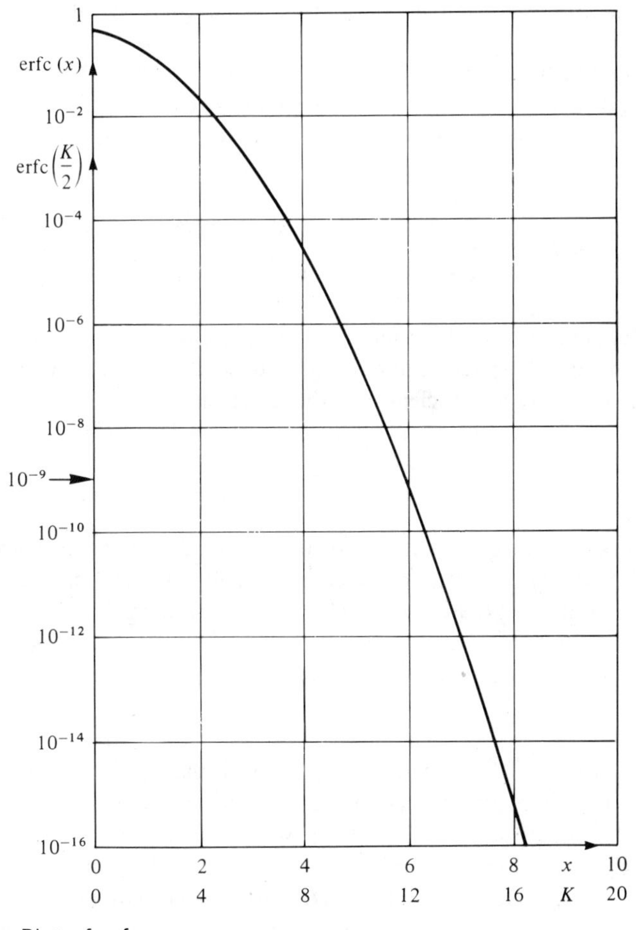

Fig. 15.8 Plot of erfc x.

theoretical system; to this is added any signal margins needed for known departures from the theoretical assumptions such as pulse spreading; and finally an unallocated margin is added for safety. It was on this basis that we suggested in Section 14.6 that a signal-to-noise ratio of $K = 20$ might be used rather than the theoretical value of $K = 12$. The difference was simply an unallocated system margin.

We are now in a position to compare the sensitivity of a receiver dominated by amplifier noise with the ideal quantum limit of Section 15.2. To simplify the comparison we will assume a high input resistance, or, in the case of a transimpedance-feedback amplifier, high input and feedback resistances. Then noise terms (a) and (d) in eqn. (14.4.10) or eqn. (14.5.14) will be negligible as well as term (c).

The required bandwidth will be taken to be one half of the bit rate so that $\Delta f = B/2$. Then by rearranging the equations we can obtain an expression for the minimum current, I_m, required to ensure a specified signal-to-noise ratio, K. This is

$$I > I_m = K\left(\frac{4\pi^2}{3M^2}(CV_A^*)^2\left(\frac{B}{2}\right)^2 + \frac{(I_A^*)^2}{M^2}\right)^{1/2}\left(\frac{B}{2}\right)^{1/2} \quad (15.3.8)$$

In turn this current can be related by the detector responsivity, $\mathcal{R}$, to the minimum received power required which depends on the wavelength of the light and the quantum efficiency of the detector:

$$\Phi_R = \frac{1}{\eta}\frac{I}{e}\frac{hc}{\lambda} = \frac{I}{\mathcal{R}} \quad (12.6.2)$$

When, on average, equal numbers of 0s and 1s are transmitted, the average current, $\bar{I}$, and the average received power, $\bar{\Phi}_R$, will be half these values: $\bar{\Phi}_R = \frac{1}{2}\Phi_R$ and $\bar{I} = \frac{1}{2}I$. Equation (15.3.8) then becomes

$$\bar{I} > \bar{I}_m = \frac{KB^{1/2}}{2\sqrt{2M}}\left(\frac{\pi^2(CV_A^*)^2B^2}{3} + (I_A^*)^2\right)^{1/2} \quad (15.3.9)$$

In Fig. 15.9, $\bar{I}_m$ is shown plotted against the bit rate, B, for $K = 12$, corresponding to the theoretical limit, and for $M = 1$, corresponding to a p–i–n photodiode. The noise parameters for a Si-BJT and Si-FET are taken to be the same as those used in Chapter 14, that is, $V_A^* = 2\ \mathrm{nV}/\sqrt{\mathrm{Hz}}$ and $4\ \mathrm{nV}/\sqrt{\mathrm{Hz}}$, and $I_A^* = 2\ \mathrm{pA}/\sqrt{\mathrm{Hz}}$ and $10\ \mathrm{fA}/\sqrt{\mathrm{Hz}}$, respectively. In order to show the importance of the input capacitance at high bit rates, values of 2 pF and 5 pF have been used. The FET can be seen to have a clear advantage at bit rates less than about 50 Mb/s. For each type of device the minimum required current varies at low bit rates as $B^{1/2}$. It then depends on I_A^* and R (or R and R_F). That is, it would increase above the levels shown if the input resistance (or the parallel combination of R and R_F) were too small. At high bit rates $\bar{I}_m$ varies in each case as $B^{3/2}$. It then depends on V_A^* and C.

We can see from Fig. 15.9 that the rule of thumb we originally suggested, 0.1–1 nW/(Mb/s), implies a quite different functional relationship for $\bar{I}_m(B)$. It can be related to the figure only when the detector responsivity is known. For example, with $\mathcal{R} = 0.5$ A/W, 1 nW/(Mb/s) implies a minimum average detector current of 0.5 nA/(Mb/s) or 3100 electrons per bit. Then, 0.1 nW/(Mb/s) corresponds to 310 electrons per bit.

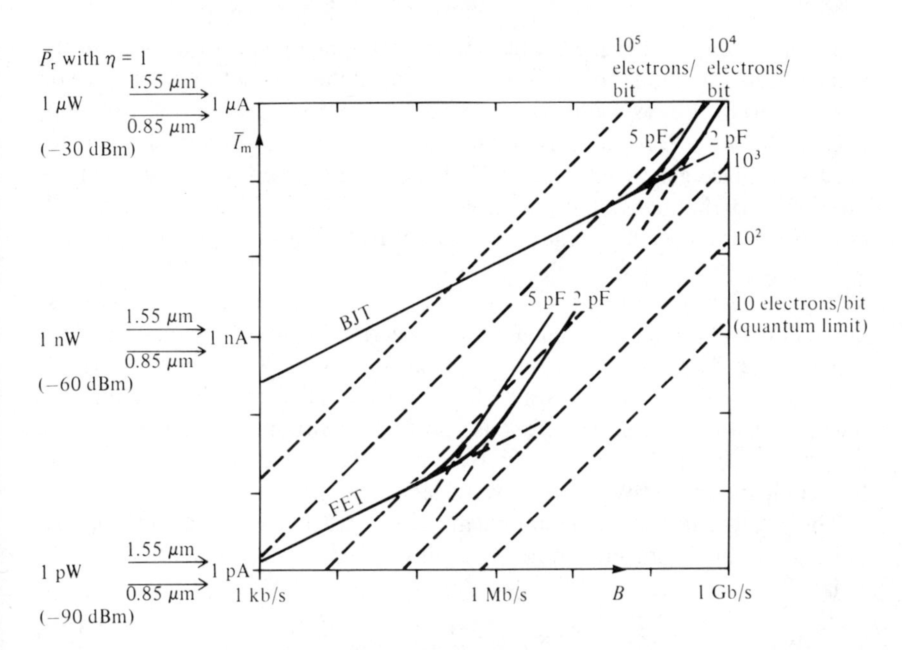

Fig. 15.9 Minimum average detector current required to ensure error probability less than 10^{-9} plotted as a function of bit-rate.

Solid lines correspond to the quoted noise parameters for Si-BJT and Si-FET with input capacitances of 2 pF and 5 pF. Dashed lines indicate constant values of electrons per bit. The corresponding power levels at $\eta = 1$ are shown on the left-hand side for different wavelengths.

15.3.2 Probability of Error When Multiplied Shot Noise is Comparable to Other Sources of Noise

It became clear in the discussion of Chapter 14 that when a p–i–n photodiode detector is used, the shot noise is normally negligible, as we have just assumed. However, the use of a good avalanche photodiode giving a gain of 20 or more is usually sufficient to ensure that the relative effect of noise from other sources is reduced to a level comparable to or less than the multiplied shot noise. The situation is exceedingly difficult to deal with theoretically to the same degree of rigor that has been possible so far. The reason for this concerns the different distribution functions to be expected for the shot noise and for the other noise sources, and the fact that the noise level may then be dependent on the signal strength.

The thermal and amplifier noise sources have been assumed to have Gaussian amplitude distributions. This has enabled us to express the total effect of a number of independent and uncorrelated sources as the sum of the mean square amplitudes of each. We have added the effect of the shot noise on the same basis. But as we have seen in Section 15.2 we should expect the shot noise to follow a Poisson distribution. The amplitude distribution of the multiplied shot noise generated in an APD further depends on the statistics of the avalanche carrier generation processes, and this is something that is not well-established theoretically. The justification for adding these various noise sources together as though they were Gaussian, as we did in Chapter 14, is simply that when the numbers involved are large, as they are here, all these distributions approximate to the Gaussian distribution in the region around the mean value. The total r.m.s. noise values so obtained are thus reasonable approximations. However, in estimating error probabilities we are dealing with the tails of the distribution functions and it is important to be aware that the simple assumption that they all approximate to the Gaussian function may lead to significant errors. We shall nevertheless continue on this basis.†

The solution for the minimum mean current, when shot noise and amplifier noise are both significant can be obtained from equations (14.4.32)–(14.4.34). It is

$$\bar{I}_{\mathrm{m}} = \tfrac{1}{2}I_{\mathrm{m}} = \tfrac{1}{2}p\{1 + (1 + q/p^2)^{1/2}\} \tag{15.3.10}$$

where

$$p = K^2 eFB/2 \tag{15.3.11}$$

and

$$q = \frac{K^2 B}{2M^2}\left(\frac{4\pi^2}{3}(CV_{\mathrm{A}}^*)^2(B/2)^2 + (I_{\mathrm{A}}^*)^2\right) \tag{15.3.12}$$

We have substituted $B/2$ for Δf and have assumed R to be large enough for noise terms (a) and (d) to be negligible. Note that

$$q/p^2 = \frac{2}{K^2 e^2 M^2 F^2}\left(\frac{\pi^2}{3}(CV_{\mathrm{A}}^*)^2 B + \frac{(I_{\mathrm{A}}^*)^2}{B}\right) \tag{15.3.13}$$

The solution for I_{m} expressed in eqn. (15.3.10) falls either into two or three parts as a function of bit rate. This is something we have already found in Section 14.4.3. The minimum value of q/p^2 occurs when

$$B = B_0 = \frac{\sqrt{3}I_{\mathrm{A}}^*}{\pi CV_{\mathrm{A}}^*} \tag{15.3.14}$$

† A simple way of representing a Poisson distribution by an equivalent Gaussian distribution so that there is a good match between the tails of the two distributions (especially in the region of Prob = 10^{-9}) is given by W. M. Hubbard, The approximation of a Poisson distribution by a Gaussian distribution, *Proc. I.E.E.E.*, **58**, 1374–5 (1970).

giving

$$\tfrac{1}{4}(q/p^2)_{\min} = \frac{\pi C V_A^* I_A^*}{\sqrt{3}(KeMF)^2} \tag{15.3.15}$$

For $B > B_0$ the voltage noise term dominates the right hand side of eqns. (15.3.12) and (15.3.13); for $B < B_0$ the current noise term is dominant. If $\frac{1}{4}(q/p^2)_{\min} \gg 1$ then at all bit rates the shot noise is negligible and $\bar{I}_m \simeq \frac{1}{2}q^{1/2}$. For $B < B_0$,

$$\bar{I}_m \simeq \tfrac{1}{2}q^{1/2} = \frac{KI_A^* B^{1/2}}{2\sqrt{2}M} = \frac{3\sqrt{2}I_A^* B^{1/2}}{M} \tag{15.3.16}$$

assuming Gaussian noise distributions with $K = 12$. For $B > B_0$,

$$\bar{I}_m \simeq \tfrac{1}{2}q^{1/2} = \frac{\pi K C V_A^* B^{3/2}}{2\sqrt{6}M} = \frac{\sqrt{6}\pi C V_A^* B^{3/2}}{M} \tag{15.3.17}$$

Even when $\frac{1}{4}(q/p^2)_{\min} \ll 1$, the current noise term still dominates at bit rates low enough to make $\frac{1}{4}q/p^2 \gg 1$, and the voltage noise term still dominates at bit rates high enough to satisfy the same condition. Thus, when

$$B \ll B_1 = \frac{(I_A^*)^2}{2K^2e^2M^2F^2} = \frac{(I_A^*)^2}{288e^2M^2F^2} \tag{15.3.18}$$

$\bar{I}_m$ is once more given by eqn. (15.3.16), and when

$$B \gg B_2 = \frac{6(KeMF)^2}{(\pi C V_A^*)^2} = \frac{864}{\pi^2}\frac{e^2M^2F^2}{(CV_A^*)^2} \tag{15.3.19}$$

$\bar{I}_m$ is again given by eqn. (15.3.17). However, now a middle range of bit rates may exist such that $B_1 \ll B \ll B_2$, and $\frac{1}{4}q/p^2 \ll 1$. Then,

$$\bar{I}_m \simeq p = \frac{K^2eFB}{2} = 72eFB \tag{15.3.20}$$

In this case the noise is dominated by the multiplied shot noise whose distribution may be non-Gaussian and even asymmetric. It may then not be appropriate to use the value $K = 12$ for the signal-to-noise ratio required to keep the error probability below 10^{-9}. It may also not be appropriate to set the threshold level midway between the 0 and 1 signal levels.

Equations (15.3.16) and (15.3.17) are, of course, the low and high bit-rate asymptotes of eqn. (15.3.9) which was plotted in Fig. 15.9 for $M = 1$. Multiplication factors greater than unity cause these curves to be shifted downwards in direct proportion to M. Equation (15.3.20) represents a degradation of the ultimate quantum-limited sensitivity derived in Section 15.2 by the factor $K^2F/20 = 7.2F$. In terms of carrier pairs generated per bit received, the sensitivity becomes $\frac{1}{2}K^2F$ or $72F$ carriers/bit. These three asymptotic solutions to eqn. (15.3.10) have been plotted for some specific examples in Fig. 15.10. We have again used the noise parameters appropriate to Si-FET and SI-BJT input stages, in which the input capacitance is 5 pF. Thus, for the FET, $(I_A^*) = 10$ fA/$\sqrt{\text{Hz}}$ and $(CV_A^*) = 2 \times 10^{-20}$ C/$\sqrt{\text{Hz}}$; for the BJT, $(I_A^*) = 2$ pA/$\sqrt{\text{Hz}}$ and $(CV_A^*) = 10^{-20}$ C/$\sqrt{\text{Hz}}$. Values used for the APD gain and noise factor are $M = 20$, $F = 5$ and $M = 100$, $F = 4$, the higher gain being an indication of what might be expected from a low-noise Si-APD. It can be seen that the FET amplifier is shot-noise limited throughout most of the useful frequency range in both cases, whereas the bipolar amplifier scarcely reaches the shot-noise limit with the lower APD gain. It is important to keep in mind that all these calculations have assumed uniform amplifier noise spectral densities at all frequencies. There is every reason to believe that in practice I_A^* and V_A^* will increase at high and low frequencies. This will tend to reduce the range of frequencies over which shot-noise limited detection can be maintained.

15.3.3 System Optimization

In Fig. 15.10 we may have implied that it is normal practice for the design of a particular system to be fixed and then its behavior examined as a function of bit rate. This, of course, is not how things happen. In practice the bit rate is

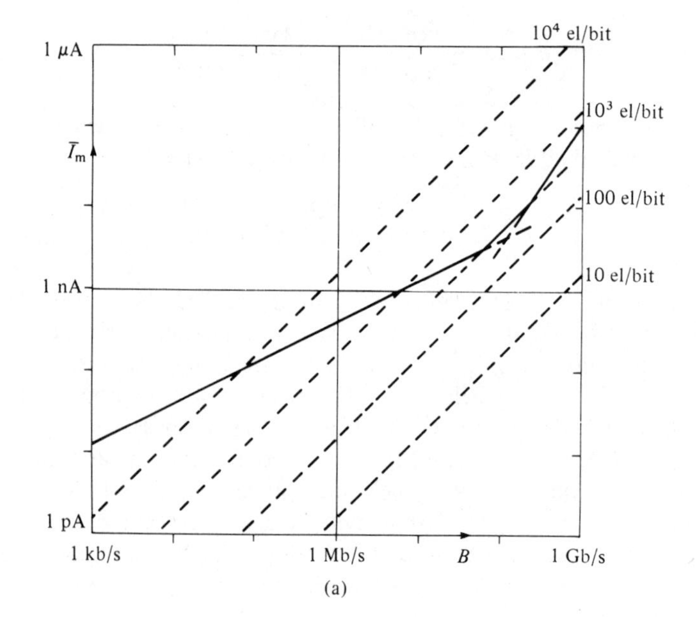

(a)

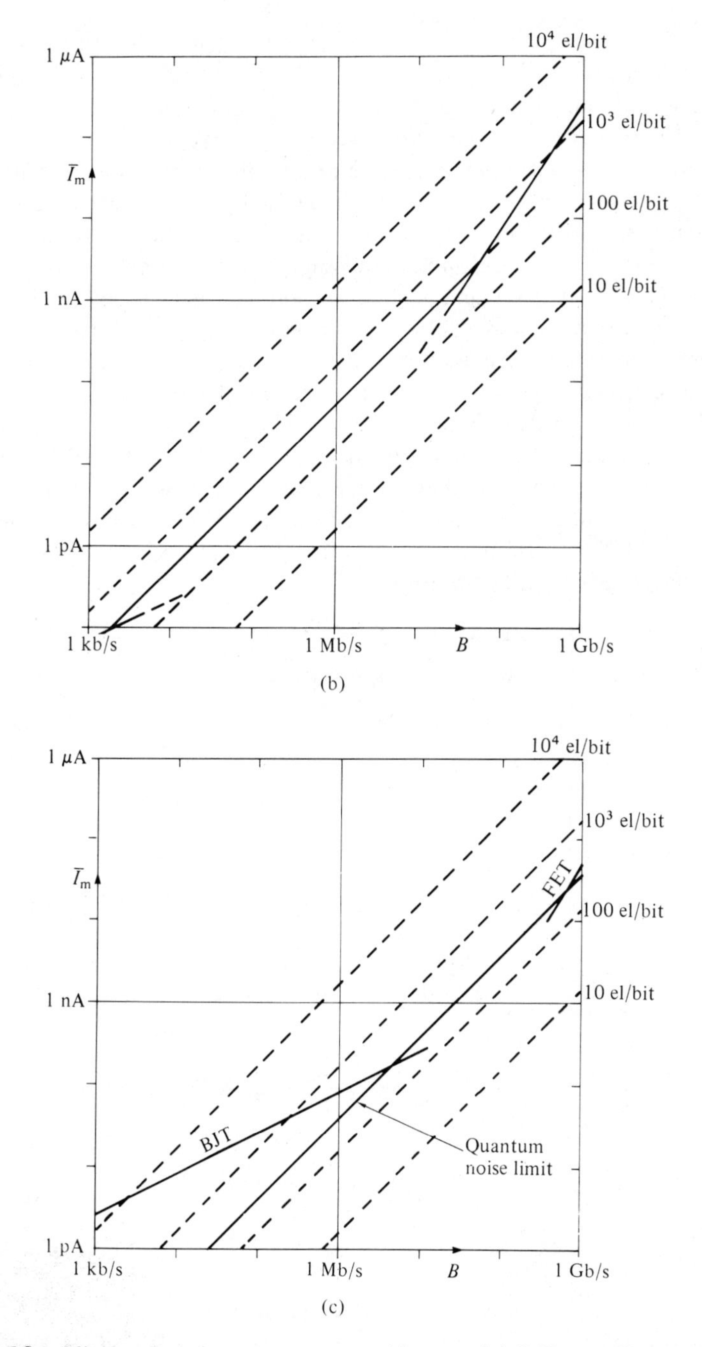

Fig. 15.10 Minimum detector current vs. bit-rate: (a) BJT amplifier, $M = 20$, $F = 5$; (b) FET amplifier, $M = 20$, $F = 5$; (c) BJT, FET amplifiers, $M = 100$, $F = 4$. The FET limit at low frequencies and the BJT limit at high frequencies lie outside the ranges covered by this graph.

normally determined at an early stage of system design and what is then required is an optimum design of receiver to suit it. We have seen that we might prefer to use a Si-FET input stage for bit rates lower than about 50 Mb/s and a Si-BJT for higher frequencies. Then, if an APD is to be used we have the freedom to choose the most appropriate multiplication factor. When the APD noise factor follows a simple law such as those expressed by eqns. (13.4.1)–(13.4.3) it is possible to find an optimum gain which minimizes the overall noise. In many cases these laws break down at a certain level of reverse voltage as microplasmas develop. Then the dark current and noise factor increase sharply if higher values of M are sought. If the optimum gain has not already been exceeded, it will occur at this threshold for microplasma breakdown. Figure 15.11 illustrates the variation of $\bar{I}_m$ with APD gain for a particular case in which we have assumed the values $K = 12$ and $F = M^{1/2}$. The bit rate is taken to be 140 Mb/s and a Si-BJT amplifier is assumed with $(CV_A^*) = 10^{-20}$ C/$\sqrt{\text{Hz}}$ and $(I_A^*) = 2$ pA/$\sqrt{\text{Hz}}$. Then as a function of APD gain eqn. (15.3.10) takes the form

$$\bar{I}_m = \frac{K^2 eFB}{4}\left[1 + \left(1 + \frac{2}{K^2 e^2 M^2 F^2}\left[\frac{\pi^2}{3}(CV_A^*)^2 B + \frac{(I_A^*)^2}{B}\right]\right)^{1/2}\right]$$

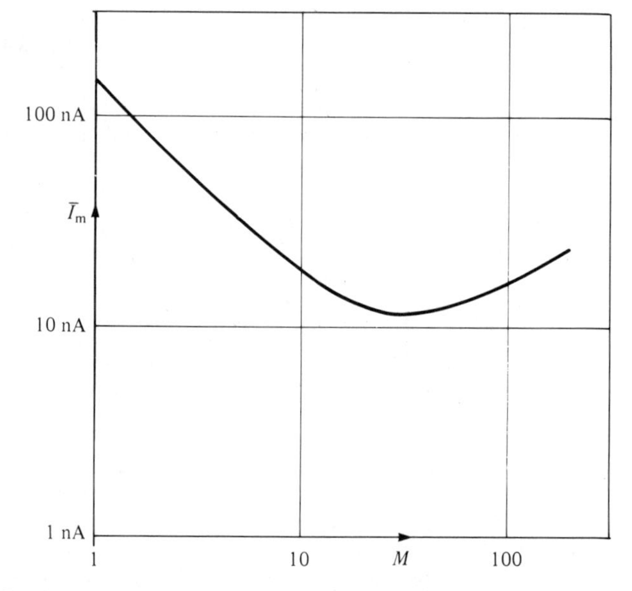

Fig. 15.11 Minimum detector current vs. APD gain; BJT amplifier at 140 Mb/s; $F = M^{0.5}$.

Thus,

$$\bar{I}_{\mathrm{m}} = 8 \times 10^{-10} M^{1/2}[1 + (1 + 4 \times 10^4 M^{-3})^{1/2}] \qquad (15.3.21)$$

when the numerical values and $F = M^{1/2}$ are substituted. The function defined by eqn. (15.3.21) has been plotted in Fig. 15.11. In this case the optimum gain is seen to be $M = 32$, giving $(\bar{I}_{\mathrm{m}})_{\mathrm{opt.}} = 11$ nA. With an unmultiplied responsivity of $\mathcal{R} = 0.55$ A/W, this would correspond to a mean received power level of $(\bar{\Phi}_{\mathrm{R}})_{\mathrm{opt}} = 20$ nW, or –47 dBm.

15.4 NOISE PENALTIES IN PRACTICAL SYSTEMS

15.4.1 Introduction

In practical optical communication systems a number of factors cause the performance to be degraded from the theoretical values calculated in the preceding sections. These were derived on the assumption that the optical energy associated with each bit would arrive as an impulse, that there would be zero energy when 0 was sent, that the receiver amplifier would be sharply band-limited, that there would be no random variation in the amplitude of the received optical power or of its arrival time within the bit period. In practice none of these is true and each violation causes an increase in the level of received signal power needed to ensure a given error probability. This additional required power we have referred to as a *power penalty*. We will examine the penalties caused by some of these effects in this section.

15.4.2 Non-zero Extinction Ratio

The ratio of the optical energy received when 0 is sent to that received when 1 is sent is defined in Section 15.1.1 as the *extinction ratio*, r_{e}:

$$r_{\mathrm{e}} = \frac{\mathcal{E}_{\mathrm{R}}(0)}{\mathcal{E}_{\mathrm{R}}(1)} \qquad (15.4.1)$$

Any dark current present in the photodiode, because it adds to the signal current at both levels, appears to increase the extinction ratio. However, with a GaAs laser source and a Si detector it is the need to bias the laser at or above the lasing threshold that is the most likely reason for r_{e} to be greater than zero. A typical value is between 0.05 and 0.1. It would be quite possible, using the statistical techniques discussed previously, to evaluate the noise penalty that this introduces in any specific system. The penalty is greatest in the case of quantum-limited detection, but often in practice it is found to be about 1–2 dB.

15.4.3 Finite Pulse Width and Timing Jitter

The inevitable fact that the received optical power has a finite pulse width and that some timing jitter is present leads to a noise penalty for two distinct reasons. The first is because non-optimal filtering is needed either to provide equalization against pulse distortion or to minimize intersymbol interference. The second is because some intersymbol interference remains and degrades the signal-to-noise ratio. In order to calculate the likely magnitude of these effects it is necessary to define the shape of the received pulses and the distribution of the jitter. We will attempt to deal only with the former.

In Section 2.4 we showed how pulses of various shape could be characterized in terms of parameters such as the r.m.s. width, σ, and the full width at half power, τ. We also indicated that when the pulse shapes approximated to Gaussian, the total mean square width of the received pulse could be obtained by adding the mean square values of the initial pulse width, the material dispersion and the mode dispersion of the fiber. The problem with the assumption of Gaussian-shaped pulses lies in the fact that the amplitude in the tails is higher than would normally be expected. While this does not affect the r.m.s. values it does make a significant difference to the calculated intersymbol interference and thus may lead to an overestimate of the power penalty. Values have been calculated for received pulses of various shape |14.5|. Results for Gaussian pulses are shown in Fig. 15.12. This figure is extremely important because it demonstrates the possible trade-off between bit rate and signal power, and thus relates the effects of fiber attenuation and fiber dispersion. It can be seen that the power penalty is less than 1 dB provided that σ remains less than about $T/5$. However, the power penalty increases sharply when the pulse width exceeds these values. Thus, in practice, a system is usually quite clearly limited either by fiber dispersion (bandwidth limited) or by fiber attenuation (power limited). With the pulse shapes encountered in practical systems the condition for the power penalty not to exceed 1 dB can usually be relaxed to σ $(\simeq \tau/2) \lesssim T/4$, and this is more or less independent of pulse shape.

15.4.4 Power Amplitude Variations: Modal Noise

In all the calculations presented up to now we have tacitly assumed the optical power incident onto the detector to be some clearly defined function of time

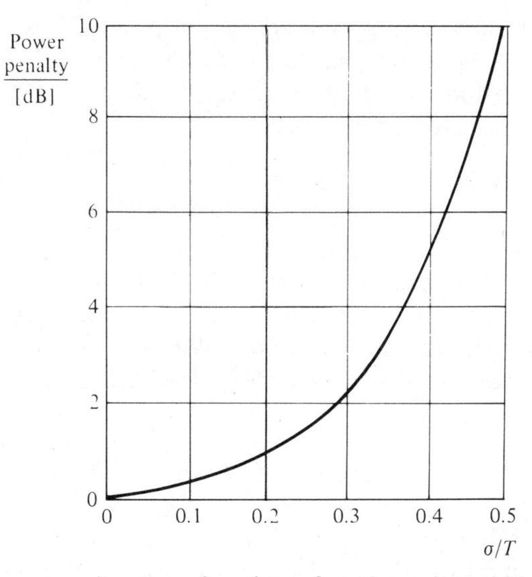

Fig. 15.12 Power penalty as a function of r.m.s. pulse-width for Gaussian pulses.

and not itself subject to statistical variation. The shot noise observed when the light is detected has been taken to be entirely caused by the statistical nature of the quantum detection process. When we reconsider the emission mechanisms at the source, this assumption clearly requires some further justification. In fact the shot-noise amplitude quoted in equations (12.6.1) and (14.2.1) represents an irreducible minimum level. Several processes, some associated with the source, some with the transmission path and some with the statistics of the detection process, may cause the observed, signal-dependent noise to be increased above this level and take on a distribution different from the Poisson distribution of eqn. (15.2.2).

We might imagine a perfect source to be one that emits a continuous wave of constant amplitude and frequency, rather like the carrier wave of a radio transmitter. An example would be the output from a stable, single-mode laser. For signalling purposes this wave would need to be chopped (by a shutter, say) into pulses which are long compared to the optical oscillation period, but short compared to the time over which the wave remains continuously in phase. This latter time is known as the *coherence time*, τ_c, and is an important property of optical radiation. Such radiation incident onto a photodiode would be expected to give rise to the levels of shot noise assumed so far.

At the other extreme we should consider the emission from a conventional, incoherent optical source. This is regarded as a sequence of independent, atomic de-excitations occurring at random times. The radiation may be thought to suffer frequent, random changes of phase. There is thus no phase correlation over long periods, and τ_c, whilst remaining long compared to the optical period, is short compared to typical signalling times, or to the response time of any ordinary detector. Thus

$$\frac{1}{f} \ll \tau_c \ll T \tag{15.4.2}$$

In this situation the random changes in the emission process average out during the response time of the detector, and again the normal level of shot noise, given by eqn. (12.6.1), is observed. With such sources there is a direct relationship between the line-width of the source and the coherence time. This is:

$$\tau_c \simeq \frac{1}{2\pi\, \Delta f} = (-)\frac{1}{2\pi c}\frac{\lambda^2}{\Delta\lambda} = \frac{\lambda}{2\pi\gamma c} \tag{15.4.3}$$

where we have used $\gamma = |\Delta\lambda/\lambda| = \Delta f/f$ for the fractional spectral spread of the source about its mean wavelength or frequency as in Chapter 2. It is also sometimes convenient to define the distance along the direction of propagation over which the radiation remains in phase as a *coherence length*, l_c. Thus

$$l_c = \frac{c}{n}\tau_c = \frac{\lambda}{2\pi\gamma n} \tag{15.4.4}$$

where we have reverted to the use of n for refractive index.

If we take $\gamma = 0.03$ to be typical of the incoherent LED sources used for optical communication, the corresponding value of $\tau_c \simeq 0.1$ ps shows condition (15.4.2) to be valid. Thus we may reasonably expect the statistical analyses given in this chapter to apply both with laser and LED sources.

There is some evidence of excess noise being generated in the source itself, particularly in lasers at frequencies near their self-resonant frequency. However, it is not clear that this leads to a significant deterioration in the overall performance of the optical communication system. An effect that is much more serious is associated with the transmission path and it occurs when narrow-band laser sources are used in conjunction with multimode fibers. This has become known as *modal noise*.

The coherent laser radiation normally excites a number of modes of propagation along a multimode fiber. As long as these retain their relative phase coherence, the radiation pattern seen at the end of the fiber takes on the form of the familiar speckle pattern produced by laser radiation. This is the result of constructive and destructive interference at any given plane. An example is shown in Fig. 15.13. After a sufficient propagation distance mode

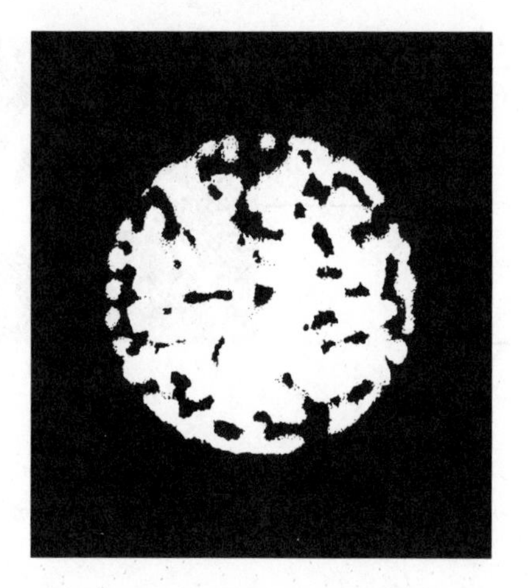

Fig. 15.13 Near-field radiation pattern for coherent light propagating in a multimode fiber. [Taken from C. P. Sandbank (ed.), *Optical Fibre Communication Systems*, John Wiley (1980) © STL Ltd.]
The number of speckles approximates the number of propagating modes. Core dia. 50 μm.

dispersion causes the relative propagation delays between the various modes to exceed the coherence time of the light in each mode. As this happens the speckle pattern becomes submerged into a uniform background radiation level.

The fact that a speckle pattern is generated is not in itself a sufficient condition for modal noise to be generated. Two further requirements have to be satisfied: there must be a point in the transmission path where the propagation is mode selective, and there must be some temporal variation in the mode propagation characteristics *before* this point. A typical cause of mode selection is a less than perfect connector, as shown in Fig. 15.14. When such mode selection occurs where there is still phase coherence, then any disturbance of the speckle pattern leads to a variation of the power transmitted. Thus any mechanical disturbance that takes place before the misaligned connector or any variation in the source wavelengths gives rise to changes in the received power and increases the apparent noise level of the system. The effect of severe modal noise on the receiver eye-diagram is shown in Fig. 15.15.

Clearly the avoidance of modal noise is of considerable importance in system design, particularly in analogue systems where large signal-to-noise ratios may be required. There is no simple recipe, however, except to restrict the use of lasers to single-mode fibers and only permit multimode fibers to be used with LED sources. This may indeed become the preferred combination in future systems, but at present it would be unduly limiting. Given a demand for laser sources to be used in conjunction with multimode fibers, it then becomes

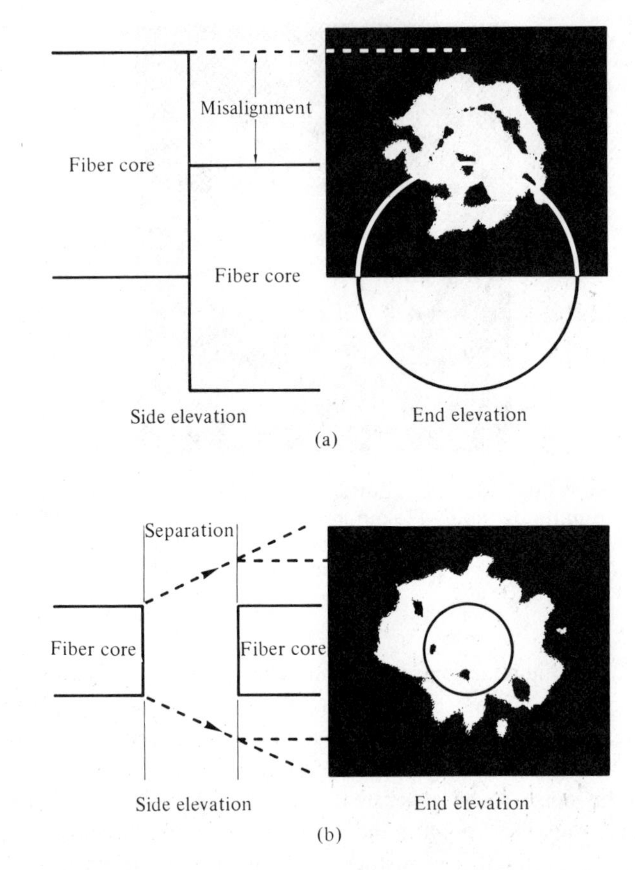

Fig. 15.14 Illustration of the effect of connector misalignment on the coupling of the fiber speckle pattern: (a) axial misalignment; (b) longitudinal misalignment. [From C. P. Sandbank (ed.), *Optical Fibre Communication Systems*, John Wiley (1980); © STL Ltd. Reprinted with permission.]

necessary to minimize mode-selective losses until mode dispersion delay differences exceed the coherence time. Mode dispersion may be increased by the use of step-index fibers with high numerical aperture or of poorly graded GI fibers. Coherence times may be reduced by using lasers which emit light in many modes simultaneously. But, of course, both of these steps work entirely contrary to all the efforts made to increase the bandwidth capability of the system! The best answer to the problem of modal noise is to limit the system bandwidth to the minimum required and to ensure that any dispersion introduced occurs as close to the source as possible. The remainder of the system should then be free of modal noise.

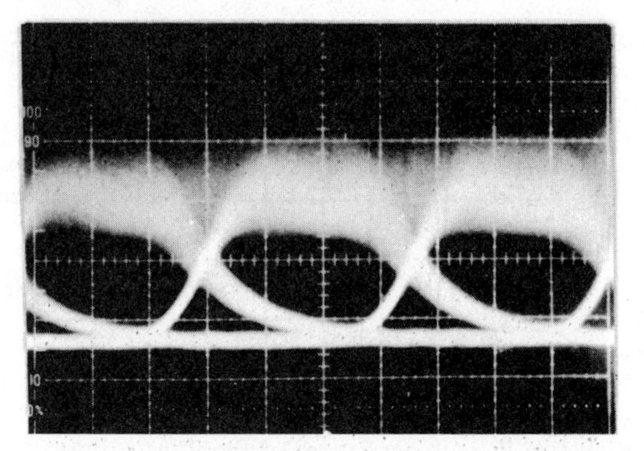

Fig. 15.15 Eye diagram for a 140 Mb/s receiver showing the effect of modal noise at a misaligned connector. [From C. P. Sandbank (ed.), *Optical Fibre Communication Systems*, John Wiley (1980); © STL Ltd. Reprinted with permission.]

PROBLEMS

15.1 On the assumption that the errors originating in each individual length of an extended communication link are generated independently, show that the error probability of the whole channel is directly proportional to its length. Take the length of each section to be l and to have a probability PE_l of generating errors, and consider a channel made up of n such lengths. Hence calculate the maximum permissible PE in sections of length 10 km and 50 km, when there is a requirement that over a loop of 2500 km PE shall be less than 2×10^{-7}.

15.2 Show that eqn. (15.1.6) is the Fourier transform of eqns (15.1.5) and that eqn. (15.1.8) is the Fourier transform of eqns (15.1.7). Use tables of Fourier transforms and make use of the normal rules for combining transforms, as required.

15.3 Show that the mean value of the Poisson probability distribution, eqn. (15.2.2), is N and that the mean square variation about the mean value is also N.

15.4 Sketch a graph showing the average number of photogenerated electrons per bit period in the absolute quantum limit of detection as a function of the required error probability.

Hence calculate the received power required per Mb/s, in this limit, when $\lambda = 0.85\ \mu m$, $\eta = 0.8$ and the required error probability is (a) 10^{-5} (b) 10^{-10} (c) 10^{-15}.

15.5 Using Fig. 15.8 compare the ratios of peak signal to r.m.s. noise required to ensure error probabilities of 10^{-5}, 10^{-10} and 10^{-15} when the noise distribution is Gaussian and noise is equally present on both binary levels.

(Note that in real systems nonstochastic processes often introduce errors at a low but limiting rate that is relatively independent of the signal-to-noise ratio.)

15.6 Expand eqns (15.3.9) and (15.3.12) so that all four amplifier noise terms (a), (b), (d) and (e) in eqn. (14.4.10) are included.

15.7 In a particular receiver $C = 10$ pF, $V_A^* = 3\ \mathrm{nV}/\sqrt{\mathrm{Hz}}$, $I_A^* = 1\ \mathrm{pA}/\sqrt{\mathrm{Hz}}$. Calculate the bit-rate B_0 above which the voltage noise term dominates the right-hand side of eqn. (15.3.13) and below which the current noise term is greater.

15.8 For the receiver of Problem 15.7 calculate values of B_1 and B_2 from eqns (15.3.18) and (15.3.19) when $K = 12$ and

(a) $M = 100$, $F = 5$
(b) $M = F = 10$
(c) $M = F = 1$.

Hence indicate in each case those of the three regimes expressed by eqns (15.3.16), (15.3.17) and (15.3.20) which may occur within the range of bit-rates 1 Mb/s to 1 Gb/s.

15.9 A 2 Mb/s optical fiber communication system working at 1.3 μm wavelength uses a Ge APD detector for which $F = M$. The input stage of the receiver is configured as a voltage amplifier. Its noise performance may be represented by an equivalent input noise voltage source of $5\ \mathrm{nV}/\sqrt{\mathrm{Hz}}$ and an equivalent input noise current source of $20\ \mathrm{fA}/\sqrt{\mathrm{Hz}}$. The total input capacitance is 5 pF. Assuming that the amplifier is to be equalized over the frequency range 0 to 1 MHz:

(a) Calculate the maximum value of the input resistance that will enable subsequent equalization to be avoided. Assume $2\pi f_{\mathrm{max}} CR = 0.3$.
(b) With this value of R compare the magnitudes of the amplifier noise sources (a), (b), (d) and (e) in eqn. (14.4.10).
(c) Calculate the value of R required to make the thermal noise term (d) negligible compared to the amplifier current noise term (e). Is term (a) then also negligible?

15.10 In order to preserve a reasonable dynamic range the input resistance of the receiver described in Problem 15.9 is made $R = 100\ \mathrm{k\Omega}$ and the necessary equalization is provided.

(a) Recalculate the magnitudes of the amplifier noise sources.
(b) Obtain an expression equivalent to eqn. (15.3.21) for the minimum average photodiode current $\bar{I}_{\mathrm{m}}$ needed to ensure a signal-to-noise ratio of $K = 12$.
(c) Sketch a graph showing the variation of $\bar{I}_{\mathrm{m}}$ with the APD gain M.
(d) Estimate the optimum gain and the corresponding value of $(\bar{I}_{\mathrm{m}})_{\mathrm{opt}}$.
(e) With $\eta = 0.8$ calculate the average received optical power required.

15.11 A 2 Gb/s optical fiber communication system working at 1.3 μm wavelength uses a p–i–n FET detector, so that $F = M = 1$. The input stage takes the form of a transimpedance amplifier in which $R = R_{\mathrm{F}} = 1\ \mathrm{k\Omega}$. Its noise performance may be represented by an equivalent input noise voltage source of $2\ \mathrm{nV}/\sqrt{\mathrm{Hz}}$ and an equivalent input noise current source of $0.1\ \mathrm{pA}/\sqrt{\mathrm{Hz}}$. The total input capacitance is 2 pF.

(a) Calculate the gain required to ensure that the amplifier is equalized over the frequency range 0 to 1 GHz.

(b) Calculate the magnitudes of the amplifier noise sources.
(c) Calculate the value of $\bar{I}_m$ required to ensure a signal-to-noise ratio of $K = 12$.
(d) With $\eta = 0.8$ calculate the average received optical power required.

15.12 Calculate the coherence time and the coherence length of the light emitted by the following sources:
(a) An InGaAsP light emitting diode having a spectral spread of 100 nm about 1.3 μm.
(b) A GaAlAs/GaAs laser with a spectral spread of 2 nm about 0.82 μm.
(c) A laser operating in a single transverse and a single longitudinal mode with a short term frequency stability of 100 MHz.

15.13 For each of the sources described in Problem 15.12 calculate the propagation distance in a multimode fiber having 0.3 ns/km mode dispersion, beyond which it is not possible for modal noise to be generated.

REFERENCE

15.1 R. Loudon, *The Quantum Theory of Light*, O.U.P. (1973).

SUMMARY

The quantum limit of detection for an error probability of 10^{-9} in an ideal digital system transmitting 0s and 1s in equal proportion is $\bar{\Phi}_R = 10\,\mathcal{E}_{ph}B$. That is, 10 photons per bit.

Real systems require some two orders of magnitude more received power in order to overcome detector and amplifier noise and because of departures from the ideal system such as noise and jitter on the received power, nonzero extinction ratio, intersymbol interference and the use of non-return-to-zero signalling.

Figures 15.9 and 15.10 give the results of calculations on the effect of detector and amplifier noise in particular systems. These assume that the noise statistics of all noise sources are Gaussian.

Each departure from the assumptions of the ideal system gives rise to a noise penalty. In particular the power penalty incurred because of the finite width of the received pulses is < 1 dB, provided that $\sigma\ (\simeq \tau/2) \lesssim T/4$.

Multimode systems using monochromatic sources such as buried heterostructure lasers may be seriously degraded by modal noise.

16

Unguided Optical Communication Systems

16.1 INTRODUCTION

In Chapter 1 the possible use of unguided transmission at optical frequencies was discussed and the likely advantages and disadvantages were set out in Table 1.1. The subsequent fourteen chapters have concentrated exclusively on components expressly designed for use with optical fiber systems. For these the physical constraints imposed by the fiber limit the practical choice of source and detector to the semiconductor types discussed in Chapters 8, 11, 12 and 13. With unguided systems there are no such constraints and, as was mentioned in Section 1.4, a much wider range of components may be considered.

Unguided optical communication systems were the subject of a great deal of research and development activity between 1960 and 1970. A brief summary of some of this work will be found in Ref. [16.1]. The effort devoted to optical fibers has entirely dominated this work in the years following 1970 and only a few highly specialized unguided applications have been pursued since then. It remains unclear just what role they may play in future telecommunications technology. Certainly they will have an advantage over guided systems when one or both of the terminals has to be mobile. They may also have a role in short links, between buildings, for example, when because of special local circumstances it is difficult to run cable across the intervening space. In either case they are then in competition with radio links. It should be noted that even modestly powered laser systems ($\sim$1 mW at 1 μm) need to be sited so that all necessary safety requirements are satisfied.

In the present chapter we start by giving a brief discussion of some of the transmission parameters that determine the received optical power and hence the performance of unguided systems. These include the divergence of the optical beam, which is discussed in Section 16.2.1, and attenuation which is discussed in Section 16.2.2. A major restriction on the outdoor terrestrial use of optical frequencies is the extreme variability of the atmospheric attenuation with changing meteorological conditions. It has to be accepted that under particular conditions of fog or heavy rain or snow the system is rendered inoperative, but this is true also of high-frequency microwave links in heavy rain. What is unfortunate is that fog and haze are, in general, much more probable in those regions of high population density where there is the greatest need for a diversity of communication systems. In Sections 16.3 and 16.4 we go on to examine some of the types of optical sources and detectors that may be used. Finally the wide range of possible systems will be illustrated in Section 16.5. These include systems for indoor use having a range of the order of 10 m, systems for terrestrial use outdoors having a range of about 1 km, and systems for use in extra-terrestrial space where the range may be required to exceed 10^5 km.

16.2 TRANSMISSION PARAMETERS

16.2.1 Beam Divergence

Figure 16.1 is a highly simplified and idealized illustration of an unguided optical communication system. No details of the optical lens systems at the transmitter and receiver are shown, and in the diagram, as in the discussion to follow, the thin-lens approximation is used for simplicity. The source is assumed to be diffuse like an LED and to have a radiating area A_S. The radiant intensity, I_0, is taken to be constant for all the light collected by the transmitter lens. This has an effective aperture A_T and a focal length f. The receiver is situated at a distance $l \gg f$. Its effective aperture is A_R and all light incident on

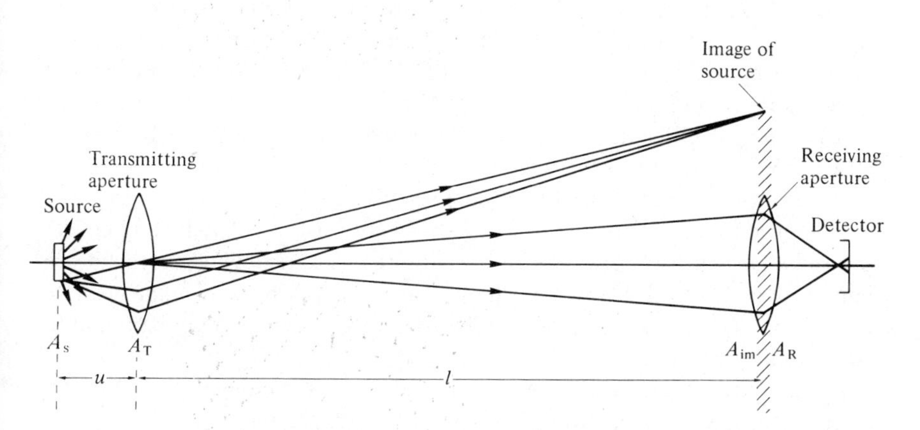

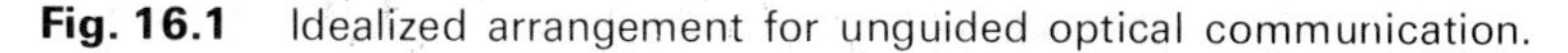
Fig. 16.1 Idealized arrangement for unguided optical communication.

to A_R is assumed to be focussed into the active region of the detector. In order to maximize the received power an image of the source should be formed in the plane of the receiver aperture. Using elementary, thin-lens, optical theory the distance, u, from the source to the center of the transmitter lens is given by

$$\frac{1}{u} = \frac{1}{f} - \frac{1}{l} \simeq \frac{1}{f} \tag{16.2.1}$$

and the image area is

$$A_{\text{im}} = \frac{A_S l^2}{u^2} \simeq A_S \frac{l^2}{f^2} \tag{16.2.2}$$

The power, Φ_T, collected by the transmitter lens is given by

$$\Phi_T = \frac{I_0 A_T}{u^2} \simeq \frac{I_0 A_T}{f^2} \tag{16.2.3}$$

Assuming that the image of the source more than fills A_R, the fraction of Φ_T which reaches the detector is simply A_R/A_{im}, so that the received power, Φ_R, is given by

$$\begin{aligned}\Phi_R = \Phi_T \frac{A_R}{A_{\text{im}}} = \Phi_T \frac{A_R u^2}{A_S l^2} \simeq \Phi_T \frac{A_R f^2}{A_S l^2} \\ = \frac{I_0 A_T A_R}{A_S l^2} = \frac{L A_T A_R}{l^2}\end{aligned} \tag{16.2.4}$$

where $L = I_0/A_S$ is the radiance of the source. Clearly a source of high radiance and apertures of large area are required. Consider as an example a system using a light-emitting diode source of radiance $L = 0.1$ W/mm^2 sr and having apertures of area $A_R = A_T = 10^{-3}$ m^2 (diameter 35 mm). At a distance of 1 km the power incident onto the receiver would be $\Phi_R = 10^5 \times 10^{-6} \div 10^6 = 10^{-7}$ W $=$ 100 nW. Although these calculations have been based on an idealized lens system, the basic conclusions apply equally if mirror or catadioptric (mixed mirror and lens) systems are used to collimate and collect the radiation. In each case the images formed in a real optical system are imperfect on account of the aberrations introduced. A brief account of these and of the optical systems that may be used at the receiver may be found in Ref. [16.2], Section 5.3.

If the source is too small, the size of its 'image' is no longer given by eqn. (16.2.2) but is determined by diffraction at the transmitter aperture. It is well known that the diffraction pattern produced by a uniformly illuminated circular aperture of diameter d_T consists of a set of concentric rings. The radial variation of power density takes the form shown in Fig. 16.2. A detailed discussion will be found in Section 8.5.2 of Ref. [2.1]. The image size may be said to be diffraction limited when the radius of the first intensity minimum, or dark ring, of the diffraction pattern becomes comparable in size with the diameter, d_{im}, of the normally focussed image, that is when

$$d_{im} = \frac{l}{u} d_S < \frac{1.22\lambda l}{d_T} \tag{16.2.5}$$

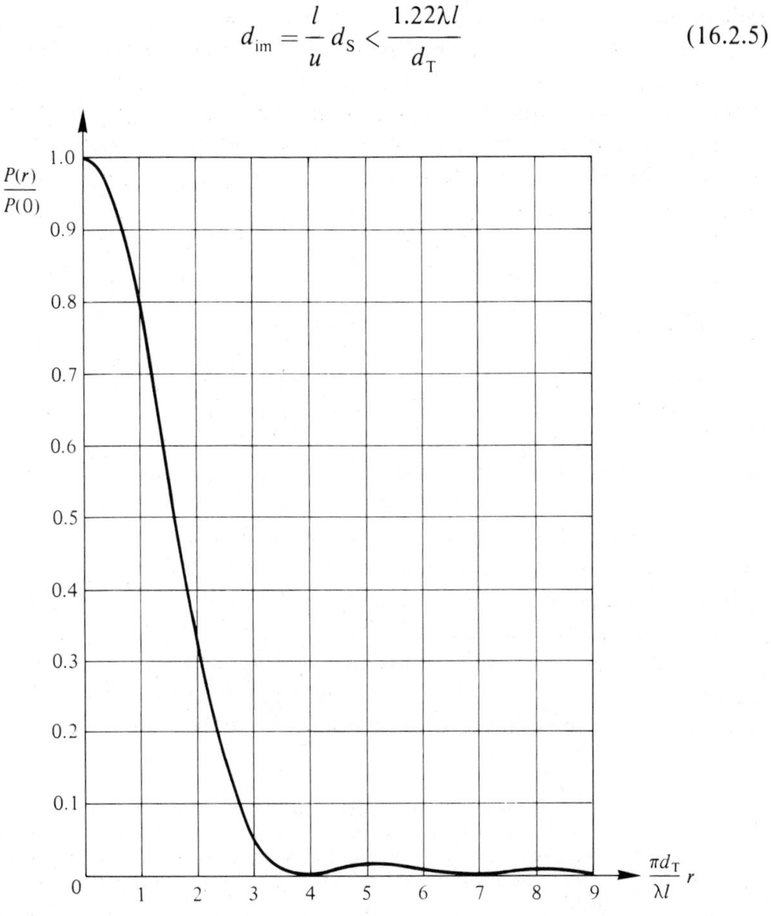

Fig. 16.2 The radial variation of power density at a distance l for a uniform beam diffracted at a circular aperture of diameter d_T.

$$\frac{P(r)}{P(0)} = \left[\frac{2\lambda l}{\pi d_T r} J_1\left(\frac{\pi d_T r}{\lambda l}\right)\right]^2$$

where $J_1(x)$ is the first-order Bessel function. The first zero of $J_1(x)$ is at $x = 3.832$ (see Table 5.1) where $r = 1.22\lambda l/d_T$.

where d_S is the diameter of the source; therefore

$$d_S < \frac{1.22u\lambda}{d_T} \simeq \frac{1.22f\lambda}{d_T} \tag{16.2.6}$$

The light from a laser source, being inherently highly collimated and coherent, normally produces a diffraction-limited image. In either case the radiant intensity at the center of the diffraction pattern is given by $\Phi_T A_T/\lambda^2$, where Φ_T is the total transmitted power. In the case of a point source Φ_T is given by eqn. (16.2.3) and in other cases it is just the product of the power density incident onto the transmitting aperture and the aperture area. The total power collected by a small receiving aperture of area A_R positioned in the center of the diffraction pattern is then

$$\Phi_R = \frac{\Phi_T A_T}{\lambda^2}\frac{A_R}{l^2} \tag{16.2.7}$$

in general, and

$$\Phi_R = \frac{I_0 A_T^2 A_R}{\lambda^2 u^2 l^2} \simeq \frac{I_0 A_T^2 A_R}{\lambda^2 f^2 l^2} \tag{16.2.8}$$

in the case of a point source of radiant intensity I_0. It should be understood that these are ideal performance figures and further losses will occur within the optical system and as a result of aberrations. A discussion of these matters from the point of view of antenna theory may be found in Ref. [16.3].

It will be useful to explore the circumstances under which the LED system discussed in the example given earlier ($A_R = A_T = 10^{-3}$ m^2, $l = 1$ km) would become diffraction limited. Assume that $\lambda = 1$ µm and that the focal length of the transmitter lens is $f = 100$ mm.† Then, eqn. (16.2.6) shows that diffraction would determine the beam divergence only if the source diameter, d_S, were less

† The ratio $X = f/d_T$ is known as the f-number of an optical system, and determines its light-gathering power. It is customarily expressed in a rather unusual way as the *aperture* or *speed* of the system: f/X. An alternative way of expressing the light-gathering power is in terms of a numerical aperture (NA) which was defined in Section 2.1. Provided that X is not too small the two terms are related,

$$(NA) \simeq \frac{d_T}{2f} = \frac{1}{2X} \tag{16.2.9}$$

The lens in the example given here ($d_T = 35$ mm, $f_T = 100$ mm) would have the speed of $f/2.8$ and a numerical aperture of 0.18. Although the focal length does not appear in eqn. (16.2.7), low f-numbers are desirable because the transmitter and receiver modules may then be made more compact.

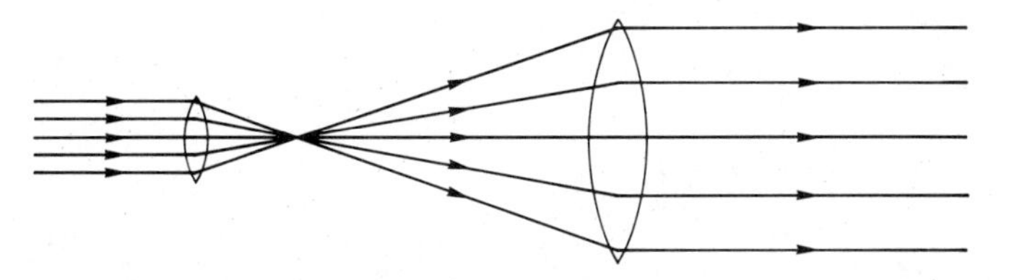

Fig. 16.3 A beam expander used to reduce the diffraction-limited divergence.

than 3 μm. Thus, an edge-emitting LED might be diffraction-limited, but a surface-emitting LED would not be.

In order to reduce the diffraction-limited divergence of a laser source, a beam expander of the type shown in Fig. 16.3 may be used. The diffracting aperture is thereby increased. We saw in Section 11.2 that the radiation from semiconductor lasers did not normally diverge with circular symmetry. In order to launch the radiation efficiently onto a fiber, a cylindrical lens may be employed to remedy this. A similar lens may be used as part of a beam expansion system, and the result obtained in eqn. (16.2.7) then applies as usual. To give an example, let us say that 10 mW of the output from a semiconductor laser is collimated by and fills the 10 mm diameter objective of a beam expander. Then, again taking $\lambda = 1\ \mu m$ and $A_R = 10^{-3}\ m^2$, the power received at a range of $l = 10$ km is

$$\Phi_R = \frac{10^{-2}(\pi/4) \times 10^{-4} \times 10^{-3}}{10^8 \times 10^{-12}} = 8\ \mu W$$

With a beam divergence as small as this, about 0.1 milliradian, the required pointing accuracy would put severe demands on the directional control and mechanical stability of any portable transmitter.

Equations (16.2.7) and (16.2.8) apply when the transmitter aperture is uniformly illuminated. Another case in which the diffraction-dominated, far-field power distribution may be readily calculated occurs when the near-field power density takes on a Gaussian distribution as illustrated in Fig. 16.4. It may be shown that the far-field, diffraction-limited beam retains the Gaussian profile, see for example Section 4.6.4. of Ref. [16.3]. This is an important case in practice because according to theory the fundamental transverse mode of oscillation of a cylindrical laser source would give rise to such a distribution of output power and this is indeed observed.

The system can best be defined in terms of cylindrical coordinates (r, ϕ, z) with origin at the center of the transmitter aperture. We shall assume radial symmetry and so ignore the azimuthal coordinate, ϕ. Let the near-field power

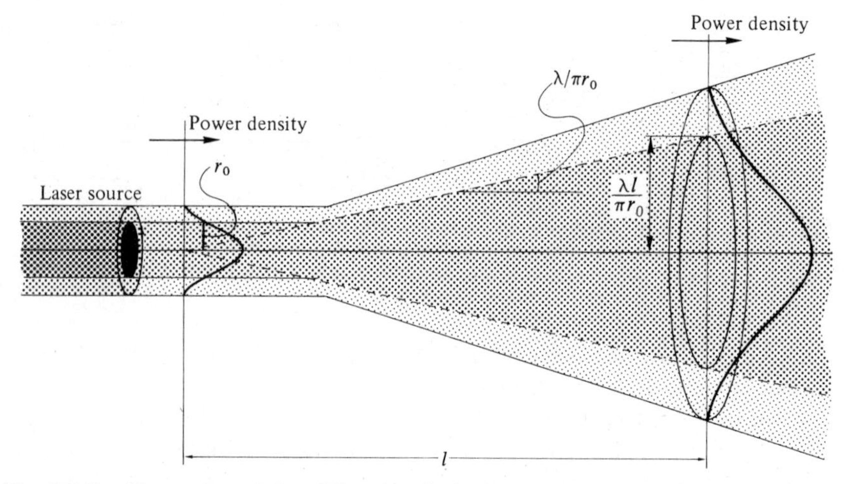

Fig. 16.4 Illustration of the diffraction-limited propagation of a Gaussian beam.

density emerging from the transmitter aperture have the form:

$$P(r, 0) = P_0 \exp\{-(r/r_0)^2\} \tag{16.2.10}$$

Then, at the plane $z = l$, provided that $l \gg \pi r_0^2/\lambda$, the power density distribution becomes

$$P(r, l) = \left(\frac{\pi r_0^2}{\lambda l}\right)^2 P_0 \exp\left\{-\left(\frac{\pi r_0 r}{\lambda l}\right)^2\right\} \tag{16.2.11}$$

If the receiving aperture is so small ($r_R \ll \lambda l/\pi r_0$, where $A_R = \pi r_R^2$) that the exponential term is approximately unity everywhere, the received power is given by

$$\Phi_R = \left(\frac{\pi r_0^2}{\lambda l}\right)^2 P_0 A_R \tag{16.2.12}$$

It may easily be shown that the total transmitted power is

$$\Phi_T = \pi r_0^2 P_0 \tag{16.2.13}$$

so that

$$\Phi_R = \frac{\Phi_T \pi r_0^2 A_R}{(\lambda l)^2} \tag{16.2.14}$$

which is identical with eqn. (16.2.7) if we regard πr_0^2 as the effective transmitter aperture area, A_T.

We might note in passing the rather strange way in which radio engineers customarily describe the directional characteristics of their antennae systems. They start with the valid assumption that but for the use of a large transmitter aperture ($A_T \gg \lambda^2$) the source power would be radiated uniformly in all directions. After a distance l it would be distributed over an area $4\pi l^2$. They then use the fact that a basic receiver, a matched, isotropic antenna, presents an effective aperture area of $\lambda^2/(4\pi)$ to the incoming radiation. The basic transmission loss is then $\mathcal{L} = (\lambda/4\pi l)^2$, being the fraction of the isotropically radiated power that would be picked up by the basic receiver. This is usually expressed in dB. If a large-aperture, directional array is used to concentrate the transmitted beam, the diffraction-limited beam intensity is, as we have seen, increased at its maximum by the factor $G_T = 4\pi A_T/\lambda^2$. Also normally expressed in dB, G_T is known as the 'gain' of the transmitter. Similarly, at the receiver the use of a wide-aperture array of collecting area A_R will increase the receiver power by the factor $G_R = 4\pi A_R/\lambda^2$. In dB, G_R is the 'gain' of the receiver. This notation brings out the reciprocity between transmitter and receiver which is valid at radio frequencies but which is not applicable in practice at optical frequencies. The overall transmission loss Φ_R/Φ_T may thus be expressed as

$$(\Phi_R/\Phi_T) = \mathcal{L} G_T G_R \tag{16.2.15}$$

$$= \frac{A_T A_R}{l^2 \lambda^2} \tag{16.2.7}$$

In our last example, with $\lambda = 1\ \mu m$, $l = 10$ km, $A_T = (\pi/4) \times 10^{-4}$ m^2, $A_R = 10^{-3}$ m^2, we would have a basic path loss $\mathcal{L} = -222$ dB, a transmitter gain $G_T = 90$ dB, a receiver gain $G_R = 101$ dB giving a net transmission loss $\Phi_R/\Phi_T = -31$ dB ($-222 + 90 + 101 = -31$ dB). Whether there is any merit in doing the calculations in this manner at optical frequencies is open to debate.

16.2.2 Atmospheric Attenuation

When we dealt with attenuation in optical fibers in Chapter 3, it was convenient to distinguish between absorption and scattering losses. The same is true for atmospheric attenuation. Absorption is caused mainly by water vapor and carbon dioxide, scattering by dust particles and water droplets. There are two further effects which disturb atmospheric transmission: refraction and scintillation.

Absorption clearly depends on the water vapor and carbon dioxide content of the air along the transmission path, and this in turn depends on the humidity and the altitude. A classic measurement of infra-red absorption at sea level was

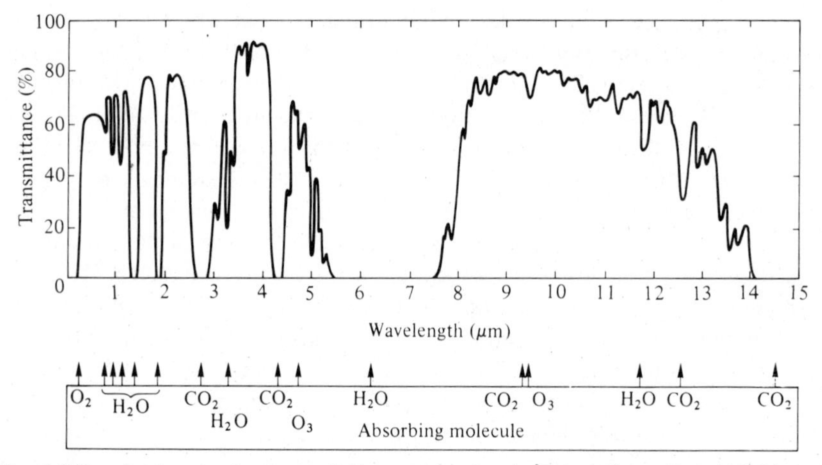

Fig. 16.5 Atmospheric transmission at sea-level. [Taken from Ref. (16.2) and based on H. A. Gebbie, Atmospheric transmission in the 1 to 14 μm region, *Proc. R. Soc.* **A 206**, 87–107 (1951).]
These results were obtained over a horizontal path of 1.8 km length. The increasing amplitude of transmittance peaks in the windows between 1 and 4 μm is evidence of residual haze scattering.

made by Gebbie† in 1951. His results are shown in Fig. 16.5. It can be seen that transmission 'windows' occur at visible wavelengths (naturally!) and in the ranges 1.5–1.8 μm, 2–2.5 μm, 3–4 μm and 8–14 μm. Within these wavelength ranges favorable transmission may be expected.

Scattering shows even greater variability than absorption. It is sometimes convenient to make a clear distinction between the terms *haze* and *fog*. In a haze the main cause of scattering is the presence of dust particles of predominantly submicron dimensions and thus small compared to the radiation wavelength. Rayleigh scattering predominates and as a result the level of attenuation reduces rapidly at longer wavelengths. In a fog scattering is mainly caused by water droplets which may range in diameter from 1 to 100 μm. Attenuation may again vary with wavelength, but the scattering becomes confused with the increased absorption in the water vapor and is usually so strong anyway that it renders the optical communication system inoperative.

Water droplets which increase in size to millimeter dimensions tend to precipitate as rain. This gives rise both to scattering and absorption. The attenuation coefficient increases with the rate of precipitation but depends also on the size distribution of the raindrops. This makes generalization difficult, but it

† H. A. Gebbie, Atmospheric transmission in the 1 to 14 μm region, *Proc. Royal Soc.*, **A206**, 87–107 (1951).

may be said that an increase in attenuation of 1–10 dB/km is not unusual. Thus while system performance may be seriously degraded by rain it is usually possible to provide sufficient power margin to allow for this.

Variation of atmospheric temperature along the transmission path gives rise to refraction of the beam. Continual changes in this refraction in the turbulence of the atmosphere give rise to scintillation, an effect commonly seen in the twinkling of stars. Refraction and scintillation make the pointing of a narrow beam difficult and effectively set a lower limit to the practical beam divergence that can be used. Scintillation also causes continuous variation in the received power level. This, along with the variability in atmospheric attenuation, rules out the use of direct analogue intensity modulation for outdoor terrestrial systems.

16.3 SOURCES

16.3.1 Introduction

We have seen in Section 16.2 that semiconductor LEDs and lasers may be adequate in good atmospheric conditions for terrestrial links of 1–10 km path length. These sources retain the advantages of being small, light, robust and easy to modulate. For longer-range systems more powerful laser sources are required. These tend to be large, fragile and inefficient, and to require complex, high-voltage supplies to drive and modulate them. Two of the more highly developed types of laser stand out as being the most suitable for optical communication. These are the neodymium laser, working at 1.06 μm or frequency doubled at 0.53 μm, and the carbon dioxide laser working at 10.6 μm. Other laser sources have been suggested from time to time, for example shorter wavelength ones for communicating between satellites and submarines, but we shall concentrate the discussion on these two. Clearly, they will give rise to two very different types of communication system.

16.3.2 Neodymium Laser Sources

Neodymium lasers are solid-state lasers in which the active material, ions of the rare-earth element neodymium (Nd^{3+}), are present as a dilute impurity in a solid host material. Two widely used host materials are:

(a) Single-crystal rods of yttrium aluminum garnet (YAG—$Y_3Al_5O_{12}$) where the Nd substitutes for the yttrium.

(b) Certain types of glasses in which up to 5% by weight of Nd_2O_3 is included among the component materials.

Because these host materials are electrically insulating, it is necessary for the lasers to be pumped optically. Usually high-power incoherent sources such as tungsten–halogen incandescent-filament lamps or krypton or xenon arc lamps are used. The host material must be transparent to the pumping radiation as well as to the laser radiation.

Because the active material is present at a relatively low concentration, the electron energy level structure of the free atoms is preserved to some extent, although the energy levels themselves are considerably modified by the presence of the host material. This is particularly the case in glass, where the Nd^{3+} concentration is higher and where there are compositional variations in the host material surrounding different neodymium ions. This gives rise to an asymmetric broadening of the energy levels. Although the electron energy-level structure is complicated, a schematic illustration of Nd^{3+} in YAG is given in Fig. 16.6. The metastable $^4F_{3/2}$ level (spontaneous lifetime 200–500 μs) forms the upper level of the normal laser transition. The lower laser level is the $^4I_{11/2}$ state. This has a short spontaneous radiative lifetime and is far enough above the ground state for it not to be thermally populated. The upper laser level may be rapidly populated by de-excitations from the higher F, G and H levels. Neodymium thus forms a 4-level laser system. Population inversion can be readily achieved by pumping with radiation in the wavelength range 500–800 nm. This is absorbed in excitations from the ground state to the higher bands of levels shown in the figure.

The population inversion can be maintained in the face of continuous laser emission. The level of output power that can be produced continuously depends on the thermal properties of the laser rod. In this respect YAG is superior to glass, so it is the preferred host material when continuous operation at high average power is required. However, much larger rods and plates may be manufactured using Nd-glass and these are more suitable for the production of very high-energy pulses at a low duty cycle. It has been possible to produce several hundred watts of continuous output by pumping with Kr-arc lamps. Then the overall efficiency from electrical input power to laser output can be greater than 1%. However, the laser under these conditions oscillates in many high-order transverse modes and, thinking in terms of practical communication systems, the stability and reliability of the arc lamp are less than desired. Several watts of continuous laser power under more stable conditions can be obtained using tungsten–halogen lamps as the pump source. Again the overall efficiency is about 1% or rather less if the laser is restricted to oscillate in its fundamental transverse mode only. By including a wavelength-selective interference plate (etalon) inside the laser cavity, as shown in Fig. 16.7, a single

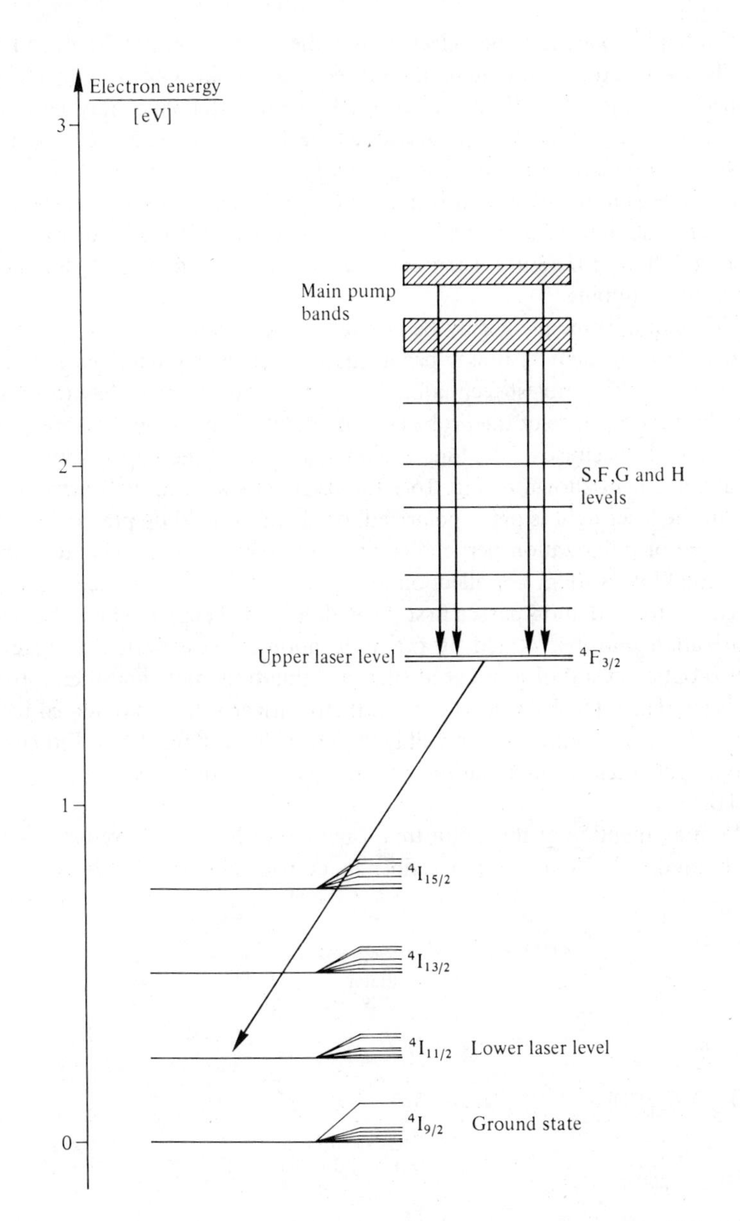

Fig. 16.6 Electron energy levels involved in the laser action of Nd^{3+}/YAG.

longitudinal mode may be selected and the frequency stabilized and tuned simply by rotating the etalon. If, further, a material with non-linear optical properties is inserted within the cavity, frequency doubling may be obtained with a relatively small loss in overall efficiency—perhaps a factor of two. A possible arrangement is also shown in Fig. 16.7. A material that has been found to be suitable is barium sodium niobate. When all these features are put together it should be perfectly feasible to produce about 1 W of single-mode laser radiation at 0.53 μm using about 1 kW of power from tungsten–halogen lamps for pumping.

Although in principle it may be possible to modulate the laser frequency by means of the intracavity tuning provided, in practice external intensity modulation remains the simplest technique. It is not possible to modulate the intensity directly, first because of the problem of modulating the optical pumping power and secondly because of the long radiative lifetime of the upper lasing level. In the arrangement shown in Fig. 16.7 the laser rods are cut at Brewster's angle so that the laser light is plane polarized: oscillations build up preferentially with the plane of polarization perpendicular to the plane of the rod end, since this radiation alone suffers no reflection losses at the rod–air interface. External to the cavity the radiation passes first through a Pockel cell, in which the plane of polarization may be rotated by the application of an electrical voltage to an electro-optic crystal of a material such as lithium niobate, and then through a polarizer. The intensity emerging from the polarizer will be modulated between 0 and 100% depending on the voltage applied to the Pockel cell. Either digital or analogue intensity modulation may be used, at modulation frequencies up to 1 GHz.

We may mention at this point that a miniature Nd-laser may be made from short lengths (1–2 cm) of optical fiber made with Nd-doped core glass. These

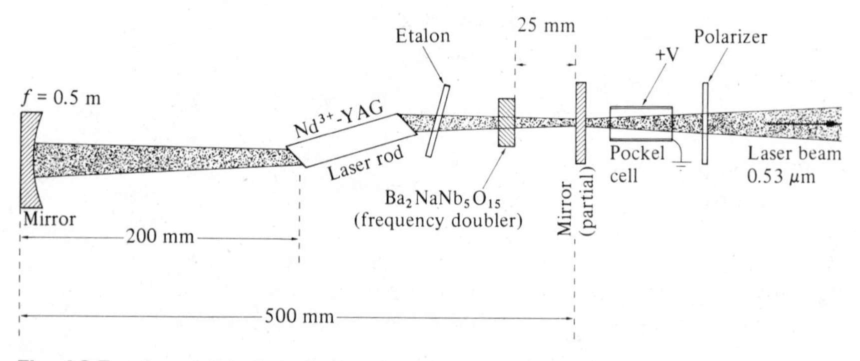

Fig. 16.7 A possible laser transmitter arrangement using a neodymium source with an external modulator.

may be end-pumped using a high-radiance LED, and have been considered as a possible 1.06 μm source for optical fiber systems.

It may be convenient in many cases to have the laser produce a continuous stream of high-powered pulses which are gated on or off by the Pockel cell. In this way the peak pulse power may be increased by ten or one hundred times the mean power level and efficient return-to-zero signalling will be available. There are three techniques which are commonly used to produce repetitive laser pulses. These are repetitive Q-switching, repetitive cavity dumping and mode-locking.

(a) Q-switching may be used to produce pulses of several microseconds duration at repetition frequencies of a few kilohertz. Optical pumping is used to build up a large population inversion without an optical cavity being formed. For example, one of the mirrors may be rotated so that it is only as the two mirrors become parallel and multiple reflections occur that a cavity is formed. This is timed to occur at the peak of the population inversion, at which point the energy stored in the laser medium is discharged rapidly and a pulse of radiation builds up and then dies away.

(b) Cavity dumping allows rather shorter pulses to be produced than does Q-switching, typically about 100 ns, thus permitting pulse repetition frequencies in the region of 1 MHz. Here the output mirror is made perfectly reflecting during the pumping period, with the result that the input energy is stored as radiation within the cavity. When this reaches its maximum, the mirror transparency is increased from zero to as high a value as possible (ideally 100%). Then during a period of a few photon transit times, τ_{ph} ($\tau_{ph} = 2l/c$, where l is the optical length of the cavity), the radiation is emitted as a single pulse. The radiation energy is 'dumped'.

(c) Mode locking may be used to produce subnanosecond pulses at pulse repetition frequencies up to 1 GHz. It is most easily achieved by including a further nonlinear optical material within the optical cavity in place of the etalon. This is a saturable absorber and produces what is known as passive mode locking. A saturable absorber is a material whose transparency increases with the radiation intensity. Intensity fluctuations occur naturally and the radiation of higher intensity experiences a higher round-trip gain within the cavity than radiation of lower intensity. The radiation energy then builds up into a single, high-power pulse which oscillates back and forth within the cavity. Emission from the partially transparent mirror occurs while the pulse is being reflected. Now it is the pulse repetition frequency that depends on the photon transit time and is simply $c/2l$. Exactly the same effect can be obtained by modulating the loss (or the gain) of the laser medium at the frequency $2l/c$. For example a polarizer and an electro-optic crystal may be included in the cavity and a voltage of this frequency applied to the crystal. This is known as active-mode locking. The pulse width, τ, depends on the range of optical frequencies over which laser action can occur, that is, on the line width of the laser transition: $\Delta f = \Delta \mathcal{E}/h$. Approximately $\tau = 1/\Delta f$. The room temperature line width of Nd^{3+} in YAG has been measured as 180 GHz, and whilst it is not possible to maintain optical gain over the whole of this line width, mode-locked pulse lengths of less than 50 ps can be produced.

16.3.3 Carbon Dioxide Laser Sources

The other laser source which can readily be made in a form suitable for use in communication is the four-level gas laser based on carbon dioxide and operating at 10.6 μm. As with most gas lasers, the upper laser level is pumped directly or indirectly by electronic excitation in a gas discharge. Either a d.c. or a radio-frequency glow discharge may be used at low pressures, say below about one tenth of an atmosphere (or 10^4 Pa). What is important is to obtain a uniform and steady discharge throughout the active volume. More sophisticated techniques have been devised for producing very short laser pulses (< 1 ns) using high-pressure discharges, and for producing very high continuous powers (> 100 kW) using continuously pumped gas-flow techniques. More suitable as a communications source would be a compact, sealed-off device capable of delivering a few watts or tens of watts of continuous power in a form that can be readily modulated. A particular type of waveguide laser has been especially developed to this end (Ref. [16.4]), and we shall concentrate the discussion on this design. A schematic illustration is shown in Fig. 16.8. We shall return to consider some of the features included in the diagram after first examining the origin of laser action in carbon dioxide.

Most types of CO_2 laser use a mixture of carbon dioxide, nitrogen and helium in the approximate proportions of 1:2:3, respectively. The essential part played by the nitrogen in promoting the efficient excitation of the upper laser level may be seen on the simplified electron energy level diagram shown in Fig. 16.9. The vibrational modes associated with the sets of levels are indicated at the top of the diagram. The vibrational levels of nitrogen are metastable, with the result that the most probable mode of de-excitation is by resonance transfer in a collision with a ground-state CO_2 molecule which is thereby excited to the 00°1 level. As a result population inversion between the 00°1 and either the 10°0 or the 02°0 level is easily achieved. It should be noted that associated with each of the vibrational levels is a large number of closely

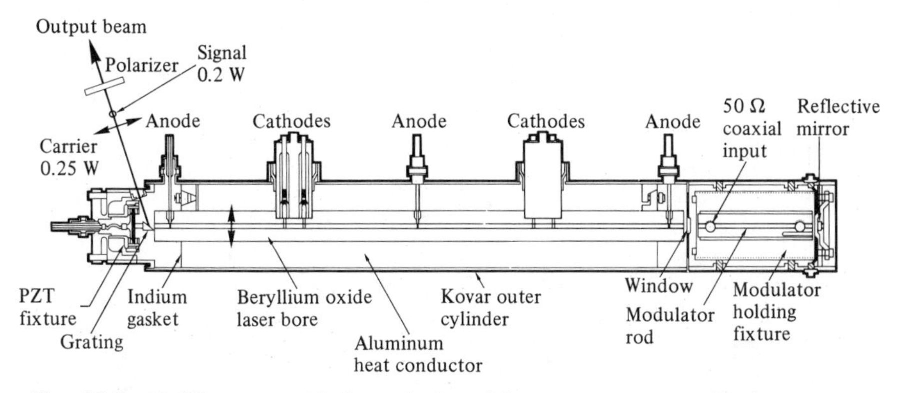

Fig. 16.8 A CO_2 waveguide laser designed for use as a communication system source. [Taken from Ref. [16.4].]

spaced, higher-energy rotational states. Thus many laser transitions in the region of 10.6 μm and 9.6 μm, as well as other nearby wavelengths, may be excited. The closeness of the laser energy levels to the ground-state, together with the high probability of excitation to the 00°1 level, make the CO_2 laser highly efficient, better than 10% overall efficiency being quite feasible. However, this is also a cause of limitation of the output power since as the gas heats up the 01^10 level becomes thermally populated and inhibits the de-excitation of the lower laser level. This is where the helium is beneficial since it aids the conduction cooling of the discharge, and by collisional de-excitation it directly increases the rate of depopulation of the lower laser level. A third beneficial effect is that by varying the helium pressure some control of the electron energy distribution in the discharge is obtained, and this can be used to produce the maximum rate of excitation into the upper laser level.

The laser shown in Fig. 16.8 has a bore of cross-section 1.5 mm square and of length 260 mm. The electrical discharges were established at 4 kV and 3 mA in a gas pressure of 1.6×10^4 Pa. With optimum output coupling 4.5 W of laser output were produced at 9% overall efficiency. The output coupling shown is via a blazed grating. This has two other functions: it acts as a polarizer and it enables the laser radiation frequency to be tuned to one of a number of transitions in the 00°1 to 10°0 sets of levels ranging from 10.467 μm to 10.788 μm in wavelength. Frequency stability is then better than ±100 kHz. Intensity modulation at up to 300 Mb/s may be obtained by means of the CdTe electro-optic crystal within the optical cavity. The voltage applied to the modulator changes the plane of polarization of the radiation within the cavity and hence the proportion coupled out by the blazed grating. It is more efficient to use intracavity modulation rather than an external modulator as described earlier, and smaller modulation voltages are required.

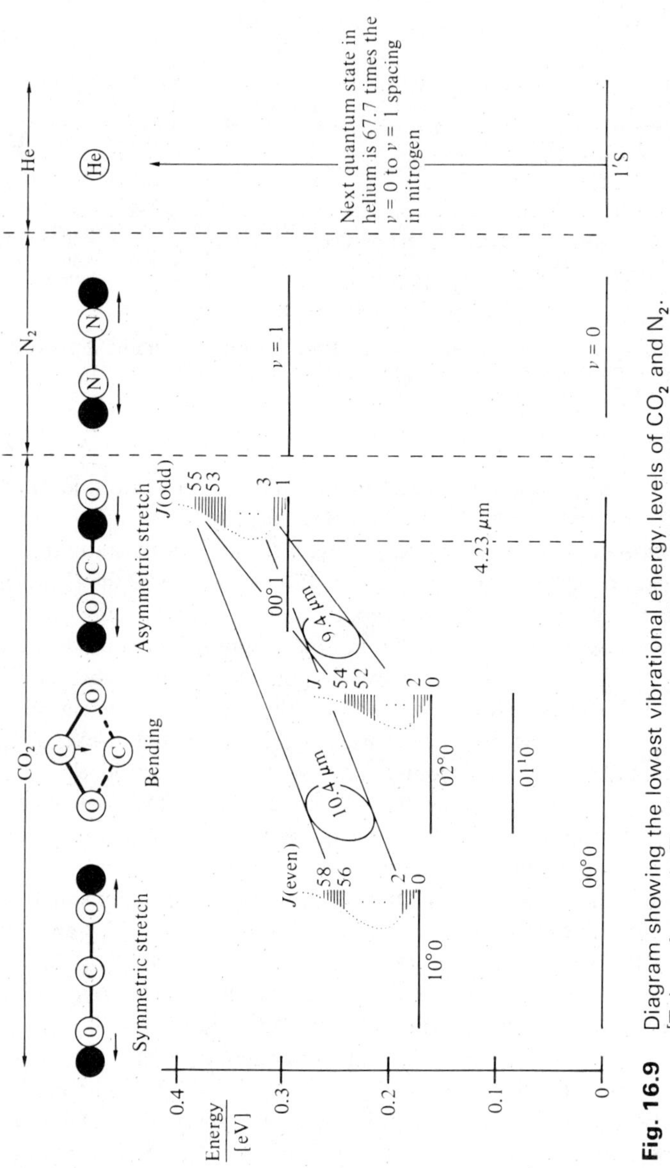

Fig. 16.9 Diagram showing the lowest vibrational energy levels of CO_2 and N_2. [Taken from J. T. Verdeyen, *Laser Electronics*, p. 279; © 1981, reprinted by permission of Prentice-Hall Inc., Englewood Cliffs, NJ.]

16.4 DETECTORS

16.4.1 Introduction

For transmission wavelengths in the range 1 to 2 μm direct detection using p–i–n or avalanche photodiodes, as described in Chapters 12 and 13, remains the most convenient method of signal recovery for unguided as well as for guided systems. The possible use of longer or shorter wavelengths, however, does bring other techniques and other types of device into consideration. At wavelengths less than 1 μm it becomes practical to use photomultiplier tube (PMT) detectors. These have the advantage of being large-area devices (up to 10 cm^2), of giving very high internal multiplication factors ($>10^6$), of introducing relatively little additional noise, and of having bandwidths in excess of 1 GHz. Their main disadvantages are a low quantum efficiency (<0.1), large size, limited life, mechanical fragility, and a need for stable, high-voltage supplies, typically of about 1 kV. At longer wavelengths, particularly at the 10 μm wavelengths associated with carbon dioxide sources, the use of heterodyne detection becomes practical, permitting increased sensitivity and the possible use of other modulation techniques.

16.4.2 The Use of Photomultiplier Tubes at Shorter Wavelengths

The basic conversion process used in photomultiplier tubes is the photoelectric effect, where the energy given up by the incoming radiation is sufficient to eject an excited electron from the surface of the material. The process is illustrated in the electron energy band diagram shown in Fig. 16.10. The photocathode is shown there as a p-type semiconductor because such material is to be preferred for a number of reasons. First note that the wavelength threshold, λ_{th}, for photoelectric emission from a p-type semiconductor is determined by the sum of the band-gap energy and the electron affinity. Thus,

$$\lambda_{th} = \frac{hc}{\mathcal{E}_{th}} = \frac{hc}{(\chi + \mathcal{E}_g)} \qquad (16.4.1)$$

A semiconductor is preferred because the incident radiation has a far higher probability of interacting with a valence-band electron than with a free electron, either in a metal or in the conduction band of an n-type semiconductor. Making the semiconductor p-type gives two further advantages. First the work-function, ϕ, is increased, with the result that thermionic emission from the photocathode in the absence of any incident radiation (dark emission) is reduced. Secondly, the surface states of a p-type semiconductor tend to develop a net positive surface charge and thus cause the bands to bend

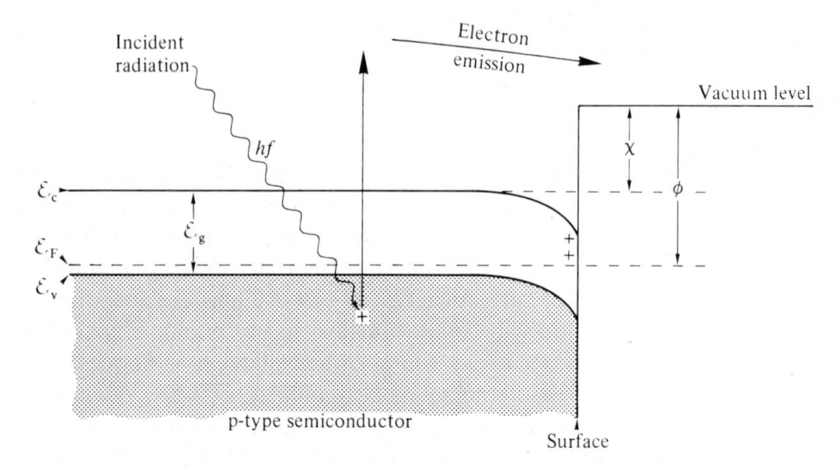

Fig. 16.10 Band diagram showing photoelectric emission from a p-type semiconductor.

downwards at the surface as shown in Fig. 16.10. In an n-type semiconductor the band structure is more likely to bend upwards at the surface, thereby increasing the effective electron affinity. This not only increases the threshold photon energy required for emission but it also means that the probability that an electron with sufficient energy to escape actually does escape from the surface is reduced.

The photoelectric quantum efficiency of a number of photocathode surfaces is shown as a function of wavelength in Fig. 16.11. The parameters of some of these are tabulated in Table 16.1. The importance of the recently developed materials with negative electron affinity (NEA) should be noted. In these the energy corresponding to the bottom of the conduction band is higher than the vacuum level.

Even the best photoelectric quantum efficiencies are substantially below those obtained in photodiode detectors. Compensating for this is the ability to use a high-gain, secondary-emission multiplier, of the type shown schematically in Fig. 16.12. The dynodes should be coated with a material having a high secondary electron emission coefficient, δ. This again could well be an NEA material. If N stages are used, each giving a gain of δ, the overall gain is

$$M = \delta^N \tag{16.4.2}$$

with $N = 10$ and $\delta = 4$, $M \simeq 10^6$.

As with the avalanche photodiodes discussed in Chapter 13, shot noise current, I_{Sh}, is generated in the quantum conversion process at the photocathode and this is multiplied together with the signal. In addition some further noise is added in the multiplication process and this can be taken into account by means of a noise factor, F, as before. According to theory, see, for

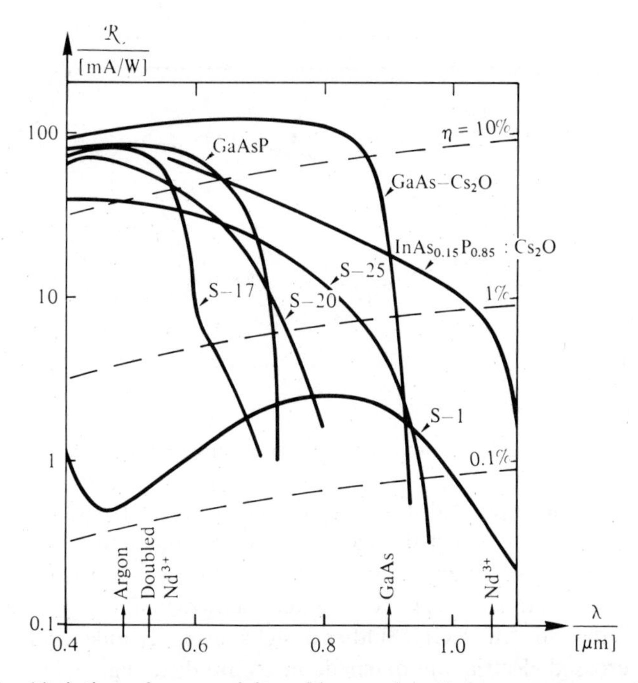

Fig. 16.11 Variation of responsivity with wavelength for some commonly used photocathode materials. [Taken from Ref. [12.1], © 1970 IEEE.)

example, Section 13.1 of Ref. [14.2] or Section 21.3 of Ref. [14.3], when $\delta \gg 1$,

$$F \simeq \frac{\delta}{\delta - 1} \tag{16.4.3}$$

Thus in our example with $\delta = 4$, $F \simeq 1.33$. Examination of eqn. (14.4.10) shows that the use of a PMT normally enables the receiver noise to be entirely

Table 16.1 Some photocathode materials

Type	Material		$\mathcal{E}_g$/[eV]	χ/[eV]
S1	Ag–O–Cs	Structure uncertain	($\mathcal{E}_g + \chi \simeq 1.0$ eV)	
S17	Cs_3Sb	p-type semiconductors	1.6	0.45
S20 (Multi-alkali)	Sb–K–Na_2–Cs		1.0	0.55
	GaAs–Cs_2O	Negative electron	1.4	–0.55
	$InAs_{0.15}P_{0.85}$–Cs_2O	affinity materials	1.1	–0.25

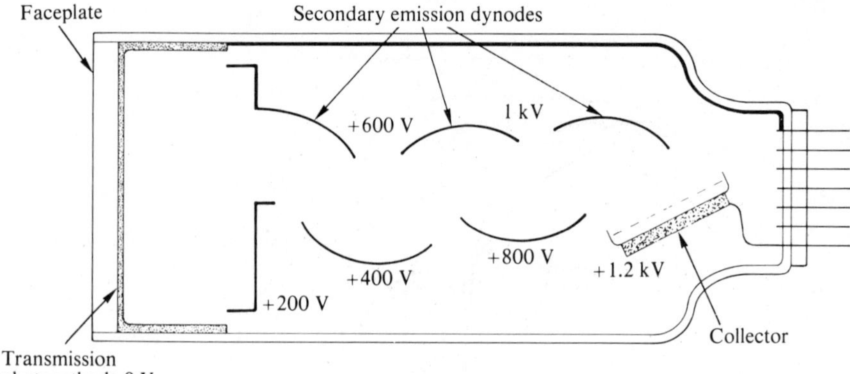

Fig. 16.12 Schematic diagram of a photomultiplier tube.

dominated by the multiplied shot noise represented by term (c) in the equation. The bandwidth over which the gain, M, can be maintained depends on the electron transit-time through the PMT. Careful attention to the electrode design in electrostatic PMTs like the one shown in Fig. 16.12 enables the bandwidth to extend to 1 GHz. Other designs, using dynodes with high values of δ and crossed electric and magnetic fields, produce high gain at still higher modulation frequencies.

16.4.3 Detectors for Longer Wavelengths

For the detection of wavelengths longer than 1 μm narrow bandgap semiconductors are required. Of the binary III–V compound semiconductors listed in Table 7.2, InSb has the smallest value of $\mathcal{E}_g$ and can be used out to about 6 μm. For the 10.6 μm radiation from CO_2 laser sources other materials are needed. In former times impurity semiconductors such as germanium doped with copper or with mercury were used at these wavelengths, acting as extrinsic photoconductors. To take Ge–Hg as an example, the mercury introduces a band of acceptor levels at an energy 0.09 eV above the top of the valence band. Of course, at quite modest temperatures these are all filled by electrons thermally excited from the valence band. But at sufficiently low temperatures, less than about 30 K, they are mostly empty, and then electrons may be excited optically. The extra holes so generated increase the electrical conductivity of the material in direct proportion to the light flux absorbed. More recently, p–n junction diodes made from the ternary II–VI semiconductor cadmium mercury telluride ($Cd_xHg_{1-x}Te$—often referred to as

CMT) have become available. Reducing the proportion of cadmium enables the room-temperature band gap of this material to be reduced from 1.8 eV to zero. With $x = 0.2$, $\mathcal{E}_g \simeq 0.1$ V at 77 K and diodes with quantum efficiency exceeding 0.25 at 10.6 μm can be produced. In order to avoid excessive dark current caused by thermal excitation, cooling to about 120 K or below is required. In terrestrial applications liquid nitrogen (77 K) may be used, but in space passive radiation coolers are adequate for these temperatures. The use of about 0.2–0.5 V reverse bias minimizes the capacitance and the response time of the device without causing excess dark current as a result of tunnelling. Bandwidths in excess of 100 MHz can then be obtained.

16.4.4 The Use of Heterodyne Detection

At 10.6 μm heterodyne detection becomes practical and because of the high sensitivity it offers it is likely to be preferred at these wavelengths. A similar principle is often employed in microwave detectors. Waves from a local oscillator source are mixed with the received, modulated waves from the transmitter in the region of the detector junction. Provided that both sources are coherent and maintain their mutual coherence over the area of the detector, the photodiode current has a component which varies at the difference frequency. In radio parlance an intermediate frequency (or i.f.) is extracted. This carries the same modulation that was present on the received optical wave. The detection sensitivity is seriously degraded if the alignment of the wavefronts of the two beams varies by more than a few percent of a wavelength over the area of the detector and also if the intensity distributions and polarization do not match. The required alignment precision is not really feasible in a practical system at 1 μm, but it does become quite possible at 10.6 μm.

Consider a photodiode that is illuminated simultaneously and uniformly across its area by a plane, modulated optical wave of power density P_R and also by a plane coherent wave from a local oscillator producing a power density P_L. In order to see what happens to the signal modulation under these circumstances, we need to recognize that the instantaneous rate of carrier generation is proportional to the square of the instantaneous electromagnetic field amplitude. By 'instantaneous' we mean a value averaged over a period that is short compared to the modulation period but includes many optical cycles. The electromagnetic field amplitudes are represented in terms of the electric field components, $E_R(t)$ and $E_L(t)$, referring respectively to the received field and the field from the local oscillator. Then

$$E_R(t) = \hat{E}_R \cos(\omega_R t + \phi_R) \qquad (16.4.4)$$

and

$$E_L(t) = \hat{E}_L \cos \omega_L t \tag{16.4.5}$$

The signal modulation $f(t)$ may be included in eqn. (16.4.4). In the case of amplitude modulation,

$$\hat{E}_R = \hat{E}_{R0}(1 + f(t)) \tag{16.4.6}$$

with intensity modulation

$$\hat{E}_R^2 = \hat{E}_{R0}^2(1 + f(t)) \tag{16.4.7}$$

With phase modulation

$$\phi_R = \phi_{R0}(1 + f(t)) \tag{16.4.8}$$

Frequency modulation has to be defined with some care. It may be expressed most easily in terms of the carrier phase (see Refs. [1.2] to [1.4]),

$$\phi_R = \phi_{R0}\left(1 + \int_0^t f(t)\,dt\right) \tag{16.4.9}$$

Then the 'instantaneous' angular frequency becomes

$$\omega = \frac{d}{dt}(\omega_R t + \phi_R) = \omega_R + \phi_{R0} f(t)$$

The optical power densities in the two waves P_R and P_L are proportional to $\hat{E}_R^2$ and $\hat{E}_L^2$, respectively. We shall therefore put

$$P_R = \tfrac{1}{2}\alpha \hat{E}_R^2 \tag{16.4.10}$$

and

$$P_L = \tfrac{1}{2}\alpha \hat{E}_L^2 \tag{16.4.11}$$

where

$$\frac{1}{\alpha} = \left(\frac{\mu_r \mu_0}{\varepsilon_r \varepsilon_0}\right)^{\frac{1}{2}} = Z \tag{16.4.12}$$

is the local space impedance ($\mu_r \mu_0$ is the local permeability and $\varepsilon_r \varepsilon_0$ is the local permittivity). In vacuum, $1/\alpha = Z_0 = 377\ \Omega$. The average photocurrent density generated by the two waves is

$$\bar{J} = \mathcal{R}(P_R + P_L) \tag{16.4.13}$$

where

$$\mathcal{R} = \frac{\eta e \lambda}{hc} \tag{12.1.2}$$

is the *responsivity* of the photodiode. We may integrate over the area, A, of the detector in order to obtain the average current, $\bar{I}$, in terms of the optical fluxes, Φ_R and Φ_L. When the illumination is uniform and the diode responsivity is uniform, $\bar{I} = \bar{J}A$, $\Phi_R = P_R A$ and $\Phi_L = P_L A$, so that

$$\bar{I} = \mathcal{R}(\Phi_R + \Phi_L) \tag{16.4.14}$$

This current will give rise to a total mean-square, shot-noise spectral density

$$(I^*_{Sh})^2 = 2e\bar{I} = 2e\mathcal{R}(\Phi_R + \Phi_L) \tag{16.4.15}$$

The time-varying photocurrent produced is given by the low-frequency (non-optical) components of

$$i(t) = \alpha\mathcal{R}A[E_R(t) + E_L(t)]^2 \tag{16.4.16}$$

$$\begin{aligned}
&= \alpha\mathcal{R}A[\hat{E}_R \cos(\omega_R t + \phi_R) + \hat{E}_L \cos\omega_L t]^2 \\
&= \alpha\mathcal{R}A[\tfrac{1}{2}\hat{E}_R^2\{1 + \cos 2(\omega_R t + \phi_R)\} \\
&\quad + 2\hat{E}_R\hat{E}_L \cos(\omega_R t + \phi_R)\cos\omega_L t + \tfrac{1}{2}\hat{E}_L^2\{1 + \cos 2\omega_L t\}] \\
&= \mathcal{R}A(P_R + P_L) + \alpha\mathcal{R}A[\tfrac{1}{2}\hat{E}_R^2 \cos 2(\omega_R t + \phi_R) + \tfrac{1}{2}\hat{E}_L^2 \cos 2\omega_L t \\
&\quad + 2\hat{E}_R\hat{E}_L\{\cos(\omega_R t + \omega_L t + \phi_R) + \cos(\omega_R t - \omega_L t + \phi_R)\}]
\end{aligned} \tag{16.4.17}$$

As we have indicated, the second harmonic and sum frequencies which appear in this classical analysis of the problem do not arise in a proper quantum-mechanical treatment. In any case the photodiode bandwidth only allows it to respond to the d.c. and difference frequency terms, and the higher frequencies average to zero. Thus,

$$i(t) = \bar{I} + 2\mathcal{R}(\Phi_R\Phi_L)^{1/2}\cos\{(\omega_R - \omega_L)t + \phi_R\} \tag{16.4.18}$$

and the r.m.s. current generated at the intermediate (or difference) frequency is

$$I_{IF} = \mathcal{R}(2\Phi_R\Phi_L)^{1/2} \tag{16.4.19}$$

In order to make comparison with previous results, let us assume that the transmitted carrier is intensity modulated with a binary signal. Thus the received flux is either zero or has the value Φ_R which provides the peak signal, I_{IF}. Then, on our previous definition, the r.m.s. signal-to-noise ratio, K, is the ratio of I_{IF} to the total r.m.s. noise generated at the photodiode and in its pre-amplifier. We will assume this to be the voltage amplifier circuit discussed in Section 14.4. The total noise comprises:

(a) The shot noise generated by the incoming radiation

$$(I^*_{Sh})^2 = 2e\mathcal{R}(\Phi_R + \Phi_L + \Phi_B) \tag{16.4.20}$$

where, to the terms included in eqn. (16.4.15), we have added Φ_B to represent any background radiation entering the detector.

(b) The random signal produced at the i.f. frequency by any background radiation that lies within the range of i.f. frequencies, Δf, about the local oscillator frequency, f. By analogy with eqn. (16.4.19) this is

$$(I_{BN})^2 = 2\mathcal{R}^2 \Phi_L \frac{d\Phi_B}{df} \Delta f \qquad (16.4.21)$$

The term $(d\Phi_B/df)\,\Delta f$ represents the flux of background radiation in the frequency range Δf about f.

(c) The noise introduced by the amplifier. Referring back to eqn. (14.4.10), this includes terms (a), (b), (d) and (e) of that equation. For convenience we shall express the sum of these as

$$(I_c^*)^2 = (V_A^*)^2 \left\{ \frac{1}{R^2} + \frac{4\pi^2}{3} C^2 (\Delta f)^2 \right\} + \frac{4kT}{R} + (I_A^*)^2 \qquad (16.4.22)$$

Putting all these factors together the overall signal-to-noise ratio takes the form:

$$K = \frac{\mathcal{R}(2\Phi_R \Phi_L)^{1/2}}{[2e\mathcal{R}(\Phi_R + \Phi_L + \Phi_B) + 2\mathcal{R}^2 \Phi_L (d\Phi_B/df) + (I_c^*)^2]^{1/2} (\Delta f)^{1/2}} \qquad (16.4.23)$$

That is

$$K^2 = \frac{\Phi_R}{\left[\dfrac{e}{\mathcal{R}\Phi_L} (\Phi_R + \Phi_L + \Phi_B) + \dfrac{d\Phi_B}{df} + \dfrac{(I_c^*)^2}{2\mathcal{R}^2 \Phi_L} \right] \Delta f} \qquad (16.4.24)$$

It can be seen from eqn. (16.4.24) that the effect of increasing the local oscillator power level is similar to the use of avalanche or secondary emission multiplication in that it reduces the importance of the amplifier noise terms. When Φ_L is large enough for these to be negligible and for Φ_R and Φ_B to be neglected in comparison,

$$K^2 = \frac{\Phi_R}{\left[\dfrac{e}{\mathcal{R}} + \dfrac{d\Phi_B}{df} \right] \Delta f} = \frac{\Phi_R}{\left[\dfrac{hf}{\eta} + \dfrac{d\Phi_B}{df} \right] \Delta f} \qquad (16.4.25)$$

Thus in theory, heterodyne detection with sufficient local oscillator power should permit the quantum limit of detection,

$$K^2 = \frac{\eta \Phi_R}{hf\,\Delta f} \tag{16.4.26}$$

to be obtained as long as background radiation is not excessive. It has been estimated that even if a receiver tuned to 10.6 μm points directly at the sun, the signal-to-noise ratio may be degraded by no more than 3 dB.

16.5 EXAMPLES OF UNGUIDED OPTICAL COMMUNICATION SYSTEMS

16.5.1 Terrestrial Systems

What is likely to be the most widely used example of unguided optical transmission is almost trivial in communication terms. It is the common and simple, infra-red, remote television controller. A hand-held unit actuates one or more conventional, infra-red LEDs to send out a sequence of optical pulses of a fraction of a milliwatt in power. A photodiode mounted on the set detects either the direct or scattered radiation. The pulse sequence can be encoded to change channels, switch off the set, raise or lower the volume, and in more sophisticated versions to select a required page of Teletext. The main competitor in this application is an ultrasonic unit, and the most important factor in deciding which is to be preferred in this very competitive field is the cost of the transducers: LEDs and photodiodes in the one case and ultrasonic units and microphones in the other. In either case the electronic signal processing required is similar and is relatively inexpensive.

A second, indoor application for which equipment has been developed enables a person moving about a room to receive high quality audio transmissions on a head-phone set. These may come from any conventional source: radio, tape-recorder or telephone. Although simple intensity modulation of an LED mounted on the source amplifier may be adequate, a better system, which is a little more complex, is illustrated by the block diagram of Fig. 16.13. Here a frequency-modulated subcarrier is generated in the form of a train of optical pulses. With standard LEDs this could be at an average power level of about 1 mW and at a frequency of about 100 kHz. The subcarrier frequency modulation and demodulation may be carried out by means of a standard, integrated-circuit, voltage-controlled oscillator as shown in the figure. This arrangement would be an alternative to the inductive loop that is often put around cinemas

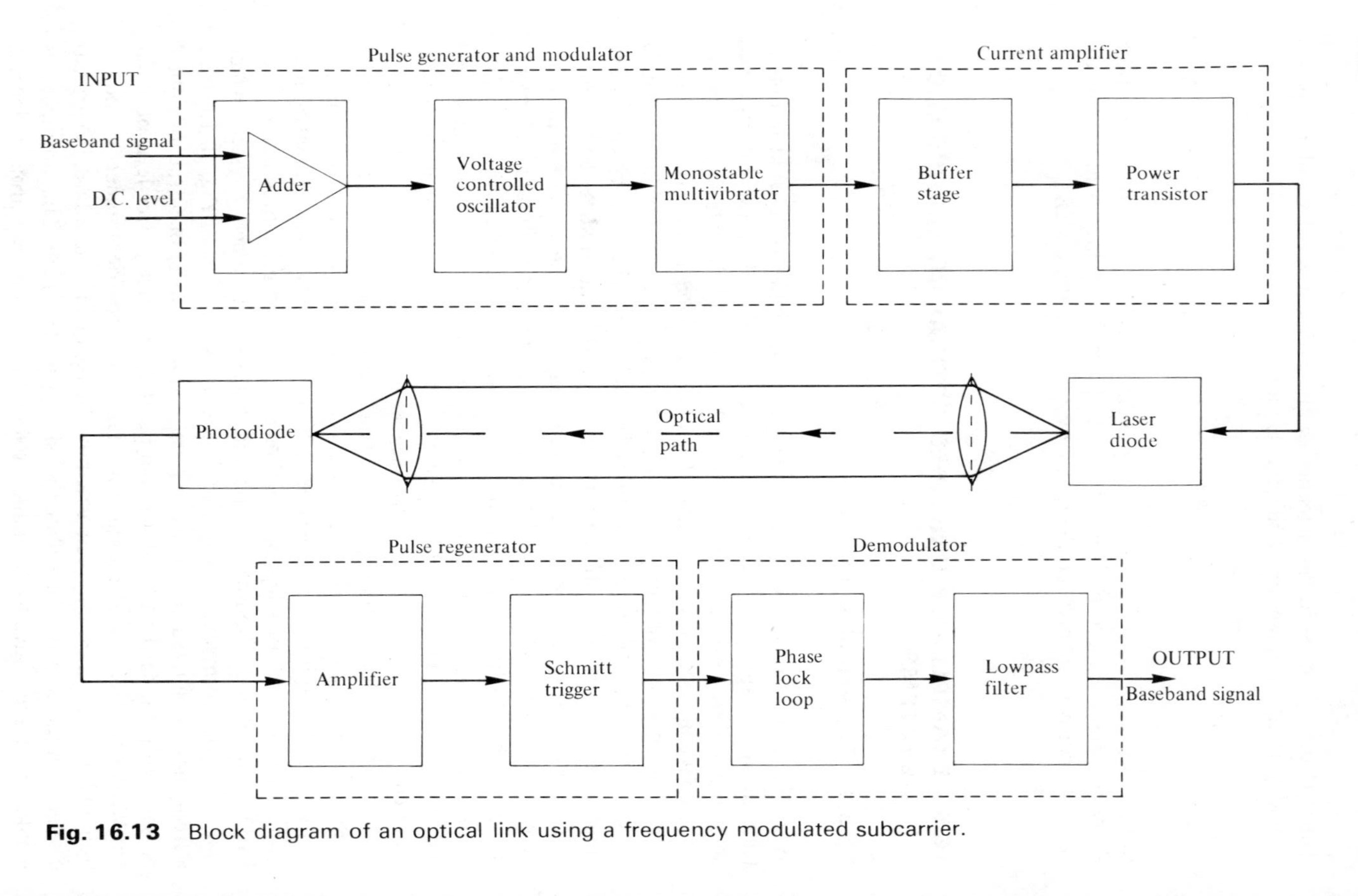

Fig. 16.13 Block diagram of an optical link using a frequency modulated subcarrier.

and lecture halls for the benefit of wearers of hearing-aids. Again, cost and convenience will determine its use, especially the power consumption required at the receiver, which for full portability must be battery powered.

The basic idea has been developed for outdoor use as a two-way, covert, field telephone. The transmitting/receiving electronics is packed into one half of a standard pair of 7 × 50 binoculars. The other half is used as a monocular for sighting purposes. The unit is connected to a lightweight head-set consisting of headphones and microphone. With a few milliwatts of transmitted, infra-red beam power from an LED, a range of the order of 1 km is easily obtained in good conditions. Under adverse weather conditions aural communication can be maintained as long as there is visual contact between operators. The advantages of the system are those that were set out in Table 1.1, but we would emphasize particularly the avoidance of normal transmission frequencies with their associated interference and the narrow beam width that can be obtained. This is typically 1° (17 milliradians) between half power points. Security is high in that the transmission is not detectable outside the narrow beam, nor even in the line of sight beyond its normal range. A similarly housed competing, microwave equipment generates about 10 mW at 30 GHz with an 8 degree beam-width, and gives a range of up to 30 km. It has to be said, however, that neither of these units shows signs of coming into general use to replace conventional, radio-frequency field telephones.

A number of similar unguided systems have found application between fixed terminals when users have needed to link computers to computers or computers to peripherals sited in scattered outbuildings. These data transmission links have usually carried binary digital data, typically at 9.6 kb/s. Mainly, they have been based on GaAs lasers or LEDs and used Si p–i–n or avalanche photodiodes for transmission over distances up to 1 km. Again the main motivation for the use of such systems has been the avoidance of electromagnetic interference and the independence they give from land lines and from the public authorities. It has been suggested that computers and peripherals distributed within a room might also be linked by infrared sources and detectors. Any one source could broadcast to all receivers through radiation scattered off the walls, floor and ceiling.

A feature of all the applications mentioned so far is that they utilize perfectly conventional components and they depend for their (limited) success on detailed engineering to meet specific requirements. We will end this section with a brief reference to a technologically rather more advanced system. This again was an experimental field communication unit and it was developed for use by the US Army in 1972 following the proven high efficiency of CO_2 lasers emitting at 10.6 μm. Systems using both intensity and frequency modulation were made and in both cases demodulation was effected by heterodyning with a local oscillator using cooled semiconductor detectors. It was demonstrated that very high receiver sensitivity, approaching the quantum limit, could be obtained. This early system formed the basis for the satellite communication links to be discussed in the next section. We might note in passing that optical systems developed for range-finding, target-identification and remote sensing utilize generation and detection techniques similar to those required for communication. It has to be said that these other applications are much more highly developed than unguided optical communication.

16.5.2 A Proposed Optical System for Communication in Near-space

In this section a much more advanced and sophisticated system will be described, one in which the high power available from laser sources may be expected to be particularly advantageous. The requirement is for a system that transmits data between a satellite in close orbit around the earth and one in a geostationary orbit, and also between two geostationary satellites. The height of the low-orbit satellite may be between 200 km and 2000 km above the earth's surface. Its purpose is to monitor the earth's surface and relay the data via one or more geostationary satellites back to a ground station. The estimated data-rate requirement is 300 Mb/s. The proposed system to be discussed here was described in detail in Ref. [16.4]. An illustration is shown in Fig. 16.14.

Clearly, in choosing a satellite communication system minimization of the overall weight and power consumption is of the utmost importance. Thus the narrow beam widths which can be obtained with optical systems using antennae of modest aperture area may be thought to give an over-riding advantage. In practice, though, at the present time, microwave (5–10 mm wavelength)

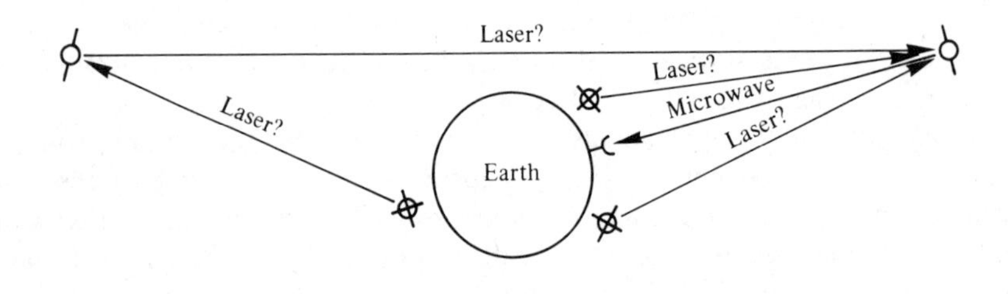

Fig. 16.14 Proposed near-space data transmission system for earth surface monitoring satellites. [Taken from Ref. [16.4] © 1977 IEEE.]

links are preferred in all parts of the proposed system because of their proven reliability, their high transmitter efficiency and their low receiver noise level. However, both frequency-double neodymium and carbon dioxide lasers have been considered for the intersatellite links and must be possible contenders for future systems. Apart from the need to establish proven long life and reliability in each case, the main problem with the neodymium system is the source efficiency and the main problem with carbon dioxide lasers lies in the modulation of the laser output. In both cases very narrow beam widths of no more than a few seconds of arc can readily be obtained, but then complex signal acquisition and stabilizing systems are called for, and accurate tracking of both transmitter and receiver is necessary. In this section we will give a brief outline of the CO_2 laser system developed under the auspices of NASA and described in Ref. [16.4], but as yet not deployed.

The transmitter laser was the one shown in Fig. 16.8. It was designed to produce 1 W of output power carrying modulation at up to 300 Mb/s. Its electrical power requirements were 55 W for the laser and 90 W for the modulator, and the total weight was 1.5 kg. The folded Gregorian optical system shown in Fig. 16.15 was chosen to collimate the transmitted beam and to collect the received signal. The diameter of the primary mirror was 185 mm giving an 'antenna gain' at 10.6 μm of 92 dB and a beam divergence of 82 μrad. The obscuration caused by the secondary mirror and the fact that the beam intensity distribution is Gaussian act to reduce the gain and increase the divergence compared to the theoretical values given in Section 16.2.

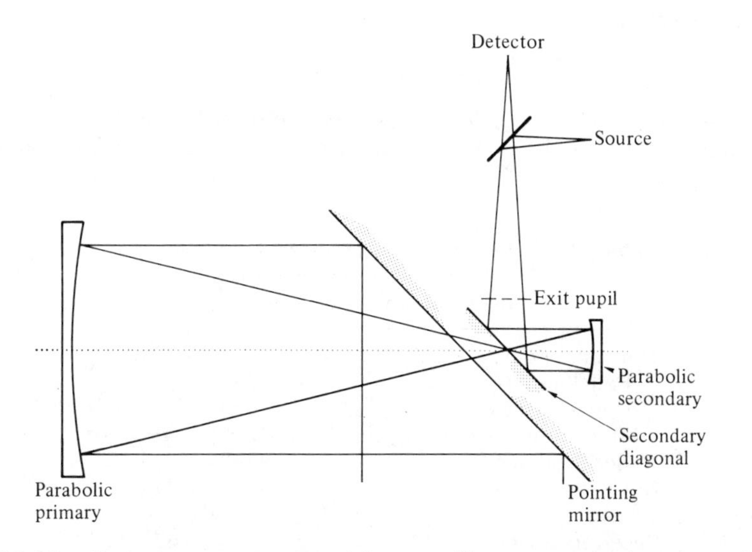

Fig. 16.15 Optical system developed for a satellite communication transceiver. [Taken from Ref. [16.4] © 1977 IEEE.]

The height of a geosynchronous orbit above the ground is 38,600 km and the maximum distance between a close orbit satellite and one in a synchronous orbit is about $l = 47{,}000$ km. Thus, the 'path loss' $(\lambda/4\pi l)^2$, is about –275 dB. The calculation set out in Table 16.2 indicates that the expected received power level is of the order of 0.2 nW, (–67 dBm). A complete optical heterodyne receiver employing a cooled cadmium mercury telluride junction photodetector has shown that this level of receiver sensitivity can be achieved in practice. At 300 Mb/s the receiver noise level was measured to be equivalent to an input noise power of –84 dBm. This is within 10 dB of the quantum limit, $hf\,\Delta f$. It was found that a signal-to-noise ratio of 14 dB was sufficient to ensure a probability of error of less than 10^{-6}. Thus a signal level of –70 dBm or 0.1 nW should be adequate, and there is an unallocated margin of 3 dB on our calculation. The two or three orders of magnitude improvement in sensitivity compared with direct detection of wavelengths around 1 μm can be ascribed partly to the reduced quantum (shot) noise at the longer wavelength and partly to the use of the heterodyne method.

As local oscillator, a smaller, externally stabilized version of the waveguide CO_2 laser used for the transmitter was developed. It produced 65 mW of output power, was tunable over ±300 MHz, and stable to 150 kHz.

One interesting feature of optical satellite systems is the Doppler frequency variation of the received carrier wave caused by relative motion between source and receiver. The relative velocity, v, between a geostationary and a close-orbit satellite can be as high as 8 km/s, giving a Doppler shift of more than ±700 MHz at 10.6 μm. If f_c is the transmitted carrier frequency and f_r the frequency received,

$$\Delta f = f_r - f_c = f_c v/c = v/\lambda \tag{16.5.1}$$

With the figures given ($v = 8$ km/s, $\lambda = 10.6$ μm) $\Delta f = 755$ MHz.

It may be seen from this one example that in future earth satellite and deep-

Table 16.2 Power account for laser communication link between a close-orbit and a geostationary satellite

Transmitter power (1 W)		+30 dBm
Path loss $(\lambda/4\pi l)^2$	−275 dB	
Optical losses	−6 dB	
Transmitter gain $(4\pi A_T/\lambda^2)$	+92 dB	
Receiver gain $(4\pi A_R/\lambda^2)$	+92 dB	
Net loss		−97 dB
Received power		−67 dBm

space communication systems laser communication may well have an important role. But it will always have to compete with microwave systems that will usually be already developed for, and proven in, terrestrial applications.

PROBLEMS

16.1 The transmitter for an unguided optical communication system consists of a diffuse source of wavelength 0.85 μm and an optical system of aperture $f/8$. Calculate the limiting source diameter below which the beam divergence becomes diffraction limited.

16.2 The source for a simple unguided optical communication system is a light-emitting diode which behaves as a diffuse source of diameter 0.1 mm and emits 10 mW into air under normal bias conditions. The photodiode detector in the system receiver has a diameter of 1 mm and the radiant power incident onto it should exceed 1 nW for satisfactory system performance.
(a) Calculate the source radiance and its normal radiant intensity.
(b) Calculate the maximum range of the system when no optical collimation or collection systems are used at the transmitter and receiver.
(c) Calculate the factors by which the range is enhanced first when a collimating lens of 50 mm diameter is used at the transmitter and further when a lens of similar diameter is used to collect the received radiation and focus it onto the detector.
(d) Calculate the pointing accuracy required when the transmitter lens mentioned in (c) has an $f/4$ aperture.

16.3 Prove eqn. (16.2.13).

16.4 The output beam from a 10 W CO_2 laser has a Gaussian intensity profile. It is to be used as a 10.6 μm wavelength source for optical communication. After beam expansion the characteristic radius, r_0, at the output aperture is 50 mm.
(a) Neglecting attenuation, calculate the power density at the center of the beam at a range of 100 km. Express the answer in nW/mm^2.
(b) Calculate the pointing accuracy required to keep the power density falling onto a small detector within a factor exp(–0.09) of the central power density, i.e. within 9%.
(c) Taking $A_T = \pi r_0^2$ and $A_R = 10^{-4}$ m^2, express the overall transmission loss in dB as the sum of the transmitter gain, the receiver gain and the basic transmission path loss.

16.5 The measurements shown in Fig. 16.5 indicate an absorption loss of about 25% over a 1.8 km length at 10.6 μm wavelength.
(a) Express this as an attenuation in dB/km.
(b) Calculate the effect of this level of attenuation on the level of received power over the 100 km path length of the example discussed in Problem 16.4.

16.6 (a) If it were possible for all the power input to a laser to be used to raise electrons from the ground state to the upper lasing level, calculate the optimum efficiency that would be obtained in terms of the energies of the upper and lower laser levels. Using Figs. 16.6 and 16.9 estimate values for these 'ideal' efficiencies for neodymium and carbon dioxide lasers.

(b) Discuss the various effects which cause the actual efficiency of lasers to fall considerably below the 'ideal' efficiency defined in (a).

16.7 The optical length of a neodymium laser cavity is 15 cm. Calculate the photon transit time. Hence calculate the approximate pulse length when cavity dumping is used, and the pulse repetition frequency to be expected when the laser is mode-locked.

16.8 Calculate the long wavelength thresholds to be expected for the photocathode materials listed in Table 16.1, and compare these with the data shown in Fig. 16.11.

16.9 Taking the solar spectral irradiance at 10.6 μm to be 0.2 W m^{-2} μm^{-1} compare the magnitudes of the background noise and quantum noise terms in eqn. (16.4.25) when a heterodyne detector operating at the quantum limit of detection is pointed directly at the sun. Take the quantum efficiency of the detector to be 0.2 and its area to be 1 mm^2. Hence comment on the statement that ends Section 16.4.

16.10 Compare the antenna gain and beam divergence figures quoted in Section 16.5.2 with the theoretical expressions obtained in Section 16.2.

REFERENCES

16.1 F. E. Goodwin, A review of operational laser communication systems, *Proc. I.E.E.E.* **58**(10), 1746–52 (1970).

16.2 R. D. Hudson, *Infrared System Engineering*, Wiley–Interscience (1969).

16.3 K. F. Sander and G. A. L. Reed, *Transmission and Propagation of Electromagnetic Waves*, C.U.P. (1978).

16.4 J. H. McElroy *et al.*, CO_2 laser communication systems for near-earth space applications, *Proc. I.E.E.E.* **65**, 221–51 (1977).

SUMMARY

Unguided optical communication systems may have a minor role to play in special applications when one of the terminals has to be mobile or difficult terrain has to be crossed. They will always have to compete with radio and microwave links, and will be susceptible to atmospheric conditions. A much wider choice of components for source and detector, and also of modulation techniques, becomes possible compared with fiber systems.

With a large, diffuse source the received power is $\Phi_R = LA_TA_R/l^2$. With a laser or a small source giving a diffraction-limited beam divergence

$$\Phi_R = \Phi_T A_T A_R/\lambda^2 l^2.$$

Possible sources are semiconductor lasers and LEDs, Nd-lasers (1.06 or 0.53 μm) and CO_2 lasers (10.6 μm), all of which operate at wavelengths where the atmosphere is relatively transparent—see Fig. 16.5. The semiconductor sources may be modulated directly, the others by means of an electro-optic cell either external to or within the laser cavity.

Photomultiplier tube detectors may be used at wavelengths shorter than about 1 μm, semiconductor junction detectors at longer wavelengths. In particular, cadmium mercury telluride may be used at 10.6 μm, where heterodyne detection becomes practical.

Several simple systems have been developed for use in the atmosphere, and an extended development study for a near-space laser communication system has demonstrated feasibility.

17

Optical Fiber Communication Systems

17.1 INTRODUCTION

It is the intention in this final chapter to give a brief review of the ways in which the various components and techniques discussed in Chapters 2–15 may be combined to form useful systems for communication and data transmission. First, it will be clear that many possible combinations of components may be conceived. The fiber may be step-index or graded-index, it may be multimode or monomode, and different glass and polymer materials may be used. The source may be a laser or an LED, the detector a p–i–n or an avalanche photodiode. Secondly, many modulation formats may be employed, both analogue and digital. At the present time all operate directly or indirectly on the optical power level emitted by the source and the most common and simplest is direct, binary, pulse-code-modulation (PCM). However, there remain many applications in which the information is both available and required in analogue form. For these the burden of analogue-to-digital and digital-to-analogue ($A \rightarrow D$ and $D \rightarrow A$) conversion may be excessive. Then, direct analogue modulation of the source power may be used, but we will show that this is unfavorable to optical systems in comparison with electrical systems. The use of a pulsed intensity modulation as a subcarrier, which in turn may be modulated in amplitude, width, position or frequency (PAM, PWM, PPM, PFM) offers some of the advantages of digital detection, whilst retaining an essentially analogue modulation format. Although other techniques employing sinusoidal subcarriers or digital subcarrier modulation have been suggested, they would seem to offer no benefits and many disadvantages, so they will not be discussed further. Finally, optical fiber systems have been shown to have application in a great range of different situations, and for many varied reasons. Bandwidth requirements may be as little as a few hertz or as much as several gigahertz, and transmission distances may be a few meters or several thousand kilometers.

Clearly it would be impossible as well as wrong to attempt to cover all combinations and all applications. Instead, in the next section we will attempt to indicate the general areas where optical fiber systems may have advantages over their competitors and to identify the critical driving factors which will lead to their adoption. Then in subsequent sections some representative examples of specific systems designed for specific applications will be used in order to illustrate the operation of some complete systems. To some extent this has been anticipated by some of the examples and power budgets given in earlier chapters. However, published results from installed and operational systems will now be used whenever possible. At the same time we will attempt to describe the range and extent of current and planned activities so that the growth of the use of optical fibers in the recent past may be followed and predictions for the future assessed.

17.2 THE ECONOMIC MERITS OF OPTICAL FIBER SYSTEMS

17.2.1 An Overview

Although in the early stages of a new technology curiosity may provide a driving force sufficient to support and encourage its development, once the level and range of its use are established only overall economic advantage will sustain it. In the case of optical fibers this economic advantage can arise in very different forms in different applications, and any valid comparison requires an evaluation of the entire system as it may be developed either with fibers or without.

In the case of long-haul, telecommunication links, for example, the cost comparison is dominated by the cost of fiber as compared to the cost of electrical cable for a given capacity. However, any cost advantage is shifted in favor of fibers by the increased repeater spacing they permit, and this is greatly enhanced if those repeaters that are required can be sited and powered within existing installations, thereby eliminating remotely powered stations. With shorter links the cost of terminal equipment, including the electrical circuitry required to drive the source and detector and to modulate and demodulate the signal, assumes much greater importance. There is, of course, no clearly defined break point between long and short links in this sense, but the changeover is likely to be somewhere in the range 1–10 km.

The importance of looking at the whole system may be illustrated by taking the example of the data transmission systems required in a sophisticated military aircraft. Merely replacing an existing electrical system with optical fibers will give little saving, if any, and the cost of the terminal equipment is likely to rise significantly. On the whole-life (say 20 years) cost of the aircraft the greatest saving will come from the weight reduction of the fiber system and the consequent reduced fuel consumption. Against this must be set the increased cost of maintenance caused by the need for more highly skilled personnel to make simple repairs to the fiber. If the aircraft is in the planning stage and can be reconfigured, even greater economies are possible. This is because the lighter weight and smaller size of the fiber system permit a smaller and lighter design. Furthermore it is possible to route the fiber system safely through areas of high electromagnetic interference or areas containing explosive materials, both of which would have been circumvented by the alternative electrical system. Of course, studies of this kind are routinely carried out by the procurement authorities. Whether or not they will lead to the general use of fiber systems depends on the reliability these demonstrate and on the confidence that may be placed on them. Also important is the facility with which they may be adapted to the requirements of the multi-access data bus which is discussed in Section 17.5.

At the most basic level the simplest cost comparison is the one between the optical fiber itself and an equivalent length of copper wire. Even this is fraught with problems because of the range of possibilities in each case and the fact that fibers have yet to reach the point where the cost benefits of the large-scale production of standardized products can be clearly established. The same is true of sources, detectors and to a lesser extent connectors. Here it should be understood that it is the growing requirement of the authorities responsible for the telecommunication networks of the world that constitute the main driving force behind the development of optical fiber systems. Only these authorities have the procurement power sufficient to produce these economies of scale. To that extent, all other fiber applications may be considered spin-offs from this. In making cost comparisons we shall express prices in U.S. dollars. It should be understood that most estimated prices have a significant margin of error on them and that the cost to any given user depends on the source of supply, the quantity required and the tolerances demanded. An effective exchange rate appropriate to these types of products should be used in making conversions to other currencies. In practice this may differ significantly from the official exchange rate because of local cost variation, transport costs, etc.

17.2.2 Telecommunications

In assessing the likely economic benefits of the use of optical fibers in telecommunications, it is helpful to separate the many different types of communication channel that exist in a telephone system into three general categories. The first consists of the lines linking individual subscribers to their local exchange, or end-office in U.S. terminology. These are normally referred to as *local loop lines*. They are invariably twisted wire pairs at present and carry a single, analogue voice channel per wire pair. In the U.K. more than 98% of these local loops are less than 5 km in length. They do, however, represent a very considerable capital investment. In the second category are links between the local exchanges. Of special importance for fibers are those in urban and suburban areas. In the U.S. these are known as metropolitan *inter-office trunks*. In the U.K. they comprise the bulk of what is known as the *junction network*. These links may take a number of different forms such as multipair or coaxial cables carrying analogue or digital signals. In the U.K. 50% of all junction routes are less than 5 km long, but within metropolitan areas in the U.K., as in the U.S. and other countries, almost all such links are less than 10 km in length. The third category of communication links comprises the inter-city trunks. For the most part these are required to be long-haul, high-capacity links. Their lengths are much more variable, extending up to 1000 km, and as well as the various types of cable, microwave-radio and satellite channels will be used competitively.

Here we will only attempt to make a comparison between electrical and optical-fiber cables used for digital transmission. Balanced wire pairs can be adapted to carry digital signals at the two lower levels of the hierarchies set out in Table 1.2, that is, at 1.5 Mb/s and 6 Mb/s in the U.S. and at 2 Mb/s and 8 Mb/s in Europe. Repeaters are required at intervals of 1 to 3 km, and are normally located at the stations prescribed in the original analogue system that used the cables. Higher levels of the hierarchy have to be carried on coaxial cables with repeaters every 1 or 2 km. Clearly, in making use of higher bit-rate levels the increased capacity of the cable has to be set against the higher cable cost, the cost of multiplexing and demultiplexing and the greater number of repeaters. The cost of repeaters increases less than proportionately with bit rate so this, together with the reducing cost of digital multiplexers, favors the maximum use of the cable bandwidth whenever the capacity is required.

In this section we will assume the costs of optical and electrical repeaters to be similar. At present this is not the case, optical repeaters being several times more expensive, especially those requiring lasers and APDs. However, there is every expectation that the greater simplicity and the reduced component count of an optical repeater will lead eventually not just to lower capital cost but to increased reliability and reduced maintenance costs as well. As order-of-magnitude estimates we might take the cost of a two-way 2 Mb/s repeater to be in the region of $100–200 and that of a two-way 140 Mb/s repeater to be in the region of $5000. The installed cost of multipair cable is in the region of $0.05 per meter per pair; that of coaxial cable in the region of $1 per meter, varying approximately with the cross-sectional area. The cost per fiber of multifiber cable might be around $1 to $2 per meter. Fiber produced in sufficient quantity by a continuous process such as the double-crucible method is clearly cheaper than graded-index or single-mode fiber which require, respectively, careful index profiling and very close control on the concentricity and diameter of the core. However, the major cost component comes from the cabling process, not from the fiber. As a result there are likely to be benefits from the use of multi-fiber cables whenever the capacity warrants. (It has been estimated that given sufficiently high volume production, say 10^4 or 10^5 km per year, the cost of producing raw graded-index and step-index fiber by a vapor deposition process might be in the region of \$100/km, and raw double-crucible fiber might be produced for as little as \$10/km. Present indications are that these estimates may prove to be optimistic.)

Cost analyses of this kind are necessarily complex. Nevertheless, there is a clear indication that the greatest cost benefit will come from the use of fibers in the high-capacity sections of the telephone system, that is, at 140 Mb/s and 274 Mb/s rather than at 1.5 Mb/s or 2 Mb/s. This is simply the result of the relative cheapness of wire pairs. With repeater costs added in, the cost of a two-way wire-pair link will be in the region of \$200/km/pair and even the cheapest fiber in repeaterless links using the simplest transmitters and receivers (LED/p–i–n) will not be competitive unless there is a major shift in the balance of costs. At the higher levels of the hierarchy the likely extra cost of the fiber cable over coaxial cable is expected to be more than offset by the saving in repeaters. These are required every 1 km or 2 km in the coaxial systems but only every 10 km or 20 km with fibers. The saving is not only in capital equipment cost but also in installation and maintenance. It is greater still if remote, power-fed repeaters can be eliminated completely. Then, metal power-supply leads are not required and a wholly nonmetallic cable may be used. This brings added benefits in electrical isolation and freedom from electromagnetic interference and ground-loop effects. A greatly simplified cost comparison is set out in Table 17.1.

Very careful cost evaluation carried out by British Telecom has confirmed the general picture we have tried to present here. Fibers are expected to have a clear cost advantage over coaxial cables at the higher levels of the digital hierarchy. There is expected to be a large increase in the rate of installation of higher-level digital links in the U.K. junction and trunk networks from 1985 onwards. It is the intention that this demand should be met with optical fiber systems. Orders for coaxial cable are gradually being phased out. The problem is that far less fiber is required for the higher levels of the hierarchy than for the lower levels. Concentrating exclusively on these will make it harder to establish production runs with the quantities required to obtain the cost advantages of large-scale production. At the second level of the hierarchy, 8 Mb/s, when the system as a whole is considered no very clear cost advantage is apparent either way. Nevertheless, a considerable length of fiber cable is being installed in the junction network for use at this level (more than 1000 km in 1984), the more efficient use of existing ducts illustrated in Fig. 1.6 being a major motivating factor. Although at 2 Mb/s cost is expected to favor the continued use of wire pairs, where links exceed 7 km in length and when it seems likely that future

Table 17.1 Comparative costs of electrical and optical digital communication links

			Electrical			Optical
Level 1 System (2 Mb/s, 30 circuits)						
Cost per km of						
two wire-pairs or of two fibers			$100			$2000
Cost of two-way repeater / Repeater spacing	$200 / 2 km	=	$100			none required
Total cost per km			$200			$2000
Level 4 System (140 Mb/s, 1920 circuits)						
Cost per km of						
two coaxial lines or of two fibers			$2000			$2000
Cost of two-way repeater / Repeater spacing	$5000 / 2 km	=	$2500	$5000 / 10 km	=	$500
Total cost per km			$4500			$2500

The quoted costs should be understood to represent order-of-magnitude estimates only. They are based on a single two-way circuit and neglect all costs common to both systems such as ducting, housings, installation and maintenance, and all terminal costs including multiplexors and de-multiplexors.

What is made clear is that the advantage of optical fibers at the higher bit rate derives primarily from the reduced repeater spacing. The advantage would hold even if the optical repeater were considerably more costly, but would be lost if the cost of fiber exceeded the cost of high quality coaxial cable by a factor of more than about two.

system up-grading will be required, fibers may be installed, even at this first level. Similarly in the trunk network some fiber systems will be installed at the third level (34 Mb/s) with a view to upgrading at a future date. In summary, the main British Telecom effort is concentrated at 140 Mb/s for the trunk system and at 8 Mb/s for the junction network. One important feature of the trunk system is the existing distribution of power feeder stations, many of which are sited about 30 km apart. An economic optical fiber system which would give a repeater spacing of 30 km at 140 Mb/s is therefore an important design goal for the second generation of fiber systems operating at 1.3 or 1.55 μm.

In the U.S., the Bell System foresees a large requirement for metropolitan inter-office trunks at the third level (T3) of the hierarchy, operating at 44.7 Mb/s. One of its principal efforts has been directed at producing repeaterless links for this service. The required bit rate × distance product can be obtained using the first generation of fiber systems (graded-index fiber with GaAs sources, $\lambda = 0.825$ μm), but the bit rate is high enough to permit the fiber system to 'prove in' over electrical systems.

An application that will demand the ultimate performance from second-generation optical fiber systems is that of undersea telecommunications. Again the types of link that will be required are many and varied, but one of the most significant that is occupying much attention is a fiber-optic transatlantic cable. In general undersea links require the longest repeater spacing that can be achieved and the highest capacity consistent with this. But the most important requirement of all is complete reliability. The most recent transatlantic cables are analogue and carry 4400 or 5500, two-way voice channels. For digital transmission two electrical cables are required, one for each direction of transmission, and the cost per circuit at present would be greater by a factor of about three. A two- or four-fiber optical fiber cable operating at the fourth level of the digital hierarchy (274 Mb/s or 140 Mb/s) would have a roughly equivalent capacity. Analogue electrical channels of this capacity require repeaters every 5–10 km. If the repeater distance could be extended to 50 km or more using long-wavelength optical fiber systems, then a significant cost advantage would be expected. Cables in general are, of course, in competition with satellite links. They are, however, cheaper per channel at present by a factor of about three, and they have the great advantage that the two countries connected retain complete political control over the link.

17.2.3 Local Distribution Services

Compared to their use in the trunk network, the rate of introduction of optical fiber systems into local loop lines is much harder to foresee. These are almost universally 3 kHz analogue voice channels carried on wire pairs. With a suitable modem low bit rate (1200 b/s) data can be sent over these lines without further modification. Only the widespread demand for many more wide-band services such as switched cable TV, videophone, teleconferencing or high-speed data transmission will cause fibers to be introduced into the subscriber's premises. Existing wire pairs would be inadequate for this traffic and fibers would be used in preference to coaxial cables. What is at present totally uncertain is the way in which the curve of demand versus price for such services may develop in the future.

Services such as teleconferencing can be provided locally and temporarily at present by means of microwave radio links to the local exchange. It may be that the merits of using this facility rather than travelling to meetings will lead to a demand for more permanent arrangements. In several areas extensive coaxial cable networks supplying television independent of the telephone system have been established for many years. For the most part these have given way to a fuller broadcasting coverage. It is possible that there may be a demand for a much wider range of television and audio program material of a higher quality than can be obtained from the broadcast services. Such services could be provided entirely by cable either together with, or independent of, the fully switched telephone network, or alternatively could be received at a community aerial by satellite transmission and relayed to subscribers by cable (CATV). Again optical fibers are likely to be competitive whenever coaxial cable has to be used.

As the main telephone system becomes increasingly digital, a key question concerns the optimum point at which to convert the analogue signals such as voice or video being sent into the digital form in which they are transmitted and switched. With the present distribution of traffic this interface would occur, if at all, at one of the exchanges on its route. With a fully digital system this would be the local exchange (end office). An expansion of wide-band services may make it desirable to shift the analogue–digital interface either into the subscriber's premises or to a more local distribution point offering a very limited switching capability. Another possibility is that video services, CATV for example, might be distributed on a fiber system using analogue modulation. The technical trade-offs involved in this decision will be discussed in Section 17.4.

17.2.4 Local Data Transmission and Telemetry

The type of information transfer we are concerned with here is that required between the various parts of a manufacturing complex, or that required within a building, a ship, an aircraft or a land vehicle. The transmission distance does not normally exceed 1 km and in many cases the frequency (bit rate) does not exceed 10 MHz (10 Mb/s). Thus a simple wire pair is often quite adequate and even the cheapest type of optical fiber system is unlikely to be competitive on a straight cost comparison. However, the example given in Section 17.2.1 showed that optical fibers might still be favored when weight or size is at a premium. There are other situations where electrical links would be expensive or impossible and where fibers may compete. These include: links between high voltage units where fibers would eliminate the need for transformers; links through areas containing explosive or inflammable materials; transmission through regions of high electromagnetic interference; links where electromagnetic compatibility is a problem. Cost is still at a premium so not only the fiber but also the terminal equipment must be cheap and simple. Thus, large diameter, high numerical aperture, step-index fiber will be favored; and polymer-clad silica (PCS) and double-crucible fiber will be the main contenders, along with LED sources and p–i–n photodiode detectors.

There are two areas which may benefit greatly from the ready availability of cheap optical fiber. The first is instrumentation, where inherently optical, and sometimes inherently digital, transducers may be developed often using the fiber itself as a sensing element. Examples might be the direct use of the fiber as a strain-gauge, or as part of an optical shaft position sensor. The second is the fast transfer of data within computer systems. The use of a parallel multiwire data bus is limited by electromagnetic compatibility problems to data rates of less than about 10 Mwords/s. A single fiber might replace a 4-, 8-, 16- or 32-word parallel data-bus with the information being transmitted in serial form at a correspondingly higher bit rate. The parallel-to-serial conversion would be provided on-chip. The ability to transfer data in this way between the central processor units (CPUs) of a computer, or from CPU to memory, may facilitate developments in distributed computer systems that would be difficult to manage otherwise.

We must emphasize again that economic advantage cannot be assessed properly simply by thinking in terms of the replacement of wires by fiber. The potential impact of fiber technology on the overall system design has to be evaluated.

17.3 OPTICAL FIBER DIGITAL TELECOMMUNICATION SYSTEMS

17.3.1 First-generation Systems

In the context of optical fiber systems for telecommunication a 'first-generation' system is one using multimode, graded-index fiber, a GaAs laser or LED source and a Si–APD detector. In many countries such systems have been installed on a trial basis as fully integrated links within the telephone system. That is, the fiber cables run through conventional telephone ducts, or are 'ploughed-in', or carried overhead on poles where this is normal practice. Many such links routinely carry live traffic. Here we will give just a few examples of such systems and the level of performance that is expected. In this working situation the system specification has to be set conservatively to allow for tolerances and component variation and for environmental degradation and 'end-of-life' performance levels. Thus, 'worst case' parameters have to be set, rather than the 'best ever' performance that is often quoted as 'typical' in a laboratory test. Nevertheless, we might reasonably expect by today's standards that the installed attenuation of the best graded-index fibers should be in the region of 3 dB/km and the r.m.s. spread caused by multipath dispersion should be no more than 0.2 ns/km. At 0.85 μm, the material dispersion parameter for doped silica is $Y_m = 0.025$. Thus, using a laser source with $\gamma_\sigma = 0.0015$, we should expect the r.m.s. material dispersion to be 0.125 ns/km, giving a total dispersion of 0.23 ns/km and a bit rate $\times$ distance product of about 1 (Gb/s)km. With an LED source $\gamma_\sigma = 0.015$ and material dispersion limits the bit rate $\times$ distance product to about 200 (Mb/s)km, if dispersion penalties are to be avoided.

In the U.S.A., Bell Laboratories have developed a system based on the third level of the U.S. digital hierarchy (DS3; 44.7 Mb/s) for manufacture by Western Electric. This is known as the FT3 Lightwave Digital Transmission System and it is intended primarily for trunk use in metropolitan areas. Following successful trial installations at Atlanta, Georgia in 1976 and in Chicago in 1977, the system is based on the following components: The graded-index fiber has a 50 μm core diameter and a 125 μm cladding diameter. It is cabled using the ribbon structure described in Section 4.2 with 12 fibers per ribbon and up to 12 ribbons per cable. A ribbon connector based on a silicon chip with etched V-grooves may be provided in the factory, otherwise individual fibers may be connected using a biconic, molded-plastic connector based on the one shown in Fig. 4.10. A GaAlAs laser source would normally be used with the bias level controlled by monitoring the back emission, as described in Section 11.2.

At least 0.5 mW of optical power at 0.825 μm may be launched onto the fiber. The detector is a standard Si–APD. Unrestricted binary coding of the transmitted information is used. A power budget is set out in Table 17.2. If the maximum acceptable fiber loss is set at 5 dB/km and a further 1 dB/km is allowed for splices (that is, 2 per km each giving a loss of 0.5 dB), the 39.5 dB power margin permits a repeater spacing of 6.6 km. With improved fiber production techniques it is likely that a maximum loss tolerance of 4 dB/km for

Table 17.2 Power budget for the FT3 system

		Laser source		LED source
Average optical power launched onto fiber		–3.0 dBm		–20 dBm
Minimum received power for required error rate		–48.5 dBm		–50 dBm
Power margin		45.5 dB		30 dB
End connector losses	3 dB		3 dB	
Dispersion penalty	0 dB		4 dB	
Unallocated margin	3 dB		3 dB	
		6.0 dB		10 dB
Allowance for installed cable loss		39.5 dB		20 dB

installed, spliced cable might be imposed without causing the cost of fiber to become excessive. This would extend the repeater spacing to 10 km. To reduce multipath dispersion to the required levels, well-graded fibers will be needed, but again it would seem that the tolerances necessary can be met at reasonable cost. At 6.6 km the bit rate × distance product is 300 (Mb/s)km, while at 10 km it becomes 450 (Mb/s)km. In neither case should multipath dispersion be a problem, nor with a laser source should material dispersion. If an LED source were used, as it might be on shorter links, a dispersion penalty may have to be included in the power budget. In the table we have set this at 4 dB and find that an overall loss of 6 dB/km would give a repeater spacing of 3.3 km (170 (Mb/s)km) while 4 dB/km would permit 5 km (220 (Mb/s)km).

Figure 17.1 shows the route of a $27\frac{1}{2}$ mile (44 km) FT3 system in New York

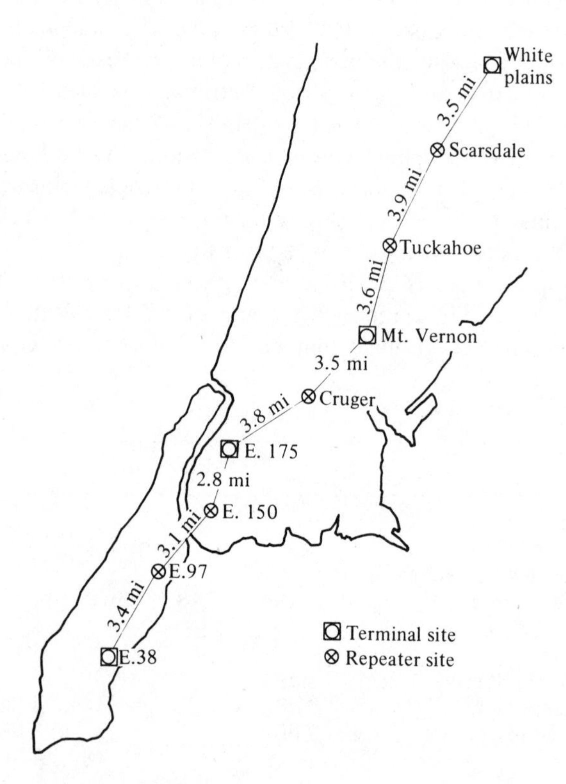

Fig. 17.1 The route of a planned FT3 lightwave system for New York City. [Taken from I. Jacobs and J. R. Stauffer, FT3—a metropolitan trunk lightwave system, *Proc. IEEE* **68**, 1286–90; © 1980 IEEE.]

City which will link White Plains to Manhattan. This is one of several similar systems being installed. In addition an inter-city trunk based on FT3 is planned to link Boston and Washington.

In the U.K. two trials of 140 Mb/s optical fiber systems were installed in standard telephone ducts in 1977. The routes were chosen as typical of the conditions likely to be met in general service. One link was set up over 9 km with two intermediate repeaters, the other was a single link of 5.75 km. Both trials used DH–GaAlAs stripe geometry laser sources, graded-index fibers and Si–APD detectors, but the components came from different sources. There were differences in the types of receiver amplifier chosen, and different binary codes were used for the purpose of limiting the maximum number of consecutive 0s and 1s that could be sent. The 9 km trial (STL/STC, Hitchin to Stevenage) used a transimpedance-feedback amplifier based on a Si-bipolar junction transistor. One parity bit was introduced after every 17 signal bits, so increasing the optical signalling rate from 139.264 to 147.456 Mb/s. Back-to-back receiver sensitivity (that is with no dispersion penalty) was –49 dBm. Over a 3 km installed link it became –46 dBm. The 5.75 km trial (BPO, Martlesham to Kesgrave) used a GaAs MESFET integrating amplifier (10 kΩ, 6 pF at input). A 7B–8B balanced disparity code was used and this increased the optical signalling rate to 159.159 Mb/s. Receiver sensitivity was better than –43 dBm. Fibers available at the time of these trials showed loss levels between 3 and 5.5 dB/km. The dispersion varied even more widely. Subsequent manufacturing improvements have reduced the average attenuation and dispersion and reduced the fiber-to-fiber variations.

On the basis of these trials the power budget of Table 17.3 may be drawn up. At the time of the trials overall losses of about 5 dB/km in the installed, spliced cables allowed satisfactory operation over unrepeatered spans of 6–7 km. This

Table 17.3 Power budget for a 140 Mb/s system using a DH laser source at 0.82–0.83 μm and a Si-APD detector

Average optical power launched onto fiber		–3 dBm
Minimum received power for required error-rate		–43 dBm
Power margin		40 dB
End connector losses	3 dB	
Unallocated margin	3 dB	
		6 dB
Allowance for installed cable loss		34 dB

was established experimentally by extending the 5.75 km link using fiber wound on drums at one of the terminals, and by omitting one of the repeaters of the 9 km link. The power budget shows that a 10 km link would require the attenuation of installed spliced fiber to be below 3.4 dB/km, and its dispersion would have to be better than 1.6 (Gb/s)km. In practice such systems are likely to be limited by multipath dispersion to unrepeatered links of 8–10 km.

In parallel with their 140 Mb/s trial the British Post Office (BPO, now British Telecom) also ran a trial 8 Mb/s system. This used the same type of cabled fiber and the same ducts except that the link was extended a further 7.25 km to the Group Switching Centre in Ipswich, giving a total length of 13 km. Both LED and laser sources were used, emitting at 0.82 and 0.86 μm, respectively. At the receiver a Si–APD was followed by a high input impedance (1 MΩ, 6 pF) amplifier. The back-to-back sensitivity was −61.5 dBm with the LED source and −60.3 dBm with the laser which had an extinction ratio of 0.1. The power budgets shown in Table 17.4 indicate that fiber with an installed loss of 4.4 dB/km permits unrepeatered link lengths of 10 km using LED sources and 12 km with lasers. The dispersion requirements of 80 and 100 (Mb/s)km are well within reasonable tolerances on index grading. With the LED source a dispersion penalty may occur if the spectral width is excessive.

Following the success of these and other trials, British Telecom ordered a further 10,000 km of fiber in 1979 and 1981, mainly in the form of 8-fiber cables. These were for installation in over 100 separate links in various parts of the telephone system, operating at 8, 34 and 140 Mb/s. Included in this program are some systems working at 1.3 μm and some which use monomode fiber. These may properly be called second-generation systems, which we will consider in the next section.

All first-generation telecommunication systems have been based on the use of APD detectors and most have required laser sources. Although the substitu-

Table 17.4 Power budgets for 8 Mb/s systems using lasers and LEDs

	Laser		LED
Average optical power launched onto fiber	0 dBm		−12 dBm
Minimum received power for acceptable error-rate	−60 dBm		−62 dBm
Power margin	60 dB		50 dB
End connector losses		3 dB	
Unallocated margin		3 dB	
	6 dB		6 dB
Allowed cable loss	54 dB		44 dB

Table 17.5 Power budget for a 34 Mb/s optical fiber system using LED sources at 0.9 μm

Power launched onto fiber		−15 dBm
Minimum received power requirement		−50 dBm
Power margin		35 dB
Connector losses	4 dB	
Dispersion penalty	5 dB	
Unallocated margin	3 dB	
		12 dB
Net margin for cable losses		23 dB

tion of LEDs and p–i–n photodiodes would give a cheaper, simpler and more reliable system, the loss of available power margin is 10–20 dB with an LED and a further 10–20 dB with a p–i–n photodiode. The transmission distance then becomes critically dependent on fiber attenuation and is otherwise limited at higher bit rates by material dispersion. A 34 Mb/s LED link over an unrepeatered distance of 7.8 km has been established as part of a more extensive trial in Rome, Italy. This has made use of the reduced attenuation and material dispersion at longer wavelengths by using a GaAlAs LED emitting at 0.9 μm with a wavelength spread of 36.5 nm. The dispersion-free sensitivity of the Si–APD/transimpedance amplifier combination was −50 dBm. The power budget shown in Table 17.5 indicates that an installed, spliced fiber loss of less than 3 dB/km was needed in order to establish the 7.8 km link.

17.3.2 Second-generation Systems

Second-generation optical fiber systems fall into two categories: those using multimode fiber and operating at the material dispersion minimum around 1.3 μm, and those using monomode fiber at one of the wavelengths of minimum attenuation. Both offer prospects for links significantly longer than those of first-generation systems, and this fact has spurred the development of long-wavelength sources and detectors and low-loss fiber. Such links will be of particular value in the long-haul parts of trunk telephone systems and in submarine applications. Here we will look briefly at some projected specifications for long-wavelength systems and at some experimental results obtained under laboratory conditions.

Figure 17.2 illustrates the power margins available in systems using different types of source and detector, and it summarizes much that has been said already. Present indications are that the power available from long-wavelength

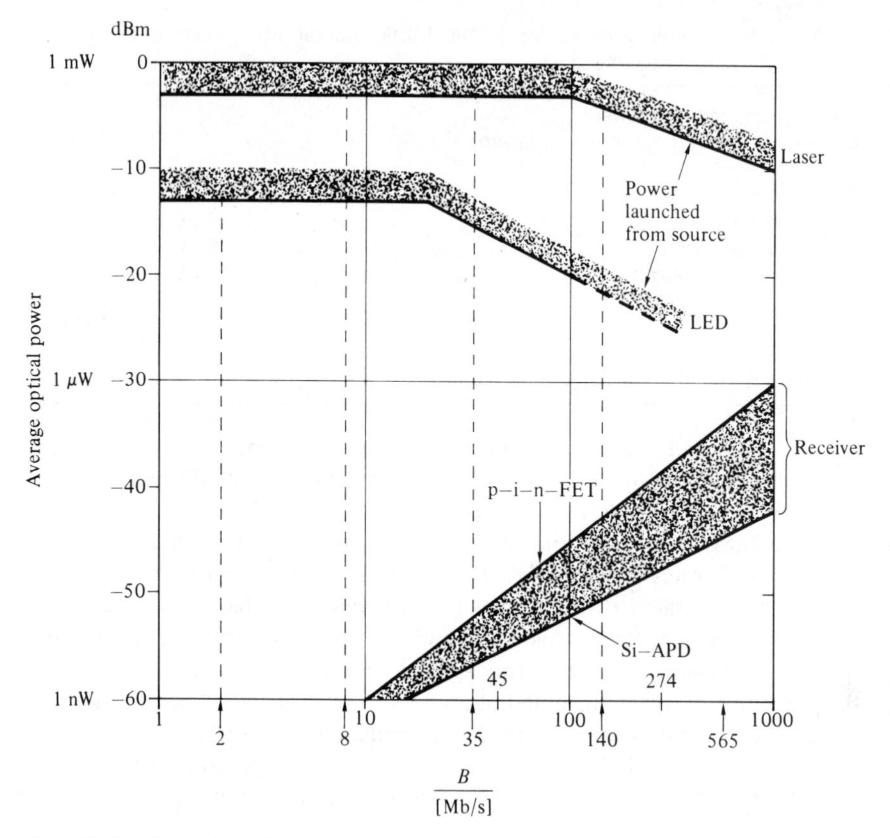

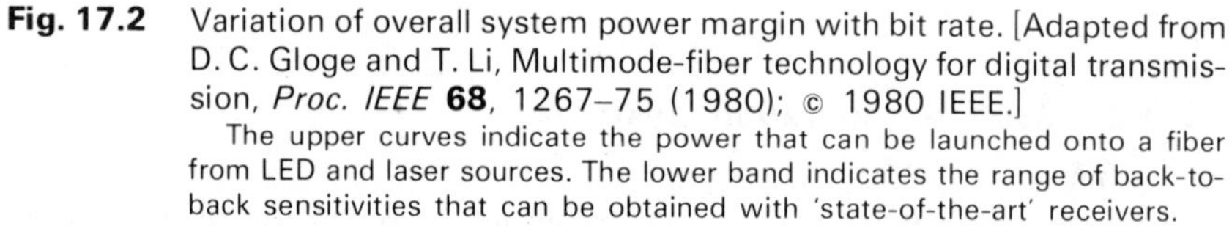

Fig. 17.2 Variation of overall system power margin with bit rate. [Adapted from D. C. Gloge and T. Li, Multimode-fiber technology for digital transmission, *Proc. IEEE* **68**, 1267–75 (1980); © 1980 IEEE.]

The upper curves indicate the power that can be launched onto a fiber from LED and laser sources. The lower band indicates the range of back-to-back sensitivities that can be obtained with 'state-of-the-art' receivers.

LEDs and lasers will be similar to that from GaAlAs sources, and the power that can be launched onto a monomode fiber is expected to be only marginally less than the power that can be launched from a laser onto a multimode fiber. The lower boundary of the range of receiver sensitivities shown in the figure is intended to represent the performance of low-noise amplifiers with Si–APD detectors. The sensitivity is then shot-noise limited and thus proportional to the bit rate. The upper boundary represents the performance of receivers using p–i–n photodiodes together with low-noise FET amplifiers. Here sensitivity at higher frequencies is proportional to $B^{3/2}$ and is highly dependent on the amplifier noise and on the total input capacitance. From the total power margin shown in the figure, about 10 dB should be subtracted to allow for connector losses, dispersion penalties, temperature variations and degradation during life, as well as providing a small (~3 dB) unallocated margin.

The success of long-wavelength, multimode systems will depend crucially on the ability to produce, at reasonable cost, precisely graded fiber having minimal multipath dispersion. The merit of such systems lies in the fact that a high performance but cheap, simple and reliable link can be set up using an LED source and a p–i–n photodiode detector. Furthermore, multimode fibers are easier to splice and interconnect, and to connect to other components than are monomode fibers. The use of laser sources would enhance the bit rates and transmission distances obtainable, although modal noise then becomes a problem. The benefit of using long-wavelength APDs is more problematical. At present they suffer from high dark current in the avalanche region and from high noise factors. They therefore offer little, if any, advantage over p–i–n photodiodes at longer wavelengths.

Fibers have been produced with attenuation levels at 1.3 μm as low as 0.5 dB/km. However, after cabling, splicing and installation it would be as well to allow at least 1.0 dB/km. The resulting attenuation-limited repeater spacings are shown in Table 17.6 for the various levels of the American and European digital hierarchies. At the dispersion minimum, LED sources typically have an r.m.s. spectral spread of $\sigma_\lambda = 40$–50 nm. This leads to a bandwidth × distance product of about 3–4 (Gb/s)km. This should not be a limitation on the requirements indicated in Table 17.6. It may be compared with the material dispersion typical for GaAlAs LEDs emitting at 0.85 μm with r.m.s. spectral spreads of $\sigma_\lambda = 15$–20 nm which give a limit of about 200 (Mb/s)km. Thus at 1.3 μm it

Table 17.6 Repeater spacings and corresponding bit rate × distance products for LED/p–i–n links with 1 dB/km fiber loss

Bit rate [Mb/s]	Overall power margin [dB]	Net power margin [dB]	=	Repeater spacing [km]	Bit rate × distance [(Mb/s) · km]
8	48		38		304
35	37		27		945
45	34		24		1080
140	22		12		1680
274	14		4*		1096*

* These values are particularly sensitive to the exact power margin available, and this must be uncertain at present to at least ±2 dB. This represents ±2 km on the repeater spacings.

will be the multipath dispersion determined by the quality of the index grading that dominates the fiber bandwidth and that may lower the limit on the repeater spacing. Bandwidths of 2 (Gb/s)km have been produced in selected fiber samples, but 1 (Gb/s)km may be a more realistic target specification. This will not be a limitation at the second and third hierarchy levels, up to 45 Mb/s, but may be at higher levels. It is clear that the target U.K. repeater spacing of 30 km can easily be reached at 8 Mb/s, may well be achieved at 35 Mb/s, but that 140 Mb/s systems will probably require two intermediate repeaters for a 30 km link. It remains an open question whether or not this will be preferable to seeking an unrepeatered link with a monomode system.

Monomode systems will find application for high-capacity, long-haul, trunk routes and for underwater links. There are three possible systems to be considered: one operating at the material dispersion minimum at 1.3 μm; one using a fiber in which the material + waveguide dispersion minimum is shifted to the attenuation minimum at 1.55 μm or thereabouts; and one operating at the attenuation minimum with a frequency-stabilized laser source. At present it would appear that from a system point of view the first two may be considered together because, as shown in Section 5.5, in order to move the dispersion minimum to 1.55 μm the fiber must have a small core diameter and a large index difference. In consequence minimum fiber losses are raised to about 0.4 dB/km,† a level which can be obtained at 1.3 μm in fibers having little or no waveguide dispersion. A target specification for either type of fiber might be 1 dB/km attenuation, cabled, spliced and installed, with 8 dB allocated to the various system overheads. Dispersion would be less than 10 ps/km, giving a bit rate × distance product in excess of 50 (Gb/s)km, indeed 250 (Gb/s)km has been claimed. The repeater spacings that should be possible in such systems are listed in Table 17.7, and they indicate that dispersion should not be a limita-

Table 17.7 Repeater spacings and corresponding bit rate × distance products for zero dispersion, monomode fibers having 1 dB/km attenuation

Bit rate [Mb/s]	Overall power margin [dB]	Net power margin [dB] = Repeater spacing [km]	Bit rate × distance [Gb/s]
35	49	41	1.4
45	47	39	1.8
140	39	31	4.3
274	32	24	6.6
565	26	18	10.2

† It may be possible to reduce this loss by a more gradual variation of the core–cladding index profile, for example by using a triangular profile, $\alpha = 1$.

tion. Significantly better results than these have been obtained in laboratory experiments, as is shown by entries (1), (2) and (4) in Table 17.8.

The third approach to monomode systems would seek to exploit the attenuation minimum between 1.5 μm and 1.7 μm where fiber losses as low as 0.2 dB/km are possible. The target specification for installed fiber might then be reduced to 0.5 dB/km. As was shown in Section 5.5, the total material dispersion parameter would be expected to lie in the range $0.005 < (Y_{TOT}) < 0.01$, so that with the normal spectral spread from a laser source, dispersion would limit the fiber to 5–10 (Gb/s)km. However, if the laser can be constrained to operate in a stable, single longitudinal mode throughout each pulse, this limit would be increased by one or two orders of magnitude. Attempts to ensure this have included the use of distributed-feedback laser structures, and the use of external optical feedback to control the lasing mode. If the dispersion limitation can be avoided in this way, the reduced attenuation will lead to repeater spacings twice as long as those given in Table 17.7. Entries (3) and (5) in Table 17.8 summarize results from recent laboratory experiments in which unrepeatered transmission over more than 100 km was achieved. An unrepeatered, monomode link of 27 km is to be installed as part of the British Telecom fiber development program.

17.3.3 Other Users

In the discussion so far we have concentrated entirely on the needs and plans of the authorities responsible for the main, fully switched, telephone network. But it is important to remember that other users may have telecommunications requirements, which for various reasons may not suit the public network, but may suit the use of optical fibers. Banks and other financial institutions may well wish to set up their own data transmission network for reasons of independence and security. Likewise the broadcasting authorities may wish to retain control over the transfer of their program material between studio and transmitter. In the U.K. both groups have been required by law to use the public, common-carrier network and there is nothing in their technical requirements that makes this unsatisfactory. However, if it were legally and economically possible for such bodies to set up their own network of links, it is likely that they would do so.

Two other public utilities have more technical reasons for seeking independence from the common-carrier telephone network. These are the electricity distribution authorities and the railways. We might note in passing that in the nineteenth century it was the need for safer railways that acted as a great spur to the development of the electric telegraph. Both these bodies depend entirely on the rapid and reliable transfer of information over long distances in a hostile electrical environment for the satisfactory operation of

Table 17.8 Results of laboratory tests on transmission through monomode fibers

Note*	Wave-length [μm]	Bit rate [Gb/s]	Fiber length *l*/[km]	Total loss [dB]	Launched power [dBm]	Received power [dBm]	Measured power/[dBm] for PE = 10^{-9} Back to back	After *l*	Power margin [dB]
(1)	1.55	2.0 0.4 (RZ)	51.5	27.7	−2.2	−29.9	−32 −39.5	−31.4	1.5 (∼9.6)
(2)	1.303	2.0 0.4 (RZ)	44.3	25.3	−1.2	−26.5	−31.9 −38.5	−29.4	2.9 (∼12)
(3)	1.52	0.140	102	34	−8	−42	−45.7	−43.1	1.1
(4)	1.303	0.420 0.274 (NRZ)	84 101	35.7 40.1	3.5 4.1	−32.2 −36.0	−33.2 −37.0	−33.2 −37.0	1.0 1.0
(5)	1.536	0.400 (RZ)	104	35	−4	−39	−39.4	−39.6 (!)	0.6
(6)	1.55	1.0 (NRZ)	100.9	27.4	1.6	−25.8	−28.2	−27.1	1.3

In (1), (2) and (4) the wavelength was close to the fiber dispersion minimum.

* See following page.

(1) Data taken from J. Yamada, A. Kawana, H. Nagai, T. Kimura and T. Miya, 1.55 μm optical transmission experiments at 2 Gbit/s using 51.5 km dispersion-free fibre, *Ets. Letters*, **18**(2), 98–100 (1982).

Fiber core diameter 4.5 μm and $\Delta = 0.007$ gave dispersion minimum at 1.55 μm. Two splices. Microlens source–fiber coupler. Ge–APD reach-through detector ($\eta = 0.8$, $x = 0.83$) feeding Si-bipolar amplifier. $Bl = 103$ (Gb/s)km.

(2) Data taken from J. I. Yamada, S. Machida and T. Kimura, 2 Gbit/s optical transmission experiments at 1.3 μm with 44 km single-mode fibre, *Ets. Letters*, **17**(13), 479–480 (1981).

Fiber core diameter 10.8 μm, $\Delta = 0.0021$. About 8 splices. Hemispherical microlens gave 6 dB increase in power launched compared with butt joint. Ge–APD ($\eta = 0.6$, $x = 0.95$) with Si-bipolar amplifier, $R_{in} = 50\ \Omega$. With $\Delta\lambda = 1.5$ nm dispersion was negligible and power penalty was due to nonzero extinction ratio. $Bl = 88.6$ (Gb/s)km.

(3) Data taken from D. J. Malyon and A. P. McDonna, 102 km unrepeated monomode fibre system experiment at 140 Mb/s with an injection locked 1.52 μm laser transmitter, *Ets. Letters*, **18**(11), 445–447 (1982).

The linewidth of the injection-locked laser was 0.5 nm. Fiber loss was estimated to be 0.31 dB/km; splice loss 0.25 dB/splice for 10 splices. The measured power penalty was ascribed to dispersion (1.6 dB) and to the nonzero extinction ratio (1.0 dB). The hybrid p–i–n FET receiver comprised an InGaAs/InP p–i–n photodiode and a GaAs FET. $Bl = 14.3$ (Gb/s)km.

(4) Data taken from M. M. Boenke, R. E. Wagner and D. J. Will, Transmission experiments through 101 km and 84 km of single-mode fibre at 274 Mb/s and 420 Mb/s, *Ets. Letters* **18**(21), 897–898 (1982).

W-profile MCVD fiber was used in these experiments, the 101 km length having 12 fusion and 3 epoxy splices. A buried heterostructure InGaAsP/InP laser and an InGaAs p–i–n photodiode were used. $Bl = 35.3$ (Gb/s) km.

(5) Data taken from K. Iwashita, K. Nakagawa, T. Matsuoka and M. Nakahara, 400 Mb/s transmission test using a 1.53 μm DFB laser diode and 104 km single-mode fibre, *Ets. Letters* **18**(22), 937–938, (1982).

VAD fiber, core diameter 10 μm, $\Delta = 0.0025$, was used. It had 9 fusion splices. The buried heterostructure, distributed feedback InGaAsP/InP laser diode operated in a single longitudinal mode. The Ge reach-through APD fed a Si bipolar transimpedance amplifier. Chromatic dispersion of −15 ps/(km . nm) was offset by pulse compression because the optical wavelength became shorter during the pulse (chirping). $Bl = 41.6$ (Gb/s) km.

(6) Data taken from R. A. Linke, B. L. Kasper, J-S. Ko, I. P. Kaminow and R. S. Vodhanel, 1 Gbit/s transmission experiment over 101 km of single-mode fibre using a 1.55 μm ridge guide C^3 laser, *Ets Letters*, **19**(9), 775–776 (1983).

The fiber dispersion was 17.5 ps/(km . nm) and its loss includes 2.9 dB from 20 epoxy splices. The modulated line width of the two-section, ridge waveguide laser was about 0.1 nm. An InGaAs mesa p–i–n photodiode drove a GaAs FET with $R = 20\ k\Omega$, $C = 1.2$ pF. High extinction ratio ($r_e = 0.28$) and nonoptimal filtering introduced power penalties ($\sim 2.8 + 1.2 = 4.0$ dB). $Bl = 101$ (Gb/s) km.

their respective systems. Both have actively experimented with optical fibers from the beginning. The track side of an electrified railway is not only subject to considerable electromagnetic interference and earth-loop problems, but is also subject to extreme temperature variations. In the case of electrical power distribution, the transmission lines form a natural path for a communication link but again the electrical isolation and the freedom from interference is a major attraction of an optical fiber link. Japanese power companies have established a number of optical fiber links for power system protection, for supervision and control, and for transferring information between computers. The links are up to 10 km in length with up to 32 Mb/s information capacity, and more are planned. Experimental links have been set up in the U.K. in which the fiber cable is either suspended from or else contained inside, the earth return of balanced, six-phase power lines. Here the ability to withstand the mechanical and vibrational stresses imposed is likely to be important. Although neither the power nor the railway authorities would be able to justify the development of fiber systems on their own, both should benefit from their more general availability.

17.4 ANALOGUE SYSTEMS

17.4.1 Advantages and Disadvantages of Analogue Modulation

We have seen that optical communication systems may be designed in which the attenuation is very low (< 1 dB/km) and the modulation bandwidth is very high (> 1 GHz km). What has not been made clear specifically is that compared with electrical transmission systems the overall power margin available in an optical system is much smaller. While this is offset by the lower transmission losses, the benefits of the optical system are greatly reduced when a high received signal-to-noise ratio (K) is required, because the extra signal power required at the receiver eats into the available loss margin. One of the features of binary pulse code modulation (PCM) is that an adequately small error probability may be obtained with a relatively low signal-to-noise ratio at the receiver. According to the theory of Section 15.3, $PE = 10^{-9}$ requires $K = 12$ (21.6 dB). The dynamic range of the encoded analogue signal, which in many cases is required to be as high as 50 or 60 dB, is determined by the number of bits per sample, and this is reflected in the bandwidth required for the PCM signal. With direct analogue transmission at baseband, the dynamic range would normally be determined by the signal-to-noise ratio at the receiver, which would therefore normally have to be much more than 21.6 dB. Thus the potential benefits of optical fiber systems are likely to be greatest when intensity-modulated binary PCM signals are to be transmitted, and are likely to be significantly degraded if direct, baseband analogue intensity modulation is required. Many users will, nevertheless, wish to transmit analogue signals in analogue form, not least to avoid the expense and complexity of providing digital encoders and decoders at the terminals. For them the use of a pulsed intensity modulation, which is used as a subcarrier and may easily be further modulated in frequency (PFM) or in pulse position (PPM), offers a compromise between the two extremes. The most common requirements for analogue fiber transmission are in simple telemetry and for the distribution of television signals. Before discussing specific examples we will examine the available power margins in optical and electrical systems in a little more detail. For this purpose we will choose systems designed to have a signal bandwidth of 100 MHz.

We have seen that the power that can reasonably be launched onto a fiber with 50 μm core diameter is in the region of $\Phi_T = 1$ mW (0 dBm). With an LED source it would be an order of magnitude less; with a larger core diameter it could be more. We have also seen in Section 14.4 that the quantum noise limit of an ideal optical receiver of bandwidth Δf is represented by

$$\Phi_R = 2(\mathcal{E}_{ph}/\eta)FK^2 \, \Delta f \qquad (14.4.14)$$

where Φ_R is the received optical power required to ensure a signal-to-noise ratio K, with a photon energy, $\mathcal{E}_{ph}$, a detector quantum efficiency η and noise factor F. For the ideal case when $\eta = F = 1$, this becomes

$$\Phi_R = 2\mathcal{E}_{ph}K^2 \, \Delta f \qquad (17.4.1)$$

We will define the overall power margin as the ratio Φ_T/Φ_R when $K = 1$. Then with an optical wavelength of 1 μm ($\mathcal{E}_{ph} = 1.24$ eV) and $\Delta f = 100$ MHz, $\Phi_R = 2\mathcal{E}_{ph}\,\Delta f = 40$ pW (−74 dBm), giving an overall power margin of 74 dB_{opt}. In a practical system with this bandwidth the additional noise introduced in the detector or the receiver amplifier may reduce this overall power margin by 10 or 20 dB.

In a 100 MHz electrical system a high-frequency power transistor might be expected to launch a 100 mW (+20 dBm) signal onto a 50 Ω line (2.2 V r.m.s.) with reasonable linearity. This could be increased by 10 or 20 dB if a vacuum transmitting tube were used. At the electrical receiver the noise power at the input of an ideal amplifier would be $kT\,\Delta f$, where k is Boltzmann's constant and T the absolute temperature. With $T = 300$ K and $\Delta f = 100$ MHz this is 0.4 pW (−94 dBm) giving an overall system power margin of 114 dB. A practical 100 MHz amplifier should have a noise figure of no more than a few decibels. Allowing 10 dB, the overall margin is reduced to 104 dB. Note that in both cases the effect of the noise is proportional to Δf. This means that although the absolute values of the power margin depend on the channel bandwidth, the relative advantage of the electrical system does not.

A consequence of these comparisons is illustrated in Fig. 17.3, which is a

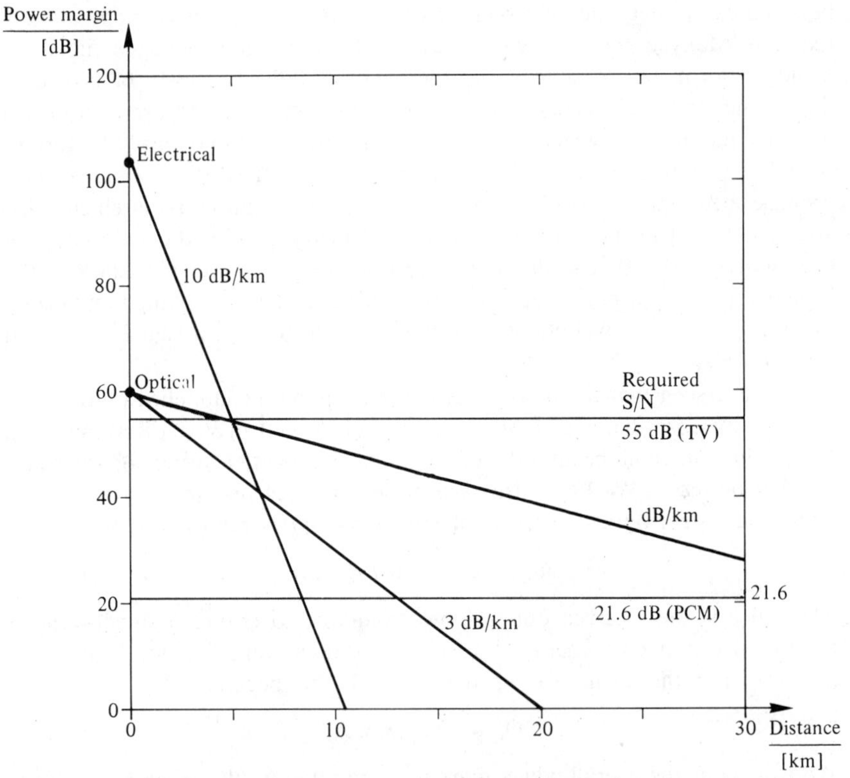

Fig. 17.3 Graph of power margin vs. repeater spacing showing the serious effect that a higher signal-to-noise ratio requirement at the receiver has on the available margin of an optical system.

plot of the receiver signal-to-noise ratio as a function of repeater spacing. The electrical system is represented as having an overall power margin of 104 dB and an attenuation at 100 MHz of 10 dB/km. This may be compared with the values shown in Fig. 4.12. The optical system is represented as having an overall power margin of 60 dB, and attenuation rates of 1 dB/km and 3 dB/km are shown. Lines corresponding to receiver signal-to-noise ratios of 21.6 dB (PCM) and 55 dB have been drawn. It will be appreciated that these results are dependent on the details of the systems chosen for comparison. Nevertheless, the overall conclusion is valid: optical systems show the greatest benefit with pulse modulation. This does not mean they are necessarily useless for analogue transmission. Optical analogue systems will be worth consideration when the fiber is bandwidth rather than attenuation limited, and when terminal costs are at a premium.

17.4.2 Direct Intensity Modulation at Baseband

As well as the demands of the increased signal-to-noise ratio, two further effects limit the use of direct intensity modulation for analogue transmission. One is the modal noise that arises when laser sources are used, which was discussed in Section 15.4. The other is source linearity, which is particularly important for frequency multiplexing, because intermodulation products give rise to interchannel interference. In addition, the transmission of color TV is sensitive to small amounts of phase distortion. A number of techniques for increasing the linearity of the transmitter have been devised. These include pre-distortion of the electrical waveform and the use of electronic feedback and feedforward circuits. The problem of pre-distortion is that once set it cannot easily be varied to adjust for changing source characteristics during life. However, a significant increase in linearity can readily be obtained. Significant reduction in second- and third-harmonic distortion can be obtained using the simple feedback circuit shown in Fig. 17.4. However, signal delay in the feedback loop presents a problem, and wide-band amplifiers are required if a good phase performance is to be achieved. Even better compensation has been claimed for the feedforward arrangement shown in Fig. 17.5, provided two identical LEDs are available. An LED, which uncompensated gave second- and third-harmonic levels -35 dB and -55 dB down on the fundamental, had these values both reduced to below -70 dB when operated with the feedforward arrangement.

In order to calculate the signal-to-noise ratio to be expected, we first define the modulation index, m, of an intensity-modulated optical signal as

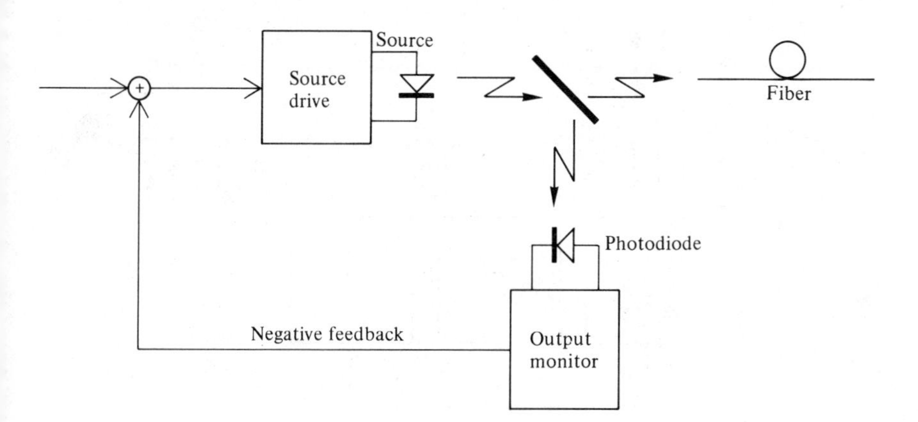

Fig. 17.4 Schematic diagram of simple feedback arrangement.

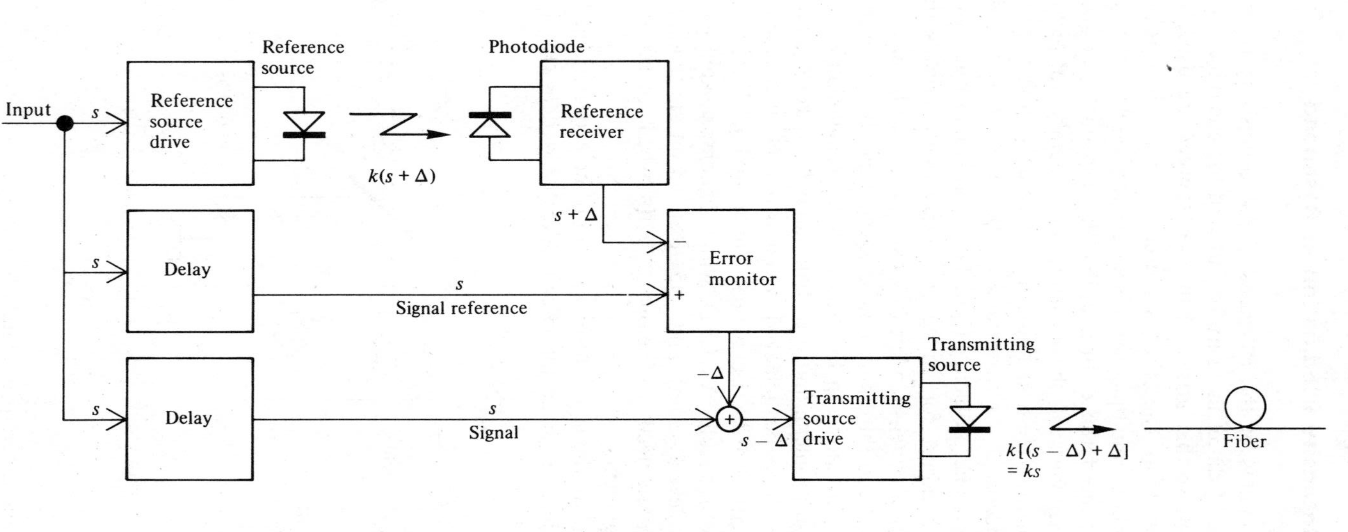

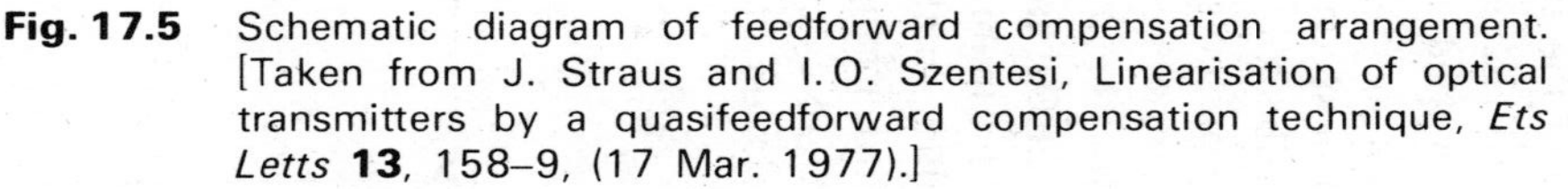

Fig. 17.5 Schematic diagram of feedforward compensation arrangement. [Taken from J. Straus and I.O. Szentesi, Linearisation of optical transmitters by a quasifeedforward compensation technique, *Ets Letts* **13**, 158–9, (17 Mar. 1977).]

$$m = \hat{\Phi}/\Phi_0 \tag{17.4.2}$$

where Φ_0 is the optical power level in the absence of modulation and $\hat{\Phi}$ is the maximum deviation of the instantaneous power from Φ_0. Clearly $0 < m < 1$, but m is further constrained in practice by the maximum permissible distortion level. The signal current generated in the detector is given by

$$I = mI_0 = m\Phi_0 \mathcal{R} \tag{17.4.3}$$

where I_0 is the current generated by the unmodulated carrier and $\mathcal{R}$ is the detector responsivity. The received signal-to-noise ratio, K, is then given by eqn. (14.4.10) or (14.5.14) with the factor I in term (c) of the denominator changed to I_0. This then represents the ratio of the peak signal to the r.m.s. noise.

We may gather terms (a), (b), (d) and (e) of eqns. (14.4.10) and (14.5.14) into a single term, I_N^*, which represents the total circuit-generated noise. Thus,

$$(I_N^*)^2 = \frac{(V_A^*)^2}{M^2}\left[\frac{1}{R^2} + \tfrac{4}{3}\pi^2 C^2(\Delta f)^2\right] + \frac{4kT}{M^2 R} + \frac{(I_A^*)^2}{M^2} \tag{17.4.4}$$

Then,

$$K = \frac{I}{[2eI_0 F + (I_N^*)^2]^{1/2}(\Delta f)^{1/2}} = \frac{m\Phi_0 \mathcal{R}}{[2e\Phi_0 \mathcal{R} F + (I_N^*)^2]^{1/2}(\Delta f)^{1/2}} \tag{17.4.5}$$

Figure 17.6 shows K plotted as a function of Φ_0 for $m = 0.5$, $\mathcal{R} = 0.5$ A/V, $F = 1$ and $\Delta f = 5$ MHz, and with $(I_N^*)^2$ as a parameter. The values of $(I_N^*)^2$ may be compared with those given in the worked example of Section 14.6. Figure 17.6 illustrates the way that systems requiring a high signal-to-noise ratio tend to be quantum-noise-limited even when p–i–n diode detectors are used.

Baseband, intensity-modulated, video transmission has been considered for a closed-circuit television (CCTV) system used to monitor a railway track (Japanese National Railway and Mitsubishi). A 1.29 μm InGaAsP/InP laser diode, multimode graded-index fiber and an InGaAsP/InP, p–i–n photodiode were used. Feedback and pre-distortion circuits improved the source linearity such that a modulation index greater than 0.5 could be used. Attenuation over 16.5 km of fiber with 7 connectors was 27.3 dB. Launched power was –7 dBm and the received power level of –34.3 dBm gave a video signal-to-noise ratio of 42.3 dB, which was adequate. As fiber bandwidth was not a limitation, it was possible to use a broad-band laser source emitting in multiple longitudinal modes in order to minimize modal noise.

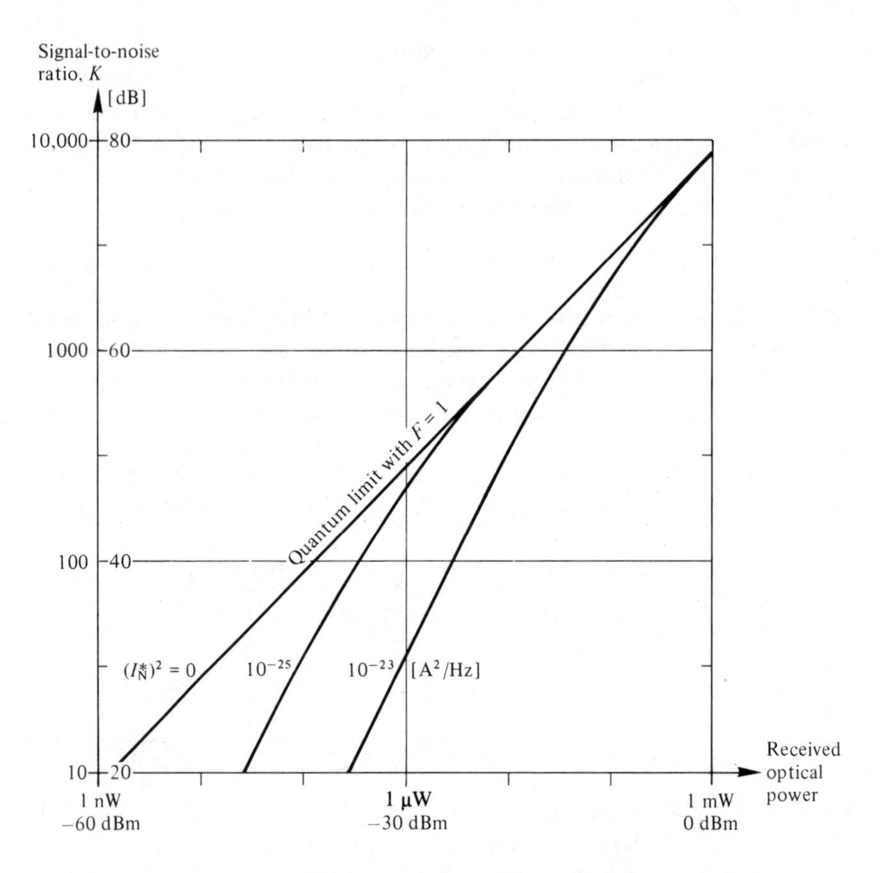

Fig. 17.6 Variation of signal-to-noise ratio with received optical power.

17.4.3 The Use of a Frequency-modulated Subcarrier

We have seen in Section 16.5 that modulating the pulse repetition frequency (p.r.f.) of a pulsed optical source enables an analogue transmission system to be realized very easily at audio frequencies. This applies, of course, whether the transmission path is guided or unguided. The technique can be extended to deal with video signals simply by using a higher subcarrier p.r.f. Systems operating with p.r.f.s of several hundred megahertz have been reported as giving successful transmission. High signal-to-noise ratios can be obtained with less received optical power than is required with baseband intensity modulation, and less channel bandwidth is needed for the transmission of any given signal compared to the use of pulse-code modulation (PCM). Whenever a PCM system is limited by dispersion rather than by attenuation, pulse frequency modulation (PFM) may offer a better performance because it enables the channel bandwidth requirement to be exchanged for signal-to-noise ratio. Frequency multiplexing may be used if the system is not band-limited by dispersion. Overall linearity depends on the linearity of the modulating and demodulating circuits. At 0.85 μm laser sources normally have to be used because material dispersion limits the transmission distance. With multimode fibers modal noise is then a serious limitation. This conflict can be eliminated either by the use of an LED source at 1.3 μm, in which case dispersion is no problem, or by the use of single-mode fibers.

Detailed analysis of a PFM communication channel is made difficult by the fact that a number of inherently nonlinear processes are involved. Furthermore, there are several possible variants in the type of modulation that may be used (frequency or position modulation using pulses; maintaining either a constant pulse-width or a constant duty-cycle as the frequency or position varies; frequency or phase modulation using a sine-wave subcarrier modulation), and different ways of effecting the modulation and demodulation. For these reasons we will not attempt a quantitative discussion of the signal-to-noise performance to be expected from an optical PFM link at this point. Suffice it to say that the general characteristics are similar to those of a conventional FM radio channel, which are discussed in most communications texts. (See, for example, Ref. [1.3].) Note that the quantity K given by eqns. (14.4.10), (14.5.14) and (17.4.5) now represents the *carrier-to-noise* ratio, when evaluated over the channel bandwidth. The use of 'wide-band' FM, in which the frequency deviation is large compared with the signal bandwidth, permits a significant reduction in the signal-to-noise ratio provided the carrier-to-noise ratio exceeds a certain threshold value which is sufficient to ensure reliable pulse regeneration.

In a number of different trials the reception of high-quality television pictures (signal-to-noise ratio $\sim$55 dB) has been found to require a received optical power level of almost 1 μW (−30 dBm). Compared with direct intensity modulation at baseband, an improvement of some 10–15 dB in signal-to-noise ratio can be obtained.

This discussion has concentrated on the transmission of video waveforms because this is thought to be the most likely application for optical PFM transmission, over distances up to about 10 km. This might be in the local distribution of networked services or as feeds to subscribers from a central reception antenna (common antenna television, CATV), perhaps one receiving satellite transmissions.

17.5 APPLICATIONS IN LOCAL DATA COMMUNICATION SYSTEMS

The increasing use of distributed, multiple computing elements in many diverse situations is establishing a need for reliable and efficient local area networks (LANs) for digital data communication. Networks containing up to one hundred nodes and capable of handling data at bit rates of 1 Mb/s or more over internodal distances up to 1 km, will be required in many types of industrial plant for purposes of process monitoring and control. A similar requirement exists in the military environment, where sophisticated weapons and communications systems, most under local computer control, need to be linked in an overall 'command, control and communications' network. In future office systems high-bandwidth links for data transfer between individual, powerful, interconnected work stations will be needed. Related data distribution needs arise within large mainframe computer systems, particularly in the transfer of data between sections of the central processor unit, or into and out of fast memory storage, or between the central processor and remote peripherals. In many of these systems the transmission paths pass through difficult chemical and electromagnetic environments where the merits of optical fiber links, listed in Table 1.1 and rehearsed several times since, will clearly be of great benefit.

It is always possible to design an optical link to replace a section of an existing electrical network so that the electrical signals received and supplied at the terminals are unchanged from those in the original system. The optical link is then said to be 'transparent' or 'invisible' in the network. Much greater benefit may accrue if the network can be designed from the start with a view to exploiting the small physical size and the large communication capacity of the fiber. However, the difficulty of producing anything other than point-to-point optical links does put a severe constraint on the use of fibers, and does make fibers less than ideally suited to the topological demands of some of the more sophisticated types of electrical network now under development.

In comparison with digital telecommunication links the bit rates and transmission distances required for local data distribution may at first sight seem trivial. Certainly internodal distances are, by definition, short, and may range from 1 m to 1 km. At present the bit rates handled are also low, lying in the range 1–10 Mb/s. But if the special characteristics of optical fibers cause changes to the architecture of computer and data distribution systems, this may result in demands for bit rates up to 1 Gb/s. There are, of course, other constraints. With such short links the cost of the optical terminals is crucial, and they have to be compatible, as far as possible, with the rest of the system. This implies as much integration as possible, and an avoidance of special supply voltages. Thus APDs will normally be avoided and lasers used only when needed for high bit rates. If, in a TTL-compatible system, only a 5 V line is available, a severe constraint is put on the design of the receiver, especially on the dynamic range that can be expected.

Five ways of interconnecting networks are illustrated in Fig. 17.7 and described below.

(a) *Mesh*
A complete interconnection of all nodes by means of point-to-point links. This arrangement is complex, expensive, and difficult to reconfigure should more nodes have to be added or some removed. Some links may never be used.
(b) *Tree or Branching Network*
This provides a limited connection of terminals to a central control or distribution unit. It is suitable in a mainframe computer and for the distribution of services.
(c) *Star*
This is relatively inflexible and expensive. Less cable is required than for a mesh, but more than for a bus or loop. The central interconnector is a vulnerable element on which the whole system depends.
(d) *Multi-access Data Bus*
(e) *Ring or Loop*
These offer the greatest flexibility and permit the greatest number of nodes to be interconnected with the least length of cable.

The connections or junctions, shown by circles in Fig. 17.7, may be purely passive or they may be active in two distinct ways. First, they may or may not permit some switching or routing of the data between the nodes they connect. Second, they may temporarily remove the data from the system and then regenerate it, if required, for onward transmission. Clearly, passive junctions should make for a simpler and more reliable system. If active junctions are used, it is desirable that they be powered centrally so that they do not have to rely on local power from the nodes they support. This requirement increases the cost and complexity of the distribution cable. Where transmission paths have to be shared, access to the network has to be controlled by some means. This may be done in one of several ways. For example, in one method one of the nodes acts as a network controller and determines which terminal may transmit at any given time. This is then a vulnerable part of the whole system. In a second method the use of the network is allocated to each terminal on a fixed, time-shared basis. This is inefficient. A third method uses *contention*. Here messages are transmitted as required. If two clash at any time this is detected, both are aborted, and the terminals try again at a randomly selected later time. For irregular traffic, and less than a certain critical level of network loading, this third technique has many advantages.

Specific examples of three well-established electrical local-area-network systems are:

(a) Ethernet: A 3 Mb/s system based on the multi-access data bus using a coaxial cable with passive taps and access by contention.
(b) Systems based on the U.S. Department of Defense avionics standard MIL-STD-1553B: This specifies a 1 Mb/s multi-access data bus using a shielded, twisted-wire pair with passive taps. Access is determined by a bus controller.
(c) Cambridge Ring: A 10 Mb/s ring using twisted-wire-pairs with active junction repeaters. As all the data on the ring has to be handled by a repeater in every node, each can access the ring during any message slot it finds to be empty.

It can be seen at once that any network based on regenerative junctions can really be regarded as no more than an organized set of point-to-point links. Substitution of optical fibers is then no problem and makes much higher bit-rates and internodal distances immediately available, if required. Indeed an optical link has been included in the Cambridge ring. The passive taps that historically have been associated with the multi-access data bus systems create a much greater problem for fibers. The reason is that optical taps are reciprocal. If power is to be fed in as well as taken out through a simple beam splitter, then a significant fraction, perhaps up to one half (–3 dB), of the circulating optical power is lost at each junction. For two-way transmission on a single fiber two such beam-splitters are needed for each terminal connected, and in addition some method of separating the received from the transmitted power is required. Unless an efficient optical switch can be developed some 5–10 dB of loss per junction, including connector losses, is inevitable. Clearly the available power margins shown in Fig. 17.2 will be able to supply few terminals even at bit rates as low as 1 Mb/s.

There are two ways out of this difficulty. Either the junctions have to be made active, as in the system shown in Fig. 17.8, or the network has to be reconfigured to make use of a passive star coupler, as shown in Fig. 17.9.

An illustration of a 7-way, transmission star coupler is given in Fig. 17.10(a). The two fiber bundles are each passed through a taper so that they form a close-packed, hexagonal array. The ends are ground flat and polished, and then fastened together inside a cladding tube using index-matching epoxy. The possible number of ports, N_P, is determined by the number of layers in the array, k:

$$N_P = k^3 - (k - 1)^3 \tag{17.5.1}$$

Thus with $k = 2$, $N_P = 7$, as shown, and with $k = 3$, $N_P = 19$; cross-sections

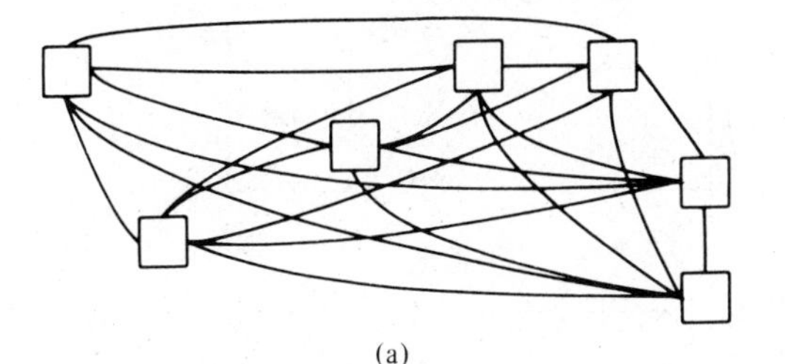

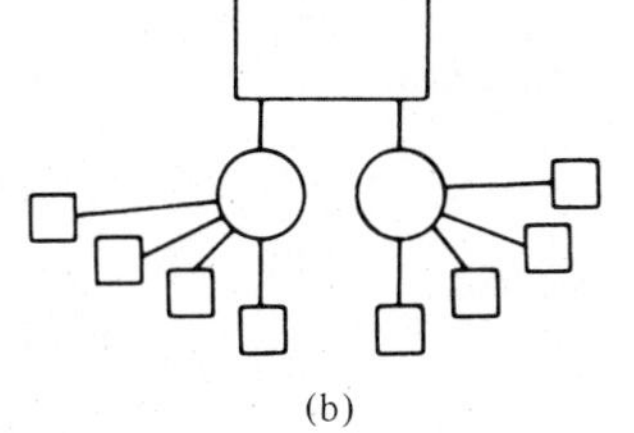

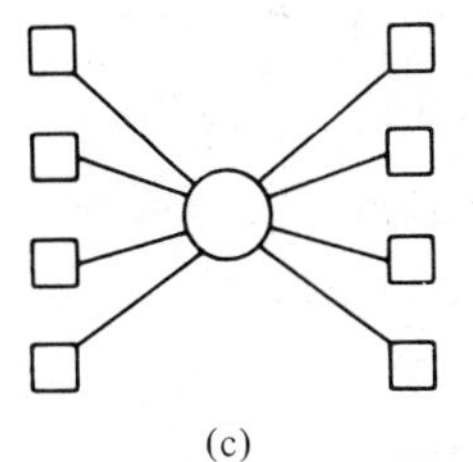

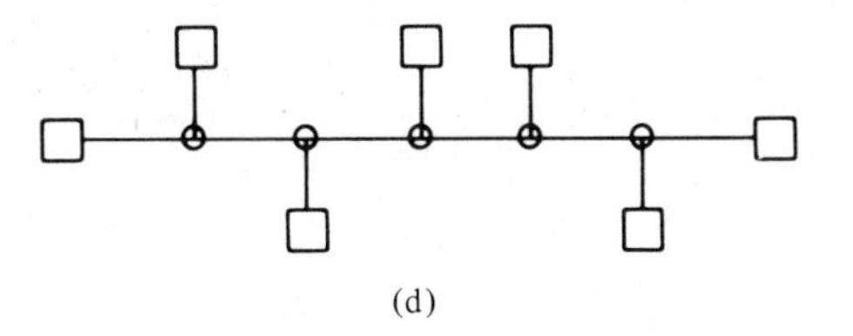

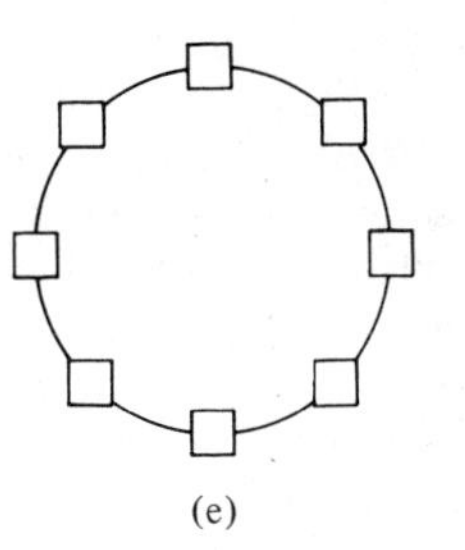

Fig. 17.7 A taxonomy of network topologies: (a) mesh; (b) tree; (c) star; (d) bus; (e) ring.

Squares □ indicate nodes, i.e. data terminals, computers, work-stations, etc. Circles ○ indicate connections or junctions, which may be either active or passive and which may or may not involve the switching (i.e. routing) of the data. With passive junctions (c) and (d) are really equivalent networks in which data is transferred directly between nodes over a shared transmission path.

are shown in Fig. 17.10(b). Star couplers of this design with 7 and 19 ports have been fabricated successfully. The average port-to-port insertion loss was some 4 dB greater than the loss arising from the even distribution of the light. There was a ± 2 dB variation, apart from the coupling between the two axial ports which was 5 dB higher.

In one proposed avionics system regenerative junctions are used on an optical multi-access data bus. In order to reduce the system vulnerability to

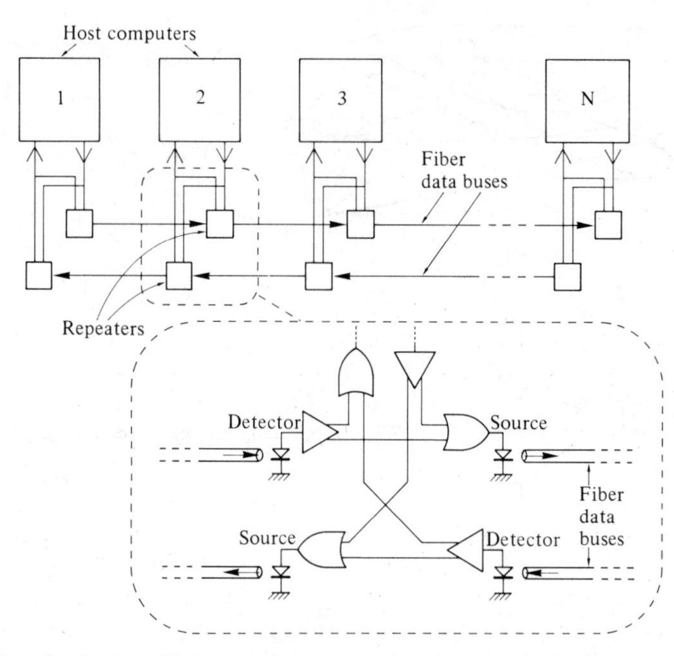

Fig. 17.8 A double fiber, two-way, multi-access optical data bus using regenerative repeaters. [Taken from E. G. Rawson and R. M. Metcalf, Fibernet: multimode optical fibers for local computer networks, *IEEE Trans. on Comm.*, **COM-26**, 983–90 (1978); © 1978 IEEE.]

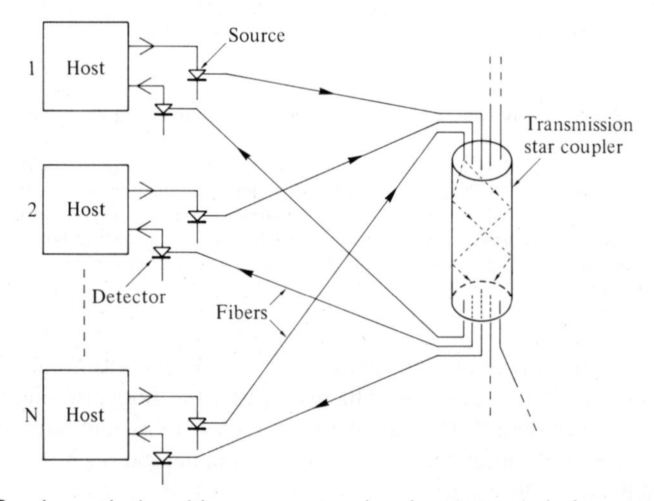

Fig. 17.9 An optical multi-access network using a transmission star coupler. [Taken from E. G. Rawson and R. M. Metcalf, Fibernet: multimode optical fibers for local computer networks, *IEEE Trans. on Comm.*, **COM-26**, 983–90 (1978); © 1978 IEEE.]

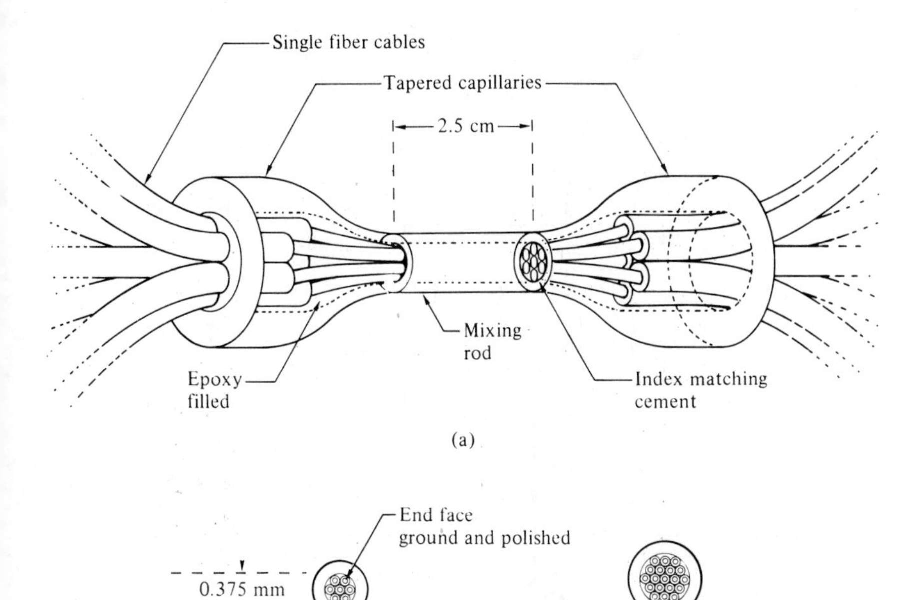

Fig. 17.10 (a) A seven-port transmission star coupler. [Taken from E. G. Rawson and R. M. Metcalf, Fibernet: multimode optical fibers for local computer networks, *IEEE Trans. on Comm.*, **COM-26**, 983–90 (1978); © 1978 IEEE]; (b) cross-sections of close-packed hexagonal arrays of 7 and 19 fibers.

failure at a single terminal, each junction is by-passed so as to provide redundant paths. One terminal in the system is illustrated schematically in Fig. 17.11. It requires a two-way, four-to-one coupler. Clearly transmission and regeneration delays have to be matched with care.

The systems described are manifestly a great deal more complex than simple point-to-point links, and apart from their direct substitution in ring systems using active junctions, the extent to which the merits of optical fibers off-set their inherent disadvantages remains an open question. The answer will depend on the future cost of optical system components, the environmental difficulties in any given situation and the demand for systems with higher bit rates.

The requirements of mainframe computer systems present a rather different set of problems, and ones to which fiber links may be more readily amenable. Usually data is transmitted between units of the system along bundles of

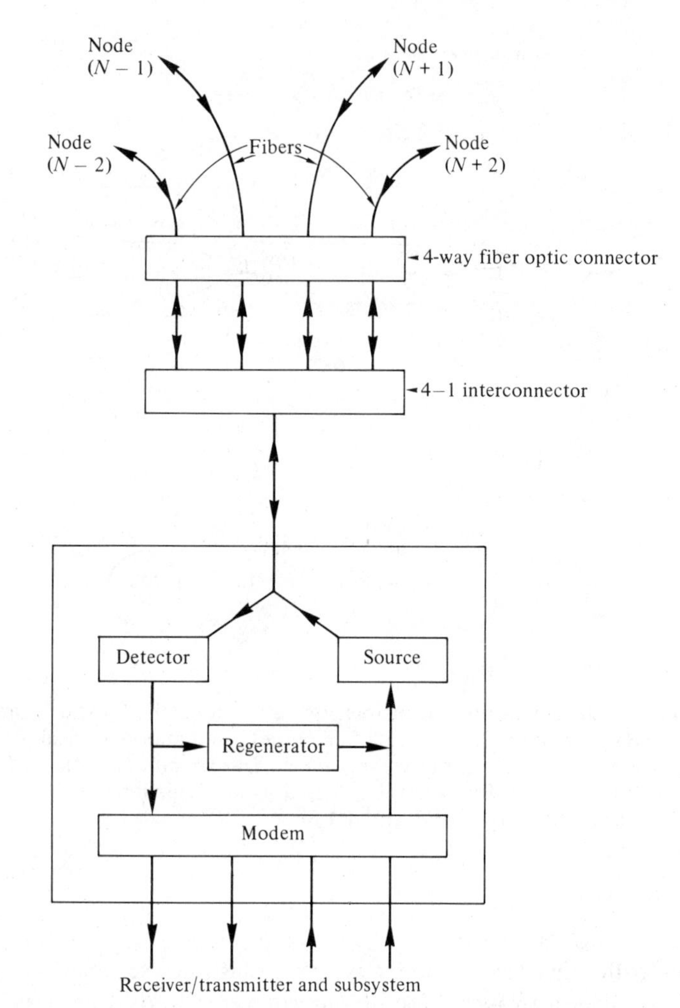

Fig. 17.11 Schematic diagram of a regenerative terminal at node N of a multi-access optical data bus. Multiple connections provide redundancy. [Adapted from figure kindly supplied by R. P. G. Collinson, Marconi Avionics Ltd., Flight Automation Research Laboratory.]

coaxial lines, as many as 72 to a bundle. Words of 8, 16 or more bits are carried in parallel. The bit rate rarely exceeds a few megabytes per second and is limited in two ways. The first is due to electromagnetic compatibility (EMC) problems, especially crosstalk between the coaxial lines. The second is the result of the error detection and correction protocols used. These usually involve the checking and acknowledgement of each word as it is received. The bit rate then becomes limited by the total, two-way, time delay of the link. The required error probability levels that impose these procedures are in the order of one per day. With optical fibers it would be possible to serialize the data using on-chip parallel-to-serial and serial-to-parallel converters, and transmit at higher bit rates, more than 100 Mb/s. A less than once-per-day error frequency at such rates implies $PE < 10^{-13}$. This should be well within the capabilities of the optical system with the power margins and bandwidths available, provided the EMC problems are eliminated. While this is not expected to be a problem in the fiber, the terminal equipment requires careful design and special screening. Then, much simpler error-correcting protocols may be used and system integrity preserved.

17.6 THE WIRED CITY

In this final section we shall take a brief and tentative look at the range of services that may become generally available when optical fibers bring wide-band transmission capacity onto the premises of the individual business and domestic subscriber. To be honest, the possibilities are limited, as much as anything, by our imagination of what the subscriber will wish to do. We may divide the services into those that are fully interactive between any pairs or combinations of terminals (like the present telephone system), and those that are selective (like Teletext) or distributive (like television). They will include the following:

Broadcast audio and video services offering a wider range and possibly a higher technical quality than the presently available radio and television services.
New interactive services such as videophone and teleconferencing.
Access to audio, video and data libraries.
Development of information services such as Viewdata and Teletext.
Electronic mail, cash transfer, business transactions.
Document transfer and business information services.
Remote metering and control.

The requirements and priorities of the business and the domestic subscriber will be different and it may be sensible to think in terms of several linked but essentially independent, fully and partially switched systems for these two communities.

If a demand for wide-band video and other services were to develop rapidly (say during the 1980s), then it would probably have to be satisfied by hybrid analogue and digital transmission from the local exchange or distribution point.

There is no doubt, however, that in the longer term (say during the 1990s) all-digital systems with coders and decoders (codecs) for analogue-to-digital conversion sited on the subscribers' premises will become a practical possibility. This will offer the greatest flexibility, and the widest range of services. There will be possibilities for expansion and development, and for the establishment of economies of scale in the production of the standardized VLSI interface circuits needed. A system arrangement that seems particularly appropriate in either case is one in which all services are made available to a local switching point serving a number of subscribers. Each subscriber may then access a limited number of services simultaneously, including the local, end-office telephone exchange, through this switching point. This is described as providing '*N* out of *M*' facilities.

Several quite elaborate and extensive experimental projects have been set up in different parts of the world to examine not only the technical problems but also the economic and social effects of offering such interactive wide-band services. The first was the Hi-Ovis project in Osaka, Japan. This supplies about 150 homes in a new suburb. Two fibers go to each subscriber, supplying analogue and digital, audio, video, and data services. Another hybrid system has been initiated at Elie, Manitoba, Canada. This is a rural area requiring longer cable runs of up to 5 km. A single fiber carries two-way transmission between each of about 150 subscribers and a local switching point. An all-digital system, which could involve several thousand subscribers, is being established in Jutland, Denmark. Here a three-fiber cable goes to each subscriber with two fibers carrying inward transmissions at 140 Mb/s and one for outgoing signals. An extensive experiment is under way in Biarritz, France. There are, no doubt, many others in various stages of planning and development. The Heinrich Hertz Institut für Nachrichtentechnik in West Berlin has for some time been engaged in extensive studies of the problems involved in these complex, interactive, subscriber networks and has set up some optical-fiber distribution systems. One is a subscriber loop or ring (Fig. 17.7(e)) operating at 280 Mb/s.

At the end of a book which is mainly concerned with the results of nearly two decades of work, one feels obliged to comment on the progress that has been made. In fact, very little that has been described here was not foreshadowed to some extent in the 1966 paper of Kao and Hockham (Ref. |1.1|). What stands out at the present level of development, effective though it may be, is its essentially hybrid nature. Only transmission is carried out in the optical domain. All modulation, signal processing and switching are done electronically. What remains to be seen is the extent to which a greater number of processes may be carried out optically, and how they may be integrated with the rest of the system. Perhaps the first step along this road will be the development of an integrated optical repeater. This will avoid the double conversion of the signal into and out of the electronic domain. Next might be the development of optical switches and modulators. These possibilities, which are outside the scope of the present book, are subject to intense research activity. A comprehensive review may be found in Ref. [17.1].

We find ourselves at a particularly interesting threshold. Optical transmission has reached the point where it can be introduced on a scale which will expand our telecommunications capacity almost without limit. As yet the full social and economic implication of this is something that can only be the subject of speculation. In terms of technology, though, the exploitation of the optical domain, the use of optically active materials, and the development of integrated optical systems are very much in their infancy.

PROBLEMS

17.1 In Fig. 17.2 the upper line limiting the range of receiver sensitivities assumes receiver noise to be dominated by the amplifier noise term (term (b) in eqn. (14.4.10)). Verify that the slope of this line agrees with the theoretical frequency variation and estimate the value of CV_A^* to which the line corresponds. Assume a detector responsivity of 1 A/W, $M = F = 1$, signal-to-noise ratio $K = 12$ and noise bandwidth $\Delta f = B/2$.

17.2 In Fig. 17.2 the lower line bounding the range of receiver sensitivities assumes that multiplied shot noise dominates the other receiver noise terms. Verify that the slope of the line is in agreement with the theoretical frequency variation and calculate the value of F implied. Assume an unmultiplied detector responsivity $\mathcal{R} = 1$ A/W and a signal-to-noise ratio $K = 12$.

17.3 Mark on Fig. 17.2 the quoted sensitivities of the various types of receiver described in the text, especially in the tables of Chapter 17. Add any others that you find that have been published more recently.

17.4 In the quantum limit (i.e. neglecting amplifier noise) obtain an expression for the received mean optical power Φ_0 required to ensure a ratio of peak signal-to-r.m.s. noise of K in an intensity modulated optical communication system, covering a bandwidth Δf at baseband.

Assume that the modulation depth is limited to m, the unmultiplied detector responsivity is $\mathcal{R}$ and the detector noise factor is F.

17.5 An optical communication link operates by direct intensity modulation over the frequency range 0–10 MHz. A receiver signal-to-noise ratio of 50 dB is required (peak signal to rms noise, $K = 316$). The source is a light emitting diode which launches 50 μW average power onto a multimode fiber and its modulation index is limited to 0.5. The fiber attenuation is 4 dB/km. The detector is an avalanche photodiode having a gain of 100, a noise factor of 5, and an unmultiplied responsivity of 0.6 A/W.

(a) Calculate the minimum optical power required at the receiver, and hence the maximum, attenuation-limited path length, assuming amplifier noise to be negligible.

(b) Calculate the magnitude of the multiplied shot noise and comment on the assumption that this dominates any amplifier noise.

17.6 Calculate by what factor the required receiver power described in the example at the end of Section 17.4.2 exceeds the power predicted from quantum limited operation. Assume $\eta = 0.6$, $M = F = 1$, $m = 0.5$, $\Delta f = 5$ MHz.

17.7 Consider possible configurations for effecting a passive tap to launch and receive signals travelling in either direction on a single fiber. Calculate the insertion loss of the complete tap, assuming that beam-splitters divide the optical power equally (–3 dB) and that each of the component parts are separated by a demountable connector—insertion loss 1 dB. Hence comment on the number of ports that could be served by such a system, used as a two-way data bus.

REFERENCE

17.1 A. Miller, D. A. B. Miller and S. D. Smith, Dynamic non-linear optical processes in semiconductors, *Adv. Phys.* **30**, 697–800 (1981).

SUMMARY

The economic advantage of optical fiber telecommunication systems over those using electrical cable is greater at higher bit rates. Then the electrical systems have to use co-axial cable or waveguide rather than twisted-wire pairs. But the principal benefit comes because greatly increased optical repeater spacings are possible. Optical links will be introduced progressively into telecommunication networks along with increased use of digital signalling. This will start at the higher system levels. The extent of fiber penetration into the local network is more speculative, although a number of trials of the kind of wide-band, interactive services that could be provided are in progress in different countries.

First-generation optical fiber telecommunication systems use GaAlAs/GaAs laser sources (0.8–0.9 μm), Si–APD detectors, and graded-index (GI) fibers. With a dispersion limit of around 1 (Gb/s)km and attenuation of 3.5 dB/km, repeater spacings of 15 km at 8 Mb/s, 11 km at 45 Mb/s and 7 km at 140 Mb/s can be achieved. The use of p–i–n photodiode detectors reduces the power margin by 10–20 dB and the repeater spacing by 3–6 km. The use of light-emitting diode (LED) sources not only reduces the power margin by 10–20 dB but also reduces the dispersion limit to about 200 (Mb/s)km.

Three longer wavelength, second generation systems are of particular interest. The first is an LED/p–i–n FET/multimode (GI) fiber system working at the wavelength of minimum material dispersion, 1.3 μm. Its performance depends critically on the minimization of multipath dispersion. Repeater spacings should exceed 10 km at 140 Mb/s and 20 km at 45 Mb/s. The second system uses laser/APD or p–i–n FET/monomode fiber and works at the attenuation minimum at 1.55 μm. Its performance depends on minimizing the laser source linewidth by operating in a single longitudinal mode. In laboratory tests 400 Mb/s signals have been received over a 100 km link. The third system combines laser/APD or p–i–n FET/monomode fiber working at the dispersion minimum which in this system can be varied by the fiber design. In laboratory tests 2 Gb/s signals have been received over 51.5 km of fiber.

The overall power margin in optical systems is small compared with that of electrical systems, but their bandwidths may be high. Digital signalling thus gives them their best advantage in that the signal-to-noise ratio required at the receiver is minimal but the required bandwidth is increased. With direct analogue intensity modulation much of this advantage is lost. The use of a frequency modulated subcarrier offers a compromise analogue system which may be particularly suited to carrying TV signals over distances up to 10 km.

The use of fibers in local data communication networks is limited if passive taps are required. No more than about five T-couplers are possible and with a greater number of nodes, one or more star couplers have to be used. If active taps are permitted then between each there is effectively a point-to-point link and fibers can be used without difficulty, although redundant paths may be thought necessary. Within mainframe computer systems the use of fibers may well enable the input/output data to be distributed in serial rather than parallel form, and will encourage the development of distributed computer systems.

Appendix 1

The Electromagnetic Wave Equation in an Isotropic Medium Subject to Cylindrical Boundary Conditions

A1.1 THE WAVE EQUATION

Maxwell's equations for a homogeneous, isotropic, nonconducting and space-charge-free region are:

$$\text{curl } \mathbf{E} = -\frac{\partial \mathbf{B}}{\partial t} \qquad \text{curl } \mathbf{H} = \frac{\partial \mathbf{D}}{\partial t} \tag{A1.1}$$

$$\text{div } \mathbf{D} = 0 \qquad \text{div } \mathbf{B} = 0$$

where

$$\mathbf{D} = \varepsilon_r \varepsilon_0 \mathbf{E} \quad \text{and} \quad \mathbf{H} = \mathbf{B}/\mu_r \mu_0$$

Then

$$\text{curl curl } \mathbf{E} = -\text{curl}\left(\frac{\partial \mathbf{B}}{\partial t}\right) = -\frac{\partial}{\partial t}(\text{curl } \mathbf{B})$$

$$= -\mu_r \mu_0 \frac{\partial}{\partial t}(\text{curl } \mathbf{H}) = -\mu_r \mu_0 \frac{\partial^2 \mathbf{D}}{\partial t^2}$$

But

$$\text{curl curl } \mathbf{E} = \nabla(\text{div } \mathbf{E}) - \nabla^2 \mathbf{E} \quad \text{and} \quad \text{div } \mathbf{E} = \text{div } (\mathbf{D}/\varepsilon_r \varepsilon_0) = 0$$

$$\therefore \quad \nabla^2 \mathbf{E} - \varepsilon_r \varepsilon_0 \mu_r \mu_0 \frac{\partial^2 \mathbf{E}}{\partial t^2} = 0 \tag{A1.2(a)}$$

and similarly

$$\nabla^2 \mathbf{H} - \varepsilon_r \varepsilon_0 \mu_r \mu_0 \frac{\partial^2 \mathbf{H}}{\partial t^2} = 0 \tag{A1.2(b)}$$

We will write ψ to represent either **E** or **H** and will seek wave solutions in cylindrical coordinates (r, ϕ, z), in which the direction of wave propagation is along the z-axis, the axis of symmetry of the boundary conditions. Thus

$$\nabla^2 \psi - \varepsilon_r \varepsilon_0 \mu_r \mu_0 \frac{\partial^2 \psi}{\partial t^2} = 0 \tag{A1.2(c)}$$

We recognize equations (A1.2) as forms of the wave equation, which in the absence of boundary conditions, give plane wave solutions. The phase velocity is

$$v_p = \frac{1}{(\varepsilon_r \varepsilon_0 \mu_r \mu_0)^{\frac{1}{2}}}$$

In vacuum $\varepsilon_r = \mu_r = 1$, so that

$$v_p = \frac{1}{(\varepsilon_0 \mu_0)^{\frac{1}{2}}} = c$$

In an isotropic medium the refractive index, n, is defined as $v_p = c/n$, so that $n = \sqrt{(\varepsilon_r \mu_r)}$ and we may therefore express eqn. (A1.2(c)) as

$$\nabla^2 \psi - \frac{n^2}{c^2} \frac{\partial^2 \psi}{\partial t^2} = 0 \tag{A1.2(d)}$$

A1.2 SOLUTIONS IN CYLINDRICAL COORDINATES

Because fiber boundary conditions have cylindrical symmetry, and we are interested in electromagnetic waves that propagate in the axial direction, we shall seek solutions having the form

$$\psi = \psi_0(r, \phi) \exp\{-j(\omega t - \beta z)\} \tag{A1.3}$$

Provided that the boundary conditions do indeed have cylindrical symmetry, the variables will separate to give

$$\psi_0(r, \phi) = \psi_r(r)\, e^{jk\phi} \tag{A1.4}$$

where k is an integer. Thus

$$\psi = \psi_r(r)\, e^{jk\phi}\, e^{j\beta z}\, e^{-j\omega t} \tag{A1.5}$$

Now in cylindrical coordinates eqn. (A1.2(d)) becomes

$$\frac{\partial^2 \psi}{\partial r^2} + \frac{1}{r}\frac{\partial \psi}{\partial r} + \frac{1}{r^2}\frac{\partial^2 \psi}{\partial \phi^2} + \frac{\partial^2 \psi}{\partial z^2} - \frac{n^2}{c^2}\frac{\partial^2 \psi}{\partial t^2} = 0 \tag{A1.6}$$

Putting (A1.5) into (A1.6) gives

$$\frac{d^2 \psi_r}{dr^2} + \frac{1}{r}\frac{d\psi_r}{dr} + \left(\frac{n^2 \omega^2}{c^2} - \beta^2 - \frac{k^2}{r^2}\right)\psi_r = 0 \tag{A1.7}$$

Equation (A1.7) defines the radial field patterns that satisfy the wave equation (A1.3) and it may be re-expressed as

$$(ur)^2 \frac{d^2 \psi_r}{d(ur)^2} + (ur)\frac{d\psi_r}{d(ur)} + \{(ur)^2 - k^2\}\psi_r = 0 \tag{A1.8}$$

in the core, where

$$u^2 = \frac{n_1^2 \omega^2}{c^2} - \beta^2 \tag{A1.9}$$

and as

$$(wr)^2 \frac{d^2 \psi_r}{d(wr)^2} + (wr)\frac{d\psi_r}{d(wr)} + j\{k^2 - (wr)^2\}\psi_r = 0 \tag{A1.10}$$

in the cladding, where

$$w^2 = \beta^2 - \frac{n_2^2 \omega^2}{c^2} \tag{A1.11}$$

Equations (A1.8) and (A1.10) are forms of Bessel's equation and their solutions include the Bessel and modified Hankel functions, in (ur) and (wr) respectively, that we described in Section 5.2. For the solutions to be well-behaved, that is to be finite at $r = 0$, and to tend to zero as $r \to \infty$, it is necessary that u and w are both real.

Appendix 2

Electromagnetic Waves in Graded-index Fiber: The WKB Approximation

Although the derivation of the wave eqn. (A1.2) in Appendix 1 depended on the assumption of a homogeneous medium, we have shown in Section 6.1 that it remains valid provided that $(1/n^2)\nabla(n^2) = (2/n)\nabla n$ is small. Here we shall assume that this condition is met and that the system has cylindrical symmetry so that the radial variation of the fields is defined by eqn. (A1.7). With n explicitly shown to be a function of radius, this becomes:

$$\frac{d^2\psi_r}{dr^2} + \frac{1}{r}\frac{d\psi_r}{dr} + \left[\frac{\omega^2}{c^2}n^2(r) - \beta^2 - \frac{k^2}{r^2}\right]\psi_r = 0 \qquad \text{(A2.1)}$$

Following the method given by Jeffreys,† approximate solutions to eqn. (A2.1) valid in certain regions, can be obtained by means of a series of substitutions. First the term in $d\psi_r/dr$ may be eliminated by writing

$$U = r^{1/2}\psi_r \qquad \text{(A2.2)}$$

Then (A2.1) becomes

$$\frac{d^2U}{dr^2} + \left[\frac{\omega^2}{c^2}n^2 - \beta^2 - \frac{(k^2 - \frac{1}{4})}{r^2}\right]U = 0 \qquad \text{(A2.3)}$$

Put

$$\left[\frac{\omega^2}{c^2}n^2 - \beta^2 - \frac{(k^2 - \frac{1}{4})}{r^2}\right] = K^2 \qquad \text{(A2.4)}$$

Next put

$$g^2 = \frac{1}{U}\frac{d^2U}{dr^2} = -K^2 \qquad \text{(A2.5)}$$

so that $g = jK$, where $j^2 = -1$.

† H. & B. S. Jeffreys, *Methods of Mathematical Physics* 3rd Ed, C.U.P. (1972), Section 17.12, pp. 519–22.

Now put

$$h(r) = \frac{1}{U}\frac{\mathrm{d}U}{\mathrm{d}r} \tag{A2.6}$$

Then

$$\ln U = \int h(r)\,\mathrm{d}r \qquad \text{or} \qquad U = \exp\{\int h(r)\,\mathrm{d}r\} \tag{A2.7}$$

and

$$\frac{\mathrm{d}h}{\mathrm{d}r} = \frac{1}{U}\frac{\mathrm{d}^2U}{\mathrm{d}r^2} - \frac{1}{U^2}\left(\frac{\mathrm{d}U}{\mathrm{d}r}\right)^2 = g^2 - h^2 \tag{A2.8}$$

so that

$$g^2(r) = h^2 + \frac{\mathrm{d}h}{\mathrm{d}r} \tag{A2.9}$$

As a first approximation, then, we may take just the first term of eqn. (A2.9), giving

$$h \simeq g \qquad \text{and} \qquad \frac{\mathrm{d}h}{\mathrm{d}r} \simeq \frac{\mathrm{d}g}{\mathrm{d}r} \tag{A2.10}$$

The size of the second term in (A2.9) may now be estimated as follows:

$$h \simeq (g^2 - \mathrm{d}h/\mathrm{d}r)^{1/2} = g\left(1 - \frac{\mathrm{d}g/\mathrm{d}r}{g^2}\right)^{1/2} \simeq g\left(1 - \frac{\mathrm{d}g/\mathrm{d}r}{2g^2}\right) \tag{A2.11}$$

So the first term alone, eqn. (A2.10), may be sufficient, provided that

$$\left|\frac{\mathrm{d}g/\mathrm{d}r}{2g^2}\right| \ll 1 \tag{A2.12}$$

Taking just the first term gives

$$\begin{aligned} U &= \{\exp \int g\,\mathrm{d}r\} = \exp\{\mathrm{j}\int K\,\mathrm{d}r\} \\ &= \exp\left\{\mathrm{j}\int\left[\frac{\omega^2}{c^2}n^2(r) - \beta^2 - \frac{(k^2 - \frac{1}{4})}{r^2}\right]^{1/2}\mathrm{d}r\right\} \end{aligned} \tag{A2.13}$$

and

$$\psi_r = \frac{U}{r^{1/2}} = \frac{1}{r^{1/2}}\exp\{\int g\,\mathrm{d}r\} = \frac{1}{r^{1/2}}\exp\{\mathrm{j}\int K\,\mathrm{d}r\} \tag{A2.14}$$

In regions where K is real and g is imaginary, ψ_r varies cyclically with radial position (exponential function with imaginary index). In regions where g is real and K imaginary, ψ_r decays monotonically with radius. The two sets of solutions must match and this restricts the allowed values of β to certain eigenvalues, β_{km}. The method followed here will be familiar to many from its wide application to the Schrödinger wave equation in quantum mechanics. It is usually referred to as the WKB (Wentzal, Kramers, Brillouin) approximation. The solutions given (A2.14) do not apply on the axis, although the method can easily be adapted to permit valid solutions at $r = 0$. It is clear from condition (A2.12) that the approximation becomes invalid if dg/dr is large and if g is small. This means that the transition regions near to r_1 and r_2 in Fig. 6.2, where $g = 0$, cannot be treated simply, and special techniques have to be employed to find the correct boundary-matching conditions.

Appendix 3

Ray Trajectories in Graded-index Fiber

A3.1 DERIVATION OF A PARAMETRIC EQUATION FOR THE RAY TRAJECTORY

We start with the general equation for a ray propagating in an inhomogeneous medium:†

$$\frac{\mathrm{d}}{\mathrm{d}s}\left(n\frac{\mathrm{d}\mathbf{r}}{\mathrm{d}s}\right) = \nabla n \tag{A3.1}$$

where $\mathbf{r}$ is the position vector of the ray, s is the distance measured along the ray trajectory and n is the refractive index of the medium. In our case n is axisymmetric so we shall choose the axis of a system of cylindrical polar coordinates to be coincident with the axis of symmetry of n so that n is a function of $\mathbf{r}$ only. The axial distance is z and the azimuthal coordinate is ϕ. At any point $\mathbf{r}$ on the ray trajectory we define unit vectors $\hat{\mathbf{i}}_r$, $\hat{\mathbf{i}}_\phi$, $\hat{\mathbf{i}}_z$ in each of the three coordinate directions. Then,

$$\mathbf{r} = r\hat{\mathbf{i}}_r + z\hat{\mathbf{i}}_z \tag{A3.2}$$

so that

$$\frac{\mathrm{d}\mathbf{r}}{\mathrm{d}s} = \frac{\mathrm{d}r}{\mathrm{d}s}\hat{\mathbf{i}}_r + r\frac{\mathrm{d}\hat{\mathbf{i}}_r}{\mathrm{d}s} + \frac{\mathrm{d}z}{\mathrm{d}s}\hat{\mathbf{i}}_z + z\frac{\mathrm{d}\hat{\mathbf{i}}_z}{\mathrm{d}s} \tag{A3.3}$$

Now

$$\frac{\mathrm{d}\hat{\mathbf{i}}_r}{\mathrm{d}s} = \hat{\mathbf{i}}_\phi\frac{\mathrm{d}\phi}{\mathrm{d}s} \tag{A3.4a}$$

$$\frac{\mathrm{d}\hat{\mathbf{i}}_\phi}{\mathrm{d}s} = -\hat{\mathbf{i}}_r\frac{\mathrm{d}\phi}{\mathrm{d}s} \tag{A3.4b}$$

† Reference [2.1], Section 3.2.1.

as illustrated in Fig. A.3.1, and

$$\frac{\mathrm{d}\hat{\mathbf{i}}_z}{\mathrm{d}s} = 0 \tag{A3.4c}$$

Therefore,

$$\frac{\mathrm{d}\mathbf{r}}{\mathrm{d}s} = \frac{\mathrm{d}r}{\mathrm{d}s}\hat{\mathbf{i}}_r + r\frac{\mathrm{d}\phi}{\mathrm{d}s}\hat{\mathbf{i}}_\phi + \frac{\mathrm{d}z}{\mathrm{d}s}\hat{\mathbf{i}}_z \tag{A3.5}$$

Now,

$$\frac{\mathrm{d}}{\mathrm{d}s}\left(n\frac{\mathrm{d}\mathbf{r}}{\mathrm{d}s}\right) = \nabla n \tag{A3.1}$$

$$\therefore \quad \frac{\mathrm{d}}{\mathrm{d}s}\left(n\frac{\mathrm{d}r}{\mathrm{d}s}\hat{\mathbf{i}}_r\right) + \frac{\mathrm{d}}{\mathrm{d}s}\left(nr\frac{\mathrm{d}\phi}{\mathrm{d}s}\hat{\mathbf{i}}_\phi\right) + \frac{\mathrm{d}}{\mathrm{d}s}\left(n\frac{\mathrm{d}z}{\mathrm{d}s}\hat{\mathbf{i}}_z\right) = \frac{\mathrm{d}n}{\mathrm{d}r}\hat{\mathbf{i}}_r \tag{A3.6}$$

since $\mathrm{d}n/\mathrm{d}\phi$ and $\mathrm{d}n/\mathrm{d}z$ are each zero.

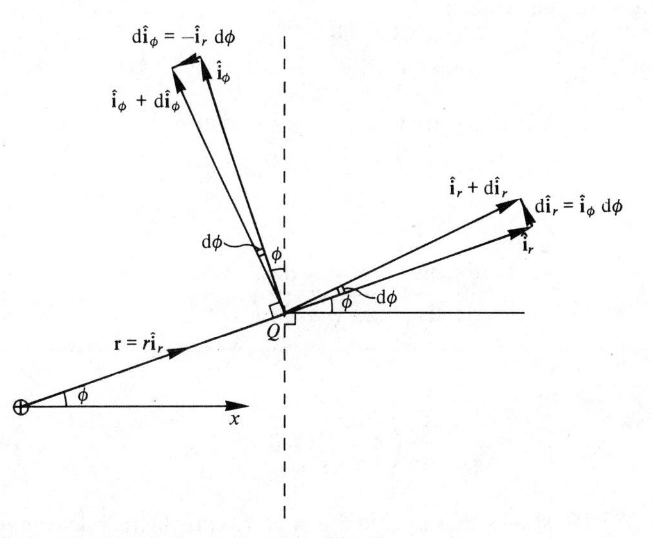

Fig. A3.1 Unit vectors at a point, $Q(r, \phi)$.

The diagram illustrates the differential relationships between unit vectors $\hat{\imath}_r$ and $\hat{\imath}_\phi$:

$$\mathrm{d}\hat{\imath}_r = \hat{\imath}_\phi \mathrm{d}\phi \tag{A3.4(a)}$$
$$\mathrm{d}\hat{\imath}_\phi = -\hat{\imath}_r \mathrm{d}\phi \tag{A3.4(b)}$$

The unit vector in the z-direction is independent of position so that

$$\mathrm{d}\hat{\imath}_z = 0 \tag{A3.4(c)}$$

Thus, performing the differentiations in eqn. (A3.6) term by term, we obtain:

$$\frac{\mathrm{d}}{\mathrm{d}s}\left(n\frac{\mathrm{d}r}{\mathrm{d}s}\right)\hat{\mathbf{i}}_r + n\frac{\mathrm{d}r}{\mathrm{d}s}\frac{\mathrm{d}\hat{\mathbf{i}}_r}{\mathrm{d}s} + \frac{\mathrm{d}}{\mathrm{d}s}\left(nr\frac{\mathrm{d}\phi}{\mathrm{d}s}\right)\hat{\mathbf{i}}_\phi + nr\frac{\mathrm{d}\phi}{\mathrm{d}s}\frac{\mathrm{d}\hat{\mathbf{i}}_\phi}{\mathrm{d}s}$$
$$+ \frac{\mathrm{d}}{\mathrm{d}s}\left(n\frac{\mathrm{d}z}{\mathrm{d}s}\right)\hat{\mathbf{i}}_z + n\frac{\mathrm{d}z}{\mathrm{d}s}\frac{\mathrm{d}\hat{\mathbf{i}}_z}{\mathrm{d}s} = \frac{\mathrm{d}n}{\mathrm{d}r}\hat{\mathbf{i}}_r$$

and hence

$$\frac{\mathrm{d}}{\mathrm{d}s}\left(n\frac{\mathrm{d}r}{\mathrm{d}s}\right)\hat{\mathbf{i}}_r + n\frac{\mathrm{d}r}{\mathrm{d}s}\frac{\mathrm{d}\phi}{\mathrm{d}s}\hat{\mathbf{i}}_\phi + \frac{\mathrm{d}}{\mathrm{d}s}\left(nr\frac{\mathrm{d}\phi}{\mathrm{d}s}\right)\hat{\mathbf{i}}_\phi - nr\left(\frac{\mathrm{d}\phi}{\mathrm{d}s}\right)^2\hat{\mathbf{i}}_r$$
$$+ \frac{\mathrm{d}}{\mathrm{d}s}\left(n\frac{\mathrm{d}z}{\mathrm{d}s}\right)\hat{\mathbf{i}}_z = \frac{\mathrm{d}n}{\mathrm{d}r}\hat{\mathbf{i}}_r \quad \text{(A3.7)}$$

using eqns. (A3.4).

Gathering terms in $\hat{\mathbf{i}}_r$, $\hat{\mathbf{i}}_\phi$ and $\hat{\mathbf{i}}_z$ and equating each separately leads directly to the following three equations:

in $\hat{\mathbf{i}}_r$,

$$\frac{\mathrm{d}}{\mathrm{d}s}\left(n\frac{\mathrm{d}r}{\mathrm{d}s}\right) - nr\left(\frac{\mathrm{d}\phi}{\mathrm{d}s}\right)^2 = \frac{\mathrm{d}n}{\mathrm{d}r} \quad \text{(A3.8)}$$

in $\hat{\mathbf{i}}_\phi$,

$$n\frac{\mathrm{d}r}{\mathrm{d}s}\frac{\mathrm{d}\phi}{\mathrm{d}s} + \frac{\mathrm{d}}{\mathrm{d}s}\left(nr\frac{\mathrm{d}\phi}{\mathrm{d}s}\right) = 0 \quad \text{(A3.9)}$$

in $\hat{\mathbf{i}}_z$,

$$\frac{\mathrm{d}}{\mathrm{d}s}\left(n\frac{\mathrm{d}z}{\mathrm{d}s}\right) = 0 \quad \text{(A3.10)}$$

Equation (A3.10) shows that $n(\mathrm{d}z/\mathrm{d}s)$ is a constant which is determined by the direction and position of the ray as it enters the fiber. These coordinates are defined in Fig. 6.3(a) as $(\alpha_0, \beta_0, \gamma_0)$ and $(r_0, \phi_0, 0)$ respectively. Thus

$$\left(\frac{\mathrm{d}z}{\mathrm{d}s}\right)_{z=0} = \cos\gamma_0 \quad \text{(A3.11)}$$

and

$$n\frac{\mathrm{d}z}{\mathrm{d}s}=n(r_0)\left(\frac{\mathrm{d}z}{\mathrm{d}s}\right)_{z=0}=E=\text{constant} \tag{A3.12}$$

where E is sometimes referred to as the 'energy' parameter of the ray.

Equation (A3.9) is easily integrated by expanding the second term and multiplying through by r. This gives:

$$nr^2\frac{\mathrm{d}^2\phi}{\mathrm{d}s^2}+2nr\frac{\mathrm{d}r}{\mathrm{d}s}\frac{\mathrm{d}\phi}{\mathrm{d}s}+r^2\frac{\mathrm{d}n}{\mathrm{d}s}\frac{\mathrm{d}\phi}{\mathrm{d}s}=\frac{\mathrm{d}}{\mathrm{d}s}\left(nr^2\frac{\mathrm{d}\phi}{\mathrm{d}s}\right)=0 \tag{A3.13}$$

Thus $nr^2(\mathrm{d}\phi/\mathrm{d}s)$ is also a constant that depends only on the manner in which the ray first enters the fiber. We therefore have

$$nr^2\frac{\mathrm{d}\phi}{\mathrm{d}s}=nr^2\frac{\mathrm{d}\phi}{\mathrm{d}z}\frac{\mathrm{d}z}{\mathrm{d}s}=Er^2\frac{\mathrm{d}\phi}{\mathrm{d}z}=El \tag{A3.14}$$

where

$$l=r_0^2\left(\frac{\mathrm{d}\phi}{\mathrm{d}z}\right)_{z=0}=r_0\sec\gamma_0(\cos\beta_0\cos\phi_0-\cos\alpha_0\sin\phi_0) \tag{A3.15}$$

as shown in Fig. A3.2. The parameter l is sometimes referred to as the 'angular momentum' parameter of the ray.

We can obtain a differential equation for $r(z)$ from eqn. (A3.8) as follows. We note that

$$\frac{\mathrm{d}r}{\mathrm{d}s}=\frac{\mathrm{d}r}{\mathrm{d}z}\frac{\mathrm{d}z}{\mathrm{d}s}=\frac{\mathrm{d}r}{\mathrm{d}z}\frac{E}{n} \tag{A3.16}$$

using eqn. (A3.12), and that

$$\left(\frac{\mathrm{d}\phi}{\mathrm{d}s}\right)^2=\left(\frac{El}{nr^2}\right)^2 \tag{A3.17}$$

using eqn. (A3.14). Therefore, eqn. (A3.8) becomes

$$\frac{\mathrm{d}}{\mathrm{d}s}\left(n\frac{E}{n}\frac{\mathrm{d}r}{\mathrm{d}z}\right)-nr\left(\frac{El}{nr^2}\right)^2=\frac{\mathrm{d}n}{\mathrm{d}r}$$

that is,

$$\frac{\mathrm{d}}{\mathrm{d}s}\left(E\frac{\mathrm{d}r}{\mathrm{d}z}\right)-\frac{l^2E^2}{nr^3}=\frac{\mathrm{d}n}{\mathrm{d}r} \tag{A3.18}$$

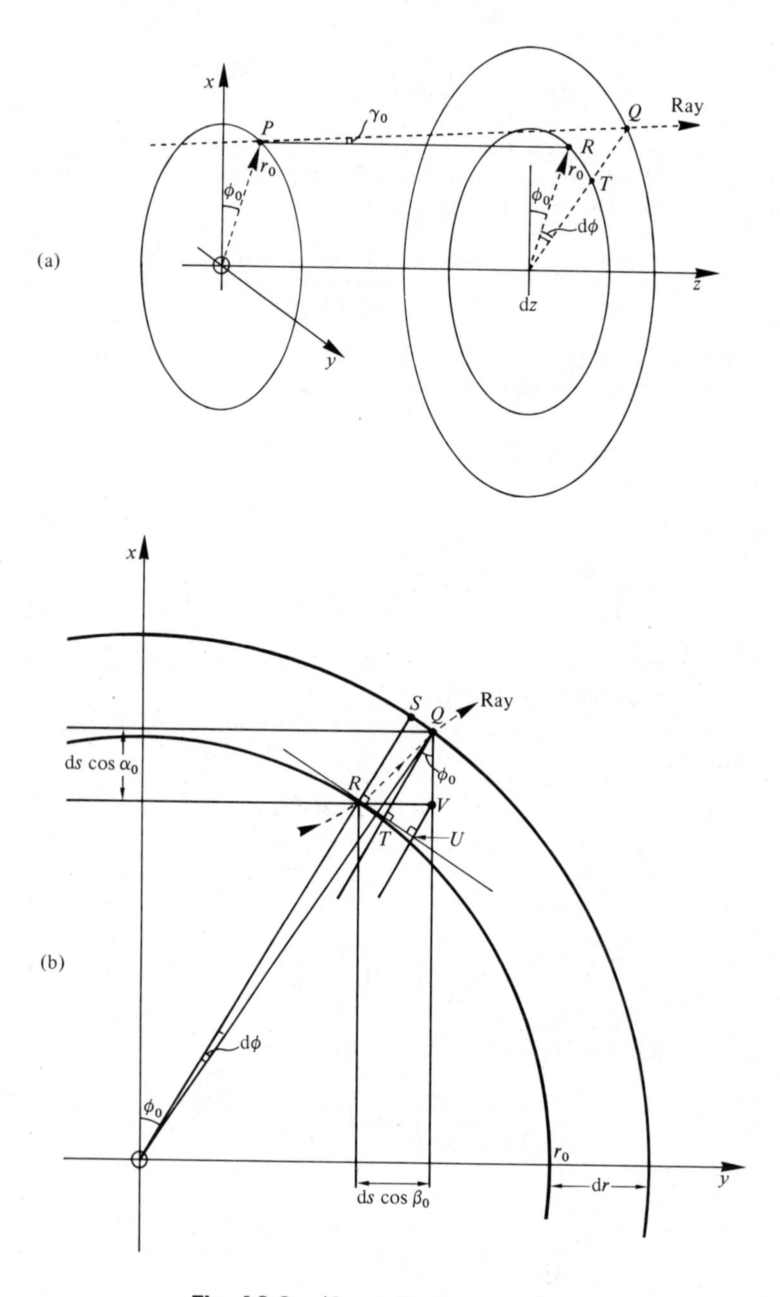

Fig. A3.2 (See following page.)

However,

$$\frac{\mathrm{d}}{\mathrm{d}s}\left(\frac{\mathrm{d}r}{\mathrm{d}z}\right)=\frac{\mathrm{d}^2 r}{\mathrm{d}z^2}\frac{\mathrm{d}z}{\mathrm{d}s}=\frac{\mathrm{d}^2 r}{\mathrm{d}z^2}\frac{E}{n} \tag{A3.19}$$

Introducing eqn. (A3.19) into eqn. (A3.18) and dividing through by E^2/n, we obtain:

$$\frac{\mathrm{d}^2 r}{\mathrm{d}z^2}=\frac{n}{E^2}\frac{\mathrm{d}n}{\mathrm{d}r}+\frac{l^2}{r^3}=\frac{1}{2E^2}\frac{\mathrm{d}}{\mathrm{d}r}(n^2)-\frac{l^2}{2}\frac{\mathrm{d}}{\mathrm{d}r}\left(\frac{1}{r^2}\right) \tag{A3.20}$$

A3.2 SOLUTIONS TO THE RAY EQUATIONS

Equations (A3.20), (A3.14) and (A3.12) determine the trajectories of every ray entering a graded-index fiber. In order to obtain more explicit expressions, we

---

Fig. A3.2 Proof of eqn. (A3.15):

(a) The ray enters the fiber at $P(r_0, \phi_0, 0)$ and after propagating an axial distance, $\mathrm{d}z$, reaches $Q(r_0 + \mathrm{d}r, \phi_0 + \mathrm{d}\phi, \mathrm{d}z)$. Thus

$$PQ = \mathrm{d}s, \quad PR = \mathrm{d}z = \mathrm{d}s \cos\gamma_0 \tag{A3.11}$$

and

$$l = r_0^2\left(\frac{\mathrm{d}\phi}{\mathrm{d}z}\right)_0 = r_0^2\left(\frac{\mathrm{d}\phi}{\mathrm{d}s}\right)_0\left(\frac{\mathrm{d}s}{\mathrm{d}z}\right)_0 = r_0^2\left(\frac{\mathrm{d}\phi}{\mathrm{d}s}\right)_0 \sec\gamma_0$$

(b) A view along the fiber axis: with

$$RS//TQ//UV,\ RT = RU - TU = SQ = (r_0 + \mathrm{d}r)\mathrm{d}\phi$$

but

$$RV = \mathrm{d}s\cos\beta_0 \quad \therefore \quad RU = \mathrm{d}s\cos\beta_0\cos\phi_0$$

and

$$QV = \mathrm{d}s\cos\alpha_0 \quad \therefore \quad TU = \mathrm{d}s\cos\alpha_0\sin\phi_0$$

$$\therefore \quad (r_0 + \mathrm{d}r)\mathrm{d}\phi = \mathrm{d}s(\cos\beta_0\cos\phi_0 - \cos\alpha_0\sin\phi_0)$$

$$\therefore \quad \text{as } (\mathrm{d}r/r_0) \to 0,\ r_0\left(\frac{\mathrm{d}\phi}{\mathrm{d}s}\right)_0 \to \cos\beta_0\cos\phi_0 - \cos\alpha_0\sin\phi_0$$

Hence $l = r_0 \sec\gamma_0 (\cos\beta_0\cos\phi_0 - \cos\alpha_0\sin\phi_0)$

(A3.15)

first multiply through eqn. (A3.20) by $2(\mathrm{d}r/\mathrm{d}z)$:

$$2\frac{\mathrm{d}r}{\mathrm{d}z}\frac{\mathrm{d}^2r}{\mathrm{d}z^2}=\frac{\mathrm{d}}{\mathrm{d}z}\left(\frac{\mathrm{d}r}{\mathrm{d}z}\right)^2=\frac{1}{E^2}\frac{\mathrm{d}}{\mathrm{d}r}(n^2)\frac{\mathrm{d}r}{\mathrm{d}z}-l^2\frac{\mathrm{d}}{\mathrm{d}r}\left(\frac{1}{r^2}\right)\frac{\mathrm{d}r}{\mathrm{d}z}$$

$$=\frac{1}{E^2}\frac{\mathrm{d}}{\mathrm{d}z}(n^2)-l^2\frac{\mathrm{d}}{\mathrm{d}z}\left(\frac{1}{r^2}\right) \tag{A3.21}$$

This integrates directly to give

$$\left(\frac{\mathrm{d}r}{\mathrm{d}z}\right)^2=\frac{n^2}{E^2}-\frac{l^2}{r^2}+A \tag{A3.22}$$

The integrating constant, A, is obtained from the initial conditions:

$$A=\left(\frac{\mathrm{d}r}{\mathrm{d}z}\right)^2_{z=0}-\frac{n_0^2}{E^2}+\frac{l^2}{r_0^2} \tag{A3.23}$$

Substituting eqns. (A3.12) and (A3.15) gives:

$$A=\left(\frac{\mathrm{d}r}{\mathrm{d}z}\right)^2_{z=0}-\left(\frac{\mathrm{d}s}{\mathrm{d}z}\right)^2_{z=0}+r_0^2\left(\frac{\mathrm{d}\phi}{\mathrm{d}z}\right)^2_{z=0}=-1 \tag{A3.24}$$

since

$$(\mathrm{d}s)^2=(\mathrm{d}r)^2+(r\,\mathrm{d}\phi)^2+(\mathrm{d}z)^2$$

Thus,

$$\frac{\mathrm{d}r}{\mathrm{d}z}=\pm\left(\frac{n^2}{E^2}-\frac{l^2}{r^2}-1\right)^{1/2} \tag{A3.25}$$

Let r_1 and r_2 be the roots of the equation

$$n^2(r)-\frac{E^2l^2}{r^2}-E^2=0 \tag{A3.26}$$

These roots are functions of the profile, $n(r)$, and the initial ray trajectory, as represented by the parameters E and l. This is shown in Fig. A3.3. When both roots are real, the two radii, r_1 and r_2, represent turning points for the trajectory, at which $\mathrm{d}r/\mathrm{d}z=0$ and about which the ray trajectory is symmetric. The radial position of the ray may therefore be represented as in Fig. A3.4 where it is seen to be periodic in z. The periodic axial distance is given by z_0,

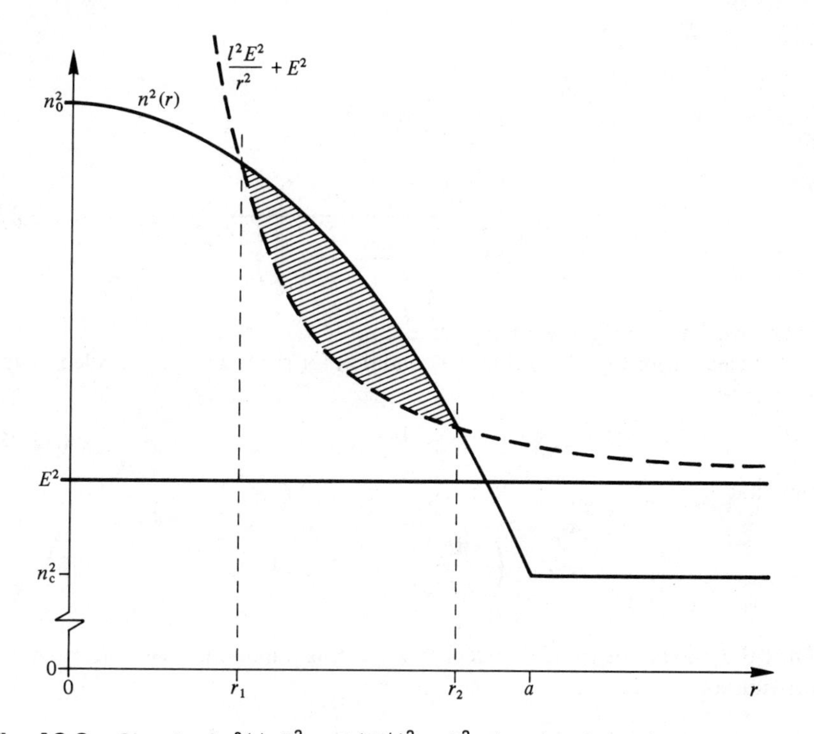

Fig. A3.3 Sketch of $n^2(r)$, E^2 and $(lE/r)^2 + E^2$, showing the region bounded by r_1 and r_2 where the roots of eqn. (A3.26) are real. The correspondence with Fig. 6.2(a) is clear.

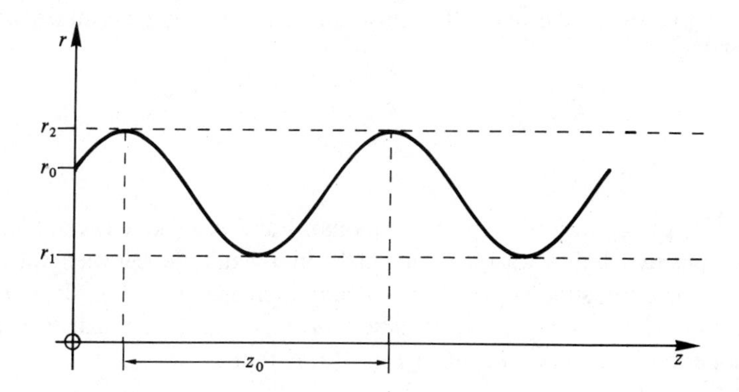

Fig. A3.4 The periodic variation of the radial position of a ray with axial distance.

where

$$z_0 = 2\int_{r_1}^{r_2} \frac{\mathrm{d}z}{\mathrm{d}r}\,\mathrm{d}r$$

$$= 2\int_{r_1}^{r_2} \frac{\mathrm{d}r}{\left(\dfrac{n^2(r)}{E^2} - \dfrac{l^2}{r^2} - 1\right)^{1/2}} \tag{A3.27}$$

where we have made use of eqn. (A3.25).

It is clear from Fig. A3.3 that the condition for rays to be fully guided by the fiber is

$$n_0 > E > n_c \tag{A3.28}$$

and that

$$r^2\left(\frac{n^2(r)}{E^2} - 1\right) > l^2 \geqslant 0 \tag{A3.29}$$

Thus the maximum value of l occurs when E is minimum, that is $E = n_c$, and depends on $n(r)$:

$$l_{\max}^2 = \frac{r^2}{n_c^2}(n^2(r) - n_c^2) \tag{A3.30}$$

For step-index fiber

$$l_{\max} = (2\Delta)^{1/2} a \tag{A3.31}$$

The variation in the azimuthal position of the ray is determined by eqn. (A3.14):

$$\frac{\mathrm{d}\phi}{\mathrm{d}z} = \frac{l}{r^2} \tag{A3.14}$$

For $l = 0$, $\phi = \phi_0$ and the ray is meridional. For $l \neq 0$ (skew rays), $(\mathrm{d}\phi/\mathrm{d}z)$ is always positive but it varies periodically in magnitude between (l/r_1^2) and (l/r_2^2). As a result, superimposed on a steady increase of ϕ with z is a periodic variation due to the periodic variation of r with z. This makes $\phi(z)$ the pseudoperiodic function shown in Fig. A3.5; that is,

$$\phi(z + z_0) - \phi(z) = \Phi = \text{constant} \tag{A3.32}$$

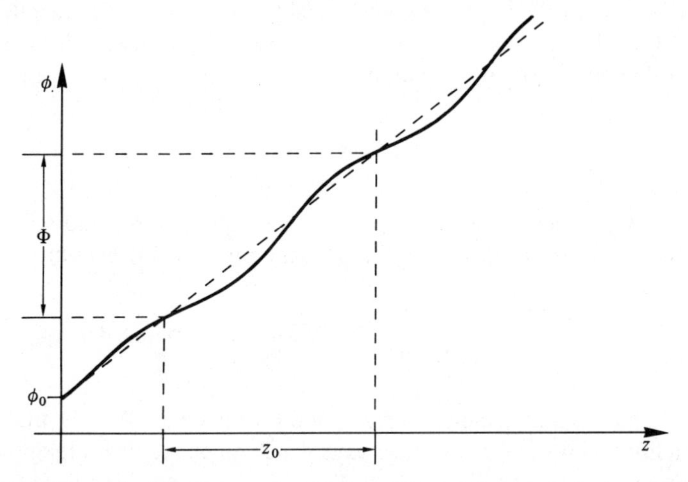

Fig. A3.5 The variation of the azimuthal position of the ray trajectory with axial distance.

where

$$\Phi = \int \frac{d\phi}{dz}\,dz = 2\int_{r_1}^{r_2} \frac{d\phi}{dz}\frac{dz}{dr}\,dr = 2\int_{r_1}^{r_2} \frac{(l/r^2)\,dr}{\left[\dfrac{n^2(r)}{E^2} - \dfrac{l^2}{r^2} - 1\right]^{1/2}} \tag{A3.33}$$

A3.3 MULTIPATH DISPERSION

Multipath dispersion is governed by the variation of z_0 and Φ over the range of possible values of E and l. The profiles which are shown by the mode analysis of Section 6.3 to minimize mode dispersion can be shown by the corresponding ray analysis to minimize multipath dispersion. The two methods are equivalent.

A3.4 LEAKY RAYS

Using eqns. (A3.11) and (A3.12) the condition for total guidance may be expressed as:

$$E^2 = n^2(r_0)\cos^2\gamma_0 > n_c^2 \tag{A3.34}$$

This is simply Snell's law again, and for step-index fibers is the equivalent of eqn. (2.1.5). Leaky rays are necessarily skew rays, which do not satisfy (A3.34) but do satisfy the condition

$$E^2 > n_c^2 - \frac{l^2 E^2}{a^2} \tag{A3.35}$$

This may be verified by comparing Fig. A3.3 with Figs. 6.2 and 6.9. In terms of the position and direction of the incident ray (A3.35) becomes

$$\cos^2 \gamma_0 > \frac{n_c^2}{n^2(r_0)} - \frac{r_0^2}{a^2}(\cos \beta_0 \cos \phi_0 - \cos \alpha_0 \sin \phi_0)^2 \tag{A3.36}$$

where we have again substituted for E and l using eqns. (A3.12) and (A3.15). To simplify this we may take those rays which enter the fiber along the radius OX as representative and put $\phi_0 = 0$. Then, the leaky rays are those for which

$$\cos \gamma_0 < n_c/n(r_0)$$

but

$$\cos^2 \gamma_0 + \frac{r_0^2}{a^2} \cos^2 \beta_0 > \frac{n_c^2}{n^2(r_0)} \tag{A3.37}$$

For step-index fibers this becomes

$$\cos^2 \gamma_0 + \frac{r_0^2}{a^2} \cos^2 \beta_0 > \frac{n_2^2}{n_1^2} \tag{A3.38}$$

and at $r_0 = a$,

$$\cos^2 \gamma_0 + \cos^2 \beta_0 > n_2^2/n_1^2 \tag{A3.39}$$

Appendix 4

Radiometry and Photometry

Radiometry is concerned with the measurement of the power and energy carried by electromagnetic radiation. The power is often referred to as the radiant flux, Φ, and at optical wavelengths, when the quantum nature of the radiation is evident in the generation and detection mechanisms, it may well be thought of as a flux of photons. The power density, **P**, in an electromagnetic wave at any point is related to the local values of the electric and magnetic fields through the Poynting vector:

$$\mathbf{P} = \mathbf{E} \wedge \mathbf{H} \qquad \text{(A4.1)}$$

The laws of radiometry are based on the principles of geometrical optics and the radiant energy is assumed to be conserved unless specific allowance is made for absorption. An entirely equivalent set of photometric laws governs the visual effect of the radiation as seen by a standard observer.

We may define the principal radiometric quantities by reference to the radiant energy emitted by the surface A in Fig. A4.1, and in particular to that part of the radiation from A which is incident on the surface S. Consider the radiant flux emanating from the element $\mathrm{d}A$ of A and passing through the element $\mathrm{d}S$ of S. The two elements are separated by the distance r and the line joining them makes an angle θ with the surface normal at $\mathrm{d}A$ and an angle ϕ with the surface normal at $\mathrm{d}S$. In order to find the total power, Φ, passing from A to S we shall have to integrate the flux travelling from each element $\mathrm{d}A$ to each element $\mathrm{d}S$ over both surfaces. We therefore write this element of the total power as $\mathrm{d}^2\Phi$. Then

$$\int_A \int_S \mathrm{d}^2\Phi = \Phi \qquad \text{(A4.2)}$$

From purely geometrical considerations we can see that the magnitude of $\mathrm{d}^2\Phi$ is proportional to:

(a) the projected area of $\mathrm{d}A$ in the direction of $\mathrm{d}S$, that is to $\mathrm{d}A \cos\theta$
(b) the projected area of $\mathrm{d}S$ in the direction of $\mathrm{d}A$, that is, to $\mathrm{d}S \cos\phi$

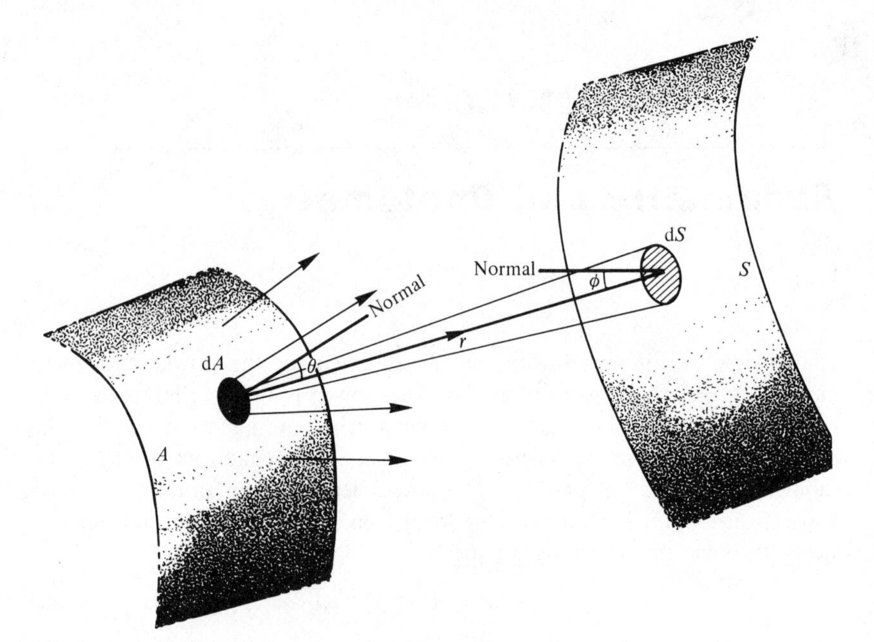

Fig. A4.1 Diagram showing the radiation emitted from a small element dA of a luminous surface A which passes through the small element dS of the surface S.

(c) the inverse square of the distance between dA and dS, that is, to $1/r^2$.

Putting the constant of proportionality as L gives

$$d^2\Phi = \frac{L\, dA \cos\theta\, dS \cos\phi}{r^2} \tag{A4.3}$$

where L is known as the *radiance* of A at dA in the direction of dS.

The quantity

$$d\Omega = \frac{dS \cos\phi}{r^2} \tag{A4.4}$$

is the solid angle, measured in steradians, subtended by dS at dA.

Thus

$$d^2\Phi = L\, d\Omega\, dA \cos\theta \tag{A4.5}$$

and the units of L are seen to be $[\mathrm{W\ m^{-2}\ sr^{-1}}]$. In the case of a diffuse emissive surface, L is found to be independent of direction, that is, independent of θ.

When the emissive region, A, is small and distant, a quantity of greater interest is the radiant power per unit solid angle emitted from the whole of the surface A in the direction of dS. This is known as the *radiant intensity*, I, of A and has the units of $[\mathrm{W\ sr^{-1}}]$. We may write the radiant intensity of the element dA as

$$\mathrm{d}I = \frac{\mathrm{d}^2\Phi}{\mathrm{d}\Omega} = L\,\mathrm{d}A\cos\theta \tag{A4.6}$$

and then integrating dI over A:

$$I = \int_A \mathrm{d}I = \int_A L\,\mathrm{d}A\cos\theta \tag{A4.7}$$

When A is small, distant and plane, θ is independent of the position of dA on A. If in addition the source is diffuse, so that L is isotropic,

$$I(\theta) = I_0 \cos\theta \tag{A4.8}$$

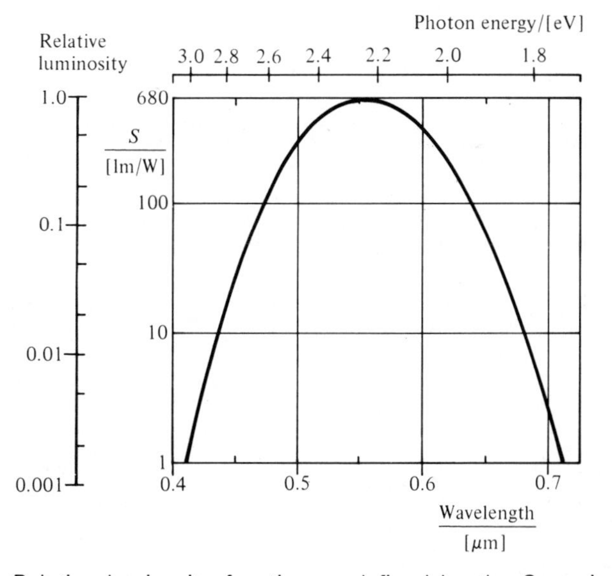

Fig. A4.2 Relative luminosity function as defined by the Commission Internationale de l'Eclairage (CIE) at a 2° view angle for normal photopic vision.

where

$$I_0 = \int_A L \, dA \tag{A4.9}$$

Such a source is known as a Lambertian radiator. In Section 2.1, when eqn. (2.1.12) was obtained for the power propagated in a step-index fiber, the source was assumed to be Lambertian.

The visual effect of electromagnetic radiation is a function of its wavelength, and the photometric quantity corresponding to optical power measured in watts is the *luminous flux* measured in *lumens*. Thus, the relative spectral sensitivity of the human eye may be calibrated directly in lumens per watt. The standard Commission Internationale de L'Eclairage (CIE) luminosity function, $S(\lambda)$, for normal (photopic) vision is plotted in Fig. A4.2. The definition of the photometric units is based on the photometric quantity corresponding to radiant intensity. The *luminous intensity* normal to an area $10/6 \text{ mm}^2$ of a black body radiator at a temperature of 2045 K (the solidification temperature of platinum at standard pressure) is defined to be 1 lumen per steradian [lm/sr] or candela [cd].

Appendix 5

Source–Fiber Coupling Efficiency

Here we wish to consider the optical power that can be usefully launched onto a fiber of numerical aperture NA and core area A_c from a diffuse source which has a light-generating region of area A_s immersed in a medium of refractive index n_s. We will assume that the Fresnel reflection losses at the interfaces can be made negligible by 'blooming', and that there is negligible absorption in any of the media.

With a butt joint between fiber and source it is only the light generated within the fiber core diameter that can couple into the fiber. Any light generated outside this area is simply wasted. It is perfectly possible to use an optical system to focus the light from a source of large area into a smaller region on the end of the fiber. However, this inevitably involves increasing the divergence of the incident light with the result that only the same total power is captured. The effect is independent of the precise optical arrangement used, and is an example of a general law of optics: that the radiance of an image formed in the same medium as its object cannot exceed the radiance of the object. This may be illustrated by Fig. A5.1 which shows a plane, axial element of a diffuse source. This has an area δA_s normal to the axis and forms an image of area δA_{im} in some kind of optical system. The power $\delta \Phi_s$ collected at

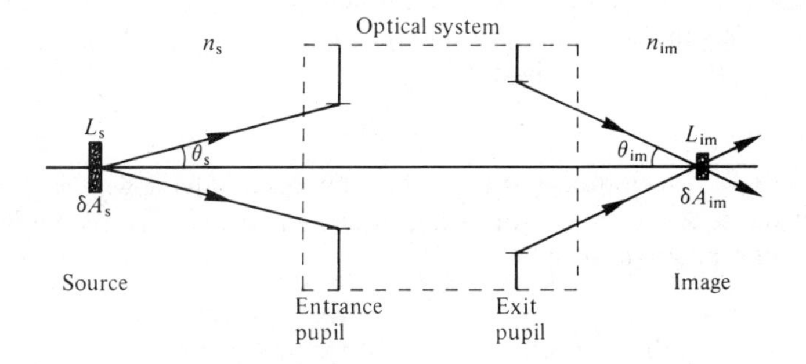

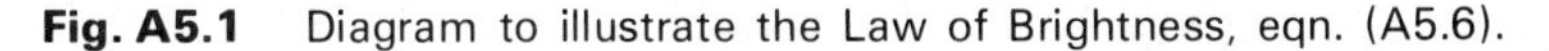
Fig. A5.1 Diagram to illustrate the Law of Brightness, eqn. (A5.6).

the entrance pupil of the optical system is given by

$$\delta\Phi_s = \pi L_s \, \delta A_s \sin^2 \theta_s \tag{A5.1}$$

This results from the same calculation that led to eqn. (2.1.13), except that here θ_s is the semi-angle subtended at the object by the entrance pupil and L_s is the radiance of the source. By an analogous argument the power leaving the exit pupil to form the image is

$$\delta\Phi_{im} = \pi L_{im} \, \delta A_{im} \sin^2 \theta_{im} \tag{A5.2}$$

where L_{im} is the image radiance and θ_{im} is the semi-angle subtended at the image by the exit pupil. Clearly energy conservation demands that

$$\delta\Phi_{im} \leqslant \delta\Phi_s \tag{A5.3}$$

Furthermore, the *sine condition* (see Ref. [2.1], Section 4.5) states that

$$n_s \sin \theta_s = m n_{im} \sin \theta_{im} \tag{A5.4}$$

where m is the linear magnification of the system. Thus,

$$\delta A_{im} = m^2 \, \delta A_s \tag{A5.5}$$

Putting eqns. (A5.1)–(A5.5) together we see that

$$L_{im} \leqslant \left(\frac{n_{im}}{n_s}\right)^2 L_s \tag{A5.6}$$

In the case of a plane, diffuse, uniformly luminescent region like the active region of an LED, the radiance is

$$L_s = I_{s0}/A_s = \Phi_s/2\pi A_s = (\eta_{int}/2\pi)(J/e)\mathcal{E}_{ph} \tag{A5.7}$$

where we have used eqns. (8.5.1) and (8.4.5) and written $J = I/A_s$ for the injection current density. The aim of the optical system should be to form an image of the source in the core of the fiber, which has an area A_c and a refractive index n_1. This image should just fill the core. In an ideal, lossless system the radiance of the image is given by

$$L_{im} = (n_1/n_s)^2 L_s \tag{A5.8}$$

The power Φ_T which propagates along the fiber is that lying within the cone of semi-angle ϕ_m, shown on Fig. 2.2, where $\sin\phi_m = (n_1^2 - n_2^2)^{1/2}/n_1$. The propagating power is then

$$\Phi_T = \pi L_{im} A_c \sin^2 \phi_m = \pi \frac{n_1^2}{n_s^2} L_s A_c \frac{(n_1^2 - n_2^2)}{n_1^2} \tag{A5.9}$$

$$= \eta_{int}(J/e)\mathcal{E}_{ph}A_c(n_1^2 - n_2^2)/2n_s^2 \quad \text{(A5.10)}$$

We may express eqn. (A5.10) in terms of the overall source-to-fiber quantum efficiency, η_f, discussed in Section 8.6. This may be defined by

$$\Phi_T = \eta_f(JA_s/e)\mathcal{E}_{ph} \quad \text{(A5.11)}$$

so that

$$\eta_f = \eta_{int}\frac{A_c}{A_s}\frac{(n_1^2 - n_2^2)}{2n_s^2} \quad \text{(A5.12)}$$

The value given by eqn. (A5.12) would be reduced by a factor of two if the radiant intensity of the active region were isotropic rather than Lambertian; it would be further reduced by any Fresnel reflection or absorption losses. We should recognize that no proper account has been taken of the coupling of skew and leaky rays especially those which enter the fiber near to the core–cladding interface. This is beyond the scope of the present treatment, but the correction is small for radiation that will propagate long distances in the fiber. We ought, however, to extend the argument to graded-index fibers, and this can best be done by an appeal to the number of modes of wave propagation excited by the source. As we shall see, this leads again, as it should, to eqn. (A5.10) in the case of step-index fiber.

A source of area A_s emitting into a solid angle Ω_s will be radiating in

$$M_s = n_s^2 A_s \Omega_s/\lambda^2 \quad \text{(A5.13)}$$

independent spatial modes of wave propagation. Each mode may be thought to occupy a solid angle

$$\Omega = \Omega_s/M_s = \lambda^2/n_s^2 A_s \quad \text{(A5.14)}$$

We have seen in Chapter 5 that the number of guided modes of propagation that can be supported by a step-index fiber is given by

$$M = 2\pi(A_c/\lambda^2)(n_1^2 - n_2^2) \quad \text{(5.3.23)}$$

Of the total number of modes radiated by the source, then, the fiber is able to collect only the fraction

$$\frac{M}{M_s} = 2\pi\frac{A_c}{A_s}\frac{(n_1^2 - n_2^2)}{n_s^2\Omega_s} \quad \text{(A5.15)}$$

and this is the fraction (Φ_T/Φ_s) of the total source power that is collected.

When $\Omega_s = 4\pi$ as in the case of an LED,

$$\frac{\Phi_T}{\Phi_s} = \frac{M}{M_s} = \frac{A_c}{A_s}\frac{(n_1^2 - n_2^2)}{2n_s^2} \tag{A5.16}$$

which, remembering that $\Phi_s = \eta_{int}(JA_s/e)\mathcal{E}_{ph}$, is in complete agreement with eqn. (A5.10).

In Chapter 6 we found that the number of modes propagated by an α-profile, graded-index fiber was

$$M = \frac{\alpha}{(\alpha + 2)}\frac{2\pi A_c}{\lambda^2}(n_0^2 - n_c^2) \tag{6.1.26–28}$$

where n_0 is the axial refractive-index, and n_c is that of the cladding, and where we have substituted $A_c = \pi a^2$. Compared to the equivalent step-index fiber ($n_1 = n_0$ and $n_2 = n_c$) the power propagated is reduced by the factor $\alpha/(\alpha + 2)$.

Appendix 6

Derivation of Frequency Response of a Laser Diode (Equation 11.2.5)

A6.1 STEADY-STATE CONDITIONS

Consider the rate of change of the electron concentration, dn/dt, and the rate of change of the optical power density, dP/dt, in the active p-region of the N–p–P double heterostructure of Fig. 10.4. Three effects contribute to dn/dt: the rate of injection of carriers, the rate of recombination of carriers by stimulated emission and the rate of recombination of carriers by spontaneous emission. We may therefore write

$$\frac{dn}{dt} = \frac{J}{ed} - A(n - n_{th})P - \frac{n}{\tau_{sp}} \tag{A6.1}$$

where J is the injected current density; d is the thickness of the active p-region; A is a constant having units [$m^2\ J^{-1}$] which relates the number of stimulated recombinations per unit volume per second to the optical power density and to the excess of the electron concentration over the threshold value, n_{th}, at which population inversion occurs; and τ_{sp} is the spontaneous recombination time.

Only a tiny fraction of the spontaneous emission is radiated into the laser mode, and if this is neglected, the rate of change of optical power density in the laser mode is given by

$$\frac{dP}{dt} = hfcA(n - n_{th})P - \frac{P}{\tau_{ph}} \tag{A6.2}$$

where τ_{ph} is the mean photon lifetime given by eqn. (11.2.6). These coupled, nonlinear equations are simplified if the optical power density, P, is expressed in terms of the effective number of photons per unit volume in the lasing mode, n_{ph}. Thus,

$$P = n_{ph}hfc \tag{A6.3}$$

Let

$$B = Ahfc \tag{A6.4}$$

where the units of B are $[\mathrm{m}^3/\mathrm{s}]$. Then,

$$\frac{\mathrm{d}n}{\mathrm{d}t} = \frac{J}{ed} - B(n - n_{\mathrm{th}})n_{\mathrm{ph}} - \frac{n}{\tau_{\mathrm{sp}}} \tag{A6.5}$$

and

$$\frac{\mathrm{d}n_{\mathrm{ph}}}{\mathrm{d}t} = B(n - n_{\mathrm{th}})n_{\mathrm{ph}} - \frac{n_{\mathrm{ph}}}{\tau_{\mathrm{ph}}} \tag{A6.6}$$

Under steady-state conditions we may put $J = J_0$, $n = n_0$, and $n_{\mathrm{ph}} = n_{\mathrm{ph0}}$ corresponding to $P = P_0$. Then $\mathrm{d}n/\mathrm{d}t$ and $\mathrm{d}n_{\mathrm{ph}}/\mathrm{d}t$ are both zero so that from eqn. (A6.6)

$$B(n_0 - n_{\mathrm{th}}) = 1/\tau_{\mathrm{ph}} \tag{A6.7}$$

and from eqn. (A6.5)

$$\frac{J_0}{ed} = \frac{n_{\mathrm{ph0}}}{\tau_{\mathrm{ph}}} + \frac{n_0}{\tau_{\mathrm{sp}}} \tag{A6.8}$$

Note that at the lasing threshold, the power in the lasing mode, P_0, is very small so we may put $n_{\mathrm{ph0}} = 0$ at this point. Thus the threshold current density, J_{th}, is given by

$$\frac{J_{\mathrm{th}}}{ed} = \frac{n_{\mathrm{L}}}{\tau_{\mathrm{sp}}} \tag{A6.9}$$

where n_{L} is the electron concentration at the lasing threshold. Above this threshold the electron concentration, which is given by eqn. (A6.7), remains unchanged in this simple model:

$$n_0 = n_{\mathrm{th}} + \frac{1}{B\tau_{\mathrm{ph}}} = n_{\mathrm{L}} \tag{A6.10}$$

and each additional injected carrier produces a photon by stimulated emission. Thus the laser power and the photon density increase linearly with the injected current:

$$n_{\mathrm{ph0}} = \frac{J_0\tau_{\mathrm{ph}}}{ed} - \frac{n_{\mathrm{L}}}{\tau_{\mathrm{sp}}}\tau_{\mathrm{ph}} = \frac{(J_0 - J_{\mathrm{th}})}{ed}\tau_{\mathrm{ph}} \tag{A6.11}$$

when the spontaneous emission term is neglected. If $n_{\mathrm{L}} \gg n_{\mathrm{th}}$, the gain coefficient is given by

$$B = \frac{1}{\tau_{ph}(n_L - n_{th})} \simeq \frac{1}{\tau_{ph} n_L} = \frac{ed}{\tau_{sp}\tau_{ph} J_{th}} \qquad (A6.12)$$

using eqns. (A6.10) and (A6.9).

A6.2 SMALL PERTURBATIONS ABOUT THE STEADY STATE

We wish to consider the small variations in these steady-state quantities caused by a small variation with time of the current density about J_0. It is therefore convenient to put $J = J_0 + J_1$, $n = n_0 + n_1$, $n_{ph} = n_{ph0} + n_{ph1}$. Equation (A6.5) then becomes

$$\frac{dn_1}{dt} = \frac{J_0}{ed} - B(n_0 - n_{th})n_{ph0} - \frac{n_0}{\tau_{sp}} + \frac{J_1}{ed} - Bn_1 n_{ph0} - B(n_0 - n_{th})n_{ph1} - \frac{n_1}{\tau_{sp}} \qquad (A6.13)$$

where we have neglected the term in $n_1 n_{ph1}$ in order to linearize the equation. The first three terms sum to zero, so that

$$\frac{dn_1}{dt} = \frac{J_1}{ed} - Bn_1 n_{ph0} - \frac{n_{ph1}}{\tau_{ph}} - \frac{n_1}{\tau_{sp}} \qquad (A6.14)$$

Equation (A6.6) becomes

$$\frac{dn_{ph1}}{dt} = \left[B(n_0 - n_{th}) - \frac{1}{\tau_{ph}}\right](n_{ph0} + n_{ph1}) + Bn_1 n_{ph0} = Bn_1 n_{ph0} \qquad (A6.15)$$

Putting eqns. (A6.14) and (A6.15) together we obtain

$$\frac{d^2 n_1}{dt^2} + \left(Bn_{ph0} + \frac{1}{\tau_{sp}}\right)\frac{dn_1}{dt} + \frac{Bn_{ph0}}{\tau_{ph}} n_1 = \frac{1}{ed}\frac{dJ_1}{dt} \qquad (A6.16)$$

and

$$\frac{d^2 n_{ph1}}{dt^2} + \left(Bn_{ph0} + \frac{1}{\tau_{sp}}\right)\frac{dn_{ph1}}{dt} + \frac{Bn_{ph0} n_{ph1}}{\tau_{ph}} = \frac{Bn_{ph0}}{ed} J_1 \qquad (A6.17)$$

These both have the form

$$\ddot{y} + \beta\dot{y} + \omega_0^2 y = f(t) \qquad (A6.18)$$

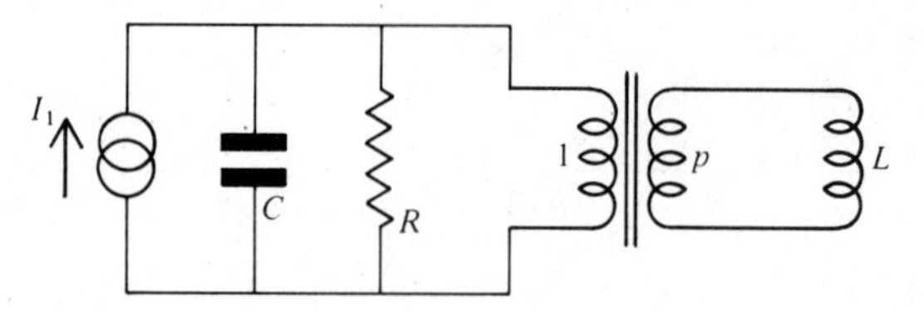

Fig. A6.1 Equivalent circuit representing the electron–photon resonance in a lasing cavity. The photon energy is represented by the energy stored in the inductance, and the transformer turns-ratio $p = \tau_{sp}/\tau_{ph}$.

where y may be either n_1 or n_{ph1}, and β and ω_0 are given by

$$\beta = Bn_{ph0} + 1/\tau_{sp} = \frac{(J_0 - J_{th})}{\tau_{sp} J_{th}} + \frac{1}{\tau_{sp}} = \frac{J_0}{\tau_{sp} J_{th}} \tag{A6.19}$$

and

$$\omega_0^2 = \frac{Bn_{ph0}}{\tau_{ph}} = \frac{(J_0 - J_{th})}{\tau_{sp}\tau_{ph} J_{th}} \tag{A6.20}$$

where we have used eqns. (A6.11) and (A6.12).

When $J_1(t)$ takes the form of a step change these equations would lead us to expect a damped harmonic oscillation in n_1 and n_{ph1} of approximate angular frequency ω_0 about their new equilibrium values. When $J_1(t)$ is a sinusoidal modulation of angular frequency ω, we shall expect a resonance response, with the amplitude inversely proportional to $(\omega_0^2 - \omega^2 + j\beta\omega)$. Thus,

$$\frac{n_{ph1}(\omega)}{n_{ph1}(0)} = \frac{P(\omega)}{P(0)} = \frac{\omega_0^2}{(\omega_0^2 - \omega^2) + j\beta\omega} \tag{A6.21}$$

This is the form quoted in eqn. (11.2.5). The coupling between the electronic carriers and the optical cavity is an inherent part of this resonance and may be emphasized by introducing a transformer into the simple circuit model of Fig. 11.8, as shown in Fig. A6.1.

Appendix 7

The Impulse Response of a Filter with Antisymmetric Frequency Response

A7.1 DEFINITION OF FREQUENCY RESPONSE

The type of frequency response under consideration is shown in Fig. A7.1. It may be defined by the function $H(\omega)$ where

$$H(0) = 1, \; H(\pi/T) = \tfrac{1}{2}, \qquad \text{and} \qquad H(\omega) = 0 \text{ for } \omega > 2\pi/T$$

We may put

$$H(\omega) = H_0 + H_1(\omega) \tag{A7.1}$$

where

$$H_0 = H(\pi/T) = \tfrac{1}{2}$$

Then the required symmetry about $\omega = \pi/T$ may be expressed by

$$H_1(\omega') = -H_1(\omega) \tag{A7.2}$$

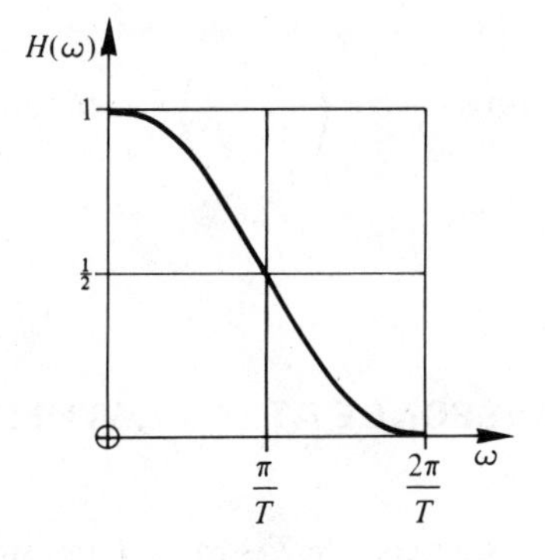

Fig. A7.1 Frequency response of filter.

where

$$\omega' = 2\pi/T - \omega \tag{A7.3}$$

A7.2 RESPONSE TO AN IMPULSE AT $t = 0$

$$h(t) = \int_0^\infty H(\omega)\cos\omega t\, d\omega = \int_0^{2\pi/T} H(\omega)\cos\omega t\, d\omega$$

$$= \int_0^{\pi/T} \{H_0 + H_1(\omega)\}\cos\omega t\, d\omega + \int_{\pi/T}^{2\pi/T} \{H_0 + H_1(\omega)\}\cos\omega t\, d\omega$$

$$= \int_0^{\pi/T} \{H_0 + H_1(\omega)\}\cos\omega t\, d\omega$$

$$+ \int_{\omega=\pi/T}^{0} \{H_0 - H_1(\omega')\}\cos\left(\frac{2\pi}{T} - \omega'\right)t\, d\left(\frac{2\pi}{T} - \omega'\right)$$

$$= \int_0^{\pi/T} \left[\{H_0 + H_1(\omega)\}\cos\omega t + \{H_0 - H_1(\omega)\}\cos\left(\frac{2\pi}{T} - \omega\right)t\right] d\omega$$

$$\therefore \quad h(t) = \int_0^{\pi/T} H_0\left(\cos\omega t + \cos\left(\frac{2\pi}{T} - \omega\right)t\right) d\omega$$

$$+ \int_0^{\pi/T} H_1(\omega)\left(\cos\omega t - \cos\left(\frac{2\pi}{T} - \omega\right)t\right) d\omega$$

$$= \int_0^{\pi/T} \left[H_0\, 2\cos\frac{\pi t}{T}\cos\left(\omega - \frac{\pi}{T}\right)t + H_1(\omega)\, 2\sin\frac{\pi t}{T}\sin\left(\frac{\pi}{T} - \omega\right)t\right] d\omega \tag{A7.4}$$

A7.3 IMPULSE RESPONSE AT $t = nT$, WHERE n IS AN INTEGER

$$h(nT) = 2\int_0^{\pi/T} |H_0\cos n\pi\cos n(\omega T - \pi) + H_1(\omega)\sin n\pi\sin n(\pi - \omega T)|\, d\omega$$

Since $\cos(n\pi) = (-1)^n$ and $\sin(n\pi) = 0$, this reduces to

$$h(nT) = (-1)^n 2H_0 \int_0^{\pi/T} \cos n(\omega T - \pi)\, d\omega$$

$$= (-1)^n \frac{2H_0}{nT} \left| \sin n(\omega T - \pi) \right|_{\omega=0}^{\pi/T} = 0$$

Appendix 8

Solutions to Numerical Problems

1.1 (a) 4.74×10^{14} Hz, 1.96 eV; (b) 2.83×10^{14} Hz, 1.17 eV; (c) 2.83×10^{13} Hz, 0.12 eV.
1.2 (a) 1.34×10^{12} Hz; (b) 1.37×10^{13} Hz.
1.3 (a) 5 m; (b) 500 m; (c) 50 km.
1.4 (a) 0.46 m^{-1}; (b) 4.6×10^{-3} m^{-1}; (c) 4.6×10^{-5} m^{-1}.
1.5 (a) 146 b/s; (b) 64 kb/s; (c) 44.3 Mb/s. Answers depend on encoding methods.

2.1 (a) 0.044, 12.1°, 8.2°, 50.5 ns/km, 19.8 (Mb/s) km.
(b) 0.172, 24.5°, 16.5°, 209 ns/km, 4.8 (Mb/s) km.
(c) $(n_1^2 - n_2^2) > 1$, (90°), 43°, 2.24 μs/km, 0.45 (Mb/s) km.
2.4 $A = 0.0000897$, $B = 0.00876961$, $C = 1.1041089$, $D = -0.00916412$, $E = -0.0000936$. At 0.82 μm, $n = 1.4530$; 1.27 μm, $n = 1.4473$; 1.55 μm, $n = 1.4440$.
2.7 0.03 ns/km, 0.13 ns/km; 0.29 ns/km. Terms in γ^3 add 1.8%, 3.6% and 5.3%, respectively. Dispersion greater at shorter wavelengths. Hence least dispersion when source center wavelength greater than λ_0.

3.2 50–80 ppb, ~150 ppb and ~25 ppb, respectively.
3.3 2 Mb/s: both limited by attenuation; 5.4 km, 9.8 km
20 Mb/s: both by dispersion (better fiber only marginally); 2.4 km, 7.0 km
100 Mb/s: both by dispersion: 0.5 km, 1.4 km.
3.5 Higher atomic mass gives lower resonance frequencies, hence minima in attenuation and dispersion occur at longer wavelengths where Rayleigh scattering is less.

4.1 9 km.
4.2 Power density higher near axis.

5.1 He_{11} :0: TE_{01}, TM_{01} :2.405; HE_{21} :approx 2.405; HE_{12}, HE_{31}, EH_{11} :3.832; HE_{41}, EH_{21} : 5.136; TE_{02}, TM_{02}, HE_{22} : 5.520.
5.2 TE_{01}, TM_{01}, HE_{21} :7.9 μm and 1.58 μm; HE_{12}, HE_{31}, EH_{11} :5.0 μm and 1.00 μm; HE_{41}, EH_{21} :3.7 μm and 0.74 μm; TE_{02}, TM_{02}, HE_{22} :3.44 μm and 0.69 μm.
5.3 At 1.55 μm, 6 and 80; at 0.85 μm, 12 and ~250.
5.5 (a) 2.405; (b) 3.401; (c) 4.165.
5.6 (a) 4.8 μm; (b) 6.8 μm; (c) 8.3 μm.
5.9 $Y_m = -0.01026$, $Y_w = 0.01045$; $Y_d = -0.0048$; $Y_T = -0.0046$; $(\Delta f)_{el} \simeq 1/8\sigma = 600$ MHz.